Reading Statistics and Research

Schuyler W. Huck
*University of Tennessee,
Knoxville*

PEARSON

Boston   Columbus   Indianapolis   New York   San Francisco   Upper Saddle River
Amsterdam   Cape Town   Dubai   London   Madrid   Milan   Munich   Paris   Montreal   Toronto
Delhi   Mexico City   Sao Paulo   Sydney   Hong Kong   Seoul   Singapore   Taipei   Tokyo

Vice President/Publisher: Kevin Davis
Editorial Assistant: Matthew Buchholz
Marketing Manager: Joanna Sabella
Production Editor: Karen Mason
Production Manager: Susan Hannahs
Senior Art Director: Jayne Conte
Cover Designer: Bruce Kenselaar
Cover Photo: Shutterstock
Photo Researcher: Annie Pickert
Full-Service Project Manager: Niraj Bhatt, Aptara®, Inc.
Composition: Aptara®, Inc.
Text and Cover Printer/Bindery: R. R. Donnelley, Harrisonburg
Text Font: Times

Credits and acknowledgments borrowed from other sources and reproduced, with permission, in this textbook appear on the appropriate page within the text.

Library of Congress Cataloging-in-Publication Data

Huck, Schuyler W.
  Reading statistics and research. — 6th ed. / Schuyler W. Huck
    p. cm.
  ISBN-13: 978-0-13-217863-1
  ISBN-10: 0-13-217863-X
1. Statistics.   2. Research.   3. Experimental design.   I. Title.
  QA276.H788 2012
  001.4'22—dc22   2010054550

10 9 8 7 6 5 4 3

www.pearsonhighered.com

ISBN-10:   0-13-217863-X
ISBN-13: 978-0-13-217863-1

This book is dedicated to two groups: those consumers of research reports who work at developing the skills needed to critically evaluate (and sometimes reject!) the claims made by researchers, and those researchers whose claims ought to be believed (and acted upon!) because they take the time to analyze carefully the data gleaned from thoughtfully designed studies that address worthy questions.

About the Author

Schuyler (Sky) Huck (Ph.D., Northwestern) is Distinguished Professor and Chancellor's Teaching Scholar at the University of Tennessee, Knoxville. His concerns for improving statistical instruction and helping consumers decipher and critique research reports show up in his books, journal articles, and convention presentations, and on his website (http://www.readingstats.com). In addition, Sky's applied/theoretical work has been cited by scholars in 337 different academic journals. Despite these achievements and other honors that have come his way, Sky takes his greatest pride in (1) the fact that two of his students have won Outstanding Dissertation Awards in stiff national competitions and (2) comments from his students that say, in essence, "You helped me learn!" Sky's hobbies include photography, puzzles, and poetry. In addition, he regularly helps prepare and serve hot meals for the homeless and makes deliveries of nonperishable groceries to those in need.

Brief Contents

Contents

17   Inferences on Percentages, Proportions, and Frequencies   404

18   Statistical Tests on Ranks (Nonparametric Tests)   434

Preface

This preface is devoted to three topics of likely concern to anyone who may be considering reading or adopting this book. These topics concern my assessment of people's need to critically evaluate research claims; the book's main objectives; and differences between the fifth and sixth editions of this book. Stated differently, one might legitimately ask:

1. For whom is this book intended?
2. In what ways will this book benefit its readers?
3. Is this simply a cosmetic revision of the fifth edition, and, if not, how does this new edition differ in significant ways from its predecessor?

People's Need to Critically Evaluate Research Claims

In the first edition of this book, I claimed that humanity could be divided into three groups: (1) those who conduct their own research studies, (2) those who do not formally engage in the research process but nonetheless encounter the results of others' investigations, and (3) those who are neither "makers" nor "consumers" of research claims. Now, nearly 40 years since I made that statement, I *still* believe that every person on the face of the Earth can be classified into one of those three groups. However, it is clear to me that the relative sizes and the needs of the three groups are different now than they were in 1974 (when the first edition of this book was published) or even in 2008 (when the fifth edition appeared).

Regarding the size of the three groups mentioned, the first group (the "doers" of research) has grown slightly larger, whereas the second group (the "consumers" of research) has expanded geometrically over the past few years. The odds are extremely high that any randomly selected person belongs to one of these two groups. The first would be populated with lots of professors, any graduate student preparing to write a master's thesis or doctoral dissertation, most employees of the many research units located in both public and private organizations, and a handful of independent researchers. Whoever isn't a member of the first group most likely is a member of the second group. That's because it is virtually impossible to avoid coming into contact with research findings.

In one way or another, almost everyone encounters the findings of empirical investigations. First, formal and full-length research reports are presented each year in thousands of technical journals and at meetings of countless international, national, regional, and local professional associations. Summaries of such studies make their way into newspaper and magazine stories, television and radio news programs, and informal conversations among coworkers, family members, and friends. Computer availability and the staggering increase in Internet websites make it possible for growing numbers of people to have access to the research "evidence" that stands behind online advice from "experts" regarding everything from arthritis to Zen Buddhism. And then there are the innumerable advertisements and commercials that bombard us on a daily basis and contain the results of so-called scientific studies that supposedly demonstrate the worth of the products or services being hawked.

Everyone in the huge second group must become a more discerning consumer of research findings and research claims. Such individuals, located on the receiving end of research summaries, cannot be *competent* consumers of what they read or hear unless they can both understand and evaluate the investigations being discussed. Such skills are needed because (1) trustworthy research conclusions come only from those studies characterized by a sound design and a careful analysis of the collected data, and (2) the screening process—if there is one in place—that supposedly prevents poor studies from being disseminated is only partially successful in achieving its objective. For these reasons, consumers must acquire the skills needed to protect themselves from overzealous or improperly trained researchers whose work leads to exaggeration, false "discoveries," and unjustified claims of "significance."

Individuals who conduct research investigations—the doers of research—also should be able to critically evaluate others' research reports. Almost every research project is built on a foundation of knowledge gleaned from previous studies. Clearly, if a current researcher cannot differentiate legitimate from unjustified research conclusions, his or her own investigation may well be doomed from the outset because it is pointed in the wrong direction or grounded in a research base made of sand. If applied researchers could more adequately critique the studies cited within their own literature reviews, they also would be able to apply such knowledge to their own investigations. The result would be better designed studies containing more appropriate statistical analyses leading to more justifiable conclusions and claims.

This edition of *Reading Statistics and Research* is targeted at two groups: those who conduct their own research investigations and those who are the recipients of research-based claims. I have tried to keep both groups in mind while working on this revision project. I hope members of *both* groups will benefit from this edition's textual discussion of statistics and research design, the many excerpts taken from published research reports, and the review questions for each chapter.

This Book's Objectives

The seven specific objectives of this edition are basically the same as those of the previous five editions. These goals include helping readers increase their ability to (1) make sense out of statistical jargon, (2) understand statistical tables and figures, (3) know what specific research question(s) can be answered by each of a variety of statistical procedures, (4) remain aware of what can and cannot be accomplished when someone sets up and tests one or more null hypotheses, (5) detect the misuse of statistics, (6) distinguish between good and poor research designs, and (7) feel confident when working with research reports.

The seven objectives just listed can be synthesized nicely in two words: *decipher* and *critique*. This book is designed to help people *decipher* what researchers are trying to communicate in the written or oral summaries of their investigations. Here, the goal is simply to distill meaning from the words, symbols, tables, and figures included in the research report. (To be competent in this arena, one must be able not only to decipher what is presented, but also to "fill in the holes"; this is the case because researchers typically assume that those receiving the research report are familiar with unmentioned details of the research process and the statistical treatment of data.) Beyond being able to decipher what is presented, I very much want readers of this book to improve their ability to *critique* such research reports. This is important because research claims are sometimes completely unjustified due to problems associated with the way studies are planned or implemented or because of problems in the way data are collected, analyzed, summarized, or interpreted.

Differences between the Fifth and Sixth Editions

In an effort to assist readers to better decipher and critique research reports, I have done my best to update, expand, and in other ways improve this edition and make it superior to the previous edition. Several of these changes are quite minor and need not be discussed. There are, however, six important ways in which this edition is different from the one published in 2008. These changes *are* worth discussing.

All New Excerpts

It is not an exaggeration to say that the boxed excerpts constitute the lifeblood of this book. I have included these tables, figures, and passages of text from published research reports to illustrate both good and not-so-good practices, to instruct via the words of others, and to demonstrate that contemporary researchers do, in fact, use the statistical procedures discussed in this text.

A total of 492 excerpts appear in this edition. All of these excerpts are new, with not even one brought forward from the fifth edition. These numbers—492 new, 0 old—can be used to back up the claim that this book contains an *extensive* array of material that illustrates what *contemporary* researchers put into their research reports.

It should be noted that the excerpts included in this edition were not chosen indiscriminately. They were not identified by students in the courses I teach, nor were they plucked from the research literature by graduate assistants. In every instance, I personally selected the excerpt. I entrusted this task to no one else because I wanted each excerpt to help readers understand a concept or practice utilized widely by applied researchers. To this end, I used five criteria in the excerpt-selection process: I wanted each item to have high-level *relevance* to the book's textual discussion; be as *short* as possible; to come from a *recent* research report; to require *no special training* in any particular academic discipline in order to be understood; and, whenever possible, to deal with content that would be *interesting* to my book's readers. I applied these criteria to *each* of this book's 492 excerpts.

New Chapters on Multivariate Tests on Means, Factor Analysis, and Structural Equation Modeling

Since the publication of the fifth edition, a slew of individuals have contacted me and asked that three topics be covered in this sixth edition: multivariate analysis of variance, factor analysis, and structural equation modeling. These advanced statistical procedures are being used by more and more applied researchers. Reference to these techniques appears regularly in the research reports presented orally at professional meetings and in print within technical journals.

In this new edition, Chapter 19 deals with multivariate analysis of variance and covariance (MANOVA and MANCOVA), Chapter 20 is focused on exploratory and confirmatory factory analysis (EFA and CFA), and Chapter 21 covers structural equation modeling (SEM). These new chapters are like others in this book in that the emphasis is on the goals of each procedure, the results typically included in research reports, and the important assumptions that should be attended to. The reader interested in the mathematical formulas associated with one or more of these statistical techniques (or how to get a computer to perform the analyses) must consult some other book or seek help from a statistician.

To make room for these three new chapters (without increasing the overall length of the book), I reduced the number of excerpts in each of the other chapters, tried to be more succinct in my discussion of topics, and eliminated the final chapter from the previous edition. That chapter was entitled, "The Quantitative Portion of Mixed Methods Studies." The absence of that chapter from this new edition is no great loss because it was (and still is) my firm belief that one's ability to decipher and critique the quantitative portions of reports of mixed-methods research endeavors is contingent on his or her ability to understand and evaluate research reports that are purely statistical in nature. As I pointed out in the previous edition,

> *Careful mixed-methods researchers do things in the same way, in the quantitative parts of their studies, as do researchers who collect only numerical information and use only statistics to analyze their data.*

It is my hope that this new edition of the book will be viewed positively by—and prove useful to—mixed-methods researchers and their intended audiences, even though there is no longer a specific chapter dedicated to that form of research. Both of these groups will be helped by a consideration here of multivariate tests on means, factor analysis, and structural equation models. These more advanced statistical procedures are being utilized more frequently in the quantitative portion of mixed-methods studies. This trend is likely to accelerate as time goes by.

Content

Several small (yet important) content changes were made as the fifth edition was transformed into this sixth edition. In addition to the three new chapters, the following items are new to this edition:

- The geometric mean
- The correlation coefficient, tau-b
- Guttman's split-half reliability procedure
- Moderated and mediated multiple regression
- sr^2 as an index of predictor variable worth in multiple regression
- Sensitivity
- Specificity
- The Sobel test
- Example of the Dunn–Sidák procedure
- The "hit rate" in logistic regression

I have made a slew of other small and large changes for the purposes of increasing clarity, updating material, emphasizing critical concepts, and improving review questions.

Diversity of Journals Represented in the Excerpts

In this sixth edition, more so than in any of its predecessors, I worked hard to select excerpts from a variety of disciplines. I did this to help readers increase their ability to cross disciplinary lines when reviewing research reports. This final point deserves a bit of explanation.

In contrast to those books that focus on a single discipline (such as psychology, education, or nursing), the manifest purpose here is to help readers feel more at ease when confronted by research claims that emanate from disciplines other than their own. To that end, the excerpts in this sixth edition come from journals such as *Journal of Criminal Justice, Body Image, Creativity Research Journal, Anti-Aging Medicine, Journal of Comparative Family Studies, Harm Reduction Journal, Measurement in Physical Education and Exercise Science, International Journal of Health Geographics, Alcohol and Alcoholism, Journal of Sex Research, Community College Review,* and *Computers in Human Behavior.*

Unless people have the ability to decipher and critique research in a multidisciplinary fashion, they become easy targets for those who inadvertently or purposefully present research "evidence" that comes from studies characterized by ill-conceived questions, poor methodology, and sloppy statistical analysis. Unfortunately, some researchers begin their studies with a strong bias as to what they would like the results to show, and the results of such biased investigations are summarized on a near daily basis in the popular press. Clearly, a person is more likely to detect such bias if he or she can decipher and critique research *across disciplines,* recognizing, for example, that the purpose of and issues related to logistic regression are the same regardless of whether the data come from sociology, ecology, or epidemiology.

Excerpts from an International Array of Authors

In an effort to honor the work of researchers located around the world, I have included a wide variety of excerpts that are international in nature. For example, Excerpt 2.4 comes from a study dealing with French people suffering from migraine headaches, Excerpt 5.1 is from a study investigating mental health in Cyprus, Excerpt 7.4 comes from a study dealing with the products purchased by adolescents in South Africa, Excerpt 8.6 comes from a study dealing with the health-care needs of elderly citizens in Japan, Excerpt 13.1 is from a study dealing with Chinese school children, Excerpt 17.20 is from a study concerned with girls suffering from anorexia nervosa in Brazil, and Excerpt 20.13 is from a study dealing with travel agents in Turkey.

Research studies are being conducted around the globe, and the Internet has made it easy for researchers to learn what others have discovered in far-off lands. Without question, the availability of shared knowledge makes it possible for fields to advance more rapidly than was the case even just a decade ago. I hope my international array of excerpts will spur researchers into looking far and wide when conducting their literature reviews and establishing communication links with others who share common research interests. Moreover, I hope these excerpts will help increase the respect researchers have for their international colleagues.

An Expanded and Updated Companion Website

The book's website (http://www.ReadingStats.com) has been updated and expanded. This website remains easy to navigate, it continues to offer different kinds of information for users with different kinds of needs, and it has been field-tested and modified on the basis of student feedback. The website and the book function to complement each other, with neither one able to successfully do alone what both can do together. This website contains more than 400 viewable pages, plus links to more than 150 carefully selected pages on other sites. The content of these pages is designed to help students learn.

The largest and most important part of the website involves information, exercises, and links carefully organized in a chapter-by-chapter format. The following items are available for each chapter:

- Chapter outlines, interactive quizzes with immediate feedback, and online resources

- Jokes, quotations, and poetry about statistics
- Statistical misconceptions
- Biographies of significant people in the field
- E-mail messages to my students that address pertinent topics
- Best passages from each chapter

It should be noted that certain features of this book's companion website provide a form of instruction that is literally impossible to duplicate either by an instructor or a book. For example, the links to other sites bring the learner into contact with interactive exercises that actually *show* statistical concepts in operation, thereby permitting a kind of presentation that no instructor or book could ever accomplish.

Five Important Similarities between the Fifth and Sixth Editions

Since the publication of the fifth edition, several individuals have contacted me with comments about this book. Most of those comments have been positive, and they have prompted me to maintain (as much as possible) five features of the fifth edition as I worked to create this new, sixth edition.

First, I kept the format the same, with excerpts from recent journal articles serving as the book's core structure. As indicated previously, I personally selected each of this book's 492 excerpts, all of which are new to the sixth edition. These excerpts are likely to be as helpful to readers as were the excerpts sprinkled throughout the fifth edition.

Second, I tried to keep the text material outside the book's excerpts as straightforward, clear, and helpful as people have said it has been in earlier editions of the book. Ever since the first edition was published nearly 40 years ago, the main compliment I've received about this book is concerned with my ability to write about statistical concepts in such a way that others can truly understand things. In preparing the new chapters for this new edition, and in revising the other chapters, I again have tried to achieve the goals of clarity, relevance, and deep understanding.

The third similarity between this edition and its predecessors is my continued effort to point out that there is often a difference—and sometimes a giant difference—between what researchers are entitled to say following their data analyses and what they actually do say. I provide many examples of statistical procedures that produced finding that were "significant" or revealed "significant differences." Such procedures inherently involved sample-to-population inferences, null hypotheses, underlying assumptions, and the possibility of inferential error. As in previous editions, I repeatedly make the point that "significance" can exist in a statistical sense but not in any practical manner.

The fourth important thing that has *not* changed is a warning. As before, I take every opportunity to point out that complex statistics do not have the magical power to create a silk purse out of a sow's ear. Unless the research questions being addressed are worthy, a study is doomed from the start. Accordingly, there is continued emphasis on critically evaluating research questions and null hypotheses as the first step in assessing the potential value of any investigation.

The final feature I have tried to maintain again concerns the excerpts. Many of these excerpts, as was the case with the excerpts in the fifth edition, have come from studies that were focused on important questions, that were designed thoughtfully, and that produced findings that may have an impact on the way you think or act. Many other excerpts came from studies focused on topics that were undoubtedly fun for the researchers to research. By considering the research questions and methodology associated with these studies, I am hoping, once again, that more than a few readers will adopt the view that research can be both fun and relevant to our daily lives.

Acknowledgments

As with most large projects, the revision of this book was made possible because of the hard work on the part of many talented people. I wish now to express my sincere appreciation to these individuals. They are not responsible, of course, for any mistakes that may have inadvertently crept into this work. They *are* responsible, however, for initiating this project, for moving it along, and for making the finished product far superior to what it would have been had they not been involved.

First and foremost, I want to thank three individuals at Pearson Publications who have supported, nurtured, and protected this gigantic book-revision endeavor. Paul Smith (my editor), Karen Mason (my production manager), and Sarah Bylund (my permissions advisor) have been enormously helpful to (and patient with) me, and I truly feel as if each one functioned, at different times, as my handler, my mentor, and my guide. Although separated by hundreds of miles, we kept in communication via telephone, e-mail, fax, and "snail mail." Without exception, the many questions I posed were answered promptly and clearly by Paul, Karen, and Sarah. They also raised important questions I never would have considered, they passed along a variety of relevant questions from others, and (most important) they offered wise counsel and moral support.

I also wish to acknowledge the help provided by Mercy Heston, my copy editor. Mercy carefully examined every page of manuscript that I submitted for this revision project, and her eagle eye caught many mistakes, both grammatical and substantive. Without question, this sixth edition is far superior to what it would have been if Mercy had not critically evaluated my initial work.

Since its publication, several students identified passages in the fifth editions that were ambiguous, contradictory, or unnecessarily repetitious. In addition, many professors at other universities and a handful of independent researchers contacted me with questions, comments, and suggestions. None of these individuals probably realizes how much I value their important roles in this revision project. Nevertheless, I am indebted to each of them for their contributions.

Three individuals deserve special recognition for their contributions to the content of this revised book. Shelley Esquivel, Amy Beavers, and Hongwei Yang provided invaluable assistance in the preparation of this edition's three new chapters. Shelley helped with Chapter 19, Amy helped with Chapter 20, and all three helped with Chapter 21. Regarding the book's final chapter on structural equation modeling, Shelley and Amy prepared the initial draft, Hongwei gave me insights that allowed me to reorganize and modify that first draft, and then I personally selected—as with every chapter in the book—all of the excerpts for Chapter 21. Although I take fully responsibility for any errors of omission or commission in Chapters 19 though 21, I am indebted to Shelley, Amy, and Hongwei for their assistance with these chapters.

Several graduate students helped in small yet non-trivial ways as this revision project unfolded, and I want to thank them for their assistance. Extensive library research was conducted by Kathy Flowers. Internet searches were conducted by Allison Biker, Gary Techer, Jordan Driver, and Jared Middleman. Draft copies of excerpts were reviewed by Andrew Hedger, Emily Young, Turner Tallman, Elle-Kate Sweeter, and S. Kylure Finn. Computer data analysis was conducted by David Kindly and Nancy Walker. Page proofs were carefully read by Rocky Alexander, Patricia Grander, Jason Traveler, Ginna Bridett, Jennifer Momminew, Josh Shutterly, and Owen Smiley. The permission file was overseen by Candace Spirit, Todd Stanford, and Keith Frisco.

Niraj Bhatt and the team at Aptara, Inc., took charge of the revision project as it moved through its production phase. I am extremely grateful to them for their work on this project. It was a pure joy working with Niraj Bhatt!

My heartfelt appreciation is extended to Ammar Safar and John Wesley Taylor, who created the original website for this book. This website (http://www.Reading Stats.com) contains extensive information and interactive exercises not contained here, and it is far more than simply a book supplement. In several respects, this companion website is equivalent in importance to the book. Having such a website would not have been possible had it not been for Ammar's and Wesley's generous contributions of their time and talent. I want to thank them for those contributions.

Finally, I want to thank my family for being supportive of my efforts to complete this revision project. At every step along the way, members of my nuclear and extended family encouraged me to consider this project to be the second highest priority (behind my students) among my many professional obligations. Had they not encouraged me to hole up in my little home office and to keep my nose to the grindstone, this revision project would have been delayed for months, if not years!

Schuyler W. Huck
Knoxville, Tennessee, 2011

Typical Format
of a Journal Article

Almost all journal articles dealing with research studies are divided into different sections by means of headings and subheadings. Although there is variation among journals with respect to the terms used for the headings and the order in which different sections are arranged, there does appear to be a relatively standard format for published articles. Readers of the professional literature will find that they can get the most mileage out of the time they invest if they are familiar with the typical format of journal articles and the kind of information normally included in each section of the article.

We are now going to look at a particular journal article that does an excellent job of illustrating the basic format that many authors use as a guide when they are writing their articles. The different sections of our model article can be arranged in outline form, as follows:

1. Abstract
2. Introduction
   a. Background
   b. Statement of purpose
   c. Hypotheses
3. Method
   a. Participants
   b. Measures
   c. Procedure
   d. Statistical plans
4. Results
5. Discussion
6. References

Let us now examine each of these items.

Abstract

An **abstract,** or *précis,* summarizes the entire research study and appears at the beginning of the article. Although it normally contains fewer than 150 words, the abstract usually provides the following information: (1) a statement of the purpose or objective of the investigation, (2) a description of the individuals who served as participants, (3) a brief explanation of what the participants did during the study, and (4) a summary of the important findings.

Excerpt 1.1 is the abstract from our model journal article. As in most articles, it was positioned immediately after the title and authors' names. This abstract was easy to distinguish from the rest of the article because it was indented and printed in a smaller font size. In some journals, the abstract is italicized to make it stand out from the beginning paragraphs of the article.

EXCERPT 1.1 • *Abstract*

Summary: A significant relationship between changes in Body Mass Index and Body Areas Satisfaction scores was found for a sample of Euro-American ($n = 97$), but not African-American ($n = 79$), women initiating a moderate exercise program. For the African-American women only, compliance with the assigned exercise regimen directly predicted changes in Body Area Satisfaction. Implications of Ethnicity for behavioral weight loss treatment were discussed.

Source: Annesi, J. J. (2009). Correlations of changes in weight and body satisfaction for obese women initiating exercise: Assessing effects of ethnicity. *Psychological Reports, 105*(3), 1072–1076.

The sole purpose of the abstract is to provide readers with a brief overview of the study's purpose, methods, and findings. Thus, most abstracts indicate *why* the study was conducted, *how* the researcher went about trying to answer the questions of interest, and *what* was discovered after the study's data were analyzed. Even though the abstract in Excerpt 1.1 is extremely brief, it addresses the how and what issues. The reason why this study was conducted was not included in the abstract, but it was articulated in the research report's first main section.

In some articles, the abstracts mention the statistical techniques used to analyze the study's data. Most abstracts, however, are like the one in Excerpt 1.1 in that they include no statistical jargon. Because of this, the abstract in the typical research report is quite "readable," even to those who do not have the same level of research expertise as the individual(s) who conducted the study.

On the basis of abstracts such as the one shown in Excerpt 1.1, you can decide that the article in front of you is a veritable gold mine, that it *may* be what you have been looking for, or that it is not at all related to your interests. Regardless of how you react to this brief synopsis of the full article, the abstract serves a useful purpose. Note, however, that it is dangerous to think you have found a gold mine after reading nothing more than an article's abstract. I elaborate on this important point near the end of this chapter.

Introduction

The **introduction** of an article usually contains two items: a description of the study's **background** and a **statement of purpose.** Sometimes, as in our model journal article, a third portion of the introduction contains a presentation of the researcher's **hypotheses.** These components of a journal article are critically important. Take the time to read them slowly and carefully.

Background

Most authors begin their articles by explaining what caused them to conduct their empirical investigations. Perhaps the author developed a researchable idea from discussions with colleagues or students. Maybe a previous study yielded unexpected results, thus prompting the current researcher to conduct a new study to see if those earlier results could be replicated. Or, maybe the author wanted to see which of two competing theories would be supported more by having the collected data conform to its hypotheses. By reading the introductory paragraph(s) of the article, you learn why the author conducted the study.

In describing the background of their studies, authors typically highlight the connection between their studies and others' previously published work. Whether this review of literature is short or long, its purpose is to show that the current author's work has been informed by, or can be thought of as an extension of, previous knowledge. Such discussions are a hallmark of scholarly work. Occasionally, a researcher conducts a study based on an idea that is not connected to anything anyone has investigated or written about; such studies, however, are rare.

Excerpt 1.2 comes from our model article. Although only two paragraphs in length, this portion of the introduction sets the stage for a discussion of the author's investigation. If you read these two paragraphs, I predict you will understand everything that the researcher presents in the way of his study's "background."

In Excerpt 1.2, note that the researcher is not presenting opinion, hope, or anecdotal experiences. Instead, he focuses his introductory remarks on what has been studied and found in earlier research investigations. This part of the researcher's report is characteristic of published articles, doctoral dissertations, master's theses, and reports from independent and government research agencies. In a very real sense, the researcher presents this information in an effort to provide a rationale for his or her spending the time and energy necessary to conduct the study that is discussed in the remaining parts of the research report.

EXCERPT 1.2 • *Background*

Approximately one-third of American women are obese (Hedley, Ogden, Johnson, Carroll, Curtin, & Flegal, 2004). Most are attempting to lose weight through either caloric restriction (dieting) alone or caloric restriction combined with increased physical activity (Powell, Calvin, & Calvin, 2007). Although the association of weight loss and reduction in health risk is acknowledged, a primary reason for women attempting weight loss is an improvement in satisfaction with their bodies (Thompson, Heinberg, Altabe, & Tantleff-Dunn, 1999). Analysis of body image as a changing process has been advocated (Gleeson, 2006), however research on the association of weight changes and changes in body satisfaction has been unclear (Houlihan, Dickson-Parnell, Jackson, & Zeichner, 1987; Foster, Wadden, & Vogt, 1997).

While obvious markers of one's body such as weight and waist circumference are readily available (through, for example, self-weighing and fit of clothes), some research suggests that feelings of competence and self-efficacy, associated with participation in an exercise program, predicts improved satisfaction with one's body even when little physiological change actually occurs (Annesi, 2000, 2006). Research also suggests ethnic differences in what is acceptable to women regarding the shapes and sizes of their bodies (Rodin & Larson, 1992; Roberts, Cash, Feingold, & Johnson, 2006; Powell, et al., 2007). For example, Euro-American women have been described as being more critical of their bodies than African-American women (Miller, Gleaves, Hirsch, Green, & Snow, 2000). Research on psychological responses to weight loss behaviors have only rarely accounted for ethnic differences. This is exemplified in some of our recent research with obese women (Annesi & Whitaker, 2008).

Source: Annesi, J. J. (2009). Correlations of changes in weight and body satisfaction for obese women initiating exercise: Assessing effects of ethnicity. *Psychological Reports, 105*(3), 1072–1076.

As in Excerpt 1.2, researchers somehow or other provide you with the information you need if you want to read any of the full research reports referred to in the "review of literature." In our model article, this is done by citing names and dates in parentheses, with this information connected to an alphabetized list of more complete citations presented at the end of the article. In many journals, footnotes are used instead of names and dates, with full citations presented (according to footnote number) at the end of the research report or at the bottom of its pages. Because it is often informative to examine primary resources rather than just second-hand summaries, take the time to read the original reports of key items referred to in the literature-review portion of any research report.

Statement of Purpose

After discussing the study's background, an author usually states the specific purpose or goal of the investigation. This statement of purpose is one of the most important parts of a research report, because in a sense, it explains what the author's "destination" is. It would be impossible for us to evaluate whether the trip was successful—in terms of research findings and conclusions—unless we know where the author was headed.

The statement of purpose can be as short as a single sentence or as long as one or two full paragraphs. It is often positioned just before the first main heading of the article, but it can appear anywhere in the introduction. Regardless of its length or where it is located, you will have no trouble finding the statement of purpose if a sentence contains the words, "the purpose of this study was to . . ." or "this investigation was conducted in order to . . ." In Excerpt 1.3, we see the statement of purpose from our model journal article.

EXCERPT 1.3 • *Statement of Purpose*

The purpose of this investigation thus was to assess the relationship of changes in Body Mass Index (kg/m^2) with changes in body satisfaction in a sample of Euro-American and African-American women with obesity who participated in a program of moderate exercise.

Source: Annesi, J. J. (2009). Correlations of changes in weight and body satisfaction for obese women initiating exercise: Assessing effects of ethnicity. *Psychological Reports, 105*(3), 1072–1076.

Hypotheses

After articulating the study's intended purpose, some authors disclose the hypotheses they had at the beginning of the investigation. Other authors do not do this, either because they did not have any firm expectations or because they consider it unscientific for the researcher to hold hunches that might bias the collection or interpretation of the data. Although there are cases in which a researcher can conduct a good study without having any hypotheses as to how things will turn out, and although it is important for researchers to be unbiased, there is a clear benefit in knowing what the researcher's hypotheses were. Simply stated, outcomes compared against hypotheses usually are more informative than are results that stand in a vacuum. Accordingly, I applaud those researchers who disclose in the introduction any a priori hypotheses they had.

Excerpt 1.4 comes from our model journal article, and it contains the researcher's hypothesis. As you can see, there really were two hypotheses, one for each group of women involved in the study. Considered together, these hypotheses say that improvements in body satisfaction can be predicted for both Euro-American and African-American women, but what is used to make these successful predictions differs across the two groups.

EXCERPT 1.4 • *Hypotheses*

It was expected that for Euro-American women, reduction in weight over 6-mo. would predict improvement in body satisfaction; while for African-American women, greater commitment to the exercise program (i.e., greater frequency of exercise) would predict improvement in body satisfaction, rather than actual weight loss. Understanding such relationships might improve weight-loss treatments by enabling them to be more sensitive to participants' ethnicities.

Source: Annesi, J. J. (2009). Correlations of changes in weight and body satisfaction for obese women initiating exercise: Assessing effects of ethnicity. *Psychological Reports, 105*(3), 1072–1076.

In most articles, the background, statement of purpose, and hypotheses are not identified by separate headings, nor are they found under a common heading. If a common heading were to be used, however, the word *introduction* would probably be most appropriate, because these three items set the stage for the substance of the article—an explanation of what was done and what the results were.

Method

In the **method** section of a journal article, an author explains in detail how the study was conducted. Ideally, such an explanation should contain enough information to enable a reader to replicate (i.e., duplicate) the study. To accomplish this goal, the author addresses three questions: (1) Who participated in the study? (2) What kinds of measuring instruments were used to collect the data? and (3) What were the participants required to do? The answer to each of these questions is generally found under an appropriately titled subheading in the method section.

Participants

Each of the individuals (or animals) who supplies data in a research study is considered to be a **participant** or a **subject.** (In some journals, the abbreviations *S* and *Ss* are used, respectively, to designate one subject or a group of subjects.) Within this section of a research report, an author usually indicates how many participants or subjects were used, who they were, and how they were selected.

A full description of the participants is needed because the results of a study often vary according to the nature of the participants used. This means that the conclusions of a study, in most cases, are valid only for individuals (or animals) who are similar to the ones used by the researcher. For example, if two different types of counseling techniques are compared and found to differ in terms of how effective they are in helping clients clarify their goals, it is imperative that the investigator indicate whether the participants were high school students, adults, patients in a mental hospital, or whatever. What works for a counselor in a mental hospital may not work at all for a counselor in a high school (and vice versa).

It is also important for the author to indicate how the participants were obtained. Were they volunteers? Were they randomly selected from a larger pool of potential participants? Were any particular standards of selection used? Did the researcher simply use all members of a certain high school or college class? As seen in Chapter 5, certain procedures for selecting samples allow results to be generalized far beyond the specific individuals (or animals) included in the study, whereas other procedures for selecting samples limit the valid range of generalization.

Excerpt 1.5 comes from our model journal article. Labeled "Participants," it was the first portion of the article's method section. The paragraph in this excerpt contains the abbreviations for three statistical concepts: *n*, *M*, and *SD*. Each of these is discussed in Chapter 2.

EXCERPT 1.5 • *Participants*

This study was based on data from the Euro-American and African American participants in an investigation published in 2008 (i.e., Annesi & Whitaker, 2008) that did not consider possible differences associated with ethnicities. Data from other ethnic groups (6% of the original sample) were not analyzed within this research. The women volunteered based on a newspaper solicitation for an exercise and nutrition education program for obese (Body Mass Index $\geq$ 30) women. Informed consent and a release form from a physician were required to participate. The Euro-American ($n = 97$) and African-American ($n = 79$) participants did not significantly differ on age (overall $M = 43.4$ yr., $SD = 10.3$) or Body Mass Index (overall $M = 36.6 \, \mathrm{kg/m^2}, SD = 5.1$).

Source: Annesi, J. J. (2009). Correlations of changes in weight and body satisfaction for obese women initiating exercise: Assessing effects of ethnicity. *Psychological Reports, 105*(3), 1072–1076.

Materials

This section of a journal article is normally labeled in one of five ways: **materials, equipment, apparatus, instruments,** or **measures.** Regardless of its label, this part of the article contains a description of the things (other than the participants) used in the study. The goal here, as in other sections that fall under the method heading, is to describe what was done with sufficient clarity so others could replicate the investigation to see if the results remain the same.

Suppose, for example, that a researcher conducts a study to see if males differ from females in the way they evaluate various styles of clothing. To make it possible for others to replicate this study, the researcher must indicate whether the participants saw actual articles of clothing or pictures of clothing (and if pictures, whether they were prints or slides, what size they were, and whether they were in color), whether the clothing articles were being worn when observed by participants (and if so, who modeled the clothes), what specific clothing styles were involved, how many articles of clothing were evaluated, who manufactured the clothes, and all other relevant details. If the researcher does not provide this information, it is impossible for anyone to replicate the study.

Often, the only material involved is the measuring device used to collect data. Such measuring devices—whether of a mechanical variety (e.g., a stopwatch), an online variety, or a paper-and-pencil variety (e.g., a questionnaire)—ought to be described very carefully. If the measuring device is a new instrument designed specifically for the study being summarized, the researcher typically reports evidence concerning the instrument's technical psychometric properties. Generally, the author accomplishes this task by discussing the reliability and validity of the scores generated by using the new instrument.[1] Even if an existing and reputable measuring instrument has been used, the researcher ought to tell us specifically what instrument was used (by indicating form, model number, publication date, etc.). One must know such information, of course, before a full replication of the study could be attempted. In addition, the researcher ought to pass along reliability and validity evidence cited by those who developed the instrument. Ideally, the authors ought to provide their own evidence as to the reliability and validity of scores used in their study, even if an existing instrument is used.

Excerpt 1.6 contains the materials section from our model article. The materials were called *measures* because the data for this study were gathered by measuring each of the study's participants in terms of three variables: body mass index (BMI), body satisfaction, and frequency of exercise.

EXCERPT 1.6 • *Materials*

Body Mass Index is the ratio of the body weight to height (kg/m^2). It was calculated using a recently calibrated scale and stadiometer. Exercise session attendance was the total number of exercise sessions completed over the 6 mo. study. Exercise sessions completed were recorded electronically through a computer attached to the cardiovascular exercise apparatus available to participants. Exercise completed outside of study facilities was recorded by participants at a kiosk near the exercise area, or through the Internet. The method was indicated to be valid through strong significant correlations previously found with changes of several measures of cardiorespiratory function (Annesi, 2000).

Body Areas Satisfaction, a scale of the Multidimensional Body-self Relations Questionnaire (Cash, 1994), is used to measure satisfaction with areas of one's body, e.g., lower torso (buttocks, hips, thighs, legs) weight. It requires responses to five

---

[1]In Chapter 4, we consider the kinds of evidence researchers usually offer to document their instruments' technical merit.

EXCERPT 1.6 • (*continued*)

items anchored by 1: Very dissatisfied and 5: Very satisfied. Internal consistency for women was reported as .73, and test–retest reliability was .74 (Cash, 1994). Internal consistency for the present sample was .79. Consistent with previous research (Jakicic, Wing, & Winters-Hart, 2002), change (mean difference) scores were calculated by subtracting scores at baseline from scores at Month 6.

Source: Annesi, J. J. (2009). Correlations of changes in weight and body satisfaction for obese women initiating exercise: Assessing effects of ethnicity. *Psychological Reports, 105*(3), 1072–1076.

This section of the research report contains several important statistical terms and numbers. To be more specific, Excerpt 1.6 contains four technical terms (valid, significant correlations, internal consistency, and test–retest reliability) and three numbers (.73, .74, and .79). In Chapter 4, we focus our attention on these and other measurement-related concepts and numerical summaries.

In most empirical studies, the **dependent variable** is closely connected to the measuring instrument used to collect data. In fact, many researchers operationally define the dependent variable as being equivalent to the scores earned by people when they are measured with the study's instrument. Although this practice is widespread (especially among statistical consultants), it is not wise to think that dependent variables and data are one and the same.

Although there are different ways to conceptualize what a dependent variable is, this simple definition is useful in most situations: a dependent variable is simply a characteristic of the participants that (1) is of interest to the researcher; (2) is not possessed to an equal degree, or in the same way, by all participants; and (3) serves as the target of the researcher's data-collection efforts. Thus, in a study conducted to compare the intelligence of males and females, the dependent variable is intelligence.

In the study associated with our model article, there are several variables of concern to the researchers: BMI, body satisfaction, frequency of exercise, and ethnicity. In one sense, all four of these variables were dependent variables. As discussed in other chapters of this book (Chapters 10 through 15), sometimes a particular statistical analysis causes a given dependent variable to assume the role of an independent variable when data are analyzed. For example, ethnicity is considered as an independent variable in our model study when the researchers analyze their data to assess their main hypothesis (about the connection between ethnicity and each of the other three variables). For now, do not worry about this "role-reversal" when dependent variables become independent variables. I assure you that this potentially confusing labeling of variables becomes fully clear in Chapters 10–15.

Procedure

How the study was conducted is explained in the **procedure** section of the journal article. Here, the researcher explains what the participants did—or what was done to them—during the investigation. Sometimes an author even includes a verbatim account of instructions given to the participants.

Remember that the method section is included to permit a reader to replicate a study. To accomplish this desirable goal, the author must outline clearly the procedures that were followed, providing answers to questions such as (1) Where was the study conducted? (2) Who conducted the study? (3) In what sequence did events take place? and (4) Did any of the subjects drop out prior to the study's completion? (In Chapter 5, we will see that subject dropout can cause the results to be distorted.)

Excerpt 1.7 is the procedure section from our model article. Even though this section is only one paragraph in length, it provides information regarding the duration of the study, how and where data were collected, and what kinds of instruction were given to the participants. In addition, the researcher points out (in the first sentence) where we can find a more expanded explanation of the study's procedure.

EXCERPT 1.7 • *Procedure*

A more detailed description of procedures is presented elsewhere (Annesi & Whitaker, 2008). Briefly, participants were provided access to YMCA wellness centers in the Atlanta, Georgia, area and given orientations to a variety of cardiovascular exercise equipment and areas for walking and running. Assignment to treatment conditions that emphasized either behavioral support or educational approaches to exercise was random. The behavioral support condition stressed the use of goal setting, progress tracking, and self-regulatory skills such as cognitive restructuring and self-reward. The educational condition stressed the need for regular exercise and knowledge of related physiological principles. All participants, however, were provided six standardized nutrition education sessions, and were assigned to three cardiovascular exercise sessions per week that progressed to 30 min. within 10 wk. Instructions on how to record exercise sessions inside the YMCA via the computer provided, and outside of the YMCA via the Internet, were given. To minimize biasing, measurements were made in a private area at baseline and Month 6 by exercise specialists unfamiliar to the participants.

Source: Annesi, J. J. (2009). Correlations of changes in weight and body satisfaction for obese women initiating exercise: Assessing effects of ethnicity. *Psychological Reports, 105*(3), 1072–1076.

Statistical Plans

Most research reports contain a paragraph (or more) devoted to the plans for statistically analyzing the study's data. In some reports, this information is presented near the end of the method section; in other reports, a discussion of the statistical plan-of-attack is positioned at the beginning of the report's results section. Excerpt 1.8,

EXCERPT 1.8 • *Statistical Plans*

An intention-to-treat design was incorporated where data missing at Month 6 was replaced by baseline scores (Gadbury, Coffey, & Allison, 2003). Statistical significance was set at $\alpha = .05$ (two-tailed). An *a priori* power analysis suggested that 64 participants per group were required to detect a medium effect size at the statistical power of .80.

Source: Annesi, J. J. (2009). Correlations of changes in weight and body satisfaction for obese women initiating exercise: Assessing effects of ethnicity. *Psychological Reports, 105*(3), 1072–1076.

which comes from our model journal article, highlights important features of the researcher's statistical plans. Here, as in most research reports, some, but not all, of those plans are delineated.

Although Excerpt 1.8 is quite brief, it contains six statistical concepts that were exceedingly important to the researcher's plan for analyzing the study's data: *statistical significance, $\alpha = .05$, two-tailed, a priori power analysis, effect size,* and *statistical power of .80.* We consider these concepts in Chapters 7 and 8. For now, let me simply say that this particular researcher deserves high marks for conducting a power analysis to determine how many participants were needed in each of the study's comparison groups.

Results

There are three ways in which the results of an empirical investigation are reported. First, the results can be presented within the text of the article—that is, with only words. Second, they can be summarized in one or more tables. Third, the findings can be displayed by means of a graph (technically called a **figure**). Not infrequently, a combination of these mechanisms for reporting results is used to help readers gain a more complete understanding of how the study turned out. In Excerpt 1.9, we see that the author of our model article presented his results in two paragraphs of text.

EXCERPT 1.9 • *Results*

Exercise attendance did not significantly differ between the Euro-American and African-American women ($t_{174} = 1.66, p = .10$). The mean number of exercise sessions attended per week was 2.07 ($SD = 0.63$). GLM mixed-model repeated measures analysis of variance indicated no significant difference in Body Mass Index scores between groups ($F_{1,174} = 3.29, p = .07; \eta^2 = .02$). Changes in Body Mass Index over 6 mo. were significant ($F_{1,174} = 41.01, p < .001; \eta^2 = .19$);

(continued)

EXCERPT 1.9 • (*continued*)

however, the change did not significantly differ by group. There was no significant difference in Body Areas Satisfaction scores between groups ($F_{1,174} = 1.03$, $p = .31$; $\eta^2 = .003$). Changes in Body Areas Satisfaction were significant ($F_{1,174} = 95.93$, $p < .001$; $\eta^2 = .35$); however, the changes did not significantly differ by group.

   For the Euro-American women, change in Body Mass Index was significantly correlated ($r = -.36$) with change in Body Areas Satisfaction, and exercise session attendance was significantly correlated ($r = -.41$) with change in Body Mass Index. There was no significant correlation between exercise session attendance and change in Body Areas Satisfaction ($r = .17$). For the African-American women, change in Body Mass Index was not significantly correlated with change in Body Areas Satisfaction ($r = -.02$). Exercise session attendance was, however, significantly correlated with both change in Body Mass Index ($r = -.46$) and change in Body Areas Satisfaction ($r = .23$).

Source: Annesi, J. J. (2009). Correlations of changes in weight and body satisfaction for obese women initiating exercise: Assessing effects of ethnicity. *Psychological Reports, 105*(3), 1072–1076.

Excerpt 1.9 contains a slew of statistical terms, abbreviations, and numerical results. If you find yourself unable, at this point, to make much sense out of the material presented in Excerpt 1.9, do not panic or think that this statistical presentation is beyond your reach. Everything in this excerpt is considered in Chapters 2, 3, 7 through 10, and 14. By the time you finish reading those chapters, you will be able to look again at Excerpt 1.9 and experience no difficulty deciphering the statistically based results of this investigation.

Although the **results** section of a journal article contains some of the most (if not *the* most) crucial information about the study, readers of the professional literature often disregard it, because the typical results section is loaded with statistical terms and notation not used in everyday communication. Accordingly, many readers of technical research reports simply skip the results section because it seems as if it came from another planet.

If you are to function as a discerning "consumer" of journal articles, you must develop the ability to read, understand, and evaluate the results provided by authors. Those who choose not to do this are forced into the unfortunate position of uncritical acceptance of the printed word. Researchers are human, however, and they make mistakes. Unfortunately, the reviewers who serve on editorial boards do not catch all of these errors. As a consequence, there is sometimes an inconsistency between the results discussed in the text of the article and the results presented in the tables. At times, a researcher uses an inappropriate statistical test. More often than you would suspect, the conclusions drawn from the statistical results extend far beyond the realistic limits of the actual data that were collected.

You do not have to be a sophisticated mathematician in order to understand and evaluate the results sections of most journal articles. However, you must become familiar with the terminology, symbols, and logic used by researchers. This text was written to help you do just that.

Look at Excerpt 1.9 once again. The text material included in this excerpt is literally packed with information intended to help you. Unfortunately, many readers miss out on the opportunity to receive this information because they lack the skills needed to decode what is being communicated or are intimidated by statistical presentations. One of my goals in this book is to help readers acquire (or refine) their decoding skills. In doing this, I hope to show that there is no reason for anyone to be intimidated by what is included in technical research reports.

Discussion

The results section of a journal article contains a technical report of how the statistical analyses turned out, whereas the **discussion** section is usually devoted to a nontechnical interpretation of the results. In other words, the author normally uses the discussion section to explain what the results mean in regard to the central purpose of the study. The statement of purpose, which appears near the beginning of the article, usually contains an underlying or obvious research question; the discussion section ought to provide a direct answer to that question.

In addition to telling us what the results mean, many authors use this section of the article to explain *why* they think the results turned out the way they did. Although such a discussion occasionally is found in articles where the data support the researchers' hunches, authors are much more inclined to point out possible reasons for the obtained results when those results are inconsistent with their expectations. If one or more of the scores turn out to be highly different from the rest, the researcher may talk about such serendipitous findings in the discussion section.

Sometimes an author uses the discussion section to suggest ideas for further research studies. Even if the results do not turn out the way the researcher anticipated, the study may be quite worthwhile in that it might stimulate the researcher (and others) to identify new types of studies that need to be conducted. Although this form of discussion more typically is associated with unpublished master's theses and doctoral dissertations, it occasionally is encountered in published forms of research reports.

It should be noted that some authors use the term **conclusion** rather than discussion to label this part of the research report. These two terms are used interchangeably. It is unusual, therefore, to find an article that contains both a discussion section and a conclusion section.

Excerpt 1.10 contains the discussion section that appeared in our model journal article. Notice how the author first provides an answer to the central research question; then, there is a lengthy discussion of possible reasons for why the results turned out as they did.

EXCERPT 1.10 • *Discussion*

As expected, change in Body Mass Index was significantly related to change in Body Areas Satisfaction for only the Euro-American women. It is possible that concern about body image in this ethnic group prompted more frequent self-weighing. Knowledge of weight change may, thus, have served as a marker for satisfaction with one's body. Although improvements in barriers and task self-efficacy were not directly measured, it is possible that the association of exercise session attendance and change in Body Areas Satisfaction for the African American women were linked to perceptions of competence (in maintaining a program of exercise) for them. It should be acknowledged that this is speculative, and direct measurement will be required for substantiation. Possibly, weight management for Euro-American women should focus on measured, manageable progress, while for African-American women the focus should be on building self-regulatory skills required to maintain weight-loss behaviors. Although this research clearly requires replication and extension, and was limited as a field design, analysis suggested that accounting for such ethnic differences when assessing psychological variables possibly related to weight loss in women is much needed.

Source: Annesi, J. J. (2009). Correlations of changes in weight and body satisfaction for obese women initiating exercise: Assessing effects of ethnicity. *Psychological Reports, 105*(3), 1072–1076.

There are two additional things to note about Excerpt 1.10, both of which are admirable features of this research report. In the next-to-last sentence of the discussion, the author suggests how his finding might be useful to practitioners. Then, in the very last sentence, the researcher points out that his research "clearly requires replication and extension." Other researchers should follow this good example of discussing implications and the need for replication.

Too often, research reports give the impression that the investigators who prepared those reports view their research as having *proven* something is true for everyone, everywhere, at all times and under all conditions. When you encounter such claims, downgrade your evaluation of the research report. However, upgrade your opinion of researchers who call for others to conduct new studies to see if initial findings can be replicated.

References

A research report normally concludes with a list of the books, journal articles, and other source material referred to by the author. Most of these items were probably mentioned by the author in the review of the literature positioned near the beginning of the article. Excerpt 1.11 is the **references** section of our model article.

The references can be very helpful to you if you want to know more about the particular study you are reading. Journal articles and convention presentations are usually designed to cover one particular study or a narrowly defined area of a subject.

EXCERPT 1.11 • *References*

Annesi, J. J. (2000). Effects of minimal exercise and cognitive-behavior modification on adherence, affective change, self-image, and physical change in obese females. *Perceptual and Motor Skills,* 91, 322–336.

Annesi, J. J. (2006). Relations of perceived bodily changes with actual changes and changes in mood in obese women initiating an exercise and weight-loss program. *Perceptual and Motor Skills,* 103, 238–240.

Annesi, J. J., & Whitaker, A. C. (2008). Weight loss and psychologic gain in obese women-participants in a supported exercise intervention. *Permanente Journal,* 12(3), 36–45. Also available at http:xnet.kp.org/permanentejournal/sum08/weight-loss.pdf.

Cash, T. F. (1994). *The multidimensional body–self relations users' manual.* Norfolk, VA: Old Dominion University.

Foster, G. D., Wadden, T. A., & Vogt, R. A. (1997). Body image in obese women before, during, and after weight loss treatment. *Health Psychology,* 16, 226–229.

Gadbury, G. L., Coffey, C. S., & Allison, D. B. (2003). Modern statistical methods for handling missing repeated measurements in obesity trial data: Beyond LOCF. *Obesity Reviews,* 4, 175–184.

Gleeson, K., & Frith, H. (2006). (De)Constructing body image. *Journal of Health Psychology, 11,* 79–90.

Hedley, A. A., Ogden, C. L., Johnson, C. L., Carroll, M. D., Curtin, L. R., & Flegal, K. M. (2004). Prevalence of overweight and obesity among US children, adolescents, and adults, 1999–2002. *Journal of the American Medical Association,* 291(23), 2847–2850.

Houlihan, M. M., Dickson-Parnell, B. E., Jackson, J., & Zeichner, A. (1987). Appearance changes associated with participation in a behavioral weight control program. *Addictive Behaviors,* 12, 157–163.

Jakicic, J. M., Wing, R. R., & Winters-Hart, C. (2002). Relationship of physical activity to eating behaviors and weight loss in women. *Medicine & Science in Sports & Exercise,* 34, 1653–1659.

Miller, K. J., Gleaves, D. H., Hirsch, T. G., Green, B. A., Snow, A. C., & Corbett, C. C. (2000). Comparisons of body image dimensions by race/ethnicity and gender in a university population. *International Journal of Eating Disorders,* 27(3), 310–316.

Powell, L., Calvin, J., & Calvin, J. (2007). Effective obesity treatments. *American Psychologist,* 62, 234–246.

Roberts, A., Cash, T., Feingold, A., & Johnson, B. (2006). Are black–white differences in females' body dissatisfaction decreasing? A meta-analytic review. *Journal of Consulting and Clinical Psychology,* 74, 1121–1131.

Rodin, J. & Larsen, L. (1992). Social factors and the ideal body shape. In Brownell, K., Rodin. J., & Wilmore. J. H. (Eds.), *Eating, body weight; and performance in athletes* (pp. 146–158): Philadelphia, PA: Lea & Febiger.

Thompson, J. K., Heinburg, L. J., Altabe, M., & Tantleff-Dunn, S. (1999). *Exacting beauty: Theory, assessment, and treatment of body disturbance.* Washington DC: American Psychological Association.

Source: Annesi, J. J. (2009). Correlations of changes in weight and body satisfaction for obese women initiating exercise: Assessing effects of ethnicity. *Psychological Reports, 105*(3), 1072–1076.

Unlike more extended writings (e.g., monographs, or books), they include only a portion of the background information and only partial descriptions of related studies that would aid the reader's comprehension of the study. Reading books and articles listed in the references section provides you with some of this information and probably gives you a clearer understanding as to why and how the author conducted the particular study you have just read. Before hunting down any particular reference item, it is a good idea to look back into the article to reread the sentence or paragraph containing the original citation to give you an idea of what is in each referenced item.

Notes

Near the beginning or end of their research reports, authors sometimes present one or more **notes.** In general, such notes are used by authors for three reasons: (1) to thank others who helped them with their study or with the preparation of the technical report, (2) to clarify something that was discussed earlier in the journal article, and (3) to indicate how an interested reader can contact them to discuss this particular study or other research that might be conducted in the future. In our model journal article, there was a single note containing the author's postal address and email address.

Two Final Comments

As we come to the end of this chapter, consider two final points. One concerns the interconnectedness among the different components of the research summary. The other concerns the very first of those components: the abstract.

In this chapter, we dissected a journal article that summarizes a research study focused on weight loss among obese women. In looking at this particular article section by section, you may have gotten the impression that each of the various parts of a research article can be interpreted and evaluated separately from the other sections that go together to form the full article. You should not leave this chapter with that thought, because the various parts of a well-prepared research report are tied together to create an integrated whole.

In our model journal article, the researchers had two principal hypotheses, shown in Excerpt 1.4. Those same hypotheses are the focus of the second paragraph in the research report's Results section (see Excerpt 1.9), the entirety of the Discussion section (see Excerpt 1.10), and the first two sentences of the abstract (see Excerpt 1.1). The author who prepared this journal article deserves high marks for keeping focused on the study's central hypotheses and for showing a clear connection between those hypotheses and his findings. Unfortunately, many journal articles display very loose (and sometimes undetectable) connections between the component parts of their articles.

My final comment takes the form of a warning. Simply stated, do not read an abstract and then think that you understand the study well enough to forgo reading the entire research report. As was stated earlier, an abstract gives you a thumbnail sketch of a study, thus allowing you to decide whether the article fits into your area of interest. If it does not, then you rightfully can move on. However, if an abstract makes it appear that the study is, in fact, consistent with your interests, you must then read the entire article for two reasons. First, the results summarized in the abstract may not coincide with the information that appears in the results section of the full article. Second, you cannot properly evaluate the quality of the results—even if they are consistently presented in the abstract, results, and discussion sections of the article—without coming to understand who or what was measured, how measurements were taken, and what kinds of statistical procedures were applied.

If you read an abstract (but nothing else in the article) and then utilize the abstract's information to bolster your existing knowledge or guide your own research projects, you potentially harm rather than help yourself, because the findings reported in some abstracts are simply not true. To be able to tell whether an abstract can be trusted, you must read the full research report. The rest of this book has been written to help make that important task easier for you.

Review Terms

Abstract	Materials
Apparatus	Measures
Background	Method
Conclusion	Notes
Dependent variable	Participant
Discussion	Procedure
Equipment	References
Figure	Results
Instruments	Statement of purpose
Introduction	Subject
Hypotheses	

The Best Items in the Companion Website

1. An important email message sent by the author at the beginning of the semester to students enrolled in his statistics and research course.
2. An interactive online quiz (with immediate feedback provided) covering Chapter 1.
3. Gary Gildner's wonderful poem entitled "Statistics."
4. Five misconceptions about the content of Chapter 1.

To access the chapter outline, practice tests, weblinks, and flashcards, visit the companion website at http://www.ReadingStats.com.

Review Questions and Answers begin on page 531.

Descriptive Statistics

The Univariate Case

In this chapter, we consider descriptive techniques designed to summarize data on a single dependent variable. These techniques are often said to be **univariate** in nature because only one variable is involved. (In Chapter 3, we look at several techniques designed for the **bivariate** case—that is, for situations in which data have been collected on two dependent variables.)

We begin this chapter by looking at several ways data can be summarized using picture techniques, including frequency distributions, stem-and-leaf displays, histograms, and bar graphs. Next, the topic of *distributional shape* is considered; here, you learn what it means when a data set is said to be normal, skewed, bimodal, or rectangular. After that, we examine the concept of *central tendency* and various methods used to represent a data set's average score. We then turn our attention to how researchers usually summarize the variability, or *spread,* within their data sets; these techniques include four types of range, the standard deviation, and the variance. Finally, we consider two types of standard scores: *z* and *T*.

Picture Techniques

In this section, we consider some techniques for summarizing data that produce a picture of the data. I use the term *picture* somewhat loosely, because the first technique really leads to a table of numbers. In any event, our discussion of descriptive statistics begins with a consideration of three kinds of frequency distributions.

Frequency Distributions

A **frequency distribution** shows how many people (or animals or objects) were similar in the sense that, measured on the dependent variable, they ended up in the same category or had the same score. Two kinds of frequency distributions are often seen in published journal articles: simple frequency distributions and grouped frequency distributions.

In Excerpt 2.1, we see an example of a **simple frequency distribution,** also called an **ungrouped frequency distribution.** The data here come from a study focused on the physical activity of school children aged 9 through 13. A brief self-report survey was completed by students in seven randomly selected schools in a large city in Florida, with one question asking the children to indicate how many days during the past week they participated in a game or sport, for at least 20 minutes, that caused them to breathe hard or sweat. This question's response options are shown in the left column of Excerpt 2.1. The numbers in the right column indicate how many students got each possible score. (Thus, there were 445 students who said they exercised vigorously seven days a week, 174 who said they did this six days a week, and so on.) The numbers in parentheses in the right-hand column indicate the percent of the full group that ended up with each possible option when answering the question about vigorous exercise.

EXCERPT 2.1 • *Simple Frequency Distribution*

TABLE 2 *Number of Days per Week Engaged in Vigorous–Intensity Physical Activity (VPA), Students (N = 1,407) in Grades 5 through 7, Sarasota County, Florida.*

No. of Days per Week Engaged in VPA	No. of Students (%)
7	445 (31.6)
6	174 (12.4)
5	214 (15.2)
4	191 (13.6)
3	145 (10.3)
2	107 (7.6)
1	75 (5.3)
0	56 (4.0)

Source: McDermott, R. J., Nickelson, J., Baldwin, J. A., Bryant, C. A., Alfonso, M., Phillips, L. M., et al. (2009). A community–school district–university partnership for assessing physical activity of tweens. *Preventing Chronic Disease.* 6(1), http://www.cdc.gov/pcd/issues/2009/jan/07_0243.htm.

Slight variations of Excerpt 2.1 appear often in research reports. The left column may correspond to a categorical variable, such as blood type, rather than anything that's quantitative. Also, the word *frequency* (or its abbreviation, *f*) might be used to label the right column of numbers. Finally, the total number of individuals is sometimes indicated by a single number, labeled *N*, positioned beneath the right column of numbers.

In Excerpt 2.2, we see an example of a **grouped frequency distribution.** This frequency distribution shows the ages of a large group of individuals who, over a 20-year period, had a coronary angiogram at the researchers' hospital. Excerpt 2.2 actually contains three frequency distributions, one for each gender and one for the combined group of males and females.

EXCERPT 2.2 • *Grouped Frequency Distribution*

TABLE 1   *Age Distribution of the Study Subjects*

Age Group (years)	Men (n)	Women (n)	Total (N)
<31	97	14	111
31–40	559	61	620
41–50	2216	345	2561
51–60	4982	1287	6269
61–70	4345	1595	5940
71–80	1251	511	1762
>80	44	16	60
Total	**13494**	**3829**	**17323**

Source: Giannoglou, G. D., Antoniadis, A. P., Chatzizisis, Y. S., & Louridas, G. E. (2010). Difference in the topography of atherosclerosis in the left versus right coronary artery in patients referred for coronary angiography. *BMC Cardiovascular Disorders, 10*(26), 1–6.

The table in Excerpt 2.2 is a *grouped* frequency distribution because the far left column has, on each row, a group of possible ages. This grouping of the patients into the seven age categories—into what are technically called *class intervals*—allows the data to be summarized in a more compact fashion. If the data in this excerpt had been presented in an ungrouped frequency distribution, with the far left column set up to reflect individual ages (e.g., 31, 32, 33), at least 52 rows would have been needed, and perhaps more than that, depending on the ages of the youngest and oldest patients.

Stem-and-Leaf Display

Although a grouped frequency distribution provides information about the scores in a data set, it carries with it the limitation of *loss of information*. The frequencies tell us how many data points fell into each interval of the score continuum, but they do not indicate, within any interval, how large or small the scores were. Hence, when researchers summarize their data by moving from a set of raw scores to a grouped frequency distribution, the precision of the original scores is lost.

A **stem-and-leaf display** is like a grouped frequency distribution that contains no loss of information. To achieve this objective, the researcher first sets up score intervals on the left side of a vertical line. These intervals, collectively called the *stem,* are presented in a coded fashion by showing the lowest score of each interval. Then, to the right of the vertical line, the final digit is given for each observed score that fell into the interval being focused on. An example of a stem-and-leaf display is presented in Excerpt 2.3. In this excerpt, the stem numbers are separated from the leaf numbers by a vertical set of asterisks rather than an actual line.

EXCERPT 2.3 • *Stem-and-Leaf Display*

Frequency	Stem	&	Leaf
1.00	2	*	2
8.00	2	*	56667899
19.00	3	*	0001122223333333444
13.00	3	*	5567788899999
10.00	4	*	0000112344
17.00	4	*	55555666677888899
9.00	5	*	000124444
4.00	5	*	5567

Stem width: 10
Each leaf: 1 case(s)

FIGURE 1 *Stem and Leaf Plot of Age*

Source: Chyung, S. Y. (2007). Age and gender differences in online behavior, self-efficacy, and academic performance. *Quarterly Review of Distance Education, 8*(3), 213–222.

To make sense of this stem-and-leaf display, we must do two things. First, we deal with the stem. Near the bottom of Excerpt 2.3, there is a note indicating the *stem width* is 10. This means we should multiple each stem number by 10 as we interpret the data. Thus, the stem numbers actually range from 20 to 50, not 2 to 5. Second, we combine stem with leaf. On the top row, there is a 2 on the left (stem) side of the column of asterisks and a 2 on the right (leaf) side. This indicates that there was one person in the group who was 22 years old. (We get 22 by adding the leaf number of 2 to the true stem number of 20.) The second row has eight digits on the leaf side, thus indicating that eight people fell into this row's interval (25–29). Using both stem and leaf from this row, we see that one of those eight people was

25 years old, three were 26, one was 27, one was 28, and two were 29. All other rows of this stem-and-leaf display are interpreted in the same way.

Notice that the individual ages are on display in the stem-and-leaf display shown in Excerpt 2.3. There is, therefore, no loss of information. Take another look at Excerpt 2.2, where a grouped frequency distribution was presented. Because of the loss of information associated with grouped frequency distributions, you cannot tell how old the youngest and oldest patients were, what specific ages appeared within any interval, or whether gaps exist inside any intervals.

Histograms and Bar Graphs

In a **histogram,** vertical columns (or thin lines) are used to indicate how many times any given score appears in the data set. With this picture technique, the baseline (i.e., the horizontal axis) is labeled to correspond with observed scores on the dependent variable whereas the vertical axis is labeled with frequencies.[1] Then, columns (or lines) are positioned above each baseline value to indicate how often each of these scores was observed. Whereas a tall bar indicates a high frequency of occurrence, a short bar indicates that the baseline score turned up infrequently.

A **bar graph** is almost identical to a histogram in both form and purpose. The only difference between these two techniques for summarizing data concerns the nature of the dependent variable that defines the baseline. In a histogram, the horizontal axis is labeled with numerical values that represent a quantitative variable. In contrast, the horizontal axis of a bar graph represents different categories of a qualitative variable. In a bar graph, the ordering of the columns is quite arbitrary, whereas the ordering of the columns in a histogram must be numerically logical.

In Excerpt 2.4, we see an example of a histogram. Notice how this graph allows us to quickly discern that about 80 percent of the 1,127 individuals in episodic migraine group were experiencing severe headaches more than once every three months but not on a weekly basis. Also notice that the columns must be arranged as they are because the variable on the baseline is clearly quantitative in nature.

An example of a bar graph is presented in Excerpt 2.5. Here, the order of the bars is completely arbitrary. The short bars could have been positioned on the left with the taller bars positioned on the right, or the bars could have been arranged alphabetically based on the labels beneath the bars.

Frequency Distributions in Words

Researchers sometimes present the information of a frequency distribution in words rather than in a picture. Excerpt 2.6 illustrates how this can be done for a grouped frequency distribution.

---

[1]Technically speaking, the horizontal and vertical axes of any graph are called the *abscissa* and *ordinate*, respectively.

EXCERPT 2.4 • *Histogram*

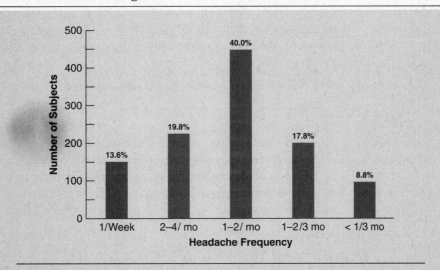

FIGURE 1 *Headache frequency in the episodic migraine group.*

Source: Radat, F., Lantéri-Minet, M., Nachit-Ouinekh, F., Massiou, H., Lucas, C., Pradalier, A., et al. (2008). The GRIM2005 study of migraine consultation in France: III: Psychological features of subjects with migraine. *Cephalalgia*, *29*, 338–350.

EXCERPT 2.5 • *Bar Graph*

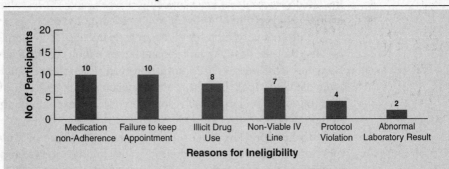

FIGURE 4 *Primary reasons for excluding participants from study after attending medical screening visit (n = 41).*

Source: Faseru, B., Cox, L. S., Bronars, C. A., Opole, I., Reed, G. A., Mayo, M. S., et al. (2010). Design, recruitment, and retention of African-American smokers in a pharmacokinetic study. *BMC Medical Research Methodology*, *10*, 1–8.

EXCERPT 2.6 • *A Frequency Distribution Described in a Sentence*

The respondents ranged in age from 21–25 (14%), 26–30 (18%), 31–35 (16%), 36–40 (20%), 41–45 (10%), 46–50 (12%), 51–56 (8%), to 56–50 years (2%).

Source: Heinrichs, J. H., & Lim, J. (2010). Information literacy and office tool competencies: A benchmark study. *Journal of Education for Business, 85*(3), 153–164.

In describing data gathered from human research participants, animal subjects, or inanimate objects, many researchers summarize their data by presenting only a mean and a standard deviation.[2] Few researchers provide descriptive summaries like we have seen in Excerpts 2.1 through 2.6. This is unfortunate because such picture techniques, as you can see, allow us to get a good "feel" for a researcher's data.

Distributional Shape

If researchers always summarized their quantitative data using one of the picture techniques just considered, then you could *see* whether the observed scores tended to congregate at one (or more) point along the score continuum. Moreover, a frequency distribution, a stem-and-leaf display, a histogram, or a bar graph allow you to tell whether a researcher's data are symmetrical. To illustrate this nice feature, take another look at Excerpt 2.1. That frequency distribution shows nicely that (1) the children varied widely in terms of how frequently they engaged in vigorous physical activity (VPA) each week, (2) over half of the children were involved in this kind of activity at least five days a week, and (3) less than 10 percent of the children engaged in VPA just once a week or not at all.

Unfortunately, pictures of data sets do not appear in journal articles very often because they are costly to prepare and because they take up a large amount of space. By using some verbal descriptors, however, researchers can tell their readers what their data sets look like. To decipher such messages, you must understand the meaning of a few terms that researchers use to describe the **distributional shape** of their data.

If the scores in a data set approximate the shape of a **normal distribution,** most of the scores are clustered near the middle of the continuum of observed scores, and there is a gradual and symmetrical decrease in frequency in both directions away from the middle area of scores. Data sets that are normally distributed are said to resemble a bell-shaped curve, because a side drawing of a bell begins low on either side and then bulges upward in the center. In Excerpts 2.2 and 2.4, we see two sets of data that resemble, somewhat, a normal distribution.

[2]We consider the *mean* and the *standard deviation* later in this chapter.

In **skewed distributions,** most of the scores end up being high or low, with a small percentage of scores strung out in one direction away from the majority. Skewed distributions, consequently, are not symmetrical. If the tail of the distribution (formed by the small percentage of scores strung out in one direction) points toward the upper end of the score continuum, the distribution is said to be **positively skewed;** if the tail points toward the lower end of the score continuum, the term **negatively skewed** applies. In Excerpt 2.7, we see an example of a negatively skewed distribution.

EXCERPT 2.7 • *Negative Skewness*

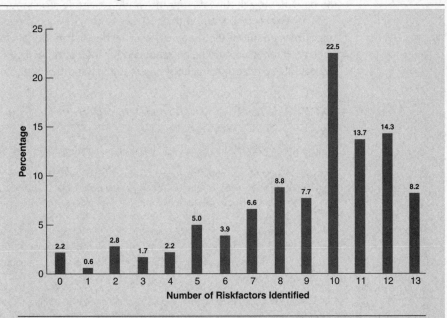

FIGURE 2 *Proportion of patients who could correctly identify different numbers of stroke/TIA risk factors.*

Source: Sloma, A., Backlund, L. G., Strender, L-E., & Skånér, Y. (2010). Knowledge of stroke risk factors among primary care patients with previous stroke of TIA: A questionnaire study. *BMC Family Practice, 11*(47), 1–10.

If the scores tend to congregate around more than one point along the score continuum, the distribution is said to be **multimodal** in nature. If there are two such places where scores are grouped together, we could be more specific and say that the data are distributed in a **bimodal** fashion. If scores are congregated at three distinct points, we use the term **trimodal.**[3] To see a bimodal distribution, take another

---

[3]Distributions having just one "hump" are said to be *unimodal* in nature.

look at the stem-and-leaf display in Excerpt 2.3. The primary mode is the 30–34 age category. There is a secondary mode, however, represented by the individuals with ages between 45 and 49.

If scores are fairly evenly distributed along the score continuum without any clustering at all, the data set is said to be **rectangular** (or **uniform**). Such a distributional shape would probably show up if someone (1) asked each person in a large group to indicate his or her birth month, and (2) created a histogram with 12 bars, beginning with January, arranged on the baseline. The bars making up this histogram would probably be approximately the same height. Looked at collectively, the bars making up this histogram would resemble a rectangle.

In Excerpts 2.8 through 2.10, we see a few examples of how researchers sometimes go out of their way to describe the distributional shape of their data sets. Such researchers should be commended for indicating what their data sets look like, because these descriptions help others understand the nature of the data collected.

EXCERPTS 2.8–2.10 • *References to Different Distributional Shapes*

Preliminary analyses revealed that the data were normally distributed.

Source: de la Sablonnière, R., Tougas, F., & Perenlei, O. (2010). Beyond social and temporal comparisons: The role of temporal inter-group comparisons in the context of dramatic social change in Mongolia. *Journal of Social Psychology, 150*(1), 98–115.

The sample's actual range of scores on the Emotional Social Support scale covered the entire theoretical range of the scale (6–30), but the distribution was extremely negatively skewed. . . .

Source: Rosenthal, B. S., Wilson, W. C., & Futch, V. A. (2010). Traumatic, Protection, and distress in late adolescence: A multi-determinant approach. *Adolescence, 44*(176), 693–703.

The histograms of the score distributions [for] both the INSPIRIT and coping religion scales [indicated] slight tendencies toward bimodal distributions.

Source: Lowis, M. J., Edwards, A. C., & Burton, M. (2010). Coping with retirement: Well-being, health, and religion. *Journal of Psychology, 143*(4), 427–448.

As we have seen, two features of distributional shape are modality and skewness. A third feature is related to the concept of **kurtosis**. This third way of looking at distributional shape deals with the possibility that a set of scores can be non-normal even though there is only one mode and even though there is no skewness in the data. This is possible because there may be an unusually large number of scores at the center of the distribution, thus causing the distribution to be overly peaked.

Or, the hump in the middle of the distribution may be smaller than is the case in normal distributions, with both tails being thicker than in the famous bell-shaped curve.

When the concept of kurtosis is discussed in research reports, you may encounter the terms **leptokurtic** and **platykurtic,** which denote distributional shapes that are more peaked and less peaked (as compared with the normal distribution), respectively. The term **mesokurtic** signifies a distributional shape that is neither overly peaked nor overly flat.

As illustrated in Excerpts 2.7 through 2.10, researchers can communicate information about distributional shape via a picture or a label. They can also compute numerical indices that assess the degree of skewness and kurtosis present in their data. In Excerpt 2.11, we see a case in which a group of researchers presented such indices in an effort to help their readers understand what kind of distributional shape was created by each set of scores that had been gathered.

EXCERPT 2.11 • *Quantifying Skewness and Kurtosis*

Internalized and externalized CB mean scores were calculated for each participant. . . . The distribution of externalized CB mean scores is much less normal (skewness = 2.43; kurtosis = 6.10) relative to the internalized CB mean scores (skewness = 0.62; kurtosis = −0.49).

Source: Field, N. P., & Filanosky, C. (2010). Continuing bonds, risk factors for complicated grief, and adjustment to bereavement. *Death Studies, 34*(1), 1–29.

To properly interpret coefficients of skewness and kurtosis, keep in mind three things. First, both indices turn out equal to zero for a normal distribution.[4] Second, a skewness value lower than zero indicates that a distribution is negatively skewed, whereas a value larger than zero indicates that a distribution is positively skewed; a kurtosis value less than zero indicates that a distribution is platykurtic, whereas a value greater than zero indicates that the distribution is leptokurtic. Finally, although there are no clear-cut guidelines for interpreting measures of skewness and kurtosis (mainly because there are different ways to compute such indices), most researchers consider data to be approximately normal in shape if the skewness and kurtosis values turn out to be anywhere from −1.0 to +1.0.

Depending on the objectives of the data analysis, a researcher should examine coefficients of skewness and kurtosis before deciding how to further analyze the data. If a data set is found to be grossly non-normal, the researcher may opt to do further analysis of the data using statistical procedures created for the non-normal

---

[4]Some formulas for computing skewness and kurtosis indices yield a value of +3 for a perfectly normal distribution. Most researchers, however, use the formulas that give values of zero (to normal distributions) for both skewness and kurtosis.

case. Or, the data can be *normalized* by means of a formula that revises the value of each score such that the revised data set represents a closer approximation to the normal. In Chapters 10–15, we consider examples of both of these options.

Measures of Central Tendency

To help readers get a feel for the data that have been collected, researchers almost always say something about the typical or representative score in the group. They do this by computing and reporting one or more **measures of central tendency.** There are three such measures that are frequently seen in the published literature, each of which provides a numerical index of the **average** score in the distribution.

The Mode, Median, and Mean

The **mode** is simply the most frequently occurring score. For example, given the nine scores 6, 2, 5, 1, 2, 9, 3, 6, and 2, the mode is equal to 2. The **median** is the number that lies at the midpoint of the distribution of earned scores; it divides the distribution into two equally large parts. For the set of nine scores just presented, the median is equal to 3. Four of the nine scores are smaller than 3; four are larger.[5] The **mean** is the point that minimizes the collective distances of scores from that point. It is found by dividing the sum of the scores by the number of scores in the data set. Thus, for the group of nine scores presented here, the mean is equal to 4.

In journal articles, authors sometimes use abbreviations or symbols when referring to their measure(s) of central tendency. The abbreviations *Mo* and *Mdn,* of course, correspond to the mode and median, respectively. The letter *M* always stands for the mean, even though all three measures of central tendency begin with this letter. The mean is also symbolized by $\overline{X}$ and μ.

In many research reports, the numerical value of only one measure of central tendency is provided. (That was the case with the model journal article presented in Chapter 1; take a look at Excerpt 1.9 to see which one was used.) Because it is not unusual for a real data set to be like our sample set of nine scores in that the mode, median, and mean assume different numerical values; researchers sometimes compute and report two measures of central tendency, or all three, to help readers better understand the data being summarized.

In Excerpt 2.12, we see a case where all three averages—the mode, the median, and the mean—were provided for each of two groups of hospital patients. It is helpful to have all three quantitative assessments of the average hospital stay. Together,

---

[5]When there is an even number of scores, the median is a number halfway between the two middle scores (once the scores are ordered from low to high). For example, if 9 is omitted from our sample set of scores, the median for the remaining eight scores is 2.5—that is, the number halfway between 2 and 3.

EXCERPT 2.12 • *Reporting Multiple Measures of Central Tendency*

It is the opinion of the authors that patients are often needlessly kept in hospital as inpatients. . . . The mean stay of the cohort patients was 12.9 days (median = 10, mode = 9), with the mean stay of patients who spent a potentially avoidable night of 13.7 days (median = 10.5, mode = 6).

Source: Forde, D., O'Connor, M. B., & Gilligan. (2009). Potentially avoidable inpatient nights among warfarin receiving patients; an audit of a single university teaching hospital. *BMC Research Notes, 2*(41), 1–5.

they paint a more complete picture of the data than would have been the case if only one had been reported.

The Relative Position of the Mode, Median, and Mean

In a true normal distribution (or in any unimodal distribution that is perfectly symmetrical), the values of the mode, median, and mean are identical. Such distributions are rarely seen, however. In the data sets typically found in applied research studies, these three measures of central tendency assume different values. As a reader of research reports, you should know not only that this happens but also how the distributional shape of the data affects the relative position of the mode, median, and mean.

In a positively skewed distribution, a few scores are strung out toward the high end of the score continuum, thus forming a tail that points to the right. In this kind of distribution, the modal score ends up being the lowest (i.e., positioned farthest to the left along the horizontal axis), whereas the mean ends up assuming the highest value (i.e., positioned farthest to the right). In negatively skewed distributions, just the opposite happens; the mode ends up being located farthest to the right along the baseline, whereas the mean assumes the lowest value. In Figure 2.1, we see a picture showing where these three measures of central tendency are positioned in skewed distributions.

After you examine Figure 2.1, return to Excerpt 2.3 and look at the stem-and-leaf display that summarizes the ages of 81 individuals. Because the distribution is not skewed very much, we should expect the mean, the median, and the mode to end up being somewhat similar. The actual values for these three measures of central tendency are 39.6, 39.9, and 33.0, respectively.

To see a case where the computed measures of central tendency turn out to be quite dissimilar, thus implying skewed data, consider again Excerpt 2.12. The mean, median, and mode for the first group of hospital patients are different from one another; in the second group, the mean and the mode are highly dissimilar (especially considering the cost of staying overnight in the hospital). To see if you can determine

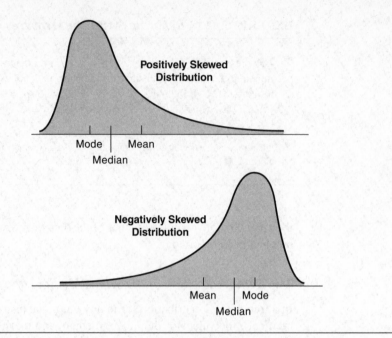

FIGURE 2.1 *Location of the Mean, Median, and Mode in Skewed Distributions*

the nature of skewness from reported values of central tendency, would you guess that the two distributions of scores reported in Excerpt 2.12 were positively skewed or negatively skewed?

In a bimodal distribution, there are two points along the score continuum where scores tend to pile up. If the distribution is symmetrical, the mean and median are located halfway between the two modes. In a symmetrical trimodal distribution, the median and mean assume a value equal to the middle of the three modes. Real data sets, however, rarely produce symmetrical bimodal or trimodal distributions. Any asymmetry (i.e., skewness) causes the median to be pulled off center toward the side of the distribution that has the longer tail—and the mean is pulled even farther in that direction.

With full-fledged rectangular distributions, the mean and median assume a value halfway between the high and low data points. In such distributions, there is no mode because all earned scores occur with equal frequency. If the distribution turns out to be only roughly rectangular, the median and mean are located close together (and close to the halfway point between the high and low scores), but the mode could end up anywhere.

Other Measures of Central Tendency

Although the mode, median, and mean are the most popular measures of central tendency, there are other techniques for summarizing the average score in a data set.

Examples include, the geometric mean, the harmonic mean, and the trimmed mean. Although you are unlikely to see these measures of central tendency reported very often, researchers may use them if their scores are extremely asymmetric or if the high or low scores are of questionable validity. In Excerpt 2.13, we see a case when the geometric mean is used for these first of these reasons.[6]

EXCERPT 2.13 • *The Geometric Mean*

Scalp hair samples, well known as a suitable specimen for monitoring human exposure to mercury, were used in this study. . . . The geometric mean (geomean) rather than the arithmetic mean was used to represent the hair mercury concentrations.

Source: Ryo, K., Ito, A., Takatori, R., Tai, Y., Tokunaga, J., Arikawa, K., et al. (2010). Correlation between mercury concentrations in hair and dental amalgam fillings. *Anti-Aging Medicine*, *7*(3), 14–17.

Measures of Variability

Descriptions of a data set's distributional shape and reports as to the central tendency value(s) help us better understand the nature of data collected by a researcher. Although terms (e.g., *roughly normal*) and numbers (e.g., $M = 67.1$) help, they are not sufficient. To get a true feel for the data that have been collected, we must also be told something about the variability among the scores. Let us consider now the standard ways that researchers summarize this aspect of their data sets.

The Meaning of Variability

Most groups of scores possess some degree of variability. That is, at least some of the scores differ (vary) from one another. A **measure of variability** simply indicates the degree of this **dispersion** among the scores. If the scores are very similar, there is little dispersion and little variability. If the scores are very dissimilar, there is a high degree of dispersion (variability). In short, a measure of variability does nothing more than indicate how spread out the scores are.

The term *variability* can also be used to pinpoint where a group of scores might fall on an imaginary homogeneous–heterogeneous continuum. If the scores are similar, they are **homogeneous** (and have low variability). If the scores are dissimilar, they are **heterogeneous** (and have high variability).

Even though a measure of central tendency provides a numerical index of the average score in a group, we must know the variability of the scores to better

[6]For any set of *N* numbers, the geometric mean is equal to the *N*th root of the product of the numbers. Thus, if our numbers are 2, 2, and 16, the product of the numbers is 64, and the cube root of this product gives us the geometric mean: 4.

understand the entire group of scores. For example, consider the following two groups of IQ scores:

Group I	Group II
102	128
99	78
103	93
96	101

In both groups the mean IQ is equal to 100. Although the two groups have the same mean score, their variability is obviously different. Whereas the scores in the first group are very homogeneous (low variability), the scores in the second group are far more heterogeneous (high variability).

The specific measures of variability that we consider next are similar in that the numerical index is zero if all of the scores in the data set are identical, a small positive number if the scores vary to a small degree, or a large positive number if there is a great deal of dispersion among the scores. (No measure of variability, no matter how computed, can ever turn out equal to a negative value.)

The Range, Interquartile Range, Semi-Interquartile Range, and Box Plot

The **range** is the simplest measure of variability. It is the difference between the lowest and highest scores. For example, in Group I of the example just considered, the range is equal to 103 − 96, or 7. The range is usually reported by citing the extreme scores, but sometimes it is reported as the difference between the high and low scores. When providing information about the range to their readers, authors usually write out the word *Range*. Occasionally, however, this first measure of variability is abbreviated as *R*.

To see how the range can be helpful when we try to understand a researcher's data, consider Excerpts 2.14 and 2.15. Notice in Excerpt 2.14 how information concerning the range allows us to sense that the participants in this study are quite heterogeneous in terms of age. In contrast, the presentation of just the mean in Excerpt 2.15 puts us in the position of not knowing anything about how much variability exists among the parents' scores. Perhaps it was a very homogeneous group, with everyone having a near-normal level of marital dissatisfaction. Or, maybe the group was bimodal, with half the parents highly distressed and the other half not distressed at all. Unless the range (or some other measure of variability) is provided, we are completely in the dark as to how similar or different the parents are in terms of their scores on marital distress.

Whereas the range provides an index of dispersion among the full group of scores, the **interquartile range** indicates how much spread exists among the middle 50 percent of the scores. Like the range, the *interquartile range* is defined as the distance between a low score and a high score; these two indices of dispersion differ,

EXCERPTS 2.14–2.15 • *Summarizing Data with and without the Range*

Participants ranged in age from 27 to 54; their average age was 42.4.

Source: Wilson, E. K., Dalberth, B. T., Koo, H. P., & Gard, J. C. (2010). Parents' perspectives on talking to preteenage children about sex. *Perspectives on Sexual & Reproductive Health, 42*(1), 56–63.

The parents' mean score of 49.9 on the Global Marital Distress (dissatisfaction) scale of the Marital Satisfaction Inventory [turned out] within the normal range, indicating that these parents were in non-distressed marriages.

Source: Kilmann, P. R., Vendemia, J. M. C., Parnell, M. M., & Urbaniak, G. C. (2009). Parent characteristics linked with daughters' attachment styles. *Adolescence, 44*(175), 557–568.

however, in that the former is based on the high and low scores within the full group of data points, whereas the latter is based on only *half* of the data—the middle half.

In any group of scores, the numerical value that separates the top 25 percent scores from the bottom 75 percent scores is the upper quartile (symbolized by Q_3). Conversely, the numerical value that separates the bottom 25 percent scores from the top 75 percent scores is the lower quartile (Q_1).[7] The interquartile range is simply the distance between Q_3 and Q_1. Stated differently, the interquartile range is the distance between the 75th and 25th percentile points.

In Excerpt 2.16, we see a case in which the upper and lower **quartiles** are presented. In this excerpt, the values of Q_1 and Q_3 give us information as to the dispersion among the middle 50 percent of the scores. Using that information along with that provided by the range, we can tell that the middle half of the individuals are between 59 and 71 years old, with the oldest one-fourth of the study's participants being between 71 and 86, and the youngest one-fourth being between 28 and 59.

EXCERPT 2.16 • *The Interquartile Range*

The median age of all participants was 66 years (interquartile range 59–71; range 28–86 years).

Source: Donaghey, C. L., McMillan, T. M., O'Neill, B. (2010). Errorless learning is superior to trial and error when learning a practical skill in rehabilitation: a randomized controlled trial. *Clinical Rehabilitation, 24*(3), 195–201.

[7]The middle quartile, Q_2, divides any group of scores into upper and lower halves. Accordingly, Q_2 is always equal to the median.

Sometimes, a researcher computes the **semi-interquartile range** to index the amount of dispersion among a group of scores. As you may guess on the basis of its name, this measure of variability is simply equal to one-half the size of the interquartile range; in other words, the semi-interquartile range is nothing more than $(Q_3 - Q_1)/2$.

With a **box plot,** the degree of variability within a data set is summarized with a picture. To accomplish this objective, a rectangle (box) is drawn such that it can be "stretched out" parallel to the axis representing the dependent variable. Some researchers position the dependent variable on a vertical axis on the left side of the graph, and in those cases the box plot stretches upward and downward. Other researchers put the dependent variable on the horizontal axis at the bottom of the graph, and box plots in these cases stretch toward the left and right.

Regardless of how a box plot is oriented, it shows graphically the variability in the data. The positions of the rectangle's ends are determined by Q_3 and Q_1, the upper and lower quartile points. Two vertical lines—sometimes called the *whiskers*—are drawn to show variability *beyond* the 75th and 25th percentiles. Researchers use different rules for drawing the whiskers, however. Sometimes the whiskers extend to the highest and lowest observed scores. In other graphs, the whiskers are drawn so they extend out to points that represent the 10th and 90th percentiles (or perhaps to the 5th and 95th percentiles). Some researchers use a rule that says that neither whisker should be longer than 1.5 times the height of the rectangle, with scores farther out than this considered to be outliers that are then indicated by small circles or asterisks.

In Excerpt 2.17, we see a case in which box plots were used to show the variability among three groups that took a test to assess their knowledge of and skills in applying evidence-based practice in physical therapy. This set of box plots shows that, on average, the faculty members scored better than either of the two student groups. The box plots also show that the faculty group was more varied in the scores they earned on the test. (Be sure to read the note beneath the box plot, as it explains the meaning of *whisker length.*)

Although box-and-whisker plots are designed to communicate information about variability, they reveal things about distributional shape. If the whiskers are of equal lengths, then we can infer that the distribution of scores is probably symmetrical. We cannot be sure of that, but such a guess is likely to be a good one. However, we should guess that the distribution of scores is skewed if the whiskers are of unequal lengths. (If the longer whisker extends toward the higher end of the score continuum, the distribution is probably positively skewed; conversely, negatively skewed distributions cause the longer whisker to extend toward lower scores.)

Standard Deviation and Variance

Two additional indices of dispersion, the **standard deviation** and the **variance,** are usually better indices of dispersion than are the first three measures of variability we considered. This is because these two measures are each based on all of the

EXCERPT 2.17 • *Box Plots*

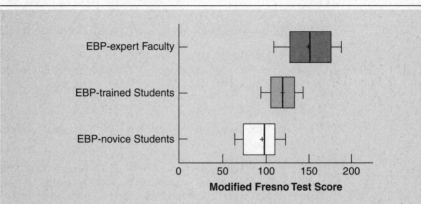

FIGURE 1 *Modified Fresno Test scores by group.*

Box and whisker plot of modified Fresno Test scores for EBP–novice PT Students ($n = 31$), EBP-trained PT Students ($n = 50$), and EBP–expert PT Faculty ($n = 27$). The central box spans from the lower to the upper quartile, the middle line represents the median, the "+" sign represents the mean, the whiskers extend from the 10th percentile to the 90th percentile of scores.

Source: Tilson, J. K. (2010). Validation of the Modified Fresno Test: Assessing physical therapists' evidence based practice knowledge and skills. *BMC Medical Education, 10*(38), 1–9.

scores (and not just the high and low scores or the upper and lower quartile points). The standard deviation is determined by (1) figuring how much each score deviates from the mean and (2) putting these deviation scores into a computational formula. The variance is found by squaring the value of the standard deviation.

In reporting their standard deviations, authors may use the abbreviation *SD*, the symbol *s* or σ, or simply write out the word **sigma.** Occasionally, authors report the standard deviation using a plus/minus format—for example, 14.83 ± 2.51, where the first number (14.83) stands for the mean and the second number (2.51) stands for the standard deviation. The *variance,* being the square of the standard deviation, is symbolized as s^2 or σ^2.

Excerpts 2.18 and 2.19 illustrate two of the ways researchers indicate the numerical value of the standard deviation. In the first of these, the abbreviation *SD* is used, whereas in the second the plus/minus format is used. These two formats for presenting the standard deviation are often seen in research reports.

Excerpt 2.20 shows how information on the standard deviation can be included in a table. In this excerpt, each row of numbers corresponds to a different variable considered important within the researchers' study. In this table, notice that the abbreviation *SD* is used to represent the term *standard deviation*.

EXCERPTS 2.18–2.19 • *Reporting on the Standard Deviation*

The mean age of the HC group was 69.79 years ($SD = 7.51$).

Source: Sapir, S., Ramig, L. O., Spielman, J. L., & Fox, C. (2010). Formant centralization ratio: A proposal for a new acoustic measure of dysarthric speech. *Journal of Speech, Language & Hearing Research, 53*(1), 114–125.

The age of the participants ranged from 28 to 73 years, with a mean age of 49.8 (±7.9) years.

Source: Faris, J. A., Douglas, K. K., Maples, D. C., Berg, L. R., & Thrailkill, A. (2010). Job satisfaction of advanced practice nurses in the Veterans Health Administration. *Journal of the American Academy of Nurse Practitioners, 22*(1), 35–44.

EXCERPT 2.20 • *Reporting the Standard Deviation in a Table*

TABLE 2 *Sample size, mean, standard deviation, and range for AVI-SOS clientele descriptive variables*

Variable	N	Mean	SD	Range
Age	105	41.6	8.5	19–61
Years lived in Victoria	103	17.3	13.4	0–55
Number of places slept last week	105	2.5	2.0	1–7
Years needle exchange client	105	7.2	5.3	0–19

Source: Exner, H., Gibson, E. K., Stone, R., Lindquist, J., Cowen, L., & Roth, E. A. (2009). Worry as a window into the lives of people who use injection drugs: A factor analytic approach. *Harm Reduction Journal, 6*(20), 1–6.

Although the standard deviation appears in research reports far more often than does any other measure of variability, a few researchers choose to describe the dispersion in their data sets by reporting the variance. Excerpt 2.21 is a case in point. This content of this excerpt illustrates nicely the danger of considering only measures of central tendency. The means made the boys and girls appear to be similar in terms of reading scores. However, the two groups differ in terms of how dispersed their scores are.

Before concluding our discussion of the standard deviation and variance, I offer this helpful hint concerning how to make sense out of these two indices of variability. Simply stated, I suggest using an article's reported standard deviation (or variance) to estimate what the range of scores probably was. Because the range

EXCERPT 2.21 • *Using the Variance to Measure Dispersion*

To explore possible gender differences in reading performance, we analysed data from [two samples]. Although the difference between the average [reading] scores of males and females in these two samples was very small, the variance of reading performance was significantly greater for males in both groups.

Source: Hawke, J. L., Olson, R. K., Willcut, E. G., Wadsworth, S. J., & DeFries, J. C. (2009). Gender ratios for reading difficulties. *Dyslexia, 15*(3), 239–242.

is such a simple concept, the standard deviation or variance can be demystified by converting it into an estimated range.

To make a standard deviation interpretable, just multiply the reported value of this measure of variability by about 4 to obtain your guess as to what the range of the scores most likely was. Using 4 as the multiplier, this rule of thumb tells you to guess that the range is equal to 20 for a set of scores in which the standard deviation is equal to 5. (If the research report indicates that the variance is equal to 9, you first take the square root of 9 to get the standard deviation, and then you multiply by 4 to arrive at a guess that the range is equal to 12.)

When giving you this rule of thumb, I said that you should multiply the standard deviation by "about 4." To guess more accurately what the range most likely was in a researcher's data set, your multiplier sometimes must be a bit smaller or larger than 4 because the multiplier number must be adjusted on the basis of the number of scores on which the standard deviation is based. If there are 25 or so scores, use 4. If N is near 100, multiply the standard deviation by 5. If N is gigantic, multiply by 6. With small Ns, use a multiplier that is smaller than 4. With 10–20 scores in the group, multiplying by 3 works fairly well; when N is smaller than 10, setting the multiplier equal to 2 usually produces a good guess as to range.

It may strike you as somewhat silly to be guessing the range based on the standard deviation. If researchers regularly included the values of the standard deviation and the range when summarizing their data (as was done in Excerpt 2.20), there would be no need to make a guess as to the size of R. Unfortunately, most researchers present only the standard deviation—and by itself, a standard deviation provides little insight into the degree of variability within a set of scores.

One final comment is in order regarding this technique of using SD to guess R. What you get is nothing more than a rough approximation, and you should not expect your guess of R to "hit the nail on the head." Using the standard deviation and range presented in Excerpt 2.20 (and using a multiplier of 5 because the N is about 100), we see that our guess of R is never perfect for any of the four rows of numbers in the excerpt. However, each of our four guesses turns out to approximate well the actual range, and it helps us understand how much spread is in a data set if only the standard deviation is presented.

Other Measures of Variability

Of the five measures of variability discussed so far, you will encounter the range and the standard deviation most often when reading researcher-based journal articles. Occasionally, you may come across examples of the interquartile range, the semi-interquartile range, and the variance, and once in a great while you may encounter some other measure of variability.

In Excerpt 2.22, we see a case where the **coefficient of variation** was used. As indicated within this excerpt, this measure of dispersion is nothing more than the standard deviation divided by the mean.

EXCERPT 2.22 • *The Coefficient of Variation*

To compare the amount of variability generated by movements performed under different experimental conditions, timing and spatial variability at movement reversals were measured using the coefficient of variation (i.e., standard deviation divided by the mean) of movement duration and movement amplitude, respectively.

Source: Shafir, T., & Brown, S. H. (2010). Timing and the control of rhythmic upper-limb movements. *Journal of Motor Behavior, 42*(1), 71–84.

The coefficient of variation is useful when comparing the variability in two groups of scores in which the means are known to be different. For example, suppose we wanted to determine which of two workers has the more consistent commuting time driving to work in the morning. If one of these workers lives 5 miles from work and the second lives 25 miles from work, a direct comparison of their standard deviations (each based on 100 days of commuting to work) does not yield a fair comparison because the worker with the longer commute is expected to have more variability. What *is* fair is to divide each commuter's standard deviation by his or her mean. Such a measure of variability is called the *coefficient of variation*.

Standard Scores

All the techniques covered thus far in this chapter describe features of the entire data set. In other words, the focus of attention is on all N scores whenever a researcher summarizes a group of numbers by using one of the available picture techniques, a word or number that reveals the distributional shape, a numerical index of central tendency, or a quantification of the amount of dispersion that exists among the scores. Sometimes, however, researchers want to focus their attention on a single score within the group rather than on the full data set. When they do this, they usually convert the raw score being examined into a **standard score**.

Although many different kinds of standard scores have been developed over the years, the ones used most frequently in research studies are called **z-scores** and **T-scores.** These two standard scores are identical in that each one indicates how many standard deviations a particular raw score lies above or below the group mean. In other words, the numerical value of the standard deviation is first looked upon as defining the length of an imaginary yardstick, with that yardstick then used to measure the distance between the group mean and the individual score being considered. For example, if you and several other people took a test that produced scores having a mean of 40 and a standard deviation of 8, and if your score on this test happened to be 52, you would be one and one-half yardsticks above the mean.

The two standard scores used most by researchers—z-scores and T-scores—perform exactly the same function. The only difference between them concerns the arbitrary values given to the new mean score and the length of the yardstick within the revised data set following conversion of one or more raw scores into standard scores. With z-scores, the mean is fixed at zero and the yardstick's length is set equal to 1. As a consequence, a z-score directly provides an answer to the question, "How many *SD*s is a given score above or below the mean?" Thus, a z-score of $+2.0$ indicates that the person being focused on is 2 standard deviations above the group mean. Likewise, a z-score of -1.2 for someone else indicates that this person scored 1.2 standard deviations below the mean. A z-score close to 0, of course, indicates that the original raw score is near the group mean.

With T-scores, the original raw score mean and standard deviation are converted to 50 and 10, respectively. Thus, a person whose raw score positioned him or her 2 standard deviations above the mean receives a T-score of 70. Someone else positioned 1.2 standard deviations below the mean receives a T-score of 38. And someone whose raw score is near the group mean has a T-score near 50.

Although researchers typically apply their statistical procedures to the raw scores that have been collected, they occasionally convert the original scores into z-scores or T-scores. Excerpts 2.23 and 2.24 provide evidence that these two standard scores are sometimes referred to in research summaries.

EXCERPTS 2.23–2.24 • *Standard Scores (z and T)*

MSFC scores were determined for all participants. . . . Standard z scores were created for each MSFC component using the baseline mean and standard deviation values. . . . Impairment on each component and composite MSFC measure was defined as a z score greater than -1.5.

Source: Drake, A. S., Weinstock-Guttman, B., Morrow, S. A., Hojnacki, D., Munschauer, F. E., & Benedict, R. H. B. (2010). Psychometrics and normative data for the Multiple Sclerosis Functional Composite: Replacing the PASAT with the Symbol Digit Modalities Test. *Multiple Sclerosis, 16*(2), 228–237.

(continued)

EXCERPTS 2.23–2.24 • (*continued*)

Maria's scored SPS indicated an overall Suicide Probability *T*-score of 82. This score was 32 points above the mean score of 50 and was more than three standard deviations above the mean.

Source: Valadez, A., Juhnke, G. A., Coll, K. M., Granello, P. F., Peters, S., & Zambrano, E. (2009). The Suicide Probability Scale: A means to assess substance abusing clients' suicide risk. *Journal of Professional Counseling: Practice, Theory & Research, 37*(1), 51–65.

A Few Cautions

Before concluding this chapter, I want to alert you to the fact that two of the terms discussed earlier—*skewed* and *quartile*—are occasionally used by researchers who define them differently than I have. I want to prepare you for the alternative meanings associated with these two concepts.

Regarding the term *skewed,* a few researchers use this word to describe a complete data set that is out of the ordinary. Used in this way, the term has nothing to do with the notion of distributional shape, but instead is synonymous to the term *atypical.* In Excerpt 2.25, we see an example of how the word *skewed* was used in this fashion.

EXCERPT 2.25 • *Use of the Term Skewed to Mean Unusual or Atypical*

The standardized instructions may have skewed these results in favor of a preponderance of "fair" ratings. The standardized instructions indicated that any street, sidewalk, public transit stop, public parks or grounds, public schools or any non-private land should be marked in "fair" condition if it showed irregular maintenance (including those with even small amounts of cracked concrete or paint or moderately overgrown vegetation) and overall the space was "in decent condition, but (rater) would recommend additional upkeep." Such instructions logically resulted in most raters ranking public spaces as being in fair condition.

Source: Parsons, J. A., Singh, G., Scott, A. N., Nisenbaum, R., Balasubramaniam, P., Jabbar, A., et al. (2010). Standardized observation of neighbourhood disorder: Does it work in Canada? *International Journal of Health Geographics, 9*, 1–19.

The formal, statistical definition of *quartile* as used in this book is "one of three points that divide a group of scores into four subgroups, each of which contains 25 percent of the full group." Certain researchers use the term *quartile* to designate the subgroups themselves. When used this way, there are four quartiles (not three), with scores falling into the quartiles (not between them). Excerpt 2.26 provides an example of *quartile* being used in this fashion.

EXCERPT 2.26 • *Use of the Term Quartile to Designate Four Subgroups*

SES was a strong, significant predictor [of students taking higher-level math courses]. This finding is of little surprise with those in the highest-SES quartile more likely to complete math courses beyond Algebra 2 when compared to students in lower-SES quartiles.

Source: Daniel T. S. (2010). Predictive factors in intensive math course-taking in high school. *Professional School Counseling, 13*(3), 196–207.

My second warning concerns the use of the term *average.* In elementary school, students are taught that (1) the average score is the mean score and (2) the median and the mode are *not* conceptually the same as the average. Unfortunately, you must undo your earlier learning if you are still under the impression that the words *average* and *mean* are synonymous.

In statistics, the term *average* is synonymous with the phrase "measure of central tendency," and either is nothing more than a generic label for *any* of several techniques that attempt to describe the typical or center score in a data set. Hence, if a researcher gives us information as to the *average score,* we cannot be absolutely sure which average is being presented. It might be the mode, it might be the median, or it might be any of the many other kinds of average that can be computed. Nevertheless, you seldom will be wrong when you see the word *average* if you guess that reference is being made to the arithmetic mean.

My final comment of the chapter concerns scores in a data set that lie far away from the rest of the scores. Such scores are called **outliers,** and they can come about because someone does not try when taking a test, does not understand the instructions, or consciously attempts to sabotage the researcher's investigation. Accordingly, researchers should (1) inspect their data sets to see if any outliers are present and (2) either discard such data points before performing any statistical analyses or perform analyses in two ways: with the outlier(s) included and with the outlier(s) excluded. In Excerpts 2.27 and 2.28, we see two cases in which data were examined for possible outliers. Notice how the researchers associated with these excerpts explain the rules they use to determine how deviant a score must be before it is tagged as an outlier; also, notice how these rules differ.

If allowed to remain in a data set, outliers can create skewness and in other ways create problems for the researcher. Accordingly, the researchers who conducted the studies that appear in Excerpts 2.27 and 2.28 deserve credit for taking extra time to look for outliers before conducting any additional data analyses.

I should point out, however, that outliers potentially can be of legitimate interest in and of themselves. Instead of quickly tossing aside any outliers, researchers would be well advised to investigate any "weird cases" within their data sets. Even if the identified outliers have come about because of poorly understood directions,

EXCERPTS 2.27–2.28 • *Dealing with Outliers*

There was only one outlier that deviated by more than three standard deviations from the variable mean. . . .

Source: de la Sablonnière, R., Tougas, F., & Perenlei, O. (2010). Beyond social and temporal comparisons: The role of temporal inter-group comparisons in the context of dramatic social change in Mongolia. *Journal of Social Psychology, 150*(1), 98–115.

Outlier individual data points . . . were excluded [if] the values exceeded the inner fences of the IQR ($<$ Q1 $-$ 1.5*IQR or $>$ Q3 $+$ 1.5*IQR).

Source: Simola, J., Stenbacka, L., & Vanni, S. (2009). Topography of attention in the primary visual cortex. *European Journal of Neuroscience, 29*(1), 188–196.

erratic measuring devices, low motivation, or effort to disrupt the study, researchers in these situations might ask the simple question, "Why did this occur?" Outliers have the potential, if considered thoughtfully, to provide insights into the genetic, psychological, or environmental factors that stand behind extremely high or low scores.

One Final Excerpt

As we finish this chapter, let us look at one final excerpt. Although it is quite short and despite the fact that it contains no tables or pictures, this excerpt stands as a good example of how researchers should describe their data. Judge for yourself. Read Excerpt 2.29 and then ask yourself: Can I imagine what the data looked like?

EXCERPT 2.29 • *A Good Descriptive Summary*

The social communication questionnaire (SCQ) for autistic spectrum disorder was previously validated in clinical populations [yet] to date there has not been a report on the use of the [40-item] SCQ in a sample representing the general population. . . . A total of 153 questionnaires completed by parents of [regular primary school children] were included in the final analysis. . . . The data had a range of 0–20, a mode of 1 and a mean of 4.22, $SD = 3.45$. The distribution was not normal, and was skewed to the right (skewness = 1.59, kurtosis = 3.94). [T]here were four outliers in the sample, each of whom had total SCQ scores more than two standard deviations above the mean total SCQ score for the sample. The corrected mean of the sample without the outlier values was 3.89, $SD = 2.77$, and the corrected sample was normally distributed (skewness = 0.67, kurtosis = -0.27).

Source: Mulligan, A., Richardson, T., Anney, R. J. L., & Gill, M. (2009). The Social Communication Questionnaire in a sample of the general population of school-going children. *Irish Journal of Medical Science, 178*, 193–199.

Review Terms

Average	Multimodal
Bar graph	Negatively skewed
Bimodal	Normal distribution
Bivariate	Outlier
Box plot	Platykurtic
Coefficient of variation	Positively skewed
Dispersion	Quartile
Distributional shape	Range
Grouped frequency	Rectangular
distribution	Semi-interquartile range
Heterogeneous	Sigma
Histogram	Simple frequency distribution
Homogeneous	Skewed distribution
Interquartile range	Standard deviation
Kurtosis	Standard score
Leptokurtic	Stem-and-leaf display
Mean	*T*-score
Measure of central	Trimodal
tendency	Ungrouped frequency
Measure of variability	distribution
Median	Univariate
Mesokurtic	Variance
Mode	*z*-score

The Best Items in the Companion Website

1. An email message from the author to his students explaining what a *standard deviation* really is (and what it is not).
2. An interactive online quiz (with immediate feedback provided) covering Chapter 2.
3. A challenging puzzle question created by the author for use with an interactive online resource called "Fun with Histograms."
4. Twelve misconceptions about the content of Chapter 2.
5. What the author considers to be the best passage from Chapter 2.

To access the chapter outline, practice tests, weblinks, and flashcards, visit the companion website at http://www.ReadingStats.com.

Review Questions and Answers begin on page 531.

Bivariate Correlation

In Chapter 2, we looked at the various statistical procedures researchers use when they want to describe single-variable data sets. We saw examples where data on two or more variables were summarized, but in each of those cases the data were summarized one variable at a time. Although there are occasions when these univariate techniques permit researchers to describe their data sets, most empirical investigations involve questions that call for descriptive techniques that simultaneously summarize data on more than one variable.

In this chapter, we consider situations in which data on two variables have been collected and summarized, with interest residing in the relationship between the two variables. Not surprisingly, the statistical procedures that we will examine here are considered to be **bivariate** in nature. Later (in Chapter 16), we consider techniques designed for situations wherein the researcher wishes to simultaneously summarize the relationships among three or more variables.

Three preliminary points are worth mentioning as I begin my effort to help you refine your skills at deciphering statistical summaries of bivariate data sets. First, the focus in this chapter is on techniques that simply summarize the data. In other words, we are still dealing with statistical techniques that are fully descriptive in nature. Second, this chapter is similar to Chapter 2 in that we consider ways to summarize data that involve both picture and numerical indices. Finally, the material covered in Chapter 4, "Reliability and Validity," draws *heavily* on the information presented here. With these introductory points now behind us, let us turn to the central concept of this chapter: correlation.

The Key Concept behind Correlation: Relationship

Imagine that each of nine adults who want to lose weight is measured on two variables: average daily time spent exercising (measured in minutes) and drop in weight over a two-week period (measured as percentage of initial weight). The data from this imaginary study might turn out as follows:

Individual	Time Spent Exercising	Weight Loss
Carol	75	2
Robert	100	3
Margaret	60	1
Tom	20	0
William	70	2
Mary	120	4
Suzanne	40	1
Craig	65	2
Jennifer	80	3

Although it would be possible to look at each variable separately and say something about the central tendency, variability, and distributional shape of the nine scores (first for exercise time, then for weight loss), the key concept of correlation requires that we look at the data on our two variables *simultaneously*. In doing this, we are trying to see (1) whether there is a **relationship** between the two sets of scores, and (2) how strong or weak that relationship is, presuming that a relationship does, in fact, exist.

On a simple level, the basic question being dealt with by **correlation** can be answered in one of three possible ways. Within any bivariate data set, it *may* be the case that the high scores on the first variable tend to be paired with the high scores on the second variable (implying, of course, that low scores on the first variable tend to be paired with low scores on the second variable). I refer to this first possibility as the *high–high, low–low* case. The second possible answer to the basic correlational question represents the inverse of our first case. In other words, it *may* be the case that high scores on the first variable tend to be paired with low scores on the second variable (implying, of course, that low scores on the first variable tend to be paired with high scores on the second variable). My shorthand summary phrase for this second possibility is *high–low, low–high*. Finally, it is possible that little systematic tendency exists in the data at all. In other words, it *may* be the case that some of the high and low scores on the first variable are paired with high scores on the second variable, whereas other high and low scores on the first variable are paired with low scores on the second variable. I refer to this third possibility simply by the three-word phrase *little systematic tendency*.

As a check on whether I have been clear in the previous paragraph, take another look at the hypothetical data presented earlier on average daily exercise time and weight loss for our nine individuals. More specifically, indicate how that bivariate relationship should be labeled. Does it deserve the label *high–high, low–low*? Or the label *high–low, low–high*? Or the label *little systematic tendency*? If you have not done so already, look again at the data presented and formulate your answer to this question.

To discern the nature of the relationship between exercise time and weight loss, one must first identify each variable's high and low scores. The top three values for the exercise variable are 120, 100, and 80, whereas the lowest three values in this same column are 60, 40, and 20. Within the second column, the top three values are 4, 3, and 3; the three lowest values are 1, 1, and 0. After identifying each variable's high and low scores, the next (and final) step is to look at both columns of data simultaneously and see which of the three answers to the basic correlational question fits the data. For our hypothetical data set, we clearly have a *high–high, low–low* situation, with the three largest exercise values being paired with the three largest weight loss values and the three lowest values in either column being paired with the low values in the other column.

The method I have used to find out what kind of relationship describes our hypothetical data set is instructive, I hope, for anyone not familiar with the core concept of correlation. That strategy, however, is not very sophisticated. Moreover, you will not have a chance to use it very often, because researchers almost always summarize their bivariate data sets by means of pictures, a single numerical index, a descriptive phrase, or some combination of these three reporting techniques. Let us now turn our attention to these three methods for summarizing the nature and strength of bivariate relationships.

Scatter Plots

Like histograms and bar graphs, a **scatter plot** has a horizontal axis and a vertical axis. These axes are labeled to correspond to the two variables involved in the correlational analysis. The **abscissa** is marked off numerically so as to accommodate the obtained scores collected by the researcher on the variable represented by the horizontal axis; in a similar fashion, the **ordinate** is labeled so as to accommodate the obtained scores on the other variable. (With correlation, the decision as to which variable is put on which axis is fully arbitrary; the nature of the relationship between the two variables is revealed regardless of how the two axes are labeled.) After the axes are set up, the next step involves placing a dot into the scatter diagram for each object that was measured, with the horizontal and vertical positioning of each dot dictated by the scores earned by that object on the two variables involved in the study.

In Excerpt 3.1, we see a scatter plot associated with a study involving a group of individuals with scoliosis, a disease characterized by an abnormal bending of the spine to the side. Each of these individuals completed two self-report questionnaires concerning their "trunk" problem. One of the instruments was the SRS-22, which provided a score on "self-image" (one of the instrument's subscales). The other

EXCERPT 3.1 • *A Scatter Plot*

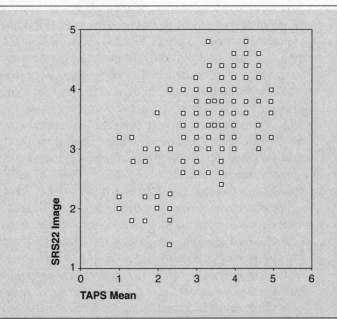

FIGURE 3 *Scatter plot between the SRS-22 self-image subscale and the mean score of TAPS.*

Source: Bago, J., Sanchez-Raya, J., Perez-Grueso, F. J. S, & Climent, J. M. (2010). The Trunk Appearance Perception Scale (TAPS): A new tool to evaluate subjective impression of trunk deformity in patients with idiopathic scoliosis. *Scoliosis*, *5*(6), 1–9.

questionnaire was a new instrument called the Trunk Appearance Perception Scale (TAPS) that yielded three scores per person, with a mean computed for each individual. On each instrument, a person's score ranged anywhere from 1 to 5, with these end-of-the-continuum scores representing bad and good posture, respectively.

The data from the scoliosis study are presented within Figure 3, with the TAPS on the abscissa, the SRS-22 on the ordinate, and each small square representing a person and his or her scores on the two instruments. Thus, the lowest small square in the scatter plot came from a person who had a score of 2.3 on the TAPS and a score of 1.4 on the SRS-22. The labeled axes allow us to know, approximately, the two scores associated with every other small square in the scatter plot. Collectively, the full set of data points show a tendency for high TAPS scores to be paired with high SRS-22 scores (and for low score on one instrument to be paired with low scores on the other instrument).

A scatter plot reveals the relationship between two variables through the pattern that is formed by the full set of dots. Researchers often insert a straight line through the data points to help us see the nature of the relationship. If the line extends from lower-left to upper-right (as it would in Excerpt 3.1), we have a high–high, low–low relationship. However, if the line extends from upper-left to lower-right, we have a high–low, low–high relationships. If the line passing through the dots has a near zero tilt, this indicates that the two variables are not related very much at all. The strength of the relationship can be determined by asking how close the dots seem to lie from the line. If the dots are close to the line, the relationship is strong; if the dots are scattered both close to and far from the line (as in Excerpt 3.1), the relationship is best thought of as moderate, or even weak.

To discern what pattern exists when there is no tilted line included, I use a simple (although not completely foolproof) two-step method. First, I draw an imaginary perimeter line, or what I think of as a "fence," around the full set of data points—and in so doing, I try to achieve a tight fit. Second, I look at the shape produced by this perimeter line and examine its tilt and its thickness. Depending on these two characteristics of the data set's scatter, I arrive at an answer to the basic correlational question concerning the nature and strength of the relationship between the two variables.

The oval produced by my perimeter line will likely be tilted, and its tilt gives me a sense for whether the relationship is of the (1) *high–high, low–low*, (2) *high–low, low–high*, or (3) *little systematic tendency* variety. The oval's thickness reveals the strength of the relationship. If the oval is elongated and thin, then I conclude that there is a *strong* relationship between the two variables. However, if the oval is not too much longer than it is wide, then I conclude that a *weak* relationship exists.

The Correlation Coefficient

Although a scatter plot has the clear advantage of showing the scores for each measured object on the two variables of interest, many journals are reluctant to publish such pictures because they take up large amounts of space. For that reason, and also because the interpretation of a scatter diagram involves an element of subjectivity, numerical summaries of bivariate relationships appear in research reports far more frequently than do pictorial summaries. The numerical summary is called a **correlation coefficient.**

Symbolized as r, a correlation coefficient is normally reported as a decimal number somewhere between -1.00 and $+1.00$. In Excerpt 3.2, we see four different correlation coefficients.

To help you learn how to interpret correlation coefficients, I have drawn a straight horizontal line to represent the continuum of possible values that result from researchers putting data into a correlational formula:

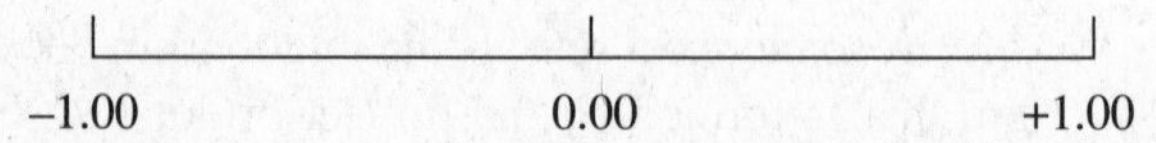

EXCERPT 3.2 • *Correlation Coefficients*

Scales to measure job demands and control were constructed for this study. . . . Correlations using the current scales closely matched those previously reported: At T1 (Time 1), for example, demands correlated at $r = .50$ with stress and at $-.40$ with job satisfaction, whereas the corresponding correlations involving control were $-.40$ and .36.

Source: Bradley, G. L. (2010). Work-induced changes in feelings of mastery. *Journal of Psychology, 144*(2), 97–119.

This correlational continuum helps you pin down the meaning of several adjectives that researchers use when talking about correlation coefficients or relationships: *direct, high, indirect, inverse, low, moderate, negative, perfect, positive, strong,* and *weak*.

First, consider the two halves of the correlational continuum. Any *r* that falls on the right side represents a **positive correlation;** this indicates a **direct relationship** between the two measured variables. (Earlier, I referred to such cases by the term *high–high, low–low*.) However, any result that ends up on the left side is a **negative correlation,** and this indicates an **indirect,** or **inverse, relationship** (i.e., *high–low, low–high*). If *r* lands on either end of our correlation continuum, the term **perfect** may be used to describe the obtained correlation. The term **high** comes into play when *r* assumes a value close to either end (thus implying a **strong** relationship); conversely, the term **low** is used when *r* lands close to the middle of the continuum (thus implying a **weak** relationship). Not surprisingly, any *r* that ends up in the middle area of the left or right sides of our continuum is called **moderate.**

In Excerpts 3.3 through 3.5, we see cases where researchers use adjectives to label their *r*s. In the first two of these excerpts, we see the concepts of *weak* and *moderate* being used to describe correlation coefficients, whereas in the third excerpt, we see the concept of *strong* used. All four of the correlations in these excerpts were positive. If each of these *r*s had turned out to be the same size as reported but negative in sign, the descriptive labels of *weak, moderate,* and

EXCERPTS 3.3–3.5 • *Use of Modifying Adjectives for the Term Correlation*

Furthermore, [the questionnaire items] "More screening if higher reimbursement" and "Proportion of male patients with alcohol problems" were weakly correlated with the use of screening instruments (Pearson's $r = 0.10$ and 0.09, respectively).

Source: Nygaard, P., Paschall, M. J., Aasland, O. G., & Lund, K. E. (2010). Use and barriers to use of screening and brief interventions for alcohol problems among Norwegian general practitioners. *Alcohol and Alcoholism, 45*(2), 207–212.

(*continued*)

EXCERPTS 3.3–3.5 • (*continued*)

Scores on MIDAS School Math subscale were moderately correlated ($r = .58$) with scores on Ohio State Math Achievement Test. . . .

Source: Shearer, C. B., & Darrell A. L. (2009). Exploring the application of multiple intelligences theory to career counseling. *The Career Development Quarterly, 58*(1), 3–13.

The strong correlation (.716) between age at graduation and pre-MBA work experience is reasonable: Older graduates are likely to have more pre-MBA work experience.

Source: Yeaple, R. N., Johnston, M. W., & Whittingham, K. L. (2009). Measuring the economic value of pre-MBA work experience. *Journal of Education for Business, 85*(1), 13–20.

strong would still have been appropriate, because these terms contrast a computed r with zero.

Before concluding our discussion of how to interpret correlation coefficients, I feel obligated to reiterate the point that when the issue of relationship is addressed, the central question being answered by r is: "To what extent are the high scores of one variable paired with the high scores of the other variable?" The term *high* in this question is considered separately for each variable. Hence, a strong positive correlation can exist even though the mean of the scores of one variable is substantially different from the mean of the scores on the other variable. As proof of this claim, consider again the data presented earlier concerning nine individuals who were measured in terms of exercise and weight loss; the correlation between the two sets of scores turns out equal to $+.96$, despite the fact that the two means are quite different (2 versus 70). This example makes clear, I hope, the fact that a correlation does *not* deal with the question of whether two means are similar or different.[1]

The Correlation Matrix

When interest resides in the bivariate relationship between just two variables or among a small number of variables, researchers typically present their *r*s within the text of their article. (This reporting strategy is shown in Excerpts 3.2 through 3.5.) When interest centers on the bivariate relationships among many variables, however, the resulting *r*s are often summarized within a special table called a **correlation matrix.**

It should be noted that *several* bivariate correlations can be computed among a set of variables, even for relatively *small* sets of variables. With eight variables,

[1]In many research studies, the focus is on the difference between means. Beginning in Chapter 10, our discussion of *t*-test and *F*-tests show how researchers compare means.

for example, 28 separate bivariate *r*s can be computed. With 10 variables, there are 45 *r*s. In general, the number of bivariate correlations is equal to $k(k - 1)/2$, where *k* indicates the number of variables.

In Excerpt 3.6, we see a correlation matrix that summarizes the measured bivariate relationships among six variables. In the study associated with this excerpt, 90 college students took a test of creativity (the Idea Generation Test) and filled out a personality survey that measured each student on the "Big 5" dimension of Openness, Conscientiousness, Extraversion, Agreeableness, and Neuroticism. As you can see, this correlation matrix contains *r*s arranged in a triangle. Each *r* indicates the correlation between the two variables that label that *r*'s row and column. For example, the value of .38 is the correlation between Openness and Agreeableness.

EXCERPT 3.6 • *A Standard Correlation Matrix*

TABLE 3 *Correlations Between the Idea Generation Test and Personality Dimensions*

Test	1	2	3	4	5	6
1. Idea Generation	1					
2. Openness	.10	1				
3. Conscientiousness	−.04	.34	1			
4. Extraversion	.22	.02	.08	1		
5. Agreeableness	.00	.38	.28	.41	1	
6. Neuroticism	−.01	.01	.01	−.39	−.10	1

Source: Ellwood, S., Pallier, G., Snyder, A., & Gallate, J. (2009). The incubation effect: Hatching a solution? *Creativity Research Journal, 21*(1), 6–14.

Two things are noteworthy about the correlation matrix shown in Excerpt 3.6. First, when a row and a column refer to the same variable (as is the case with the top row and the left column, the second row and the second column, etc.), there is a "1" positioned at the intersection of that row and column. Clearly, the correlation of any variable with itself is perfect. Thus, the correlation coefficients (each equal to 1) in the diagonal are not informative; the "meat" of the correlation matrix lies elsewhere.

Second, there are no correlation coefficients above the diagonal. If correlations appear there, they would be a mirror image of the *r*s positioned below the diagonal. The value .10 would appear on the top row in the second column, −.04 would appear on the top row in the third column, and so on. Such *r*s, if they were put into the correlation matrix, are fully redundant with the *r*s that already are present; accordingly, they add nothing.

In Excerpt 3.6, the correlation matrix is set up with the 15 bivariate correlation coefficients positioned below the diagonal of 1s. At times, you will come across a correlation matrix in which (1) the values of the correlation coefficients are positioned above rather than below the diagonal, or (2) each diagonal element has either a dash or nothing at all. Such alternative presentations should not cause you any difficulty, because they still contain all possible bivariate correlations that are interpreted in the same way that we interpreted the *r*s in Excerpt 3.6.

Excerpt 3.7 illustrates how two correlation matrices can be combined into one table. In the study associated with this table, 572 Dutch adults filled out three personality inventories (dealing with self-efficacy, intention, and action planning) and also answered questions about their current and past consumption of fruit. A similar group of 585 individuals did the same thing, except their consumption questions dealt with snacks, not fruit. After collecting the data, the researchers computed, separately for each group, bivariate correlation coefficients among the five variables. Using the note beneath the correlation matrix as a guide, we can look to see how the correlation between any two variables compares across the two groups. The two correlations between current and past consumption were quite similar (0.76 and 0.60). However, the correlations between self-efficacy and consumption were quite different (0.57 for the fruit group; −0.36 for the snack group).

EXCERPT 3.7 • *Two Correlation Matrices Shown in a Single Table*

TABLE 1   *Pearson Correlations between Cognitions, Past Behavior and Current Outcome Behaviors*[a,b]

	1	*2*	*3*	*4*	*5*
1. Self-efficacy	—	0.58	0.40	0.63	0.57
2. Intention	0.38	—	0.48	0.42	0.36
3. Action planning	0.17	0.57	—	0.31	0.33
4. Past fruit/snack consumption	−0.42	−0.31	−0.18	—	0.76
5. Fruit/snack consumption	−0.36	−0.29	−0.22	0.60	—

[a]All correlations between variables in the fruit consumption study are depicted above the diagonal; correlations between variables in the snack consumption study are depicted below the diagonal.
[b]All correlations are significant at the 0.001 level (two-tailed).

Source: van Osch, L., Beenackers, M., Reubsaet, A., Lechner, L., Candel, M., & de Vries, H. (2009). Action planning as predictor of health protective and risk behavior: An investigation of fruit and snack consumption. *International Journal of Behavior Nutrition and Physical Activity, 69*(6), 1–10.

I should point out that researchers sometimes put things in the diagonal other than 1s (as in Excerpt 3.6), dashes (as in Excerpt 3.7), or empty spaces. Occasionally, you may see numbers in the diagonal that are not correlation coefficients but rather reliability coefficients. (We consider the topic of reliability in Chapter 4.) Also, you are likely to come across a correlation matrix in which the row or column that contains no correlation coefficients has been deleted. For example, in Excerpt 3.6, the top row (for Test 1) could be eliminated without altering the amount of statistical information being communicated, as could the sixth column. When this is done, correlation coefficients appear in the table's diagonal. (This practice of eliminating a row or a column, however, never takes place in conjunction with a correlation matrix such as that shown in Excerpt 3.7, wherein two correlation matrices are combined into a single table.)

Different Kinds of Correlational Procedures

In this section, we take a brief look at several different correlational procedures that have been developed. As you will see, all these techniques are similar in that they are designed for the case in which data have been collected on two variables.[2] These bivariate correlational techniques differ, however, in the nature of the two variables. In light of this important difference, you must learn a few things about how variables differ.

The first important distinction is between quantitative and qualitative characteristics. With a **quantitative variable,** the targets of the measuring process vary as to how much of the characteristic is possessed. In contrast, a **qualitative variable** comes into play when the things being measured vary from one another in terms of the categorical group to which they belong relative to the characteristic of interest. Thus, if we focus our attention on people's heights, we have a quantitative variable (because some people possess more "tallness" than others). If, however, we focus our attention on people's favorite national park, we would be dealing with a qualitative variable (because people simply fall into categories based on which park they like best).

From the standpoint of correlation, quantitative variables can manifest themselves in one of two ways in the data a researcher collects. Possibly, the only thing the researcher wants to do is order individuals (or animals, or objects, or whatever) from the one possessing the greatest amount of the relevant characteristic to the one possessing the least. The numbers used to indicate ordered position usually are assigned such that 1 goes to the person with the greatest amount of the characteristic, 2 goes to the person with the second greatest amount, and so on. Such numbers are

---

[2]Some authors use the term *zero-order correlation* when referring to bivariate correlations to distinguish this simplest kind of correlation—that involves data on just two variables—from other kinds of correlations that involve data on three or more variables (e.g., partial correlations, multiple correlations, canonical correlations).

called **ranks** and are said to represent an **ordinal** scale of measurement. A researcher's data is also ordinal in nature if each person or thing being measured is put into one of several ordered categories, with everyone who falls into the same category given the same score (e.g., the numbers 1, 2, 3, and 4 used to represent freshmen, sophomores, juniors, and seniors).

With a second kind of quantitative variable, measurements are more precise. Here, the score associated with each person supposedly reveals how much of the characteristic of interest is possessed by that individual—and it does this without regard for the standing of any other measured person. Whereas ranks constitute data that provide relative comparisons, this second (and more precise) way of dealing with quantitative variables provide absolute comparisons. In this book, I use the term **raw score** to refer to any piece of data that provides an absolute (rather than relative) assessment of one's standing on a quantitative variable.[3]

Qualitative variables come in two main varieties. If the subgroups into which people are classified truly have no quantitative connection with each other, then the variable corresponding to those subgroups is said to be **nominal** in nature. Your favorite academic subject, the brand of jelly you most recently used, and your state of residence exemplify this kind of variable. If there are only two categories associated with the qualitative variable, then the variable of interest is said to be dichotomous in nature. A **dichotomous variable** actually can be viewed as a special case of the nominal situation, with examples being "course outcome" in courses where the only grades are pass and fail (or credit and no credit), gender, party affiliation during primary elections, and graduation status following four years of college.

In Excerpts 3.8 through 3.11, we see examples of different kinds of variables. The first two of these excerpts illustrate the two kinds of quantitative variables we have discussed: ranks and raw scores. The last two of these excerpts exemplify qualitative variables (the first being a four-category nominal variable, the second being a dichotomous variable).

EXCERPTS 3.8–3.11 • *Different Kinds of Data*

The questionnaire also asked about perceptions of work-related stress. . . . Nurses ranked the following stressors in order of importance: critical illness or acuity of patients, dying patients, rotating shifts or schedule, short staffing, long work hours, and demanding families.

Source: Gallagher, R., & Gormley, D. K. (2009). Perceptions of stress, burnout, and support systems in pediatric bone marrow transplantation nursing. *Clinical Journal of Oncology Nursing, 13*(6), 681–685.

---

[3]Whereas most statisticians draw a distinction between interval and ratio measurement scales and between discrete and continuous variables, readers of journal articles do not need to understand the technical differences between these terms in order to decipher research reports.

EXCERPTS 3.8–3.11 • (*continued*)

A comprehension test consisting of 12 multiple choice and fill-in-the-blank questions was used to assess participants' comprehension and retention of the textbook lesson. . . . Means were computed for male and female students in each condition based on the number of correct answers out of 12 total questions.

Source: Good, J. J., Woodzicka, J. A., & Wingfield, L. C. (2010). The effects of gender stereotypic and counter-stereotypic textbook images on science performance. *Journal of Social Psychology, 150*(2), 132–147.

We combined Hispanic/Latino ethnicity and race to create a race/ethnicity variable with the following four levels: (1) Hispanic/ Latino; (2) non-Hispanic Black/African American; (3) non-Hispanic White; and (4) Other.

Source: Duncan, D. T., Johnson, R. M., Molnar, B. E., & Azrael, D. (2009). Association between neighborhood safety and overweight status among urban adolescents. *BMC Public Health, 9,* 289–297.

Gender is represented by a dichotomous variable (male = 1 and female = 0).

Source: Fernandes, D. C., & Neves, J. A. (2010). Urban bias in development and educational attainment in Brazil. *Journal of Developing Areas, 43*(2), 271–288.

Researchers frequently derive a raw score for each individual being studied by combining that individual's responses to the separate questions in a test or survey. As Excerpts 3.12 and 3.13 show, the separate items can each be ordinal or even dichotomous in nature, and yet the sum of those item scores is looked upon as being what I call a *raw score*. Although theoretical statistical authorities argue back and forth as to whether it is prudent to generate raw scores by combining ordinal or dichotomous data, doing so is an extremely common practice among applied researchers.

EXCERPTS 3.12–3.13 • *Combining Ordinal or Dichotomous Data to Get Raw Scores*

A five-point Likert-type scale ranging from strongly disagree (1) to strongly agree (5) was used to rate eight items for each of the five sources of support, yielding a total of 40 items. A total score was obtained by summing the amount of support perceived from the five sources on each of the eight items. Possible scores ranged from 40–200.

Source: Sammarco, A., & Konecny, L. M. (2010). Quality of life, social support, and uncertainty among Latina and Caucasian breast cancer survivors: A comparative study. *Oncology Nursing Forum, 37*(1), 93–99.

(*continued*)

EXCERPTS 3.12–3.13 • (*continued*)

> The final MSKQ consists of 25 multiple-choice statements and takes about 20 minutes
> to complete. The score is obtained by summing the number of correct answers and
> ranges from 0 to 25.
>
> *Source:* Giordano, A., Uccelli, M. M., Pucci, E., Martinelli, V., Borreani, C., et al. (2010). The
> Multiple Sclerosis Knowledge Questionnaire: A self-administered instrument for recently
> diagnosed patients. *Multiple Sclerosis, 16*(1), 100–111.

One final kind of variable must be mentioned briefly. Sometimes, a researcher
begins with a quantitative variable but then classifies individuals into two categories
on the basis of how much of the characteristic of interest is possessed. For example, a researcher conceivably could measure people in terms of the quantitative variable of height, place each individual into a tall or short category, and then disregard
the initial measurements of height (that took the form of ranks or raw scores).
Whenever this is done, the researcher transforms quantitative data into a two-
category qualitative state. The term **artificial dichotomy** is used to describe the
final data set. An example of this kind of data conversion appears in Excerpt 3.14.

EXCERPT 3.14 • *Creating An Artificial Dichotomy*

> General racism was assessed with the question "Thinking about your race or ethnicity,
> how often have you felt treated badly or unfairly because of your race or ethnicity?"
> In descriptive analysis, response categories for general racism were dichotomized
> (never or rarely versus sometimes, often, or all the time).
>
> *Source:* Shariff-Marco, S., Klassen, A. C., & Bowie, J. V. (2010). Racial/ethnic differences in
> self-reported racism and its association with cancer-related health behaviors. *American Journal
> of Public Health, 100*(2), 364–374.

Pearson's Product–Moment Correlation

The most frequently used bivariate correlational procedure is called **Pearson's
product–moment correlation,** and is designed for the situation in which (1) each
of the two variables is quantitative in nature and (2) each variable is measured so
as to produce raw scores. The scatter diagram presented in Excerpt 3.1 provides a
good example of the kind of bivariate situation that is dealt with by means of Pearson's technique.

Excerpts 3.15 and 3.16 illustrate the use of this extremely popular bivariate
correlational technique. Note, in the first of these excerpts, that the phrase
product–moment is used without the label *Pearson*. In the second excerpt, note that

EXCERPTS 3.15–3.16 • *Pearson's Product–Moment Correlation*

The simple product-moment correlation between faculty interaction and openness to diversity is positive and statistically significant ($r = 0.22$).

Source: Pike, G. R. (2009). The differential effects on on- and off-campus living arrangements on students' openness to diversity. *NASPA Journal, 46*(4), 629–45.

However, unlike previous studies that reveal positive correlations among the workaholism factors, the present study showed that the WI-3 related negatively to WI-E ($r = -.24$) and D-3 ($r = -.12$).

Source: Huang, J., Hu, C., & Wu, T. (2010). Psychometric properties of the Chinese version of the Workaholism Battery. *Journal of Psychology, 144*(2), 163–183.

only the symbol *r* is presented, and there is no adjective such as *Pearson's, Pearson's product–moment,* or *product–moment.* (In cases like this, where the symbol *r* stands by itself without a clarifying label, it is a good bet you are looking at a Pearson product–moment correlation coefficient.)

Spearman's Rho and Kendall's Tau

The second most popular bivariate correlational technique is called **Spearman's rho.** This kind of correlation is similar Pearson's in that it is appropriate for the situation in which both variables are quantitative in nature. With Spearman's technique, however, each of the two variables is measured in such a way as to produce ranks. This correlational technique often goes by the name **rank–order correlation** (instead of Spearman's rho). The resulting correlation coefficient, if symbolized, is usually referred to as r_s or ρ.

To illustrate how raw scores can be converted into ranks, I created some hypothetical raw scores for the height (in inches) and weight (in pounds) for five men. Those data are located in the first two columns of numbers. In the third and fourth columns, I show the ranks the five men would have if we got ready to do a Spearman's rank–order correlation. For these data, Spearman's correlation = .50.

Individual	Height Raw Score	Weight Raw Score	Height Rank	Weight Rank
Alex	67	175	4	4
Claude	73	210	1	2
David	70	225	2	1
Lee	66	180	5	3
Andrew	68	155	3	5

Only rarely does a researcher display the actual ranks used to compute Spearman's rho. Most of the time, the only information you are given is (1) the specification of the two variables being correlated and (2) the resulting correlation coefficient. Excerpts 3.17 and 3.18, therefore, are more typical of what you see in published journal articles than the four columns of data I created for the five men.

EXCERPTS 3.17–3.18 • *Spearman's Rank–Order Correlation*

A correlation of $\rho = .557$ was found between self-reported activity intensity and direct observation.

Source: Belton, S., & Donncha, C. M. (2010). Reliability and validity of a new physical activity self-report measure for younger children. *Measurement in Physical Education & Exercise Science, 14*(1), 15–28.

A correlation of $r_s = 0.57$ between the sum of near fall incidents and the sum of falls during the study period was found.

Source: Nilsagård, Y., Lundholm, C., Denison, E., & Gunnarsson, L.-G. (2009). Predicting accidental falls in people with multiple sclerosis—A longitudinal study. *Clinical Rehabilitation, 23*(3), 259–269.

Kendall's tau is very similar to Spearman's rho in that both these bivariate correlational techniques are designed for the case in which each of two quantitative variables is measured in such a way as to produce data in the form of ranks. The difference between rho and tau is related to the issue of ties. To illustrate what I mean by ties, suppose six students take a short exam and earn these scores: 10, 9, 7, 7, 4, and 3. These raw scores, when converted to ranks, become 1, 2, 3.5, 3.5, 5, and 6, where the top score of 10 receives a rank of 1, the next-best score (9) receives a rank of 2, and so on. The third- and fourth-best scores tied with a score of 7, and the rank given to each of these individuals is equal to the mean of the separate ranks they would have received if they had not tied. (If the two 7s had been 8 and 6, the separate ranks would have been 3 and 4, respectively; the mean of 3 and 4 is 3.5, and this rank is given to each of the persons who actually earned a 7.)

Kendall's tau is simply a bivariate correlational procedure that does a better job of dealing with tied ranks than does Spearman's rho. Recently, I came across a journal article that contained two sets of ranks, each of which went from 1 to 13. However, there were three pairs of ties within each array of ranks. (Of the 13 objects being evaluated, nine had a tied rank on the X variable or on the Y variable or on both variables.) I correlated the ranks twice, once to get Spearman's rho and once to get Kendall's tau. The resulting correlation coefficients were .89 and .75, respectively. Many statisticians consider tau to be the more accurate of these two correlation coefficients due to the ties.

In Excerpts 3.19 and 3.20, we see two cases where Kendall's tau was used. In the second of these excerpts, note that the correlation was called **tau-b**. When researchers use the labels tau and tau-b, they are most likely referring to the same thing. (Actually, there are three versions of Kendall's tau: tau-a, tau-b, and tau-c. Only tau-b and tau-c contain a built-in correction for tied ranks.)

EXCERPT 3.19 • *Kendall's Tau*

To check whether extraneous bodily movements and irrelevant vocalizations were fewer among the older children, we treated these two sets of behaviours separately. In each case we examined whether their frequency correlated with the children's age. The appropriate statistic here was Kendall's tau [which revealed] a decrement with age in the case of extraneous bodily movements [$\tau = -.316$] but not in the case of irrelevant vocalizations.

Source: Ellis, S. A. Turner; Miles, T. R.; Wheeler, T. J.; Thomson, Michael. (2009). Extraneous bodily movements and irrelevant vocalizations by dyslexic and non-dyslexic boys during calculation tasks. *Dyslexia, 15*(2), 155–163.

EXCERPT 3.20 • *Kendall's Tau-b*

A pilot-study including 42 nursing-home patients investigated inter-rater reliability of the Cornell Scale, by having three raters, one geriatric psychiatrist (GP) and two of the research nurses, a registered nurse (RN) and a nurse specialized in psychiatry (NP), score the Cornell Scale on the basis of responses from the primary carers. . . . The correlation coefficients (Kendall's tau-b) of the Cornell sum scores were 0.964 (GP vs RN), 0.961 (GP vs NP) and 0.957 (RN vs NP).

Source: Barca, M. L., Engedal, K., Laks, J., & Selbaek, G. (2010). A 12 months follow-up study of depression among nursing-home patients in Norway. *Journal of Affective Disorders, 120*, 141–148.

Point Biserial and Biserial Correlations

Sometimes a researcher correlates two variables that are measured so as to produce a set of raw scores for one variable and a set of 0s and 1s for the other (dichotomous) variable. For example, a researcher might want to see if a relationship exists between the height of basketball players and whether they score any points in a game. For this kind of bivariate situation, a correlational technique called **point biserial** has been designed. The resulting correlation coefficient is usually symbolized r_{pb}.

If a researcher has data on two variables where one variable's data are in the form of raw scores whereas the other variable's data represent an artificial dichotomy, then the relationship between the two variables are assessed by means of a technique called **biserial correlation.** Returning to our basketball example, suppose a researcher wanted to correlate height with scoring productivity, with the second of these variables dealt with by checking to see whether each player's average is less than 10 points or some value in double digits. Here, scoring productivity is measured by imposing an artificial dichotomy on a set of raw scores. Accordingly, the biserial technique is used to assess the nature and strength of the relationship between the two variables. This kind of bivariate correlation is usually symbolized by r_{bis}.

In Excerpt 3.21, we see a case where the point-biserial correlation was used in a published research article. In this study, gender was the binary (i.e., dichotomous) variable. Grade-point average (GPA) and satisfaction-with-major were not dichotomous, as GPA was measured on a traditional 0 to 4.0 scale, whereas the satisfaction variable was measured on a 1 to 7 Likert scale.

EXCERPT 3.21 • *Point-Biserial Correlation*

The first phase of the analysis involved splitting the sample [of undergraduate business students] into roughly equal subsamples ($n = 229$; $n = 222$). . . . Gender had significant correlations with GPA [$r_{pb} = .16$ and $r_{pb} = .18$] and Satisfaction-With-Major [$r_{pb} = .24$ and $r_{pb} = .18$] in both samples (point-biserial).

Source: Wefald, A. J., & Downey, R. G. (2009). Construct dimensionality of engagement and its relation with satisfaction. *Journal of Psychology, 143*(1), 91–112.

Phi and Tetrachoric Correlations

If both of a researcher's variables are dichotomous in nature, then the relationship between the two variables is assessed by means of a correlational technique called **phi** (if each variable represents a true dichotomy) or a technique called **tetrachoric correlation** (if both variables represent artificial dichotomies). An example calling for the first of these situations involves correlating, among high school students, the variables of gender (male/female) and car ownership (yes/no). For an example of a place where tetrachoric correlation is appropriate, imagine that we measure each of several persons in terms of height (with people classified as tall or short depending on whether they measure over 5′8″) and weight (with people classified as "OK" or "Not OK" depending on whether they are within 10 pounds of their ideal weight). Here, both height and weight are forced into being dichotomies.

Excerpt 3.22 illustrates the use of phi, demonstrating nicely how the two variables involved in a correlation can each represent a true dichotomy. The notion of left versus right is involved in all three of the variables mentioned.

EXCERPT 3.22 • *Phi*

Phi correlation was used to evaluate the relationship between the side of the abdomen in which ETAP [exercise-related transient abdominal pain] was reported and both the direction of lateral curvature of the spine, if present, and side of the spine raised in the case of roto-scoliosis. . . . The site in which ETAP was reported within the abdomen was not related to the direction of scoliosis.

Source: Morton, D. P., & Callister, R. (2009). Influence of posture and body type on the experience of exercise-related transient abdominal pain. *Journal of Science and Medicine in Sport, 13*(5), 485–488.

Cramer's V

If a researcher collects bivariate data on two variables where each variable is nominal in nature, the relationship between the two variables can be measured by means of a correlational technique called **Cramer's V.** In Excerpt 3.23, we see a case where Cramer's *V* was used. In this study focused on a group of alcohol abusers, two of the variables of interest were abstinence and the stage of change. As indicated in the excerpt, Cramer's *V* measured the strength of the relationship between these variables. This correlational technique, unlike most others we have considered, yields coefficients that must lie somewhere between 0 and 1.

EXCERPT 3.23 • *Cramer's V*

There was a positive association [Cramer's $V = 0.15$] between the stage of change and abstinence in Month 12 after detoxification. . . . Half of the individuals initially in Action were abstinent (55.2%, $n = 179$) compared to 37.7% of the Contemplators ($n = 23$) and 25.0% ($n = 2$) of the Precontemplators.

Source: Freyer-Adam, J., Coder, B., Ottersbach, C., Tonigan, J. S., Rumpf, H., John, U., & Hapke, U. (2009). The performance of two motivation measures and outcome after alcohol detoxification. *Alcohol & Alcoholism, 44*(1), 77–83.

Warnings about Correlation

At this point, you may be tempted to consider yourself a semi-expert when it comes to deciphering discussions about correlation. You now know what a scatter diagram is, you have looked at the correlational continuum (and know that correlation coefficients typically extend from −1.00 to +1.00), you understand what a correlation matrix is, and you have considered several different kinds of bivariate correlation. Before you assume that you know everything there is to know about measuring the relationship between two variables, I want to provide you with six warnings that deal with causality, the coefficient of determination, the possibility of outliers, the assumption of linearity, the notion of independence, and criteria for claims of high and low correlations.

Correlation and Cause

It is important for you to know that a correlation coefficient does not speak to the issue of **cause and effect.** In other words, whether a particular variable has a causal impact on a different variable cannot be determined by measuring the two variables simultaneously and then correlating the two sets of data. Many recipients of research reports (and even a few researchers) make the mistake of thinking that a high correlation implies that one variable has a causal influence on the other variable. To prevent yourself from making this mistake, I suggest that you memorize this simple statement: correlation $\neq$ cause.

Competent researchers often collect data using strategies that allow them to address the issue of cause. Those strategies are typically complex and require a consideration of issues that cannot be discussed here. In time, however, I am confident that you will understand the extra demands that are placed on researchers who want to investigate the potential causal connections between variables. For now, all I can do is ask that you believe me when I say that bivariate correlational data alone cannot be used to establish a cause-and-effect situation.

Coefficient of Determination

To get a better feel for the strength of the relationship between two variables, many researchers square the value of the correlation coefficient. For example, if r turns out equal to .80, the researcher squares .80 and obtains .64. When r is squared like this, the resulting value is called the **coefficient of determination.**

The coefficient of determination indicates the proportion of variability in one variable that is associated with (or explained by) variability in the other variable. The value of r^2 lies somewhere between 0 and $+1.00$, and researchers usually multiply by 100 so they can talk about the *percentage* of explained variability. In Excerpt 3.24, we see an example from a stress/eyewitness study where r^2 has

EXCERPT 3.24 • r^2 and Explained Variation

Pearson's correlation coefficient between the change in heart rate (labyrinth mean heart rate–baseline mean heart rate) and state anxiety score showed a reliable association, $r = .76$ [and] $r^2 = .58$. Change in heart rate accounted for 58% of the variance in state anxiety score.

Source: Valentine, T., & Mesout, J. (2009). Eyewitness identification under stress in the London Dungeon. *Applied Cognitive Psychology, 23*(2), 151–161.

been converted into a percentage. As this excerpt indicates, researchers some-times refer to this percentage as the amount of variance in one variable that is ac-counted for by the other variable, or they sometimes say that this percentage indicates the amount of *shared variance.*

As suggested by the material in Excerpt 3.24, the value of r^2 indicates how much (proportionately speaking) variability in either variable is explained by the other variable. The implication of this is that the raw correlation coefficient (i.e., the value of r when not squared) exaggerates how strong the relationship really is between two variables. Note that r must be stronger than .70 for there to be at least 50 percent explained variability. Or, consider the case where $r = .50$; here, only one-fourth of the variability is explained.

Outliers

My third warning concerns the effect on r of one or more data points located away from the bulk of the scores. Such data points are called **outliers,** and they can cause the size of a correlation coefficient to understate or exaggerate the strength of the relationship between two variables. In Excerpt 3.25, we see a case where the researchers were aware of the danger of outliers, so they examined their scatter plots before making claims based on their correlation coefficients.

EXCERPT 3.25 • *Outliers*

[C]orrelations were run for the whole sample and for married and custodial fathers separately. . . . [W]e examined the scatterplots to ensure that the relations were not attributable to one or a few outliers. The plots show clear group tendencies not inflated by extreme data points.

Source: Bernier, A., & Miljkovitch, R. (2009). Intergenerational transmission of attachment in father–child dyads: The case of single parenthood. *Journal of Genetic Psychology, 170*(1), 31–52.

In contrast to the good example provided in Excerpt 3.25, most researchers fail to check to see if one or more outliers serve to distort the statistical summary of the bivariate relationships they study. There are not many scatter plots in journal articles, and thus you cannot examine the data yourself to see if outliers were present. Almost always, only the correlation coefficient is provided. Give the researcher some extra credit, however, whenever you see a statement to the effect that the correlation coefficient was computed after an examination of a scatter plot revealed no outliers (or revealed an outlier that was removed prior to computing the correlation coefficient).

Linearity

The most popular technique for assessing the strength of a bivariate relationship is Pearson's product–moment correlation. This correlational procedure works nicely if the two variables have a linear relationship. Pearson's technique does not work well, however, if a curvilinear relationship exists between the two variables.

A **linear** relationship does *not* require that all data points (in a scatter plot) lie on a straight line. Instead, what *is* required is that the *path* of the data points be straight. The path itself can be very narrow, with most data points falling near an imaginary straight line, or the path can be very wide—so long as the path is straight. (Regardless of how narrow or wide the path is, the path to which we refer can be tilted at any angle.)

If a **curvilinear** relationship exists between two variables, Pearson's correlation underestimates the strength of the relationship present in the data. Accordingly, you can place more confidence in any correlation coefficient you see when the researcher who presents it indicates that a scatter plot was inspected to see whether the relationship was linear before Pearson's r was used to summarize the nature and strength of the relationship. Conversely, add a few grains of salt to the rs that are thrown your way without statements concerning the linearity of the data.

In Excerpt 3.26, we see an example where a pair of researchers checked to see if their bivariate data sets were linear. These researchers deserve high praise for taking the time to check out the linearity assumption before computing Pearson's r. Unfortunately, most researchers collect their data and compute correlation coefficients without ever thinking about linearity.

EXCERPT 3.26 • Linearity

Examination of the scatter plots provided further information on linearity [and] no evidence of curvilinear relationship was identified.

Source: Tam, D. M. Y., & Coleman, H. (2009). Construction and validation of a professional suitability scale for social work practice. *Journal of Social Work Education, 45*(1), 47–63.

Correlation and Independence

In many empirical studies, the researcher either builds or uses different tests in an effort to assess different skills, traits, or characteristics of the people, animals, or objects from whom measurements are taken. Obviously, time and money is wasted if two or more of these tests are redundant. Stated differently, it is desirable (in many studies) for each measuring instrument to accomplish something unique compared to the other measuring instruments being used. Two instruments that do this are said to be **independent**.

The technique of correlation often helps researchers assess the extent to which their measuring instruments are independent. Independence exists to the extent that *r* turns out to be close to zero. In other words, low correlations imply independence, whereas high positive or negative correlations signal lack of independence.

Excerpt 3.27 illustrates how authors sometimes refer to the *independence* of variables. The full journal article from which this excerpt was taken presented only one of the three correlation coefficients referred to here. Perhaps those unseen correlation were quite close to zero. What we know for sure, however, is that the other two *r*s turned out somewhere between −.26 and +.26. (This segment of the correlational continuum seems to me to be a bit wide in order for variables to be considered to be independent. I would make that segment only one-fourth as wide as these researchers did.)

EXCERPT 3.27 • *Independence*

[A] Pearson's correlation test was performed on [each pair of] the three variables. The highest correlation between any two variables was *r* = .26, indicating independence among the three dependent variables.

Source: Ivanov, B., Pfau, M., & Parker, K. A. (2009). Can inoculation withstand multiple attacks? An examination of the effectiveness of the inoculation strategy compared to the supportive and restoration strategies. *Communication Research, 36*(5), 655–676.

Relationship Strength

My final warning concerns the labels that researchers attach to their correlation coefficients. There are no hard and fast rules that dictate when labels such as *strong* or *moderate* or *weak* should be used. In other words, there is subjectivity involved in deciding whether a given *r* is *high* or *low*. Not surprisingly, researchers are sometimes biased (by how they *wanted* their results to turn out) when they select an adjective to describe their obtained *r*s. Being aware that this happens, you must realize that you have the full right to look at a researcher's *r* and label it however you wish, even if your label is different from the researcher's.

Consider Excerpt 3.28. In this passage, the researchers assert that two of the obtained *r*s indicate "a moderate correlation." Knowing now about how the coefficient of determination is computed and interpreted, you ought to be a bit hesitant to swallow the researchers' assertion that −.241 and −.186 are moderate correlations. If squared and then turned into percentages, these *r*s indicate that income explained less than 6 percent of the variability in either of the two divorce determinants mentioned.

EXCERPT 3.28 • *Questionable Labels Used to Describe Relationships*

Income also showed moderate correlations with divorce determinants such as incompetence in supporting family (−.241) and economic bankruptcy (−.186).

Source: Chun, Y., & Sohn, T. (2009). Determinants of consensual divorce in Korea: Gender, socio-economic status, and life course. *Journal of Comparative Family Studies, 40*(5), 775–789.

Review Terms

Abscissa	Outlier
Biserial correlation	Pearson's product–moment correlation
Bivariate	Perfect
Cause and effect	Phi
Coefficient of determination	Point biserial correlation
Correlation coefficient	Positive correlation
Correlation matrix	Qualitative variable
Cramer's *V*	Quantitative variable
Curvilinear	Rank–order correlation
Dichotomous variable	Ranks
Direct relationship	Raw score
High	Relationship
Independent	Scatter plot
Indirect relationship	Spearman's rho
Inverse relationship	Strong
Kendall's tau	Tetrachoric correlation
Linear	Weak
Low	r
Moderate	r_s
Negative correlation	r^2
Nominal	r_{pb}
Ordinal	r_{bis}
Ordinate	ρ

The Best Items in the Companion Website

1. An interactive online quiz (with immediate feedback provided) covering Chapter 3.
2. Ten misconceptions about the content of Chapter 3.
3. The author's poem "True Experimental Design."

4. An email message from the author to his students in which he asks an honest question about Pearson's *r* and Spearman's rho.
5. Two jokes about statistics, the first of which concerns a student in a statistics course.

To access the chapter outline, practice tests, weblinks, and flashcards, visit the companion website at http://www.ReadingStats.com.

Review Questions and Answers begin on page 531.

Reliability and Validity

Empirical research articles focus on data that have been collected, summarized, and analyzed. The conclusions drawn and the recommendations made in such studies can be no better than the data on which they are based. As a consequence, most researchers describe the quality of the instruments used to collect their data. These descriptions of instrument quality usually appear in the method section of the article, either in the portion that focuses on materials or in the description of the dependent variables.

Regardless of where it appears, the description of instrument quality typically deals with two measurement-related concepts—reliability and validity. In this chapter, I discuss the meaning of these two concepts, various techniques employed by researchers to assess the reliability and validity of their measuring instruments, and numerical indices of instrument quality that are reported. My overall objective here is to help you refine your skills at deciphering and evaluating reports of reliability and validity.

Reliability

This discussion of reliability is divided into three sections. We begin by looking at the core meaning of the term *reliability.* Next, we examine a variety of techniques that researchers use to quantify the degree to which their data are reliable. Finally, I provide five cautionary comments concerning reports of reliability that will help you as you read technical research reports.

The Meaning of Reliability and the Reliability Coefficient

The basic idea of **reliability** is summed up by the word **consistency.** Researchers can and do evaluate the reliability of their instruments from different perspectives, but the basic question that cuts across these various perspectives (and techniques) is always the same: "To what extent can we say that the data are consistent?"

As you will see, the way in which reliability is conceptualized by researchers can take one of three basic forms. In some studies, researchers ask, "To what degree does a person's measured performance remain consistent across repeated testings?" In other studies, the question of interest takes a slightly different form: "To what extent do the individual items that go together to make up a test or an inventory consistently measure the same underlying characteristic?" In still other studies, the concern over reliability is expressed in the question, "How much consistency is there among the ratings provided by a group of raters?" Despite the differences among these three questions, the notion of consistency is at the heart of the matter in each case.

Different statistical procedures have been developed to assess the degree to which a researcher's data are reliable, and we will consider some of the more frequently used procedures in a moment. Before doing that, however, I want to point out how the different procedures are similar. Besides dealing, in one way or another, with the concept of consistency, each of the reliability techniques leads to a single numerical index. Called a **reliability coefficient,** this descriptive summary of the data's consistency normally assumes a value somewhere between 0.00 and +1.00, with these two "end points" representing situations where consistency is either totally absent or totally present.

Different Approaches to Reliability

Test–Retest Reliability.   In many studies, a researcher measures a single group of people (or animals or things) twice with the same measuring instrument, with the two testings separated by a period of time. The interval of time may be as short as one day or it can be as long as a year or more. Regardless of the length of time between the two testings, the researcher simply correlates the two sets of scores to find out how much consistency is in the data. The resulting correlation coefficient is simply renamed the **test–retest reliability coefficient.**[1]

With a test–retest approach to reliability, the resulting coefficient addresses the issue of consistency, or stability, over time. For this reason, the test–retest reliability coefficient is frequently referred to as the **coefficient of stability.** As with other forms of reliability, coefficients of stability reflect high reliability to the extent that they are close to 1.00.

In Excerpts 4.1 and 4.2, we see two examples of test–retest reliability. In the first of these excerpts, there is no indication of the statistical procedure used to compute the reliability coefficient. In cases like these, you will probably be right if you guess that the stability coefficient came from Pearson's correlation. (This is a safe

---

[1]As you recall from Chapter 3, correlation coefficients can assume values anywhere between -1.00 and $+1.00$. Reliability, however, cannot logically turn out to be negative. Therefore, if the test–retest correlation coefficient turns out to be negative, it will be changed to 0.00 when relabeled as a *reliability coefficient.*

EXCERPTS 4.1–4.2 • *Test–Retest Reliability*

The total score [of the Harvey Developmental Scale] shows a test–retest reliability of .95 after nine months and .96 after three years.

Source: Tremblay, K. N., Richer, L., Lachance, L., & Cote, A. (2010). Psychopathological manifestations of children with intellectual disabilities according to their cognitive and adaptive behavior profile. *Research in Developmental Disabilities, 31*(1), 57–9.

The first 20 participants were assessed twice for test–retest reproducibility. Their 1-wk test–retest intraclass correlation coefficient (ICC) ranged from 0.92 to 0.99 across the six muscles assessed.

Source: Lee, M. J., Kilbreath, S. L., Singh, M. F., Zeman, B., & Davis, G. M. (2010). Effect of progressive resistance training on muscle performance after chronic stroke. *Medicine and Science in Sports and Exercise, 42*(1), 23–34.

bet, because test–retest reliability is typically estimated via *r*.) As shown in Excerpt 4.2, another technique used to estimate test–retest reliability is called the **intraclass correlation.** The intraclass correlation procedure is quite versatile, and it can be used for a variety of purposes (e.g., to see if ratings from raters are reliable). In the case of a test–retest situation, all you need to know is that the intraclass correlation yields the same estimate of reliability as does Pearson's correlation.

With most characteristics, the degree of stability decreases as the interval between test and retest increases. For this reason, high coefficients of stability are more impressive when the time interval is longer. If a researcher does not indicate the length of time between the two testings, then the claims made about stability must be taken with a grain of salt. Stability is not very convincing if a trait remains stable for only an hour.

Alternate-Forms Reliability.[2]   Instead of assessing stability over time, researchers sometimes measure people with two forms of the same instrument. The two forms are similar in that they supposedly focus on the same characteristic (e.g., intelligence) of the people being measured, but they differ with respect to the precise questions included within each form. If the two forms do in fact measure the same thing (and if they are used in close temporal proximity), we would expect a high degree of consistency between the scores obtained for any examinee across the two testings. With **alternate-forms reliability,** a researcher is simply determining the degree to which this is the case.

---

[2]The terms *parallel-forms reliability* and *equivalent-forms reliability* are synonymous (as used by most applied researchers) with the term *alternate-forms reliability*.

To quantify the degree of alternate-forms reliability that exists, the researcher administers two forms of the same instrument to a single group of individuals with a short time interval between the two testings.[3] After a score becomes available for each person on each form, the two sets of data are correlated. The resulting correlation coefficient is interpreted directly as the alternate-forms reliability coefficient.[4] Many researchers refer to this two-digit value as the **coefficient of equivalence.**

To see an example where this form of reliability was used, consider Excerpt 4.3. Notice how the researcher points out that her alternate-forms reliability coefficient (that turned out equal to .91) was found by simply correlating the participants' scores on the two forms of the Situational Judgment Test (SJT). Most likely, this correlation was a Pearson's r. Note that the two forms of the SJT were equated in terms of number of items, item difficulty, and the leadership dimensions dealt with. Clearly, a concerted effort was made to make the two forms equivalent. Together, the test development process and the extremely high correlation of .91 provide strong evidence that the two forms were equivalent.

EXCERPT 4.3 • *Alternate-Forms Reliability*

Alternate forms reliability was assessed in Study 2. . . . Two forms of the SJT were developed with items equated on difficulty and dimension representation. The resulting forms [had] 72 items each. A coefficient of equivalence showed a strong, positive correlation between the two forms ($r = .91$), which indicates that the two forms are equivalent measures and can used to similarly measure leadership ability.

Source: Grant, K. L. (2009). The validation of a situational judgment test to measure leadership behavior. Unpublished master's thesis. Western Kentucky University. Bowling Green, Kentucky.

Internal Consistency Reliability. Instead of focusing on stability across time or on equivalence across forms, researchers sometimes assess the degree to which their measuring instruments possess internal consistency. When this perspective is taken, reliability is defined as consistency across the parts of a measuring instrument, with the "parts" being individual questions or subsets of questions. To the extent that these parts "hang together" and measure the same thing, the full instrument is said to possess high **internal consistency reliability.**

To assess internal consistency, a researcher need only administer a test (or questionnaire) a single time to a single group of individuals. After all responses

---

[3]The two forms will probably be administered in a *counterbalanced* order, meaning that each instrument is administered first to one-half of the examinees.

[4]As is the case with test–retest reliability, any negative correlation would be changed to 0. Reliability by definition has a lower limit of 0.

have been scored, one of several statistical procedures is then applied to the data, with the result being a number between 0.00 and +1.00. As with test–retest and equivalent-forms procedures, the instrument is considered to be better to the extent that the resulting coefficient is close to the upper limit of this continuum of possible results.

One of the procedures that can be used to obtain the internal consistency reliability coefficient involves splitting each examinee's performance into two halves, usually by determining how the examinee did on the odd-numbered items grouped together (i.e., one half of the test) and the even-numbered items grouped together (i.e., the other half). After each person's total score on each half of the instrument is computed, these two sets of scores are correlated. Once obtained, the r is inserted into a special formula (called **Spearman-Brown**). The final numerical result is called the **split-half reliability coefficient.**

Use of the split-half procedure for assessing internal consistency can be seen in Excerpts 4.4 and 4.5. Note in the first excerpt what is in parentheses immediately after the reliability coefficient is presented. This information clarifies that the Spearman-Brown formula was used to "correct" the correlation (between the two seven-item halves) in order to make the reliability estimate appropriate for a test of equal length to the 14-item test being evaluated. The Spearman-Brown correction is needed because reliability tends to be higher for longer tests (and lower for shorter tests). The Spearman-Brown correction formula "boosts" the correlation coefficient upwards so as to undo the "damage" caused by splitting the test in half. In Excerpt 4.5, we see an example of **Guttman's split-half reliability.** The Guttman procedure is like the normal split-half procedure, except that the former technique does not require the two halves to have equal variances whereas the latter technique does.

EXCERPTS 4.4–4.5 • *Split-Half Reliability*

Reliability of the questionnaire was quantified using all 184 child responses. . . . Additionally, split-half internal consistency method was employed to determine reliability. The reliability of the 14-item questionnaire was .56 (equal-length Spearman-Brown, $n = 184$).

Source: Geller, K. S., Dzewaltowski, D. A., Rosenkranz, R. R., & Karteroliotis, K. (2009). Measuring children's self-efficacy and proxy efficacy related to fruit and vegetable consumption. *Journal of School Health, 79*(2), 51–57.

The internal consistency of the scale was assessed [via] Guttman's split-half method. . . . Following the removal of six out of 22 items, [the] Guttman split-half value for the remaining 16 items was 0.77.

Source: Maijala, H., Åstedt-Kurki, P., & Åstedt-Kurki, P. (2009). From substantive theory towards a family nursing scale. *Nurse Researcher, 16*(3), 29–44.

A second approach to assessing internal consistency is called **Kuder-Richardson #20 (K-R 20).** This procedure, like the split-half procedure, uses data from a single test that has been given just once to a single group of respondents. After the full test is scored, the researcher simply puts the obtained data into a formula that provides the K-R 20 reliability coefficient. The result is somewhat like a split-half reliability, but better, because the split-half approach to assessing internal consistency yields a result that can vary depending on which items are put in the odd-numbered slots and which are placed in the even-numbered slots. In other words, if the items that go together to make up a full test are "scrambled" in terms of the way they are ordered, this likely affects the value of the split-half reliability coefficient. Thus, whenever the split-half procedure is used to assess the reliability of a measuring instrument, we do not know whether the resulting reliability coefficient is favorable (i.e., high) or unfavorable (i.e., low) as compared with what would have been the case if the items had been ordered differently.

With K-R 20, the result is guaranteed to be neither favorable nor unfavorable, because the formula for K-R 20 was designed to produce a result equivalent to what you would get if you (1) scramble the order of the test items over and over again until you have all possible orders, (2) compute a split-half reliability coefficient for each of these forms of the test, and (3) take the mean value of those various coefficients. Of course, the researcher who wants to obtain the K-R 20 coefficient does not have to go to the trouble to do these three things. A simple little formula is available that brings about the desired result almost instantaneously.

In Excerpts 4.6 and 4.7, we see an example of how Kuder-Richardson reliability results are often reported in published research reports. In the first of these excerpts, K-R 20 is used. In Excerpt 4.7, we see the use of K-R 21, a procedure that once had the advantage (before computers) of being easier to compute. K-R 21 is no longer used as much because its results are not quite as good as those provided by K-R 20.

EXCERPTS 4.6–4.7 • *Ruder-Richardson 20 and Kuder-Richardson 21 Reliabilities*

The reliability of internal consistency (KR-20) was .926.

Source: Chen, L., Ho, R., & Yen, Y. (2010). Marking strategies in metacognition-evaluated computer-based testing. *Journal of Educational Technology & Society, 13*(1), 246–259.

- -

The instrument [demonstrated] a Kuder-Richardson-21 (KR-21) reliability coefficient of .70.

Source: Gallo, M. A., & Odu, M. (2009). Examining the relationship between class scheduling and student achievement in college algebra. *Community College Review, 36*(4), 299–325.

A third method for assessing internal consistency is referred to as **coefficient alpha, Cronbach's alpha,** or simply **alpha.** This technique is identical to K-R 20 whenever the instrument's items are scored in a dichotomous fashion (e.g., "1" for correct, "0" for incorrect). However, alpha is more versatile because it can be used with instruments made up of items that can be scored with three or more possible values. Examples of such a situation include (1) a four-question essay test, where each examinee's response to each question is evaluated on a 0–10 scale; or (2) a Likert-type questionnaire where the five response options for each statement extend from "strongly agree" to "strongly disagree" and are scored with the integers 5 through 1. Excerpts 4.8 and 4.9 show two instances in which Cronbach's alpha was used to evaluate internal consistency. These excerpts demonstrate the versatility of Cronbach's technique for assessing internal consistency, because the two instruments used very different scoring systems. Whereas the instrument in Excerpt 4.8 had "1" or "0" scores (to indicate correct or incorrect responses to test questions), the instrument in Excerpt 4.9 utilized a 5-point Likert-type response format aimed at assessing the respondents' attitudes.

EXCERPTS 4.8–4.9 • *Coefficient Alpha Reliability*

Comprehension test. It consisted of 32 items and responses were given using four multiple-choice alternatives. . . . For each correct answer to a comprehension item participants received one point, otherwise they received zero points. . . . Cronbach's alpha for the overall comprehension test was .76.

Source: de Koning, B. B., Tabbers, H. K., Rikers, R. M. J. P., & Paas, F. (2010). Attention guidance in learning from a complex animation: Seeing is understanding? *Learning and Instruction, 20,* 111–122.

Students completed the empathic concern subscale [consisting] of 7 items (sample item: "I often have tender, concerned feelings for people less fortunate than me"). Items were rated on a 5-point scale ranging from 1 (*does not describe me*) to 5 (*describes me very well*). . . . Cronbach's alpha for [this] subscale in the current study was .76.

Source: McGinley, M., Carlo, G., Crockett, L. J., Raffaelli, M., Torres Stone, R. A., & Iturbide, M. I. (2010). Stressed and helping: The relations among acculturative stress, gender, and prosocial tendencies in Mexican Americans. *Journal of Social Psychology, 150*(1), 34–56.

Interrater Reliability

Researchers sometimes collect data by having raters evaluate a set of objects, pictures, applicants, or whatever. To quantify the degree of consistency among the raters, the researcher computes an index of **interrater reliability.** Five popular procedures for doing this include a percentage-agreement measure, Pearson's correlation, Kendall's coefficient of concordance, Cohen's kappa, and the intraclass correlation.

The simplest measure of interrater reliability involves nothing more than a percentage of the occasions where the raters agree in the ratings they assign to whatever is being rated. In Excerpt 4.10, we see an example of this approach to interrater reliability. This excerpt is instructional because it contains a clear explanation of how the two reliability figures were computed.

EXCERPT 4.10 • *Percentage Agreement as Measure of Interrater Reliability*

An agreement was recorded if both observers identically scored the answer as correct or incorrect. A disagreement was recorded if questions were not scored identically. Percent agreement for each probe was calculated by dividing the number of agreements by the number of agreements plus disagreements and multiplying by 100. Interrater reliability for the first dependent variable was 94.6%. . . . Interrater reliability for the second dependent variable was 97.7%.

Source: Mazzotti, V. L., Wood, C. L., Test, D. W., & Fowler, C. H. (2010). Effects of computer-assisted instruction on students' knowledge of the self-determined learning model of instruction and disruptive behavior. *Journal of Special Education*, in press.

The second method for quantifying interrater reliability uses Pearson's product-moment correlation. Whereas the percentage-agreement procedures can be used with data that are categorical, ranks, or raw scores, Pearson's procedure can be used only when the raters' ratings are raw scores. In Excerpt 4.11, we see an example of Pearson's correlation being used to assess the interrater reliability among two raters.

EXCERPT 4.11 • *Using Pearson's r to Assess Interrater Reliability*

Two high school English teachers, blind to students and to the study's hypothesis, rated each paper independently [on a 7-point Likert scale]. . . . Interobserver agreement for holistic quality for the essays (using Pearson *r*) was .95.

Source: Jacobson, L. T., & Reid, R. (2010). Improving the persuasive essay writing of high school students with ADHD. *Exceptional Children, 76*(2), 156–174.

Kendall's procedure is appropriate for situations where each rater is asked to rank the things being evaluated. If these ranks turn out to be in complete agreement across the various evaluators, then the **coefficient of concordance** will turn out equal to +1.00. To the extent that the evaluators disagree with one another, Kendall's procedure will yield a smaller value. In Excerpt 4.12, we see a case in which Kendall's coefficient of concordance was used.

EXCERPT 4.12 • *Kendall's Coefficient of Concordance*

A standardized measure of neurological dysfunction specifically designed for TBI currently does not exist and the lack of assessment of this domain represents a substantial gap. To address this, the Neurological Outcome Scale for Traumatic Brain Injury (NOS-TBI) was developed. . . . Overall interrater agreement between independent raters (Kendall's Coefficient of Concordance) for the NOS-TBI total score was excellent ($W = .995$).

Source: McCauley, S. R., Wilde, E. A., Kelly, T. M., Weyand, A. M., Yallampalli, R., Waldron, E. J., et al. (2010). The Neurological Outcome Scale for Traumatic Brain Injury (NOS-TBI): II. Reliability and convergent validity. *Journal of Neurotrauma, 27*(6), 991–997.

Kendall's coefficient of concordance establishes how much interrater reliability exists among ranked data. **Cohen's kappa** accomplishes the same purpose when the data are nominal (i.e., categorical) in nature. In other words, kappa is designed for situations where raters classify the items being rated into discrete categories. If all raters agree that a particular item belongs in a given category, and if there is a total agreement for all items being evaluated (even though different items end up in different categories), then kappa assumes the value of $+1.00$. To the extent that raters disagree, kappa assumes a smaller value.

To see a case in which Cohen's kappa was used, consider Excerpt 4.13. In the study that provided this excerpt, the researchers examined 225 social networking Web pages created by 17- to 20-year-olds. Each had a reference to alcohol. Two raters classified each Web page in terms of preset categories (e.g., explicit or figurative reference to alcohol). These categories were fully nominal in nature. Cohen's kappa was used to see how consistently the two raters assigned the Web pages to the categories.

EXCERPT 4.13 • *Cohen's Kappa*

We evaluated 400 randomly selected public MySpace profiles of self-reported 17- to 20-year-olds. . . . Two authors (L.R.B. and M.M.) conducted the initial evaluation to identify profiles with alcohol references. . . . 225 profiles contained references to alcohol and were included in all analyses (56.3%). . . . Cohen's Kappa statistic was used to evaluate the extent to which there was agreement in the coding of the web profiles before discussion [resolved differences of opinion]. The Kappa value for the identification of references to alcohol use was 0.82.

Source: Moreno, M. M., Briner, L. R., Williams, A., Brockman, L., Walker, L., & Christakis, D. A. (2010). A content analysis of displayed alcohol references on a social networking web site. *Journal of Adolescent Health, 47*(2), 168–175.

The final method for assessing interrater reliability to be considered here is called **intraclass correlation (ICC),** a multipurpose statistical procedure, as it can be used for either correlational or reliability purposes. Even if we restrict our thinking to reliability, ICC is still versatile. Earlier in this chapter, we saw a case where the intraclass correlation was used to estimate test–retest reliability. Now, we consider how ICC can be used to assess interrater reliability.

Intraclass correlation is similar to the other reliability procedures we have considered in terms of the core concept being dealt with (consistency), the theoretical limits of the data-based coefficient (0 to 1.00), and the desire on the part of the researcher to end up with a value as close to 1.00 as possible. It differs from the other reliability procedures in that there are several ICC procedures. The six most popular of these procedures are distinguished by two numbers put inside parentheses following the letters ICC. For example, ICC (3,1) designates one of the six most frequently used versions of intraclass correlation. The first number indicates which of three possible statistical models has been assumed by the researchers to underlie their data. The second number indicates whether the researchers are interested in the reliability of a single rater (or, one-time use of a measuring instrument) or in the reliability of the mean score provided by a group of raters (or, the mean value produced by using a measuring instrument more than once). The second number within the parentheses is a 1 for the first of these two cases; if interest lies in the reliability of means, the second number is a value greater than 1 that designates how many scores are averaged together to generate each mean.

I will not attempt to differentiate any further among the six main cases of ICC. Instead, I simply want to point out that researchers should explain in their research reports (1) which of the six ICC procedures was used and (2) the reason(s) behind the choice made. You have a right to expect clarity regarding these two issues because the ICC-estimated reliability coefficient can vary widely depending on which of the six available formulas is used to compute it.

In Excerpts 4.14 and 4.15, we see two examples where the intraclass correlation was used to assess interrater reliability. Notice that the researchers associated with the second of these excerpts indicate which of the six main types of ICC they used—model 2 for a single rater. Because the coefficient provided by ICC can vary

EXCERPTS 4.14–4.15 • *Intraclass Correlation*

All participants wrote their personal vision, and [then] three raters rated the vision statements according to definitions of challenge and imagery. . . . In our study, interrater reliability was intraclass correlation coefficient (ICC) = .93 for challenging and ICC = .87 for imagery.

Source: Masuda, A. D., Kane, T. D., Stoptaugh, C. F., & Minor, K. A. (2010). The role of a vivid and challenging personal vision in goal hierarchies. *Journal of Psychology, 144*(3), 221–242.

(*continued*)

EXCERPTS 4.14–4.15 • (*continued*)

The interrater reliability of the BESTest total scores was excellent, with an ICC (2,1) of .91.

Source: Horak, F. B., Wrisley, D. M., & Frank, J. (2009). The Balance Evaluation Systems Test (BESTest) to differentiate balance deficits. *Physical Therapy, 89*(5), 484–498.

---

widely depending on which of the six main formulas are used to obtain the intra-class correlation, we have a right to think more highly about the information in the second excerpt. It would have been even nicer if the authors of Excerpt 4.15 had explained why they chose ICC (2,1) instead of other variations of this reliability procedure.

The Standard Error of Measurement

Some researchers, when discussing reliability, present a numerical value for the **standard error of measurement (SEM)** that can be used to estimate the range within which a score would likely fall if a given measured object were to be re-measured. To illustrate, suppose an intelligence test is administered to a group of children, and also suppose that Tommy ends up with an IQ score of 112. If the SEM associated with the IQ scores in this group were equal to 4, then we would build an interval for Tommy (by adding 4 to 112 and subtracting 4 from 112) extending from 108 to 116. This interval, or **confidence band,** helps us interpret Tommy's IQ because we can now say that Tommy would likely score somewhere between 108 and 116 if the same intelligence test were to be re-administered and if Tommy did not change between the two testings.[5]

In a very real sense, the SEM can be thought of as an index of consistency that is inversely related to reliability. To the extent that reliability is high, the SEM is small (and vice versa). There is one other main difference between these two ways of assessing consistency. Reliability coefficients are tied to a scale that extends from 0 to 1.00, and in this sense they are completely "metric free." In contrast, an SEM is always tied to the nature of the scores generated by a test, and in this sense it is not "metric free." Simply stated, the continuum for reliability coefficients has no units of measurement, whereas the SEM is always "in" the same measurement units as are the scores around which confidence bands are built.

In Excerpt 4.16, we see a case in which the SEM was used in study dealing with stroke victims. In part of their investigation, the researchers compared four different ways of measuring upper-extremity mobility in patients: the Fugl-Meyer

---

[5]By creating an interval via the formula "score±SEM," we end up with a 68 percent confidence band. If we double or triple the SEM within this little formula, we end up with a 95 percent or a 99 percent confidence band, respectively.

Motor Test (UE-FM), the Stroke Rehabilitation Assessment of Movement (UE-STREAM), the Action Research Arm Test (ARAT), and the Wolf Motor Function Test (WMFT). In this excerpt, the number 6 associated with the UE-FM came about by doubling that test's SEM. (Similarly, the numbers 3, 4, and 12 are twice as large as the other tests' SEMs.) We know this because the researchers report that a change of more than 6 points "can be interpreted by clinicians as a real change with 95% confidence."

EXCERPT 4.16 • *Standard Error of Measurement*

We quantified random measurement errors with the standard error of measurement (SEM). . . . Our findings suggest that changes of more than 6 points, 3 points, 4 points, and 12 points in the total scores on the UE-FM (highest possible score: 66), UE-STREAM (20), ARAT (57), and WMFT (75), respectively, for each patient assessed by an individual rater are not likely to be attributable to chance variation or measurement error and can be interpreted by clinicians as a real change with 95% confidence.

Source: Lin, J., Hsu, M., Sheu, C., Wu, T., Lin, R., Chen, C., et al. (2009). Psychometric comparisons of 4 measures for assessing upper-extremity function in people with stroke. *Physical Therapy, 89*(8), 840–850.

Warning about Reliability

Before we turn to the topic of validity, there are five important warnings about reliability to which you should become sensitive. It would be nice if all researchers were also aware of these five concerns; unfortunately, that is not the case.

First, keep in mind that different methods for assessing reliability consider the issue of consistency from different perspectives. Thus, a high coefficient of stability does not necessarily mean that internal consistency is high (and vice versa). Even within the internal consistency category, a high value for split-half reliability does not necessarily mean that K-R 20 would be equally high for the same data. The various methods for assessing reliability accomplish different purposes, and the results do not necessarily generalize across methods. Because of this, I like to see various approaches to reliability used within the same study.

My second warning concerns the fact that reliability coefficients really apply to data and not to measuring instruments. To understand the full truth of this claim, imagine that a test designed for a college-level class in physics is administered twice to a group of college students, producing a test–retest reliability coefficient of .90. Now, if that same test is administered on two occasions to a group of first-grade students (with the same time interval between test and retest), the coefficient of stability would not be anywhere near .90. (The first graders would probably guess at all questions, and the test–retest reliability for this younger group most likely would end up close to 0.00.) Try to remember, therefore, that reliability is conceptually and computationally connected to the data produced by the *use* of a measuring instrument, not to the measuring instrument as it sits on the shelf.

Excerpts 4.17 and 4.18 illustrate the fact that reliability is a property of data, not of the instrument that produces the data. Reliability can vary across groups that vary in gender, age, health status, profession, or any number of other characteristics. As Excerpt 4.18 illustrates, reliability can also vary depending on the timing or conditions under which the test is administered.

EXCERPTS 4.17–4.18 • *Different Reliabilities from Different Samples and Days*

The [test–retest] correlation for grades 2–5 was .51; for the eighth graders, the correlation was .62; and for the adults, the correlation was .68, indicating increased reliability among older students.

Source: Pitts, J. (2009). Identifying and using a teacher-friendly learning-styles instrument. *Clearing House, 82*(5), 225–232.

An interesting finding with the INV-R was the variation in coefficient alpha according to day of measurement. . . . On day 1 of the cycle, the alpha was 0.71; on day 2, it rose to 0.75, and on day 3, it was 0.84.

Source: Ingersoll, G. L., Wasilewski, A., Haller, M. E., Pandya, K., Bennett, J., He, H., et al. (2010). Effect of Concord Grape Juice on chemotherapy-induced nausea and vomiting: Results of a pilot study. *Oncology Nursing Forum, 37*(2), 213–221.

Some researchers realize that reliability is a property of scores produced by the administration of a measuring instrument (rather than a property of the printed instrument itself). With this in mind, they not only cite reliability coefficients obtained by previous researchers who used the same instrument, but also gather reliability evidence *within* their own investigation. This practice is not widely practiced, unfortunately. Most of the time, researchers simply reiterate the reliability evidence gathered earlier by previous researchers who developed or used the same instrument. Those researchers who take the extra time to assess the reliability of the data gathered in their own investigation deserve credit for knowing that reliability ought to be reestablished in any current study.

My next warning calls on you to recognize that any reliability coefficient is simply an estimate of consistency. If a different batch of examinees or raters is used, you should expect the reliability coefficient to be at least slightly different—even if the new batch of examinees or raters contains people who are highly similar to the original ones. If the groups are small, there would probably be more fluctuation in the reliability coefficient than if the groups are large. Accordingly, place more faith in the results associated with large groups. Regardless of how large the group of examinees or raters is, however, give credit to researchers who use the word *estimated* in conjunction with the word *reliability.*

Another important warning concerns estimates of internal consistency. If a test is administered under great time pressure, the various estimates of internal consistency—split-half, K-R 20, and coefficient alpha—will be spuriously high (i.e., too big). Accordingly, do not be overly impressed with high internal consistency reliability coefficients if data have been collected under a strict time limit or if there is no mention as to conditions under which the data were collected.

Finally, keep in mind that reliability is not the only criterion that should be used to assess the quality of data. A second important feature of the data produced by measuring instruments (or raters) has to do with the concept of validity. The remaining portion of this chapter is devoted to a consideration of what validity means and how it is reported.

Validity

Whereas the best one-word synonym for reliability is consistency, the core essence of **validity** is captured nicely by the word **accuracy.** From this general perspective, a researcher's data are valid to the extent that the results of the measurement process are accurate. Stated differently, a measuring instrument is valid to the extent that it measures what it purports to measure.

In this portion of the chapter, we first consider the relationship between reliability and validity. Next, we examine several of the frequently used procedures for assessing validity. Finally, I offer a few warnings concerning published claims that you may see about this aspect of data quality.

The Relationship between Reliability and Validity

It is possible for a researcher's data to be highly reliable even though the measuring instrument does not measure what it claims to measure. However, an instrument's data must be reliable if they are valid. Thus, high reliability is a necessary but not sufficient condition for high validity. A simple example may help to make this connection clear.

Suppose a test is constructed to measure the ability of fifth-grade children to solve arithmetic word problems. Also suppose that the test scores produced by an administration of this test are highly reliable. In fact, let's imagine that the coefficient of stability turns out equal to the maximum possible value, +1.00. Even though the data from our hypothetical test demonstrate maximum consistency over time, the issue of accuracy remains unclear. The test may be measuring what it claims to measure—math ability applied to word problems; however, it may be that this test really measures reading ability.

Now, reverse our imaginary situation. Assume for the moment that all you know is that the test is valid. In other words, assume that this newly designed measuring instrument does, in fact, produce scores that accurately reflect the ability of fifth graders to solve arithmetic word problems. If our instrument produces scores that are valid, then those scores, of necessity, must also be reliable. Stated differently, accuracy requires consistency.

Different Kinds of Validity

In published articles, researchers often present evidence concerning a specific kind of validity. Validity takes various forms because there are different ways in which scores can be accurate. To be a discriminating reader of the research literature, you must be familiar with the purposes and statistical techniques associated with the popular validity procedures. The three most frequently used procedures are content validity, criterion-related validity, and construct validity.

Content Validity.   With certain tests, questionnaires, or inventories, an important question concerns the degree to which the various items collectively cover the material that the instrument is supposed to cover. This question can be translated into a concern over the instrument's **content validity.** Usually, an instrument's standing with respect to content validity is determined simply by having experts carefully compare the content of the test against a syllabus or outline that specifies the instrument's claimed domain. Subjective opinion from such experts establishes—or does not establish—the content validity of the instrument.

In Excerpt 4.19, we see a case in which content validity is discussed by a team of researchers. As you will see, these researchers were extremely thorough in their effort to assess—and improve—the content validity of the new measuring instrument they had developed.

EXCERPT 4.19 • *Content Validity*

In order to establish the content validity, the questionnaire was sent to a panel of six experts (three sports management faculty members and three martial arts instructors). The panel members were given the conceptual definitions of each motivation factor and instructed to retain items based on their relevance and representation of the factors and clarity of wording. Based on the feedback received, items that were unclear, irrelevant or redundant (four items) were eliminated. Additional modifications were made on a number of items, mainly to clarify wording. The questionnaire was then field-tested on a group of college students ($N = 10$) who are currently involved in some form of martial arts. They were asked to further analyse the clarity, wording and relevance of the items.

Source: Ko, Y. J., Kim, Y. K., & Valacich, J. (2010). Martial arts participation: Consumer motivation. *International Journal of Sports Marketing & Sponsorship, 11*(2), 105–123.

Criterion-Related Validity. Researchers sometimes assess the degree to which their new instruments provide accurate measurements by comparing scores from the new instrument with scores on a relevant criterion variable. The new instrument under investigation might be a short, easy-to-give intelligence test, and in this case the criterion would probably be an existing reputable intelligence test (possibly the Stanford-Binet). Or, maybe the new test is an innovative college entrance examination; hence, the criterion variable would be a measure of academic success in college (possibly grade point average). The validity of either of those new tests would be determined by (1) finding out how various people perform on the new test and on the criterion variable, and (2) correlating these two sets of scores. The resulting *r* is called the **validity coefficient,** with high values of *r* indicating high validity.

There are two kinds of criterion-related validity. If the new test is administered at about the same time that data are collected on the criterion variable, then the term **concurrent validity** is used. Continuing the first example provided in the preceding paragraph, if people were given the new and existing intelligence tests with only a short time interval between their administrations, the correlation between the two data sets would speak to the issue of concurrent validity. If, however, people were given the new test years before they took the criterion test, then *r* would be a measure of **predictive validity.**

In Excerpts 4.20 and 4.21, we see cases where the expressed concern of the researchers was with concurrent and predictive validity. In these excerpts, note that bivariate correlation coefficients—most likely Pearson's *rs*—were used to evaluate the **criterion-related validity** of the tests being investigated.

EXCERPTS 4.20–4.21 • *Concurrent and Predictive Validity*

The LEP is intended to assess learners' level of cognitive development [and] has demonstrated moderate concurrent validity with other developmental measures designed to measure similar concepts (.46 to .57 with the MID: Measure of Intellectual Development).

Source: Granello, D. H. (2010). Cognitive complexity among practicing counselors: How thinking changes with experience. *Journal of Counseling & Development, 88*(1), 92–100.

The predictive validation study of the GMAT was [assessed] via Pearson correlation analysis. . . . Its correlation with GPA at the end of the MBA program was .60.

Source: Koys, D. (2010). GMAT versus alternatives: Predictive validity evidence from Central Europe and the Middle East. *Journal of Education for Business, 85*(3), 180–185.

You may come across a research report in which the generic term *criterion-related validity* is used. When you do, you may have to make a guess as to whether reference is being made to concurrent validity or to predictive validity. Usually, you will have no difficulty figuring this out. Just ask yourself: Were the criterion data collected at the same time as or later than the test data associated with the test that is being validated?

Construct Validity.   Many measuring instruments are developed to reveal how much of a personality or psychological construct is possessed by the examinees to whom the instrument is administered. To establish the degree of **construct validity** associated with such instruments, the test developer (as well as the test's users) ought to do one or a combination of three things: (1) provide correlational evidence showing that the construct has a strong relationship with certain measured variables *and* a weak relationship with other variables, with the strong and weak relationships conceptually tied to the new instrument's construct in a logical manner; (2) show that certain groups obtain higher mean scores on the new instrument than other groups, with the high- and low-scoring groups being determined on logical grounds *prior to* the administration of the new instrument; or (3) conduct a factor analysis on scores from the new instrument.

Excerpt 4.22 provides an example of the first of these approaches to construct validity. This excerpt deserves your close attention because it contains a clear explanation of how correlational evidence is examined for the purpose of establishing **convergent validity** and **discriminant** (divergent) **validity.**

EXCERPT 4.22 • *Construct Validity Using Correlations*

To estimate construct validity of the questionnaires measuring social anxiety disorder, we used the following criteria. . . . (a) The intercorrelations of LSAS, SIAS, and SPS should be large and significantly differ from zero (convergent validity). (b) The correlations between these measures should be larger than their respective correlations to the other questionnaires (MADRS-S, BAI, and QOLI) because of the latters' intention to measure other constructs (discriminant validity). [Results] showed high convergent validity and fully met the criterion of correlations being strong among social anxiety disorder questionnaires. Regarding discriminant validity, the social anxiety disorder measures were uncorrelated or correlated to a lesser degree with QOLI than with each other, indicating discriminant validity.

Source: Hedman, E., Ljótsson, B., Rück, C., Furmark, T., Carlbring, P., Lindefors, N., et al. (2010). Internet administration of self-report measures commonly used in research on social anxiety disorder: A psychometric evaluation. *Computers in Human Behavior, 26,* 736–740.

It is not always easy to demonstrate that a measuring instrument is involved in a network of relationships where certain of those relationships are strong whereas others are weak. However, claims of construct validity are more impressive when evidence regarding both convergent *and* discriminant validity is provided. Of course, not all measuring instruments are created to deal with personality or psychological constructs, and even those that can be validated with non-correlational evidence. When you *do* encounter validation evidence like that illustrated in Excerpt 4.22, give some "bonus points" (in your evaluation) to those researchers who have utilized the two-pronged approach.[6]

In Excerpt 4.23, we see an example of the "known group" approach to construct validity. In the study associated with this excerpt, the researchers investigated the validity of a new test designed to assess an examinee's knowledge of coronary heart disease. In this study, the evidence for construct validity came from showing that a group of experienced cardiovascular nurses could perform better on the test than a group of patients admitted to the hospital.

EXCERPT 4.23 • *Construct Validity Using Comparison Groups*

Forty-nine women who had no prior CHD and presented to a primary care office for their routine medical care were recruited for the study by the office staff as the control group (Group 1). . . . Twenty-three female registered cardiovascular nurses who routinely cared for cardiac patients in a Midwest hospital were recruited by one of the researchers as a known group (Group 2) for the known group validity test. . . . Based on their education and clinical expertise, these cardiovascular nurses were expected to have higher CHD knowledge scores than laywomen without any organized educational program. . . . Statistical analyses were [conducted] to evaluate the construct validity of the new CHD knowledge tool by [investigating] the tool's ability to differentiate the control group (Group 1) from a known group (Group 2) in the known group validity test. . . . The significantly lower CHD knowledge scores of the control group in comparison to that of cardiovascular nurses provides a known group validity of the tool.

Source: Thanavaro, J. L., Thanavaro, S., & Delicath, T. (2010). Coronary heart disease knowledge tool for women. *Journal of the American Academy of Nurse Practitioners, 22*(2), 62–69.

The third procedure frequently used to assess construct validity involves a sophisticated statistical technique called **factor analysis.** I discuss the details of factor analysis in Chapter 20. For now, I simply want you to see an illustration of

---

[6]Unfortunately, some researchers make the mistake of thinking that discriminant validity is supported by moderate or strong negative correlations. When evidence for construct validity is gathered via convergent and discriminant correlations, the appropriate two "opposites" are not *r*s that are positive and negative, but rather ones that are high and low.

how the results of such an investigation are typically summarized. I do not expect you to understand everything in Excerpt 4.24; my only purpose in presenting it is to alert you to the fact that construct validity is often assessed statistically using technique called *factor analysis*.

EXCERPT 4.24 • *Construct Validity Using Factor Analysis*

Construct validity was established for the noncognitive scales (i.e., affective measures and self-reported data) through factor analyses using the data from the sixth- and eighth-grade samples in this study. The results of the factor analyses for these scales of the MSLES revealed that the one factor model for each scale was the best fit, confirming that each scale was unidimensional.

Source: McBeth, W., & Volk, T. L. (2010). The National Environmental Literacy Project: A baseline study of middle grade students in the United States. *Journal of Environmental Education*, *41*(1), 55–67.

Warnings about Validity Claims

Before concluding our discussion of validity, I want to sensitize you to a few concerns regarding validity claims. Because researchers typically have a vested interest in their studies, they are eager to have others believe that their data are accurate. Readers of research literature must be on guard for unjustified claims of validity and for cases where the issue of validity is not addressed at all.

First, remember that reliability is a necessary but not sufficient condition for validity. Accordingly, do not be lulled into an unjustified sense of security concerning the accuracy of research data by a technical and persuasive discussion of consistency. Reliability and validity deal with different concepts, and a presentation of reliability coefficients—no matter how high—should not cause one's concern for validity to evaporate.

Next, keep in mind that validity (like reliability) is really a characteristic of the data produced by a measuring instrument and not a characteristic of the measuring instrument itself.[7] If a so-called valid instrument is used to collect data from people who are too young or who cannot read or who lack any motivation to do well, then the scores produced by that instrument will be of questionable validity. The important point here is simply this: The people used by a researcher and the conditions under which measurements are collected must be similar to the people and conditions involved in validation studies before you should accept the researcher's claim that his or her research data are valid because those data came from an instrument having "proven validity."

---

[7]This is true for all varieties of validity except content validity.

My third warning concerns content validity. Earlier, I indicated that this form of validity usually involves a subjective evaluation of the measuring instrument's content. Clearly, this evaluation ought to be conducted by individuals who possess (1) the technical expertise to make good judgments as to content relevance and (2) a willingness to provide, if necessary, negative feedback to the test developer. When reporting on efforts made to assess content validity, researchers should describe in detail who examined the content, what they were asked to do, and how their evaluative comments turned out.

With respect to criterion-related and construct validity, a similar warning seems important enough to mention. With these approaches to assessing validity, scores from the instrument being validated are correlated with the scores associated with one or more "other" variables. If the other variables are illogical or if the validity of the scores associated with such variables is low, then the computed validity coefficients conceivably could make a truly good instrument look as if it is defective (or vice versa). Thus, regarding the predictive, concurrent, or construct validity of a new measuring instrument, the researcher should first discuss the quality of the data that are paired with the new instrument's data.

My next warning concerns the fact that the validity coefficients associated with criterion-related or construct probes are simply estimates, not definitive statements. Just as with reliability, the correlation coefficients reported to back up claims of validity would likely fluctuate if the study were to be replicated with a new batch of examinees. This is true even if the test-takers in the original and replicated studies are similar. Such fluctuations can be expected to be larger if the validity coefficients are based on small groups of people; accordingly, give researchers more credit when their validity investigations are based on large groups.

Finally, keep in mind that efforts to assess predictive and concurrent validity utilize correlation coefficients to estimate the extent to which a measuring instrument can be said to yield accurate scores. When construct validity is dealt with by assessing an instrument's convergent/discriminant capabilities or by conducting a factor analysis, correlation again is the vehicle through which validity is revealed. Because correlation plays such a central role in the validity of these kinds of investigations, it is important for you to remember the warnings about correlation that were presented near the end of Chapter 3. In particular, do not forget that r^2 provides a better index of a relationship's strength than does r.

Final Comments

Within this discussion of reliability and validity, I have not addressed a question that most likely passed through your mind at least once as we considered different procedures for assessing consistency and accuracy: "How high must the reliability and validity coefficients be before we can trust the results and conclusions of the study?" Before leaving this chapter, I want to answer this fully legitimate question.

For both reliability and validity, it would be neat and tidy if there were some absolute dividing point (e.g., .50) that separates large from small coefficients. Unfortunately, no such dividing point exists. In evaluating the reliability and validity of data, the issue of "large enough" has to be answered in a *relative* manner. The question that the researcher (and you) should ask is, "How do the reliability and validity of data associated with the measuring instrument(s) used in a given study compare with the reliability and validity of data associated with other available instruments?" If the answer to this query about relative quality turns out to be "pretty good," then you should evaluate the researcher's data in a positive manner—even if the absolute size of reported coefficients leaves lots of room for improvement.

My next comment concerns the possible use of multiple methods to assess instrument quality. Because there is no rule or law that prohibits researchers from using two or more approaches when estimating reliability or validity, it is surprising that so many research reports contain discussions of one and (if validity is discussed at all) *only* one kind of validity. That kind of research report is common because researchers typically overlook the critical importance of having good data to work with and instead seem intent on quickly analyzing whatever data have been collected. Give credit to those few researchers who present multiple kinds of evidence when discussing reliability and validity.

My third point is simple: Give credit to a researcher who indicates that he or she considered the merits of more than one measuring instrument before deciding on which test or survey to use. Too many researchers, I fear, decide first that they want to measure a particular trait or skill and then latch on to the first thing they see or hear about that has a name that matches that trait or skill. In Excerpt 4.25, we see an example of a better way of going about instrument selection. (In this excerpt, note that the third criterion includes a consideration of reliability and validity.)

EXCERPT 4.25 • *Reasons Provided as to Why a Given Instrument Was Selected*

Eight survey instruments that measured trust were evaluated using predetermined criteria.

The criteria described an instrument that (1) measures trust on a continuum scale; (2) has a short completion time (<10 min); (3) is available, reliable, and valid; and (4) has the ability to measure multiple dimensions of trust. The Organizational Trust Index (OTI) survey instrument was selected for use in this study as it best met the established criteria, including the ability to measure the five dimensions of trust identified during the literature review.

Source: Alston, F., & Tippett, D. (2009). Does a technology-driven organization's culture influence the trust employees have in their managers? *Engineering Management Journal, 21*(2), 3–10.

My last general comment about reliability and validity is related to the fact that data quality, by itself, does not determine the degree to which a study's results can be trusted. It is possible for a study's conclusions to be totally worthless even though the data analyzed possess high degrees of reliability and validity. A study can go down the tubes despite the existence of good data if the wrong statistical procedure is used to analyze data, if the conclusions extend beyond what the data legitimately allow, or if the design of the study is deficient. Reliability and validity are important concepts to keep in mind as you read technical reports of research investigations, but other important concerns must be attended to as well.

Review Terms

Accuracy	Equivalent-forms reliability
Alpha	Factor analysis
Alternate-forms reliability	Guttman's split-half reliability
Coefficient alpha	Internal consistency reliability
Coefficient of concordance	Interrater reliability
Coefficient of equivalence	Intraclass correlation
Coefficient of stability	Kuder-Richardson #20 (K-R 20)
Cohen's kappa	Parallel-forms reliability
Concurrent validity	Predictive validity
Confidence band	Reliability
Consistency	Reliability coefficient
Construct validity	Spearman-Brown
Content validity	Split-half reliability coefficient
Convergent validity	Standard error of measurement (SEM)
Criterion-related validity	Test–retest reliability coefficient
Cronbach's alpha	Validity
Discriminant validity	Validity coefficient

The Best Items in the Companion Website

1. An interactive online quiz (with immediate feedback provided) covering Chapter 4.
2. Ten misconceptions about the content of Chapter 4.
3. An online resource entitled "Multitrait–Multimethod."
4. An email message about convergent and discriminant validity sent from the author to his students to help them understand these two measurement concepts.
5. Chapter 4's best paragraph.

To access the chapter outline, practice tests, weblinks, and flashcards, visit the companion website at http://www.ReadingStats.com.

Review Questions and Answers begin on page 531.

CHAPTER 5

Foundations of Inferential Statistics

In Chapters 2 through 4, we considered various statistical procedures that are used to organize and summarize data. At times, the researcher's sole objective is to describe the people (or things) in terms of the characteristic(s) associated with the data. When that is the case, the statistical task is finished as soon as the data are displayed in an organized picture, are reduced to compact indices (e.g., the mean and standard deviation), are described in terms of distributional shape, are evaluated relative to the concerns of reliability and validity, and, in the case of a bivariate concern, are examined to discern the strength and direction of a relationship.

In many instances, however, the researcher's primary objective is to draw conclusions that extend beyond the specific data that are collected. In this kind of study, the data are considered to represent a sample—and the goal of the investigation is to make one or more statements about the larger group of which the sample is only a part. Such statements, when based on sample data but designed to extend beyond the sample, are called *statistical inferences*. Not surprisingly, the term **inferential statistics** is used to label the portion of statistics dealing with the principles and techniques that allow researchers to generalize their findings beyond the actual data sets obtained.

In this chapter, we consider the basic principles of inferential statistics. We begin by considering the simple notions of sample, population, and scientific guess. Next, we take a look at eight of the main types of samples used by applied researchers. Then we consider certain problems that crop up to block a researcher's effort to generalize findings to the desired population. Finally, a few tips are offered concerning specific things to look for as you read professional research reports.

Statistical Inference

Whenever a statistical inference is made, a **sample** is first extracted (or is considered to have come from) a larger group called the **population.** Measurements are then taken on the people or objects that compose the sample. Once these measurements are summarized—for example, by computing a correlation coefficient—an educated guess is made as to the numerical value of the same statistical concept (which, in our example, would be the correlation coefficient) in the population. This educated guess as to the population's numerical characteristic is the **statistical inference.**

If measurements could be obtained on all people (or objects) contained in the population, statistical inference would be unnecessary. For instance, suppose the coach of the girls' basketball team at a local high school wants to know the median height of 12 varsity team members. It would be silly for the coach to use inferential statistics to answer this question. Instead of the coach making an educated guess as to the team's median height (after seeing how tall a few of the girls are), it would be easy to measure the height of each member of the varsity team and then obtain the precise answer to the question.

In many situations, researchers cannot answer their questions about their populations as easily as could the coach in the basketball example. Two reasons seem to account for the wide use of inferential statistics. One of these explanations concerns the measurement process, whereas the other concerns the nature of the population. Because inferential statistics are used so often by applied researchers, it is worthwhile to pause for a moment and consider these two explanations as to why only portions of populations are measured, with educated guesses being made on the basis of the sample data.

First of all, it is sometimes too costly (in dollars or time) to measure every member of the population. For example, the intelligence of all students in a high school cannot be measured with an individual intelligence test because (1) teachers would be upset by having each student removed from classes for two consecutive periods to take the test and (2) the school's budget would not contain the funds needed to pay a psychologist to do this testing. In this situation, it would be better for the principal to make an educated guess about the average intelligence of the high school students than to have no data-based idea whatsoever as to the students' intellectual capabilities. The principal's guess about the average intelligence is based on a sample of students taken from the population made up of all students in the high school. In this example, the principal is sampling from a **tangible population** because each member of the student body could end up in the sample and be tested.

The second reason for using inferential statistics is even more compelling than the issue of limited funds and time. Often, the population of interest extends into the future. For example, the high school principal in the previous example probably would like to have information about the intellectual capabilities of the school's student body so improvements in the curriculum could be made. Such changes are made on the assumption that next year's students will not be dissimilar from this year's students. Even if the funds and time could be found to administer an individual intelligence test to every student in the school, the obtained data would be viewed as coming from a *portion* of the population of interest. That population is made up of students who attend the school now *plus* students who follow in their footsteps. Clearly, measurements cannot be obtained from all members of such a population because a portion of the population has not yet "arrived on the scene." In this case, the principal creates an **abstract population** to fit an existing sample.

Several years ago, I participated as a subject in a study to see if various levels of consumed oxygen have an effect, during strenuous exercise, on blood composition. The researcher who conducted this study was interested in what took place physiologically during exercise on a stationary bicycle among non-sedentary young men between the ages of 25 and 35. That researcher's population was not just active males who were 25–35 years old at the time of the investigation. The population was defined to include active males who *would be* in this age range at the time the research summary got published—approximately 18 months following the data collection. Inferential statistics were used because the research participants of the investigation were considered to be a representative sample of a population of similar individuals that extended into the future.

To clarify the way statistical inference works, consider the two pictures in Figure 5.1. These pictures are identical in that (1) measurements are taken only on the people (or objects) that compose the sample; (2) the educated guess, or inference, extends *from* the sample *to* the population; and (3) the value of the population characteristic is not known (nor can it ever be known as a result of the inferential process). Although these illustrations show that the inference concerns the mean, the pictures could have been set up to show that the educated guess deals with the median, the variance, the product–moment correlation, or any other statistical concept.

As you can see, the only differences between the two pictures involve the solid versus dotted nature of the larger circle and the black arrows. In the top picture, the population is tangible in nature, with each member within the larger circle available for inclusion in the sample. When this is the case, the researcher actually begins with the population and then ends up with the sample. In Figure 5.1, the lower picture is meant to represent the inferential setup in which the sequence of events is reversed. Here, the researcher begins with the sample and then creates an abstract population that is considered to include people (or objects) like those included in the sample.

Excerpts 5.1 and 5.2 illustrate the distinction between tangible and abstract populations. In the first of these excerpts, the population was made up all residents of Cyprus who were listed in the phone directory. This was a tangible population because (1) every person in the population had a unique name and phone number and (2) any of those individuals could have ended up in the sample. In Excerpt 5.2, the population was abstract. The people who composed the sample in this study were not "pulled from" (i.e., drawn out of) a larger group; instead, they got into the sample because they voluntarily responded to recruitment advertisements. Because

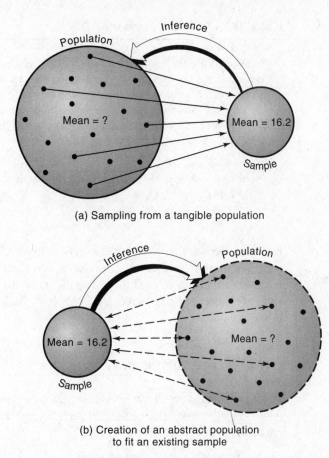

(a) Sampling from a tangible population

(b) Creation of an abstract population
to fit an existing sample

FIGURE 5.1 *Two Kinds of Sample/Population Situations*

EXCERPT 5.1–5.2 • *Tangible and Abstract Populations*

This study is one of the very first research projects on Mental Health (MH) issues in Cyprus and, to the author's best knowledge, the first community assessment with the public at large as evaluator. . . . All study participants were randomly selected from the phone directory, which lists citizens' phone numbers in alphabetical order. Specifically, all directory pages were numbered, and then the table of random digits was consulted to select a page at random. All the names on the selected page were again numbered, and the table of random digits was once again consulted to select prospective participants.

Source: Georgiades, S. (2009). Mental health in Cyprus: An exploratory investigation. *International Journal of Mental Health, 38*(2), 3–20.

(*continued*)

EXCERPTS 5.1–5.2 • (*continued*)

Individuals were recruited for a large online study using a wide variety of different methods (e.g., posters, business cards, online and magazine ads, e-mail listserv announcements, and snowball sampling from existing participants). All advertising directed potential participants to a Web site that described the study, eligibility, and incentives. Interested participants completed a brief demographic questionnaire and provided contact information.

Source: Holmberg, D., Blair, K. L., & Phillips, M. (2010). Women's sexual satisfaction as a predictor of well-being in same-sex versus mixed-sex relationships. *Journal of Sex Research*, *47*(1), 1–11.

the researchers associated with Excerpt 5.2 used inferential statistics with the data collected from their research participants, it is clear that they wanted to generalize the study's findings beyond those specific individuals. Thus, the population in this study was abstract because it existed only hypothetically as a larger "mirror image" of the sample.

The Concepts of Statistic and Parameter

When researchers engage in inferential statistics, they must deal with four questions *before* they can make their educated guess, or inference, that extends from the sample to the population:

1. What is/are the relevant population(s)?
2. How will a sample be extracted from each population of interest, presuming the population(s) is/are tangible in nature?
3. What characteristic of the sample people, animals, or objects will serve as the target of the measurement process?
4. What will be the study's statistical focus?

The first of these four questions is completely up to the researcher and is dictated by the study's topical focus. The second question is considered in detail in the next section. The third question, of course, is answered when the researcher decides what to study.[1] The notion of a measurement process is also involved in this question, thus making the issues of reliability and validity (covered in Chapter 4) important to consider when judging whether the researcher did an adequate job in measuring the

---

[1]You may, at times, disagree with the researcher as to whether the characteristic of the people, animals, or objects in the population is important. Nevertheless, I doubt that you will ever experience difficulty determining what variables were examined. A clear answer to this question is usually contained in the article's title, the statement of purpose, or the discussion of dependent variables.

research participants. This brings us to the fourth question, a concern for the statistical focus of the inference.

After the researcher has measured the sample on the variable(s) of interest, there are many alternative ways in which the data can be summarized. The researcher could compute, for example, a measure of central tendency, a measure of variability, a measure of skewness, or a measure of relationship. However, even within each of these broad categories, the researcher has alternatives as to how the data will be summarized. With central tendency, for example, the researcher might decide to focus on the median rather than on the mean or the mode. If relationship is the issue of interest, a decision might be made to compute Pearson's product–moment correlation coefficient rather than other available correlational indices. The term **statistical focus** is used simply to indicate the way in which the data are summarized.

Regardless of how a researcher decides to analyze the sample data, there will always be two numerical values that correspond to the study's statistical focus. One of these is "in" the sample—and it can be computed as soon as the sample is measured. This numerical value is called the **statistic.** The second value that corresponds to the study's statistical focus is "in" the population, and it is called the **parameter.** The parameter, of course, can never be computed because measurements exist for only a portion of the people, animals, or objects that make up the population.

Because researchers often use symbols to represent the numerical values of their statistics (and sometimes use different symbols to represent the unknown values of the corresponding parameters), it is essential that you become familiar with the symbols associated with inferential statistics. Table 5.1 shows the most frequently used symbols for the statistic and parameter that correspond to the same statistical focus. As you can easily see, Roman letters are used to represent statistics, whereas Greek letters stand for parameters.

TABLE 5.1   *Symbols Used for Corresponding Statistics and Parameters*

Statistical Focus	*Statistics* *(in the sample)*	*Parameter* *(in the population)*
Mean	$\bar{X}$ *or* M	μ
Variance	s^2	σ^2
Standard deviation	s	σ
Proportion	p	P
Product–moment correlation*	r	ρ
Rank–order correlation	r_s	ρ_s
Size of group[†]	n	N

*Unfortunately, the symbol ρ is used to designate the value of the product–moment correlation in the relevant population. This is the letter rho from the Greek alphabet. In Chapter 3, we saw that Spearman's rank–order correlation is also referred to as rho.

[†]In many articles, the symbol N is used to indicate the size of the *sample*. It would be better if the symbol n could be used instead of N when researchers give us information about their sample sizes.

Now that I have clarified the notions of statistic and parameter, I can be a bit more parsimonious in my definition of inferential statistics. When engaged in inferential statistics, a researcher uses information concerning the known value of the sample statistic to make an educated guess as to the unknown value of the population parameter. If, for example, the statistical focus is centered on the mean, then information concerning the known value of M is used to make a scientific guess as to the value of μ.

Types of Samples

The nature of the sample used by a researcher as a basis for making an educated guess as to the parameter's value obviously has an influence on the inferential process. To be more specific, the nature of the sample influences either (1) the accuracy of the inferential guess or (2) the definition of the population toward which the inferential guess is directed. To help you understand the way in which the sample can affect the inferential process in these two ways, I must distinguish among eight kinds of samples that fall into two main categories: probability samples and nonprobability samples.

Probability Samples

If all members of the population can be specified prior to drawing the sample, if each member of the population has at least some chance of being included in the sample, and if the probability of any member of the population being drawn is known, then the resulting sample is referred to as a **probability sample.** The four types of probability samples considered here are *simple random samples, stratified random samples, systematic samples,* and *cluster samples.* As you read about each of these samples, keep in mind the illustration presented in Figure 5.1a.

Simple Random Samples.   With a **simple random sample,** the researcher, either literally or figuratively, puts the names of all members of the population into a hat, shuffles the hat's contents, and then blindly selects a portion of the names to determine which members of the total group are or are not included in the sample. The key feature of this kind of sample is an equal opportunity for each member of the population to be included in the sample. It is conceivable, of course, that such a sample could turn out to be grossly unrepresentative of the population (because the sample turns out to contain the population members who are, for example, strongest or most intelligent or tallest). It is far more likely, however, that a simple random sample will lead to a measurement-based statistic that approximates the value of the parameter. This is especially true when the sample is large rather than small.

In Excerpt 5.3, we see an example of simple random samples being used in applied research studies. Because there are different kinds of random samples that can be drawn from a tangible population, these researchers deserve credit for using the word *simple* to clarify exactly what type of random sampling procedure was

EXCERPT 5.3 • *Simple Random Sample*

The population of interest was all college students at a major southeastern university. The entire student body, except those under age eighteen who were legally minors, formed the sampling frame. A simple random sample of 15,000 individuals from the population of 50,701 students age eighteen or older was e-mailed an invitation to anonymously participate in a Web-based survey.

Source: Patton, C. L., Nobles, M. R., & Fox, K. A. (2010). Look who's stalking: Obsessive pursuit and attachment theory. *Journal of Criminal Justice*, *38*(3), 282–290.

used in their studies. In this excerpt, you see the term **sampling frame.** Generally speaking, a sampling frame is simply a list that enumerates the things—people, animals, objects, or whatever—in the population. In a very real sense, there must be a sampling frame for simple random samples (or, more generally, for any probability sample) to be drawn from a population.

Stratified Random Samples. To reduce the possibility that the sample might turn out to be unrepresentative of the population, researchers sometimes select a **stratified random sample.** To do this, the population must first be subdivided into two or more parts based on the knowledge of how each member of the population stands relative to one or more stratifying variables. Then, a sample is drawn that mirrors the population percentages associated with each segment (or *stratum*) of the population. Thus, if a researcher knows that the population contains 60 percent males and 40 percent females, a random sample stratified on gender should contain six males for every four females.

An example of a stratified random sample is presented in Excerpt 5.4. This is a good example of a well-described stratified random sample because it answers the question, "Stratified on what?" Too often, researchers either make no mention

EXCERPT 5.4 • *Stratified Random Sample*

A target sample size of 40 was obtained through a stratified random sample method. Students were first grouped according to the final grade (high distinction, distinction, credit, pass, fail). Students were randomly selected so that the proportion of students with each grade in the final sample was equal to the proportion of the grades in the class as a whole. This was done to ensure that the sample was representative of the full range of abilities in the class.

Source: Neumann, D. L., Neumann, M. M., & Hood, M. (2010). The development and evaluation of a survey that makes use of student data to teach statistics. *Journal of Statistics Education*, *18*(1), 1–18.

whatsoever of the variable used to create the strata, or terms like "age-stratified" or "region-stratified" are used without any specification of how many strata were set up or what the numerical or geographic boundaries were between the strata.

In some studies using stratified random samples, researchers make the size of the sample associated with one or more of the strata larger than that strata's proportionate slice of the population. This **oversampling** in certain strata is done for one of three reasons: (1) anticipated difficulty in getting people in certain strata to participate in the study, (2) a desire to make comparisons between strata (in which case there are advantages to having equal strata sizes in the sample, even if those strata differ in size in the population), and (3) a need to update old strata sizes, when using archival data, because of recent changes in the characteristics of the population. In Excerpt 5.5, we see an example of a stratified random sample that involved oversampling for the first of these three reasons.

EXCERPT 5.5 • *Oversampling*

Computer-assisted self-interviewing was used to collect data from [a stratified] sample of household residents in four cities (Baltimore; Durham, NC; St. Louis; and Seattle) and the U.S. census-defined county subdivisions immediately adjacent to them. . . . Within the four study sites, we stratified segments by the percentage of population who were black and oversampled segments with high minority concentrations. This procedure yielded a large enough sample of couples in which one or both partners were black to provide stable estimates of both their behaviors and the antecedents of those behaviors.

Source: Billy, J. O. G., Grady, W. R., & Sill, M. E. (2009). Sexual risk-taking among adult dating couples in the United States. *Perspectives on Sexual & Reproductive Health, 41*(2), 74–83.

Systematic Samples.   A third type of probability sample, called a **systematic sample**, is created when the researcher goes through an ordered list of members of the population and selects, for example, every fifth entry on the list to be in the sample. (Of course, the desired size of the sample and the number of entries on the list determine how many entries are skipped following the selection of each entry to be in the sample.) So long as the starting position on the list is determined randomly, each entry on the full list has an equal chance of ending up in the sample. Thus, if the researcher decides to generate a sample by selecting every fifth entry, the first entry selected for the sample should not arbitrarily be the entry at the top of the list (or the one positioned in the fifth slot); instead, a random decision should determine which of the first five entries goes into the sample.

Excerpt 5.6 exemplifies the use of a systematic sample. As indicated in this excerpt, pages out of census data recorded on microfilm reels were the things being sampled. Every tenth page on a reel ended up in the sample, with the first of those

EXCERPT 5.6 • *Systematic Sampling*

The manuscripts from each census are stored on several thousand microfilm reels. Most reels contain several hundred pages. Each of these pages contains between 40 and 50 lines, with each line containing information on one person. . . . The sampling strategy is based on the census page. We generate a random starting point for each microfilm reel between 1 and 5, and then designate every 10th page thereafter as a sample page. Thus, for example, if the starting point is 3, we designate the 3rd, 13th, and 23rd pages, continuing in that fashion until the end of the reel.

Source: Davern, M. (2009). Drawing statistical inferences from historical Census data, 1850–1950. *Demography*, *46*(3), 589–603.

pages being selected at random from among pages 1 and 5. (I think the excerpt's fifth sentence would be clearer if the words "that page and" had appeared between the words "designate" and "every.")

Cluster Samples. The last of the four kinds of probability sampling to be discussed here involves what are called **cluster samples.** When this technique is used to extract a sample from a population, the researcher first develops a list of the clusters in the population. The clusters might be households, schools, litters, car dealerships, or any other groupings of the things that make up the population. Next, a sample of these clusters is randomly selected. Finally, data are collected from each person, animal, or thing that is in each of the clusters that has been randomly selected, or data are collected from a randomly selected subset of the members of each cluster.

Excerpt 5.7, we see an example of a cluster sample in which each "cluster" was a Head Start school in the state of New Hampshire. As indicated in the excerpt, 27 schools were randomly selected. Then, every student in each selected school was examined (unless their parents declined the opportunity) by a dentist who checked for cavities. This technique of cluster sampling made it much easier for

EXCERPT 5.7 • *Cluster Samples*

We conducted the survey at 27 of the 45 New Hampshire Head Start sites. . . . We used a simple random one-stage cluster sample design, in which all children at each selected site would be surveyed. . . . Four volunteer dentists provided oral examinations and determined the presence of untreated dental caries, caries experience, and treatment urgency.

Source: Anderson, L., Martin, N. R., Burdick, A., Flynn, R. T., & Blaney, D. D. (2010). Oral health status of New Hampshire Head Start children, 2007–2008. *Journal of Public Health Dentistry*, *70*(3), 245–248.

the researchers to collect their study's data than would have been the case if a simple random sample of children had been taken from all 45 Head Start schools.

As indicated in Excerpt 5.7, the Head Start dental evaluation study used a one-stage cluster sample design. You are likely to encounter research reports that refer to two- or three-stage cluster samples. In these multistage cluster samples, clusters of one kind are embedded inside clusters of different kind, with the sampling process beginning with the larger clusters and then continuing down to the smaller clusters. For example, a three-stage cluster sample of homes in a given state (perhaps to assess their painted color) might involve selecting a sample of counties first, then a sample of cities within the selected counties, and finally a sample of residential neighborhoods within the selected cities. Collecting the study's data from the resulting sample of homes, grouped in clusters, is far more convenient than if a simple random sample of homes were selected from all homes in the state.

Nonprobability Samples

In many research studies, the investigator does *not* begin with a finite group of persons, animals, or objects in which each member has a known, nonzero probability of being plucked out of the population for inclusion in the sample. In such situations, the sample is technically referred to as a **nonprobability sample.** Occasionally, an author indicates directly that one or more nonprobability samples served as the basis for the inferential process. Few authors do this, however, and so you must be able to identify this kind of sample from the description of the study's participants.

Although inferential statistics can be used with nonprobability samples, extreme care must be used in generalizing results from the sample to the population. From the research write-up, you probably will be able to determine who (or what) was in the sample that provided the empirical data. Determining the larger group to whom such inferential statements legitimately apply is usually a much more difficult task.

We next consider four of the most frequently seen types of nonprobability samples: purposive samples, convenience samples, quota samples, and snowball samples.

Purposive Samples.   In some studies, the researcher starts with a large group of potential participants. To be included in the sample, however, members of this large group must meet certain criteria established by the researcher because of the nature of the questions to be answered by the investigation. Once these screening criteria are employed to determine which members of the initial group wind up in the sample, the nature of the population at the receiving end of the "inferential arrow" is different from the large group of potential persons with whom the researcher started. The legitimate population associated with the inferential process is either (1) the portion of the initial group that satisfied the screening criteria, presuming that only a subset of these acceptable people (or objects) were actually measured; or (2) an abstract population made up of people (or objects) similar to those included in the sample, presuming that each and every "acceptable" person (or object) was measured. These two notions of the population, of course, are meant to parallel the two situations depicted earlier in Figure 5.1.

Excerpt 5.8 illustrates the way researchers sometimes use and describe their **purposive samples.** In this passage, notice how the researchers set up inclusion criteria for both members of the adolescent/parent dyads recruited into the study. As you can see, only younger adolescents and their parents were allowed into the study.

EXCERPT 5.8 • *Purposive Samples*

A purposive sample of 94 adolescents/parent dyads was recruited from eight middle schools within a single school district in southern California. The inclusion criteria for adolescents are as follows: between the ages of 12 and 15 years; able to read and speak English; signed informed assent; and signed informed parent consent. The inclusion criteria for parent participants are as follows: able to read and speak English; have legal custody of adolescent participant; and signed consent form for participation for self and child.

Source: Rutkowski, E. M., & Connelly, C. D. (2010). Obesity risk knowledge and physical activity in families of adolescents. *Journal of Pediatric Nursing, 26*(1), 51–57.

The full research report from which Excerpt 5.8 was taken included a highly detailed description of the people who composed the sample (including demographic information on the adolescents' grade point average (GPA), height, weight, and year in school, as well as information of the parents' age, educational level, and marital status). Such descriptions, along with a clear articulation of the inclusion criteria, are essential in research reports based on purposive samples. The reason for this is simple—the relevant populations associated with purposive samples are abstract rather than tangible. As pointed out earlier, the nature of an abstract population is determined by who or what is in the sample. Clearly, you cannot have a good sense of the population toward which the inference is directed unless you have a good sense for who was in the sample.

Convenience Samples. In some studies, no special screening criteria are set up by the researchers to make certain that the individuals in the sample possess certain characteristics. Instead, the investigator simply collects data from whoever is available or can be recruited to participate in the study. Such data-providing groups, if they serve as the basis for inferential statements, are called **convenience samples.**

The population corresponding to any convenience sample is an abstract (i.e., hypothetical) population. It includes individuals (or objects) similar to those included in the sample. Therefore, the sample–population relationship brought about by convenience samples is always like that pictured earlier in Figure 5.1b.

Excerpt 5.9 illustrates the use of convenience samples. In this excerpt, the researchers clearly label the kind of sample they used. Not all researchers are so forthright.

EXCERPT 5.9 • *Convenience Samples*

Our survey targeted a cross-section of students enrolled in various courses in the College of Business Administration across all levels (e.g., introductory courses to senior level courses) and areas of study (e.g., introduction to business, finance, marketing, strategic management). We derived this convenience sample from classes that comprise a part of our pre-business and business core curriculum. We selected the specific classes on the basis of the availability of the researchers to administer the survey in person and the flexibility of the faculty member in the classroom.

Source: Sipe, S., Johnson, C. D., & Fisher, D. K. (2009). University students' perceptions of gender discrimination in the workplace: Reality versus fiction. *Journal of Education for Business, 84*(6), 339–349.

It should be noted that the statements presented in Excerpt 5.9 do not constitute the full description of the convenience sample used in this study. The researchers provided information on the students' age, year in school, ethnicity, political orientation, work experience, and GPA. Unfortunately, many researchers put us in a quandary by not providing such descriptions. Unless we have a good idea of who is in a convenience sample, there is no way to conceptualize the nature of the abstract population toward which the statistical inferences are aimed.

Quota Samples.   The next type of nonprobability sample to be considered is called a **quota sample.** Here, the researcher decides that the sample should contain *X* percent of a certain kind of person (or object), *Y* percent of a different kind of person (or object), and so on. Then, the researcher simply continues to hunt for enough people/things to measure within each category until all predetermined sample slots have been filled.

In Excerpt 5.10, we see an example of a quota sample. In this investigation, the researchers wanted to have an overall sample size of 4,000, with a quota for each of four age categories. These quotas were determined by first examining census data

EXCERPT 5.10 • *Quota Samples*

Participants were identified using a national quota-sampling procedure. . . . Quota sampling in the context of this study refers to a sampling method in which the first of a pre-determined number of participants are selected, the number being 4000 for this survey. After the number of participants exceeded approximately 4000, the survey was terminated. The quota-sampling strategy stratified on four age groups: 25–29, 30–34, 35–39 and 40–45 to match 2006 U.S. Census data.

Source: Kronenfeld, L. W., Reba-Harrelson, L., Von Holle, A., Reyes, M. L., & Bulik, C. M. (2010). Ethnic and racial differences in body size perception and satisfaction. *Body Image, 7*(2), 131–136.

to find out what proportion of women ages 25–45 was in each 5-year category; then, each of those proportions was multiplied by 4,000 to arrive at the needed number of women in each of the study's age strata.

On the surface, quota samples and stratified random samples seem to be highly similar. There is, however, a big difference. To obtain a stratified random sample, a finite population is first subdivided into sections and then a sample is selected randomly from each portion of the population. When combined, those randomly selected groups make up the stratified random sample. A quota sample is also made up of different groups of people that are combined. Each subgroup, however, is not randomly extracted from a different stratum of the population; rather, the researcher simply takes whoever comes along until all vacant sample slots are occupied. As a consequence, it is often difficult to know to whom the results of a study can be generalized when a quota sample serves as the basis for the inference.

Snowball Samples. A **snowball sample** is like a two-stage convenience or purposive sample. First, the researcher locates a part of the desired sample by turning to a group that is conveniently available or to a set of individuals who possess certain characteristics deemed important by the researcher. Then, those individuals are asked to help complete the sample by going out and recruiting family members, friends, acquaintances, or coworkers who might be interested (and who possess, if a purposive sample is being generated, the needed characteristics). Excerpt 5.11 illustrates how this technique of snowballing is sometimes used in research studies.

EXCERPT 5.11 • *Snowball Samples*

We recruited participants for the present study [by] using a snowball sampling technique. Specifically, we e-mailed a Web link to an online survey to approximately 45 individuals who were employed full-time and whom we knew personally or professionally. . . . We invited individuals to participate [and] we asked individuals to forward the link to other fulltime employees.

Source: Culbertson, S. S., Huffman, A. H., & Alden-Anderson, R. (2010). Leader–member exchange and work–family interactions: The mediating role of self-reported challenge- and hindrance-related stress. *Journal of Psychology, 144*(1), 15–36.

The Problems of Low Response Rates, Refusals to Participate, and Attrition

If the researcher uses a probability sample, there will be little ambiguity about the destination of the inferential statement that is made—as long as the researcher clearly defines the *population* that supplied the study's subjects. Likewise, there will be little ambiguity associated with the target of inferential statements based upon nonprobability samples—as long as the *sample* is fully described. In each case, however, the inferential process becomes murky if data are collected from less than 100 percent of the individuals (or objects) that make up the sample. In this section, we consider three frequently seen situations in which inferences are limited because only a portion of the full sample is measured.

Response Rates

In many studies, the research data are collected by sending a survey, questionnaire, or test to a group of people by means of a mailed letter or an e-mail message. Usually, only a portion of the individuals who receive these mailed or e-mailed measurement probes furnishes the researcher with the information that was sought. In many cases, the recipient of the mailed or e-mailed survey, questionnaire, or test simply chooses not to open the envelope or the electronic message (or the attachment to the e-mail message). In other instances, the recipient looks at the research instrument(s) but decides not to take the time to read and respond to the questions. In any event, the term **response rate** has been coined to indicate the percentage of sample individuals who supply the researcher with the requested information.

In Excerpts 5.12 and 5.13, we see two cases in which response rates were reported in recent studies. In the first two of these excerpts, the response rate was far below the optimum value of 100 percent. Response rates like these are not uncommon. Some researchers attempt to justify their low response rates by saying that "It is normal to have a low response rate in mailed surveys" or that some so-called research authority says that "A response rate of 30 percent or more is adequate." You should be wary of such attempts to justify low response rates. Clearly, the statistical inferences in these studies extend only to individuals who are similar to those who returned completed surveys.

EXCERPTS 5.12–5.13 • Response Rates

Of 14,939 mailed questionnaires, 5,381 were returned—a response rate of 36 percent.

Source: Günther, O. H., Kürstein, B., Riedel-Heller, S. G., & König, H. (2010). The role of monetary and nonmonetary incentives on the choice of practice establishment: A stated preference study of young physicians in Germany. *Health Services Research, 45*(1), 212–229.

Of the 3289 email surveys delivered, 2010 unique responders submitted completed surveys (61% response rate).

Source: Faris, J. A., Douglas, M. K., Maples, D. C., Berg, L. R., & Thrailkill, A. (2010). Job satisfaction of advanced practice nurses in the Veterans Health Administration, *Journal of the American Academy of Nurse Practitioners, 22*(1), 35–44.

Adequate response rates rarely show up in studies where the researcher simply sits back and waits for responses from people who have been mailed just once a survey, questionnaire, or test. As Excerpts 5.14, 5.15, and 5.16 show, researchers can do certain things both before and after the measuring instrument is mailed in an effort to achieve a high response rate. Researchers (like those associated with these three excerpts) who try to get responses from everyone in the target sample deserve credit for their efforts; on the other hand, you ought to downgrade your evaluation of those studies in which little or nothing is done to head off the problem of ending up with a poor response rate.

EXCERPTS 5.14–5.16 • *Working to Get a Good Response Rate*

As in previous mailings, a self-addressed, stamped envelope was provided with the questionnaire. Two weeks following the mail-out a reminder phone call was made to those who had not returned their survey. Reminder post cards were mailed out approximately four weeks following the conclusion of the tournament to those spectators who had not yet returned their survey.

Source: Bee, C. C., & Havitz, M. E. (2010). Exploring the relationship between involvement, fan attraction, psychological commitment and behavioural loyalty in a sports spectator context. *International Journal of Sports Marketing & Sponsorship, 11*(2), 140–157.

Data were collected by mail survey from three groups of low-income individuals. . . . A five-contact survey approach was [employed], including two $1 bills attached to the survey cover letter, to increase the response rate.

Source: Loibl, C., Grinstein-Weiss, M., Zhan, M., & Red Bird, B. (2010). More than a penny saved: Long-term changes in behavior among savings program participants. *Journal of Consumer Affairs, 44*(1), 98–126.

To increase response rates [to the mailed survey], a small lottery incentive involving four Amazon.com gift certificates was included.

Source: Risley-Curtiss, C. (2010). Social work practitioners and the human-companion animal bond: A national study. *Social Work, 55*(1), 38–46.

Most researchers who collect data through the mail or via the Internet want their findings to generalize to individuals like those in the *full* group to whom the measuring instrument was originally sent, not just individuals like those who send back completed instruments. To get a feel for whether less-than-perfect response rates ought to restrict the desired level of generalizability, researchers sometimes conduct a midstream mini-study to see whether a **nonresponse bias** exists. As indicated in Excerpts 5.17, 5.18, and 5.19, there are different ways to check on a possible nonresponse bias. The methods exemplified by the first two of these excerpts are easier to execute, but they provide the least impressive evidence as to the existence of any nonresponse bias. In contrast, the method illustrated in Excerpt 5.19 is difficult to accomplish; it is, however, the best approach for investigating possible nonresponse bias.

EXCERPT 5.17–5.19 • *Checking for Nonresponse Bias*

In order to assess whether this substantial nonresponse rate [46.2%] was systematically biased, we performed a nonresponse analysis that compared the characteristics of respondents and complete nonrespondents. . . . The nonresponse analysis of demographic variables consisted of a series of bivariate comparisons of the two subgroups (respondents and nonrespondents) on key variables: town size, exam quality, socioeconomic status, sector, and gender.

Source: Negev, M., Garb, Y., Biller, R., Sagy, G., & Tal, A. (2010). Environmental problems, causes, and solutions: An open question. *Journal of Environmental Education*, *41*(2), 101–115.

Invitations to complete the survey were sent to 4,000 individuals who were known Internet shoppers and resided in the United States. . . . In total, 493 responded. . . . To check for possible response bias, early and late respondents were compared on demographic variables. No statistical differences were found.

Source: Milne, G. R., Labrecque, L. I., & Cromer, C. (2009). Toward an understanding of the online consumer's risky behavior and protection practices. *Journal of Consumer Affairs*, *43*(3), 449–473.

We gained data on 7 per cent of non-responders ($n = 103$) through telephone interviews. We compared responders and non-responders and found no difference in demographic and professional characteristics. In terms of 'therapeutic capacity' to care for patients who use illicit drugs, we found no difference in the level of educational adequacy, role legitimacy, motivation or self-esteem in the role. . . . These results suggest that, based on the variables measured in the study, the bias caused by the high non-response rate was not substantial.

Source: Ford, R., & Bammer, G. (2009). A research routine to assess bias introduced by low response rates in postal surveys. *Nurse Researcher*, *17*(1), 44–53.

Refusal to Participate

In studies where individuals are asked to participate, some people may decline. Such **refusals to participate** create the same kind of problem that is brought about by low response rates. In each case, valid inferences extend only to individuals similar to those who actually supplied data, not to the larger group of individuals who were *asked* to supply data. In Excerpt 5.20, we see a case in which nearly one-fourth of the potential members of the sample chose not to participate.

EXCERPT 5.20 • *Refusals to Participate*

[The researcher] contacted 150 general practitioners working in private practice in a large city in southern France. She explained the study to them, asked them to participate, and, if they agreed, arranged when to administer the experiment. Of these 150, 115 (78%) agreed to participate.

Source: Mas, C., Albaret, M., Sorum, P. C., & Mullet, E. (2010). French general practitioners vary in their attitudes toward treating terminally ill patients. *Palliative Medicine*, *24*(1), 60–67.

Just as some researchers perform a check to see whether a less-than-optimal response rate affects the generalizability of results, certain investigators compare those who agree to participate with those who decline. If no differences are noted, a stronger case exists for applying inferential statements to the full group of individuals invited to participate (and others who are similar) rather than simply to folks similar to those who supplied data. Researchers who make this kind of comparison in their studies deserve bonus points from you as you critique their investigations. Conversely, you have a right to downgrade your evaluation of a study if the researcher overlooks the possible problems caused by refusals to participate.

Attrition

In many studies, less than 100 percent of the participants remain in the study from beginning to end. In some instances, this problem arises because the procedures or data-collection activities of the investigation are aversive, boring, or costly to the participant. In other cases, forgetfulness, schedule changes, or residential relocation explain why certain individuals become dropouts. Regardless of the causal forces that bring about the phenomenon of **attrition,**[2] it should be clear why attrition can affect the inferential process.

When attrition occurs in a study, it may be possible for the researcher to check for an attrition bias. The purpose and procedures in doing this mirror the goals and techniques used in checking for a response bias. In Excerpt 5.21, we see an example in which the researchers checked to see if their attrition rate (about 27 percent) was potentially damaging to the study. Notice that the check of a possible attrition bias involved a comparison of those who remained in the study versus those who dropped out, with this done separately on nine different variables.

---

[2]The problem of attrition is sometimes referred to as **mortality.**

EXCERPT 5.21 • *Checking for Attrition Bias*

Of these teachers [in the study's sample], 987 returned a completed questionnaire at Time 1 (T1), and 719 did so at Time 2 (T2). . . . Additional analyses examining the pattern of participant attrition revealed that the T2 sample did not differ significantly from those who responded at T1 on nine dimensions (gender, age, marital status, school sector, years of teaching experience, number of teachers in the school, number of students in the school, school location, and socioeconomic status of school area).

Source: Bradley, G. L. (2010). Work-induced changes in feelings of mastery. *Journal of Psychology, 144*(2), 97–119.

A Few Warnings

As we approach the end of this chapter, I offer a handful of warnings about the inferential connection between samples and populations. I highly suggest that you become sensitive to these issues, because many professional journals contain articles in which the researcher's conclusions seem to extend far beyond what the inferential process legitimately allows. Unfortunately, more than a few researchers get carried away with the techniques used to analyze their data—and their technical reports suggest that they gave little or no consideration to the nature of their samples and populations.

My first warning has to do with *a possible mismatch between the source of the researcher's data and the destination of the inferential claims.* Throughout this chapter, I have emphasized the importance of a good match between the sample and the population. Be on guard when you read or listen to research reports, because the desired fit between these two groups may leave much to be desired. Consider, for example, the information presented in Excerpt 5.22.

EXCERPT 5.22 • *Mismatch Between Sample and Intended Population*

We sought to develop a dual-process model of sexual aggression by examining the relationships between men's implicit power–sex association and explicit power–sex beliefs, rape myth acceptance, and rape proclivity. . . . In Study 1, we developed and validated an explicit measure of power–sex beliefs. In Study 2, we used this measure of explicit power–sex beliefs, and an implicit measure of a power–sex association, to compare two alternative dual-process models of rape proclivity. . . . Participants [in Study 1] were 131 college students from a Midwestern, Catholic university [who were] enrolled in an upper-level psychology course or an introductory anthropology course. . . . Participants [in study 2] were 108 men from a

(continued)

EXCERPT 5.22 • (*continued*)

Midwestern, Catholic university who were recruited through an introductory psychology course and given course extra credit for their participation. The mean age for this sample was 19.1 years (SD = 1.3 years).

Source: Chapleau, K. M., & Oswald, D. L. (2010). Power, sex, and rape myth acceptance: Testing two models of rape proclivity. *Journal of Sex Research, 47*(1), 66–78.

The major concern you ought to have with the passage in Excerpt 5.22 is the vast difference between the declared population of interest and the actual sample. The researcher's intention was to use sample data and inferential statistics as a basis for making claims about men's beliefs. However, data were collected from a sample of male students attending a single, religion-oriented university, with students selected from just two kinds of courses (psychology and anthropology). These features of the sample made it impossible for the study's findings to be generalized to the stated population of interest.

My next warning has to do with the *size of the sample*. If you do not know much about the members of the sample or how the researcher obtained the sample, then the inferential process cannot operate successfully—no matter how large the sample might be. Remember, therefore, that it is the quality of the sample (rather than its size) that makes statistical inference work. Proof of this claim can be seen during national elections when pollsters regularly predict with great accuracy who will win elections even though the samples used to develop these predictions are relatively small.

My third warning concerns the term *random*. Randomness in research studies is usually considered to be a strong asset, but you should not be lulled into thinking that an investigation's results can be trusted simply because the term *random* shows up in the method section of the write-up. Consider, for example, the material presented in Excerpts 5.23 and 5.24.

EXCERPTS 5.23–5.24 • *The Word Random*

[T]he aim of the present study was to evaluate the patterns of rapid weight loss in a large sample of competitive judo athletes. . . . During the [judo] competitions, the participants were approached randomly and invited to participate in the study.

Source: Artigli, G. G., Gualano, B., Franchini, E., Scagliusa, F. B., Takesian, M., Fuchs, M., et al. (2010). Prevalence, magnitude, and methods of rapid weight loss among judo competitors. *Medicine and Science in Sports and Exercise, 42*(3), 436–442.

Surveys, including the School Counselor Self-Efficacy Scale (Bodenhorn & Skaggs, 2005), questions regarding the school counseling program, achievement

(*continued*)

EXCERPTS 5.23–5.24 • (*continued*)

gap information, and demographics, were sent to a random sample of 1,600 ASCA members. Through random selection of these participants, half of the surveys were sent through postal mail and half through e-mail/Internet. . . . The overall response rate was 54% (860 individuals responded).

Source: Bodenhorn, N., Wolfe, E. W., & Airen, O. E. (2010). School counselor program choice and self-efficacy: Relationship to achievement gap and equity. *Professional School Counseling*, *13*(3), 165–174.

In Excerpt 5.23, we are told that the judo competitors were "approached randomly" and asked if they would like to participate in the researchers' study. I am not sure what this means. It is possible, of course, that a subset of all the judo competitors were randomly selected. However, it is my hunch that the word *casually,* if substituted for the word *randomly,* would more accurately describe how the judo athletes were approached.

In Excerpt 5.24, we see the word *randomly* used twice in the statistical sense of the term. A large, random sample of members of the ASCA professional association was drawn, and then the sample was randomly divided into two halves. However, this randomness was undermined when nearly half of the study's intended participants did not return the survey. The sample size of 860 may seem large; however, there is no way to know if the survey results were tainted by a response bias. Although the gender split (85 percent female) and ethnicity (89 percent European American) among the 860 respondents were reported to be "similar to the demographic characteristics of school counselors found in most national studies," the low response rate destroyed the randomness of the initial random sample.

Regarding another matter, researchers should describe the procedures used to extract samples from their relevant populations. They should do this because the question of whether a sample is a random sample can be answered only by considering the procedure used to select the sample. As indicated earlier in the chapter, one not-too-sophisticated procedure for getting a random sample is to draw slips of paper from a hat. Random samples can also be produced by flipping coins or rolling dice to determine which members of the population end up in the sample.

Most contemporary researchers do not draw their random samples by rolling dice, flipping coins, or drawing slips of paper from a hat. Instead, they utilize either a printed **table of random numbers** or a set of **computer-generated random numbers.** To identify which members of the population get into the sample, the researcher first assigns unique ID numbers (e.g., 1, 2, 3) to the members of the population. Then, the researcher turns to a table of random numbers (or a set of computer-generated random numbers) where the full set of ID numbers appear in a scrambled order. Finally, the ID numbers that appear at the top of the list (e.g., 27, 4, 9) designate which members of population get into the sample.

In Excerpt 5.25, we see how easy it is for a researcher to indicate that a random sample was selected via a table of random numbers or computer-generated random numbers. These authors deserve credit for clarifying exactly how their random samples were created. All researchers should follow this good example!

EXCERPT 5.25 • *Using a Table of Random Numbers*

In Trinidad there are seventy nine (79) health centres, most of which have walk-in clinics (where patients can present without an appointment for any medical problem). These clinics were stratified to represent all regional health authorities (administrative regions) and to capture rural and urban populations. Clinics [$n = 16$] were then selected using a table of random numbers [to match] the proportion of clinics per administrative region.

Source: Maharaj, R. G., Alexander, C., Bridglal, C. H., Edwards, A., Mohammed, H., Rampaul, T., et al. (2010). Abuse and mental disorders among women at walk-in clinics in Trinidad: A cross-sectional study. *BMC Family Practice, 11*(26), 1–21.

The final warning is really a repetition of a major concern expressed earlier in this chapter. Simply stated, an empirical investigation that incorporates inferential statistics is worthless unless there is a detailed description of the population or the sample. No matter how carefully the researcher describes the measuring instruments and procedures of the study, and regardless of the levels of appropriateness and sophistication of the statistical techniques used to analyze the data, the results are meaningless unless we are given a clear indication of the population from which the sample was drawn (in the case of probability samples) or the sample itself (in the case of nonprobability samples). Unfortunately, too many researchers get carried away with their ability to use complex inferential techniques when analyzing their data. I can almost guarantee that you will encounter technical write-ups in which the researchers emphasize their analytical skills to the near exclusion of a clear explanation of where their data came from or to whom the results apply. When you come across such studies, give the authors *high* marks for being able to flex their "data analysis muscles"—but *low* marks for neglecting the basic inferential nature of their investigations.

To see an example of a well-done description of a sample, consider Excerpt 5.26. Given this relatively complete description of the 136 children who formed this study's sample, we have a much better sense of the population to which the statistical inferences can be directed. It would be nice if all researchers described their samples with equal care.

EXCERPT 5.26 • *Detailed Description of a Sample*

We examined the summer employment and community participation experiences of 136 youth with severe disabilities. To be included in this study, students had to (a) be receiving special education services under a primary or secondary disability category of cognitive disability, autism, or multiple disabilities; (b) be eligible for the state's alternate assessment; and (c) provide assent and parental consent to participate. . . . Youth participating in our study ranged in age from 13.9 to 21.8 ($M = 18.2$; $SD = 1.8$), and slightly more than half were male (52.9%). The majority (85.3%) was European American, 11.8% were African American, and 2.9% reported other races/ethnicities (i.e., Asian American, American Indian). Twenty-six students (19.1%) were in 9th grade, 18 (13.2%) were in 10th grade, 36 (26.5%) were in 11th grade, 37 (27.2%) were in 12th grade, and 19 (14.0%) received services in 18 to 21 programs. More than one quarter (28.7%) of students were eligible for free/reduced lunch (FRL). Most youth were reported to be served under the primary disability category of cognitive disabilities (85.3%), followed by autism (10.3%) and orthopedic impairments (4.4%); 61.0% were reported to have one or more secondary disabilities.

Source: Carter, E. W., Dutchman, N., Ye, S., Trainor, A. A., Swedeen, B., & Owens, L. (2010). Summer employment and community experiences of transition-age youth with severe disabilities. *Exceptional Children, 76*(2), 194–212.

Review Terms

Abstract population
Attrition
Cluster sample
Computer-generated
    random numbers
Convenience sample
Inferential statistics
Mortality
Nonprobability sample
Nonresponse bias
Oversampling
Parameter
Population
Probability sample
Purposive sample

Quota sample
Refusals to participate
Response rate
Sample
Sampling frame
Simple random sample
Snowball sample
Statistic
Statistical inference
Statistical focus
Stratified random sample
Systematic sample
Table of random numbers
Tangible population

The Best Items in the Companion Website

1. An interactive online quiz (with immediate feedback provided) covering Chapter 5.
2. Ten misconceptions about the content of Chapter 5.

3. An email message sent from the author to his students to help them understand the difference between tangible and abstract populations.
4. A poem about questionnaires (written by a famous statistician).
5. The best passage from Chapter 5 (selected by the author).

To access the chapter outline, practice tests, weblinks, and flashcards, visit the companion website at http://www.ReadingStats.com.

Review Questions and Answers begin on page 531.

6

Estimation

In Chapter 5, we laid the foundation for our consideration of inferential statistics. We did this by considering the key ingredients of this form of statistical thinking and analysis: population, sample, parameter, statistic, and inference. In this chapter, we now turn our attention to one of the two main ways in which researchers use sample statistics to make educated guesses as to the values of population parameters. These procedures fall under the general heading **estimation.**

This chapter is divided into three main sections. First, the logic and techniques of *interval estimation* are presented. Next, we examine a second, slightly different way in which estimation works; this approach is called *point estimation*. Finally, I offer a few tips to keep in mind as you encounter research articles that rely on either of these forms of estimation.

Before beginning my discussion of estimation, I want to point out that the two major approaches to statistical inference—estimation and hypothesis testing—are similar in that the researcher makes an educated guess as to the value of the population parameter. In that sense, both approaches involve a form of guesswork that might be construed to involve estimation. Despite this similarity, the term *estimation* has come to designate just one of the two ways in which researchers go about making their educated guesses about population parameters. The other approach, hypothesis testing, is discussed in Chapters 7 and 8.

Interval Estimation

To understand how **interval estimation** works, you must become familiar with three concepts: sampling errors, standard errors, and confidence intervals (CIs). In addition, you must realize that a CI can be used with just about any statistic that is computed on the basis of sample data. To help you acquire these skills, we begin with a consideration of what is arguably the most important concept associated with inferential statistics: sampling error.

Sampling Error

When a sample is extracted from a population, it is conceivable that the value of the computed statistic will be identical to the unknown value of the population parameter. Although such a result is possible, it is far more likely that the statistic will turn out to be different from the parameter. The term **sampling error** refers to the magnitude of this difference.

To see an example of sampling error, flip a coin 20 times, keeping track of the proportion of times the outcome is heads. Consider your 20 coin flips to represent a sample of your coin's life history of flips, with that total life history being the population. I will assume that your coin is unbiased and that your flipping technique does not make a heads outcome more or less likely than a tails outcome. Given these two simple assumptions, I can assert that the parameter value is known to be .50. Now, stop reading, take out a coin, flip it 20 times, and see how many of your flips produce a heads outcome.

I do not know, of course, how your coin-flipping exercise turned out. When *I* flipped my coin (a quarter) 20 times, however, I *do* know what happened. I ended up with 13 heads and 7 tails, for a statistic of .65. The difference between the sample's statistic and the population's parameter is the *sampling error*. In my case, therefore, the sampling error turned out to be .15.[1]

If the 20 coin flips produce 10 heads, the sampling error is equal to zero. Such a result, however, is not likely to occur. Usually, the sample statistic contains sampling error and fails to mirror exactly the population parameter. Most of the time, of course, the size of the sampling error is small, thus indicating that the statistic is a reasonably good approximation of the parameter. Occasionally, however, a sample yields a statistic that is quite discrepant from the population's parameter, such as if we get 19 or 20 heads (or tails) when flipping a fair coin 20 times.

It should be noted that the term *sampling error* does *not* indicate that the sample has been extracted improperly from the population or that the sample data have been improperly summarized. (I ended up with a sampling error of .15 even though I took a random sample from the population of interest and even though I carefully summarized my data.) When sampling error exists, it is attributable not to any mistake being made, but rather to the natural behavior of samples. Samples generally do not turn out to be small mirror images of their corresponding populations, and statistics usually do not turn out equal to their corresponding parameters. Even with proper sampling techniques and data analysis procedures, sampling error ought to be expected.

In my example dealing with 20 coin flips, we knew what the parameter's value was equal to. In most inferential situations, however, the researcher knows

---

[1] I computed the sampling error by subtracting .50 from .65.

the numerical value of the sample's statistic, but not the value of the population's parameter. This situation makes it impossible for the researcher to compute the precise size of the sampling error associated with any sample, but it does not alter the fact that sampling error should be expected. For example, suppose I gave you a coin that was known *only by me* to be slightly biased. Imagine that it would turn up heads 55 percent of the time over its life history. If I asked you to flip this coin 20 times and then make a guess as to the value of the coin's parameter value, you should expect sampling error to occur. Hence, not knowing the parameter value (and thus not being able to compute the magnitude of any sample's sampling error) should not affect your expectation that the statistic and the parameter are at least slightly unequal.[2]

Sampling Distributions and Standard Errors

Most researchers extract a single sample from any population about which they want to make an educated guess. Earlier, for example, I asked you to take a sample of 20 flips of your coin's coin-flipping life history. It is possible, however, to *imagine* taking more than one sample from any given population. Thus, I can imagine taking multiple samples from the coin I flipped that gave me, in the first sample, an outcome of .65 (i.e., 65 percent heads).

When I imagine taking multiple samples (each made up of 20 flips) from that same coin, I visualize the results changing from sample to sample. In other words, whereas I obtained a statistic of .65 in my first sample, I would not be surprised to find that the statistic turns out equal to some other value for my second set of 20 flips. If a third sample (of 20 flips) were to be taken, I would not be surprised to discover that the third sample's statistic assumes a value different from the first two samples' statistics. If I continued (in my imagination) to extract samples (of 20 flips) from that same coin, I would eventually find that values of the statistic (1) would begin to repeat, as would be the case if I came across another sample that produced 13 heads; and (2) would form a distribution resembling a bell-shaped curve centered over the value of .50.

The distribution of sample statistics alluded to in the preceding paragraph is called a **sampling distribution,** and the standard deviation of the values that make up such a distribution is called a **standard error (SE).** Thus, an SE is nothing more than an index of how variable the sample statistic is when multiple samples of the same size are drawn from the same population. As you recall from Chapter 2, variability can be measured in various ways; the SE, however, is always conceptualized as being equal to the standard deviation of the sampling

---

[2]If a population is perfectly homogenous, the sampling error is equal to 0. If the population is heterogeneous but an enormously large sample is drawn, here again, the statistic turns out equal to the parameter once that statistic is rounded to one or two decimal places. Both of these situations, however, are unrealistic. Researchers typically are involved with heterogeneous populations and base their statistical inferences on small samples where $n < 50$.

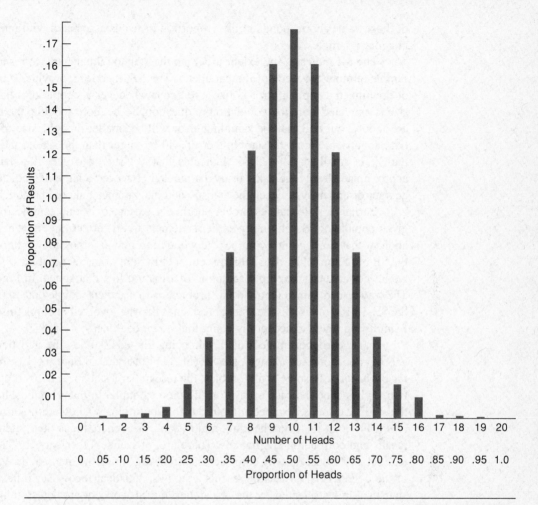

FIGURE 6.1 *Sampling Distribution of Number/Proportion of Heads in 20 Flips of a Fair Coin*

distribution of the statistic (once we imagine that multiple samples are extracted and summarized).[3]

Figure 6.1 contains the sampling distribution that we would end up with if we took many, many samples (of 20 flips per sample) from a fair coin's population of potential flips, with the statistical focus being the proportion of heads that turn up within each sample. The standard deviation of this sampling distribution is equal to about .11. This SE provides a numerical index of how much dispersion exists among the values on which the standard deviation is computed; in this case, each

---

[3]Even though the concepts of standard deviation and SE are closely related, they are conceptually quite different. A standard deviation indicates the variability inside a single set of actual data points; an SE, in contrast, indicates how variable the sample statistic is from sample to sample.

of those values corresponds to the proportion of heads associated with one of our imaginary samples.

The SE indicates the extent to which the statistic fluctuates, from sample to sample, around the value of the parameter. The SE, therefore, provides a measure of how much sampling error is likely to occur whenever a sample of a particular size is extracted from the population in question. To be more specific, the chances are about 2 out of 3 that the sampling error will be smaller than the size of the SE (and about 1 in 3 that the sampling error will be larger than the size of the SE). If the SE is small, therefore, this indicates that we should expect the statistic to approximate closely the value of the parameter. However, a large SE indicates that a larger discrepancy between the statistic and the parameter should be anticipated.

Earlier, I said that researchers normally extract only one sample from any given population. Based on my earlier statement to that effect (and now my reiteration of that same point), you may be wondering how it is possible to know what the SE of the sampling distribution is equal to in light of the fact that the researcher would not actually develop a sampling distribution like that shown in Figure 6.1. The way researchers get around this problem is to use their sample data to estimate the SE. I will not discuss the actual mechanics that are involved in doing this; rather, I simply want you to accept my claim that it *can* be done.[4]

In my earlier example about a coin being flipped 20 times, the statistical focus was a proportion. Accordingly, the SE (of .11) illustrated in Figure 6.1 is the SE *of the proportion*. In some actual studies, the researcher's statistical focus is a proportion, as has been the case in my coin-flipping example. In many studies, however, the statistical focus is something other than proportion. When reading journal articles, I find that the overwhelming majority of researchers focus their attention on means and correlation coefficients. There are, of course, other ways to "attack" a data set, and I occasionally come across articles in which the median, the variance, or the degree of skewness represents a study's statistical focus. Regardless of the statistical focus selected by the researcher, the SE concept applies so long as the study involves inferential statistics.

Consider, for example, the short passage contained in Excerpt 6.1. As you can see, this excerpt comes from a study focused on the possible influence of a criminal's age and health status on the length of the jail sentences meted out. As you can see, the data in Excerpt 6.1 summarize the sentences given to offenders who were sick and those who were healthy. The mean for each group is provided, as is each group's SE. Because the mean is the statistical focus, and because sample data only allow us to make educated guess as to population characteristics, the more complete name for each SE number is *estimated standard error of the mean*.

---

[4]For example, when I use my single sample of 20 coin flips (13 heads, 7 tails) to estimate the SE of the theoretical sampling distribution, I obtain the value of .1067. This estimated SE of the proportion approximates the true value, .1118, that corresponds to the full sampling distribution shown in Figure 6.1.

EXCERPT 6.1 • *Estimated Standard Error of the Mean*

Offenders in good health were on average given a significantly longer sentence of 62.90 months (SE = 5.49), compared to offenders in poor health who received on average sentences of 57.92 months (SE = 4.97).

Source: Mueller-Johnson, K. U., & Dhami, M. K. (2010). Effects of offenders' age and health on sentencing decisions. *Journal of Social Psychology, 150*(1), 77–97.

By providing (in Excerpt 6.1) information as to the estimated SE associated with each group's mean sentence, the researchers were alerting their readers to the fact that their data allowed them to compute sample statistics, not population parameters. In other words, each SE in this short passage cautions us not to consider each mean to be equal to its corresponding μ. If a different sample of healthy offenders (like the ones used in this study) were to be considered, the mean sentence length for the new group would probably turn out equal to some value other than 62.90.

Excerpt 6.2 contains another example of information being presented as to the SE. This time, however, the SE is connected to a percentage rather than the mean. In the study associated with this excerpt, 75 women with stage II cervical cancer were treated with a combination of chemotherapy and radiation. These women were then observed for several years regarding survival and recurrence of the disease. By presenting an SE for the five-year survival rate and for the five-year "progression free rate," the researchers warn us that these two percentages (80.6 and 71.3) should be looked upon as sample statistics, not population parameters.

EXCERPT 6.2 • *Estimated Standard Error of a Percentage*

The 5-year overall survival rate was 80.6% (standard error, 4.9%) and 5-year PFS [progression free survival] rate was 71.3% (standard error, 5.3%).

Source: Lee, D. W., Kim, Y. T., Kim, J. H., Kim, S., Kim, S. W., Nam, E. J., et al. (2010). Clinical significance of tumor volume and lymph node involvement assessed by MRI in stage IIB cervical cancer patients treated with concurrent chemoradiation therapy. *Journal of Gynecological Oncology, 21*(1), 18–23.

The SE values in Excerpts 6.1 and 6.2 give us a feel for how much variability we should expect to see if these studies were to be replicated, with the new samples in the replication studies being pulled out of the same abstract populations as were the criminals (in Excerpt 6.1) or the women with cervical cancer (in Excerpt 6.2). Consider, for example, the progression-free rate (PFR) for the 75 women involved in the second of these excerpts. Because the SE for the survival rate was

5.3 percent, we might expect the new sample's PFR to be within about 5 percentage points of 71.3. If we wanted to be more confident of our prediction, we might double the SE, round off, and predict that the new sample's survival rate most likely would be within 10–11 percentage points of 71.3.

In Excerpt 6.3, we see a case where SE values are presented in a bar graph. In this interesting study, each bar corresponds to a mean on the Graduate Record

EXCERPT 6.3 • *Estimated Standard Error of the Mean in a Bar Graph*

This research examined the benefits of interpreting physiological arousal as a *challenge* response on [the] actual Graduate Record Examination (GRE) scores. Participants who were preparing to take the GRE reported to the laboratory for a practice GRE study. Participants assigned to a reappraisal condition were told arousal improves performance, whereas control participants were not given this information. . . . One to three months later, participants returned to the lab and provided their score reports from their actual GRE.

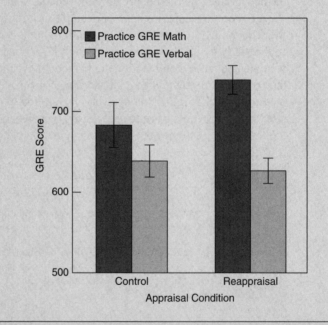

FIGURE 1   *Practice GRE performance as a function of appraisal condition and test section. Scores could range from 200 to 800. Error bars represent ± standard error of the mean.*

Source: Jamieson, J. P., Mendes, W. B., Blackstock, E., & Schmade, T. (2010). Turning the knots in your stomach into bows: Reappraising arousal improves performance on the GRE. *Journal of Experimental Social Psychology, 46*(1), 208–212.

Exam (GRE). The study's research participants, all of whom were planning to take the GRE, had a practice session first. Just prior to the practice test, the students in the experimental group were led to believe that recent research shows test anxiety to be beneficial (not harmful) to test performance. Those in the control group were not given this information. After taking the real GRE, students reported how they did on the Math and Verbal subtests. In Excerpt 6.3, the vertical line that overlaps the top of each column functions as a graphical indication of the SE of the mean. Those vertical lines are called *error bars*.

Confidence Intervals

Researchers who report SEs along with their computed sample statistics deserve to be commended. This practice helps to underscore the fact that sampling error is very likely to be associated with any sample mean, with any sample percentage, with any sample correlation coefficient, and with any other statistical summary of sample data. By presenting the numerical value of the SE (as in Excerpts 6.1 and 6.2) or by putting a line segment through the statistic's position in a graph (as in Excerpt 6.3), researchers help us to remember that they are only making educated *guesses* as to parameters.

Although SEs definitely help us when we try to understand research results, a closely related technique, **confidence intervals,** helps us even more. My four-fold objective here is to show what a confidence interval looks like, explain how confidence intervals are built, clarify how to interpret confidence intervals properly, and point out how confidence intervals carry with them a slight advantage over SEs.

Confidence Intervals: What They Look Like. A *confidence interval (CI)* is simply a finite interval of score values on the dependent variable. Such an interval is constructed by adding a specific amount to the computed statistic (thereby obtaining the upper limit of the interval) and by subtracting a specific amount from the statistic (thereby obtaining the lower limit of the interval). In addition to specifying the interval's upper and lower limits, researchers always attach a percentage to any interval that is constructed. The percentage value selected by the researcher is invariably a high number, such as 90, 95, or 99 percent.[5]

We next look at four CIs that have come from research reports. These CIs are instructive, as each one has a lesson connected to it. If you wish, look first at Excerpts 6.4 through 6.7 and see if you can determine why these particular CIs have been included here. Then, return to the following paragraphs to see if you noticed the special feature of each excerpt.

---

[5]The vast majority of researchers set up 95 percent CIs. If you read enough research reports, you may come across 90 percent and 99 percent CIs; however, you will likely consider them to be exceptions to the rule because they are used so infrequently.

EXCERPTS 6.4–6.5 • *Confidence Intervals Around a Mean and a Percentage*

ABC scores were obtained from 46 children, the mean score was 51 and 95% CI was 46–56 (SD 18).

Source: Fernell, E., Hedvall, A., Norrelgen, F., Eriksson, M., Höglund-Carlsson, L., Barnevik-Olsson, M., et al. (2010). Developmental profiles in preschool children with autism spectrum disorders referred for intervention. *Research in Developmental Disabilities, 31*(3), 790–799.

- -

Of the 18–30 year olds, 49.8% (95% confidence interval: 48.5%–51.2%) were men and 50.2% (48.8%–51.5%) were women.

Source: Cavazos-Rehg, P. A., Spitznagel, E. L., Krauss, M. J., Schootman, M., Bucholz, K. K., Cottler, L. B., et al. (2010). Understanding adolescent parenthood from a multisystemic perspective. *Journal of Adolescent Health, 46*(6), 525–531.

In Excerpt 6.4, we see a CI that has been built around a sample mean. Notice that this excerpt contains the standard deviation for the 46 test scores. This measure of variability (coupled with the sample size) ought to make us guess that range of individual scores probably extended from about 10 to about 92. This range is much larger than the distance between the end points of the CI. These "ranges" differ because the SD measures the variability of raw scores whereas the CI provides an estimate of the variability of the sample mean. Although they both deal with variability, SDs and CIs are conceptually different and will never be equivalent in size.

Excerpt 6.5 contains two CIs built around percentages. Notice how narrow each CI is, with the end points positioned quite close to each other. In each case, the upper and lower values are less than 3 percentage points different from each other. The small width of these CIs came about because of the large samples sizes used in the study. There were 3,475 men and 4,462 women. If these sample sizes had been smaller, the end points would have been further apart. This inverse relationship between sample size and interval width ought to make sense, because the size of the sample statistics (in this excerpt, a percentage) fluctuates less across different samples if those samples are large rather than small.

In Excerpt 6.6, we see a CI that has been built around a correlation coefficient. There are two things to note here. First, the researchers chose to set up a 99% CI rather than the more popular 95% kind of CI. Had they created a 95% CI, the end points would have been closer together. (Had they created a 90% CI, it would have been even narrower.) Second, notice that the value of Pearson's r does not lie precisely in the middle of the CI. Unless the value of r is 0.00, or unless the sample size is gigantic, a CI built around r will extend further in the negative direction if r is positive, or further in the positive direction if r is negative. As in Excerpt 6.6, the two ends of the CI can have the same sign and yet the CI can extend one direction farther than the other direction.

EXCERPTS 6.6–6.7 • *Confidence Intervals Around a Correlation and an Odds Ratio*

A Pearson's *r* coefficient of .62 [having] a 99% confidence interval of .495 and .72 was obtained.

Source: Smith, S., & Chonody, J. M. (2010). Peer-driven justice: Development and validation of the Teen Court Peer Influence Scale. *Research on Social Work Practice, 20*(3), 283–292.

Patients younger than age 65 had almost three times the odds of engaging in regular exercise activities as compared to patients 65 years and older (OR = 2.82, 95% CI: 1.44, 5.53).

Source: Bleich, S. N., Huizinga, M. M., Beach, M. C., & Cooper, L. A. (2010). Patient use of weight-management activities: A comparison of patient and physician assessments. *Patient Education and Counseling, 79*(3), 344–350.

Excerpt 6.7 contains the last of our four CI examples. In this excerpt, a CI has been built around something called an *odds ratio (OR),* and in this case, the OR is equal to 2.82. It was this number that prompted the researchers to use the phrase "almost three times" that appears in the excerpt. In Chapter 6, we consider in detail the concept of odds ratios. This excerpt is worth looking at now, however, because it exemplifies the way researchers often present their CIs. Instead of stating the CI as if it were a range of values extending from a low number to a high number (as in Excerpts 6.4 and 6.5), the CI in Excerpt 6.7 is summarized via its end points (1.44 and 5.53) separated by a comma.

The CIs in Excerpts 6.4 through 6.7 are the same in the sense that they appear in the text of their respective research reports. CIs may be presented in two other ways: tables and figures. I do not include here any such tables or figures, for you will find it easy to decipher such presentation so long as you understand the material we have covered about CIs as well as the material to which we now turn.

The Construction of Confidence Intervals. The end points of a CI are not selected by the researcher magically making two values appear out of thin air. Rather, the researcher first makes a decision as to the level of confidence that is desired (usually 95 or 99). Then, the end points are computed by means of a joint process that involves (1) the analysis of sample data so as to obtain the estimated SE of the statistic and (2) the multiplication of that estimated SE by a tabled numerical value.[6]

[6]For example, to build a 95 percent CI around the mean in Excerpt 6.4 from the ABC scores from the 46 children, I multiply the estimated SE of 2.65 by 2.01, with the second of these numbers coming from a *t*-table. The product, 5.33, is then added to and subtracted from the mean of 51 to establish the ends of the CI.

Although you do not need to know the various formulas used to construct CIs, you should be cognizant of the fact that a scientific approach is taken to the creation of any CI. Moreover, you should be aware of the fact that short (i.e., narrow) intervals that have a high level of confidence associated with them are more helpful in inferential statistics. In making an educated guess as to the unknown value of a population parameter, it is much better to have that guess be precise.

It should be noted that the length of a CI is also affected by the nature of the statistic computed on the basis of sample data. For example, CIs built around the mean are shorter than those constructed for the median. The same situation holds true for Pearson's product–moment correlation coefficient as compared with Spearman's rho. This may explain, in part, why Ms and rs are seen so frequently in the published literature.

The Proper Interpretation of Confidence Intervals.   CIs are often misinterpreted to designate the probability that population parameters lie somewhere between the intervals' upper and lower limits. For example, many people (including more than a few researchers) would look at the CI presented in Excerpt 6.4 and conclude that there is a .95 probability (i.e., a 95 percent chance) that the population mean of ABC scores lies somewhere between 46 and 56. CIs should *not* be interpreted in this fashion.

After a sample has been extracted from a population and then measured, the CI around the sample's statistic either does or does not "cover" the value of the parameter. Hence, the probability that the parameter lies between the end points of a CI is either 0 or 1. Because of this fact, a CI should never be considered to specify the chances (or probability) that the parameter is "caught" by the interval.

The proper way to interpret a CI is to *imagine* that (1) many, many samples of the same size are extracted from the same population and (2) a 95 percent CI is constructed separately around the statistic computed from each sample's data set. Some of these intervals would capture the parameter—that is, the interval's end points are such that the parameter lies within the interval. However, some of these CIs do *not* capture the parameter. Looked at collectively, 95 percent of these 95 percent CIs would contain the parameter. Accordingly, when you see a 95 percent CI, you should consider that the chances are 95 out of 100 that the interval you are looking at is one of those that does, in fact, capture the parameter. Likewise, when you encounter a 99 percent CI, you can say to yourself that the chances are even higher (99 out of 100) that the interval in front of you is one of the many possible intervals that would have caught the parameter.

The Advantage of Confidence Intervals over Estimated Standard Errors.   As stated previously, a CI is determined by first computing and then using the value of the estimated SE. Researchers should be commended for providing either one or the other of these inferential aids to their readers, for it is unfortunately true that many researchers supply their readers with neither SEs nor CIs for any of the sample statistics that are reported. Nevertheless, CIs carry with them a slight advantage that is worth noting.

When a CI is computed, it is labeled as to its level of confidence. (As exemplified by Excerpts 6.4 through 6.6, researchers usually build 95 percent CIs.) In contrast, SE intervals rarely are labeled as to their level of confidence. Given the fact that SE intervals usually have a confidence level of about 68 percent, they are apt to be misinterpreted and thought to be better than they really are.

Consider, for example, the information in Excerpt 6.1 regarding the length of sentences given to offenders in poor health. You can, if you wish, add 4.97 to 57.92 and subtract 4.97 from 57.92 to create the ends of an interval, with the result extending from 52.95 to 62.89 months. However, that interval is *not* a 95 percent CI. You could approximate a 95 percent CI by *doubling* the SE number and then moving above and below the sample mean by that amount to establish your guess as to the end points of the CI. However, doing this works accurately only when the sample *n* is at least 30.

Point Estimation

When engaged in interval estimation, a researcher will (1) select a level of confidence (e.g., 95 percent), (2) analyze the sample data, (3) extract a number out of a statistical table, and (4) scientifically build an interval that surrounds the sample statistic. After completing these four steps, the researcher makes an educated guess as to the unknown value of the population parameter. In making this guess, the researcher ends up saying, "My data-based interval extends from _____ to _____, and the chances are _____ out of 100 that this interval is one of the many possible intervals (each based on a different sample) that would, in fact, contain the parameter between the interval limits."

A second form of estimation is called **point estimation,** and here again, an educated guess is made, on the basis of sample data, as to the unknown value of the population parameter. With this second kind of estimation, however, the activities and thinking of the researcher are much simpler. With point estimation, no level of confidence must be selected, no statistical table must be consulted, and no interval must be created. Instead, the researcher simply computes the statistic on the basis of the sample data and then posits that the unknown value of the population parameter is the same as the data-based number. Thus, the researcher who uses this guessing technique ends up saying, "Because the sample-based statistic turned out equal to _____, my best guess is that the value of the parameter is also equal to that particular value."

Point estimation, of course, is likely to produce statements that are incorrect. Because of the great likelihood of sampling error, the value of the statistic rarely matches the value of the parameter. For this reason, interval estimation is generally considered to represent a more logical way of making educated guesses as to parameter values than is point estimation.

Despite the fact that point estimation disregards the notion of sampling error, many researchers can be seen making pinpoint guesses as to parameter values.

Consider, for example, Excerpt 6.8. In the research report from which this excerpt was drawn, the researchers looked at the relationship between life satisfaction and tolerance.

EXCERPT 6.8 • *Point Estimates Not Labeled as Such*

[We examined] the relationships between life satisfaction and two types of openness: tolerance in relation to gays and lesbians and tolerance in relation to racial and ethnic minorities. Across all countries, there is a positive correlation between life satisfaction and both types of tolerance (.78, .63, gays and racial minorities, respectively).

Source: Florida, R., Mellander, C., & Rentfrow, P. J. (2010). Socioeconomic Structures and Happiness. Working Paper Series, Martin Prosperity Institute, Rotman School of Management, University of Toronto, REF. 2010–MPIWP-002.

In Excerpt 6.8, two correlation coefficients are presented. Each of these *r*s is a point estimate, even though this label does not appear in the excerpt (or in the full research report). Despite the fact that these correlations were based on large samples—approximately 1,000 adults in each of 150 countries who participated in the Gallop Organization's World Poll—the *r*s reported in Excerpt 6.8 are still point estimates. If different people had been interviewed, the correlations most likely would have been at least slightly different. Because the phrase *point estimate* or simply the word *estimate* is nowhere to be seen, I am afraid readers of this research report might mistakenly consider these correlations to be population parameters rather than sample statistics.

In Excerpt 6.9, we see an example of researchers pointing out that their point estimates were just that, point estimates. The final sentence of this excerpt provides a warning to readers of the research report not to interpret the numbers

EXCERPT 6.9 • *Point Estimates Referred to as Point Estimates*

Adult smoking prevalence for African Americans was 19.3% compared with 15.4% for all Californians. The health care cost of smoking was $626 million for the African American community. Although African Americans account for 6% of the California adult population, they account for over 8% of smoking-attributable expenditures and fully 13% of smoking-attributable mortality costs. [However], our estimates are point estimates and do not account for the sampling variability in smoking prevalence, relative risks, or health care expenditure estimates.

Source: Max, W., Sung, H., Tucker, L., & Stark, B. (2010). The disproportionate cost of smoking for African Americans in California. *American Journal of Public Health, 100*(1), 152–158.

in the excerpt's first three sentences as fully accurate indices of smoking prevalence, smoking-related health-care expenditures, or smoking-related mortality
rates. Unfortunately, appropriate warnings such as this appear infrequently in
research reports.

Point estimates are connected with many of the statistical summaries included
in research reports: percentages, means, medians, standard deviations, to name but
a few. Researchers also engage in point estimation in the discussion of the measuring instruments used to collect data. As was indicated in Chapter 4, these discussions often involve the presentation of reliability and validity coefficients.

Give yourself a pat on the back if you recall, from what I said in Chapter 4,
that such coefficients are only estimates. If a different sample of examinees were to
provide the data for the assessment of reliability or validity, the obtained coefficients
most likely would fluctuate. Sampling error accounts for such fluctuation.

Although it is possible to build CIs around reliability and validity coefficients, researchers rarely do this. Instead, point estimates are typically provided.
This is a common practice, even in cases where the researcher uses inferential
statistical procedures in other parts of the research report. Consider, for example, Excerpts 6.10 and 6.11. The reliability coefficient in the first of these
excerpts and the validity coefficient in second excerpt are point estimates, not
parameters.

EXCERPTS 6.10–6.11 • *Point Estimates of Reliability and Validity*

Reliability analyses showed that the scale had high reliability in this study (alpha = .93).

Source: Kwo, S. Y. C. L., & Shek, D. T. L. (2010). Personal and family correlates of suicidal
ideation among chinese adolescents in Hong Kong. *Social Indicators Research, 95*(3),
407–419.

- -

Concurrent validity coefficient between this scale and ENRICH Marital Satisfaction
Questionnaire was 0.83.

Source: Rajabi, G. R. (2010). Factorial structure of marital satisfaction scale in married staff
members of Shahid Chamran University. *Iranian Journal of Psychiatry and Clinical Psychology, 15*(4), 351–358.

Occasionally, you may come across a research report that exemplifies the
good practice of building CIs around reliability and validity coefficients. Consider,
for example, Excerpts 6.12 and 6.13. The researchers who conducted these studies
deserve high praise for recognizing that their reliability and validity coefficients
were sample statistics, not population parameters.

EXCERPTS 6.12–6.13 • *Confidence Intervals Built Around Reliability and Validity Coefficients*

Intraclass correlation coefficients (ICCs) (95% confidence interval [CI]) for inter-rater reliability were .90 (.71–.97), .92 (.77–.97), and .85 (.64 –.95) for time, number of steps, and smoothness, respectively.

Source: Hess, R. J., Brach, J. S., Piva, S. R., & VanSwearingen, J. M. (2010). Walking skill can be assessed in older adults: Validity of the Figure-of-8 Walk Test. *Physical Therapy*, *90*(1), 89–99.

To measure concurrent validity [of the CCAM] with the COVS, the Pearson correlation (*r*) between scores on the CCAM and COVS was calculated. . . . The Pearson correlation coefficient between the CCAM total score and the COVS total score was very high (*r* = .96, 95% CI = .91 − 1.00) for the measure as a whole.

Source: Huijbregts, M. P. J., Teare, G. F., McCullough, C., Kay, T. M., Streiner, D., Wong, S. K. C., et al. (2009). Standardization of the Continuing Care Activity Measure: A multicenter study to assess reliability, validity, and ability to measure change. *Physical Therapy*, *89*(6), 546–555.

Although the likelihood of sampling error causes the practice of point estimation to seem quite ill-founded, this form of statistical inference deserves to be respected for two reasons. These two supportive arguments revolve around (1) the role played by point estimation in interval estimation and (2) the reliance on point estimation by more advanced scientific disciplines (such as physics). Let's consider briefly each of these reasons why it is unwise to look on point estimation with complete disrespect.

When engaged in interval estimation, the researcher builds a CI that surrounds the sample statistic. Point estimation is relied on in two ways when such intervals are constructed. First, the pinpoint value of the sample statistic is used as the best single estimate of the population parameter. The desired interval is formed by adding a certain amount to the statistic and subtracting a certain amount from the statistic. Hence, the value of the statistic, as a *point estimate* of the parameter, serves as the foundation for each and every CI that is constructed.

Interval estimation draws on point estimation in a second manner. To be more specific, the amount that is added to (and subtracted from) the statistic in order to obtain the interval's upper and lower limits is based on a point estimate of the population's variability. For example, when a CI is constructed around a sample mean, the distance between the end points of the interval is contingent on, among other things, a *point estimate* of the population standard deviation. Likewise, whenever a CI is built around a sample proportion, the length of the interval cannot be specified until the researcher first uses *point estimation* to guess how variable the population is.

From a totally different perspective, the practice of point estimation deserves to be respected. Certain well-respected scientists assert that as a discipline advances and becomes more scientifically rigorous, point estimation is turned to with both increased frequency and greater justification.

Warnings Concerning Interval and Point Estimation

As we wrap up this chapter, I want to give you four cautionary comments concerning the techniques of estimation. The first three comments concern interval estimation, whereas the fourth is relevant to both kinds of estimation techniques: point and interval. You will be a better consumer of the research literature if you keep these final points in mind.

First, be aware that the second of two numbers separated by a plus-and-minus sign can represent any of three things. In other words, if you see the notation 63 ± 8, be careful before you guess what the 8 signifies. It might be the standard deviation, it might be an estimated SE, or it might be half the distance between the end points of a CI. Researchers almost always clarify the meaning of such statements within a table or figure, or in the text of the research article. Take the time to look and read before jumping to any conclusions.

A second warning concerns the fact that sample data allow a researcher to *estimate* the SE of the statistic, not to *determine* that SE in a definitive manner. Excerpts in this chapter illustrate how researchers sometimes forget to use the word *estimated* prior to the phrase *standard error.* Keep in mind that the researcher never knows for sure, based on the sample data, how large the SE is; it can only be estimated.

The third warning concerns, once again, CIs. The sample statistic, of course, is always located between the upper and lower limits of the CI—but it is *not* always be located halfway between the interval's end points. When CIs are built around a sample mean, it is true that M turns out to be positioned at the midpoint of the interval. When CIs are constructed for many other statistics, however, one "side" of the interval is longer than the other side.[7] We saw an example of this in Excerpt 6.6, where a CI was built around a correlation coefficient. Whenever a CI is built around a proportion (or percentage), the same thing happens unless the value of the statistic is .50 (i.e., 50 percent).

My final warning applies to both interval estimation and point estimation—and this is by far the most important of my end-of-chapter cautionary comments. Simply stated, the entire process of estimation requires that the data used to form the inference come from a *random* sample. For the techniques of estimation to work properly, therefore, there must be a legitimate connection between the sample and

---

[7]The degree to which such CIs appear to be lopsided is inversely related to sample size. If *n* is large enough, the statistic will be positioned in the middle of the interval.

population such that either (1) the former is actually extracted, randomly, from the latter (with no refusals to participate, attrition, or response rate problems); or (2) the population, if hypothetical, is conceptualized so as to match closely the nature of the sample. Without a strong link between sample and population, neither form of estimation can be expected to function very well.

Review Terms

Confidence interval	Sampling distribution
Estimation	Sampling error
Interval estimation	Standard error
Point estimation	

The Best Items in the Companion Website

1. An interactive online quiz (with immediate feedback provided) covering Chapter 6.
2. Ten misconceptions about the content of Chapter 6.
3. An online resource entitled "Sampling Distributions."
4. An email message about a "dead even" political race sent from the author to his students.
5. Two jokes, one about probability and the other about statisticians screwing in light bulbs.

To access the chapter outline, practice tests, weblinks, and flashcards, visit the companion website at http://www.ReadingStats.com.

Review Questions and Answers begin on page 531.

Hypothesis Testing

In Chapter 6, we saw how the inferential techniques of estimation can assist researchers when they use sample data to make educated guesses about the unknown values of population parameters. Now, we turn our attention to a second way in which researchers engage in inferential thinking. This procedure is called **hypothesis testing**.

Before we examine the half-dozen elements of hypothesis testing, let me reiterate something I said near the beginning of Chapter 5. In order for inferential statistics to begin, the researcher must first answer four preliminary questions: (1) What is/are the relevant population(s)? (2) How will a sample be extracted from the population(s) of interest? (3) What characteristic(s) of the sample people, animals, or objects will serve as the target of the measurement process? (4) What is the study's statistical focus—or stated differently, how will the sample data be summarized so as to obtain a statistic that can be used to make an inferential statement concerning the unknown parameter? In this chapter, I assume that these four questions have been both raised and answered by the time the researcher starts to apply the hypothesis testing procedure.

To help you understand the six-step version of hypothesis testing, I first simply list the various steps in their proper order (i.e., the order in which a researcher ought to do things when engaged in this form of statistical inference). After presenting an ordered list of the six steps, I then discuss the function and logic of each step.

An Ordered List of the Six Steps

Whenever researchers use the six-step version of the hypothesis testing procedure, they do the following:

1. State the null hypothesis.
2. State the alternative hypothesis.

3. Select a level of significance.
4. Collect and summarize the sample data.
5. Refer to a criterion for evaluating the sample evidence.
6. Make a decision to discard/retain the null hypothesis.

It should be noted that there is no version of hypothesis testing that involves fewer than six steps. Stated differently, it is outright impossible to eliminate any of these six ingredients and have enough left to test a statistical hypothesis.

A Detailed Look at Each of the Six Steps

As indicated previously, the list of steps just presented is arranged in an ordered fashion. In discussing these steps, however, we now look at these six component parts in a somewhat jumbled order: 1, 6, 2, 4, 5, and then 3. My motivation in doing this is not related to sadistic tendencies! Rather, I am convinced that the function and logic of these six steps can be understood far more readily if we purposely chart an unusual path through the hypothesis testing procedure. Please note, however, that the six steps are rearranged here for pedagogical reasons only. If I were asked to apply these six steps in an actual study, I would use the ordered list as my guide, not the sequence to which we now turn.

Step 1: The Null Hypothesis

When engaged in hypothesis testing, a researcher begins by stating a **null hypothesis**. If there is just one population involved in the study, the null hypothesis is a pinpoint statement as to the unknown quantitative value of the parameter in the population of interest. To illustrate what this kind of null hypothesis might look like, suppose that (1) we conduct a study in which our population contains all full-time students enrolled in a particular university, (2) our variable of interest is intelligence, and (3) our statistical focus is the mean IQ score. Given this situation, we could set up a null hypothesis to say that $\mu = 100$. This statement deals with a population *parameter*, it is *pinpoint* in nature, and *we* made it.

The symbol for null hypothesis is H_0, and is usually followed by (1) a colon, (2) the parameter symbol that indicates the researcher's statistical focus, (3) an equal sign, and (4) the pinpoint numerical value that the researcher has selected. Accordingly, we specify the null hypothesis for our imaginary study by stating H_0: $\mu = 100$.

If our study's statistical focus involves something other than the mean, we must change the parameter's symbol to make H_0 consistent with the study's focus. For example, if our imaginary study is concerned with the variance among students' heights, the null hypothesis must contain the symbol σ^2 rather than the symbol μ. Or, if we are concerned with the product–moment correlation between the students' heights and weights, the symbol ρ must appear in H_0.

With respect to the pinpoint numerical value that appears in the null hypothesis, researchers have the freedom to select any value that they wish to test. Thus, in our example dealing with the mean IQ of university students, the null hypothesis could be set up to say that $\mu = 80$, $\mu = 118$, $\mu = 101$, or $\mu =$ any specific value of our choosing. Likewise, if our study focuses on the variance, we could set up H_0, the null hypothesis, to say that $\sigma^2 = 10$ or that $\sigma^2 =$ any other positive number of our choosing. And in a study having Pearson's product–moment correlation coefficient as its statistical focus, the null hypothesis could be set up to say that $\rho = 0.00$ or that $\rho = -.50$ or that $\rho = +.92$ or that $\rho =$ any specific number between -1.00 and $+1.00$.

The only statistical restrictions on the numerical value that appears in H_0 are that it (1) must lie somewhere on the continuum of possible values that correspond to the parameter and (2) cannot be fixed at the upper or lower limit of that continuum, presuming that the parameter has a lowest or highest possible value. These restrictions rule out the following null hypotheses:

$$H_0: \sigma^2 = -15 \qquad H_0: \rho = +1.30$$
$$H_0: \sigma^2 = 0 \qquad H_0: \rho = -1.00$$

because the variance has a lower limit of 0 whereas Pearson's product–moment correlation coefficient has limits of ± 1.00.

Excerpts 7.1 and 7.2 show how researchers sometimes talk about their null hypotheses. In the first of these excerpts, the statistical focus of the null hypothesis is the mean, as is made clear by the inclusion of the symbol μ. As you can see, there are two μs in this null hypothesis, because there are two populations involved in this study, boys and girls. The symbol μ, of course, corresponds to the mean score

EXCERPTS 7.1–7.2 • *The Null Hypothesis*

$H_0: \mu_1 - \mu_2 = 0$, the null hypothesis [is] that the population mean of boys is equal to the population mean of the girls on valuing of reading.

Source: Sturtevant, E. G., & Kim, G. S. (2010). Literacy motivation and school/non-school literacies among students enrolled in a middle-school ESOL program. *Literacy Research and Instruction, 49*(1), 68–85.

To compare whether there is a significant correlation between age and the anthropometric dimensions of the Malaysian elderly, [we set up] the null-hypothesis (H_0: $\rho = 0$). . . .

Source: Rosnah, M. Y., Mohd Rizal, H., & Sharifah-Norazizan, S. A. R. (2009). Anthropometry dimensions of older Malaysians: Comparison of age, gender and ethnicity. *Asian Social Science, 5*(6), 133–140.

on some variable of interest. As indicated in the excerpt, the researchers wanted to compare the two groups in terms of how much they valued reading.

Previously, I indicated that every null hypothesis must contain a pinpoint numerical value. From what is stated in Excerpt 7.1, it is clear that the pinpoint number in this excerpt's H_0 is zero. This pinpoint number would still be in the null hypothesis (but slightly hidden from view) if the researchers had said $H_0: \mu_1 = \mu_2$. If two things are equal, there is no difference between them, and the notion of no difference is equivalent to saying that a zero difference exists.

Although researchers have the freedom to select any pinpoint number they wish for H_0, a zero is often selected when the samples from two populations are being compared. When this is done, the null hypothesis becomes a statement that there is no difference between the populations. Because of the popularity of this kind of null hypothesis, people sometimes begin to think that a null hypothesis *must* be set up as a "no difference" statement. This is both unfortunate and wrong. When two populations are compared, the null hypothesis can be set up with any pinpoint value the researcher wishes to use. (For example, in comparing the mean height of men and women, we could set up a legitimate null hypothesis that stated $H_0: \mu_{men} - \mu_{women} = 2$ inches.) When the hypothesis testing procedure is used with a single population, the notion of "no difference," applied to parameters, simply does not make sense. How could there be a difference, zero or otherwise, when there is only one μ (or only one ρ, or only one σ^2, etc.)?

Excerpt 7.2 contains a null hypothesis that involves a correlation. The symbol ρ in this H_0 represents the Pearson product–moment correlation in the study's population. As you can see, ρ is set equal to zero in this null hypothesis. Theoretically, the value of ρ in H_0 could have been set equal to any pinpoint number, such as $+.20$, $-.55$, or any other value between $+1.00$ and -1.00. However, it is almost always the case that researchers set ρ equal to 0.00 in H_0 when using the hypothesis testing procedure in a correlational study.

In Excerpts 7.3 and 7.4, we see two additional null hypotheses. In the first of these excerpts, the null hypothesis stated that two percentages were equal.[1] Because of the wording in this excerpt, you might think that this H_0 stipulates that the percentage of intangible outputs from the 10 members of the T work group is identical to

EXCERPTS 7.3–7.4 • *Two Additional Null Hypotheses*

This research utilized a set of subjects, 19 in total (ten from T-work group and nine from the WT-work group). . . . $H_0: P_{wto} = P_{to}$, this means the percentage of "intangible" outputs by WT-work is equal to the percentage of "intangible" outputs by T-work.

Source: Waters, N. M., & Beruvides, M. G. (2009). An empirical study analyzing traditional work schemes versus work teams. *Engineering Management Journal, 21*(4), 36–43.

---

[1]In Chapter 17, we consider in depth statistical tests that focus on percentages.

EXCERPTS 7.3–7.4 • (*continued*)

More specifically, the article explores the perceived influence of adolescents on the purchase of various product groups across four family communication types, namely laissez-faire, protective, pluralistic, and consensual families. The following null hypothesis was formulated for the purposes of the study:

$$H_0: \mu_1 = \mu_2 = \mu_3 = \mu_4.$$

Source: Tustin, D. (2009). Exploring the perceived influence of South African adolescents on product purchases by family communication type. *Communicatio: South African Journal for Communication Theory & Research*, *35*(1), 165–183.

the percentage of such outputs from the 9 members of the WT work group. This is an incorrect conceptualization of this study's H_0, because the two percentages represented in the null hypothesis refer to the *population* of people in T work groups and the *population* of people in WT work groups. Without exception, null hypotheses always are focused on populations (and this is true for studies that focus on means, correlations, percentages, or anything else).

Excerpt 7.4 shows a null hypothesis that involves four population means. The μs in this null hypothesis correspond to the means of the four populations represented by the different kinds of families involved in this study. In the stated H_0, μ_1 is connected to families with a laissez-faire style of communication, μ_2 is connected to families having a protective style of communication, and so on. The data of the study came from the parents in each of the four samples who were asked to indicate how much influence their adolescent children had on their (the parents') decisions to purchase various items (e.g., cell phones, fast food, clothing).

Before we leave our discussion of the null hypothesis, it should be noted that H_0 does *not* always represent the researcher's personal belief, or hunch, as to the true state of affairs in the population(s) of interest. In fact, the vast majority of null hypotheses are set up by researchers in such a way as to *disagree* with what they actually believe to be the case. We return to this point later in the chapter; for now, however, I want to alert you to the fact that the H_0 associated with any given study probably is *not* an articulation of the researcher's personal belief concerning the involved population(s).

Step 6: The Decision Regarding H_0

At the end of the hypothesis testing procedure, the researcher does one of two things with H_0. One option is for the researcher to take the position that the null hypothesis is probably false. In this case, the researcher **rejects** H_0. The other option available to the researcher is to refrain from asserting that H_0 is probably false. In this case, a **fail-to-reject** decision is made.

If, at the end of the hypothesis testing procedure, a conclusion is reached that H_0 is probably false, the researcher communicates this decision by saying one of four things: (1) H_0 was rejected, (2) a statistically significant finding was obtained, (3) a **reliable difference** was observed, or (4) p is less than a small decimal value (e.g., $p < .05$). In Excerpts 7.5 through 7.7, we see examples of how researchers sometimes communicate their decision to disbelieve H_0.

EXCERPTS 7.5–7.7 • *Rejecting the Null Hypothesis*

The authors were able to reject the null hypothesis that the program would have no effect on knowledge.

Source: Rethlefsen, M. L., Piorun, M., & Prince, D. (2009). Teaching Web 2.0 technologies using Web 2.0 technologies. *Journal of the Medical Library Association*, 97(4), 253–259.

ESPN Internet articles included a significantly higher proportion of descriptors about the positive skill level/accomplishments and family roles/personal relationships than CBS SportsLine articles.

Source: Kian, E. T. M., Mondello, M., & Vincent, J. (2009). ESPN—The women's sports network? A content analysis of Internet coverage of March Madness. *Journal of Broadcasting & Electronic Media*, 53(3), 477–495.

Participants generated more original analogies of time following exposure to dual cultures or a fusion culture (vs. control) ($t = 2.08, p < .05$).

Source: Leung, A. K., & Chiu, C. (2010). Multicultural experience, idea receptiveness, and creativity. *Journal of Cross-Cultural Psychology, 41*(5–6), 723–741.

Just as there are different ways for a researcher to tell us that H_0 is considered to be false, there are various mechanisms for expressing the other possible decision concerning the null hypothesis. Instead of saying that a fail-to-reject decision has been reached, the researcher may tell us (1) H_0 was tenable, (2) H_0 was **accepted**, (3) no reliable differences were observed, (4) no significant difference was found, (5) the result was not significant (often abbreviated as *ns* or *NS*), or (6) p is greater than a small decimal value (e.g., $p > .05$). Excerpts 7.8 through 7.10 illustrate these different ways of communicating a fail-to-reject decision.

EXCERPTS 7.8–7.10 • *Failing to Reject the Null Hypothesis*

Hence, this null hypothesis was accepted.

Source: Vinodh, S., Sundararaj, G., & Devadasan, S. R. (2010). Measuring organisational agility before and after implementation of TADS. *International Journal of Advanced Manufacturing Technology, 47*(5–8), 809–818.

EXCERPTS 7.8–7.10 • (*continued*)

The male participants were evenly split, with 51% choosing the true crime book and 49% choosing the war book, $(1, N = 259) = 0.04$, *ns*.

Source: Vicary, A. M., & Fraley, R. C. (2010). Captured by true crime: Why Are women drawn to tales of rape, murder, and serial killers? *Social Psychological and Personality Science*, *1*(1), 81–86.

No significant age variance was found between Jewish and Muslim participants $(t(215) = 1.89, p > .05)$.

Source: Winstok, Z. (2010). The effect of social and situational factors on the intended response to aggression among adolescents. *Journal of Social Psychology*, *150*(1), 57–76.

It is especially important to be able to decipher the language and notation used by researchers to indicate the decision made concerning H_0, because most researchers neither articulate their null hypotheses nor clearly state that they used the hypothesis testing procedure. Often, the only way to tell that a researcher has used this kind of inferential technique is by noting what happened to the null hypothesis.

Step 2: The Alternative Hypothesis

Near the beginning of the hypothesis testing procedure, the researcher must state an **alternative hypothesis.** Referred to as H_a (or as H_1), the alternative hypothesis takes the same form as the null hypothesis. For example, if the null hypothesis deals with the possible value of Pearson's product–moment correlation in a single population (e.g., $H_0: \rho = .00$), then the alternative hypothesis must also deal with the possible value of Pearson's correlation in a single population. Or, if the null hypothesis deals with the difference between the means of two populations (perhaps indicating that $\mu_1 = \mu_2$), then the alternative hypothesis must also say something about the difference between those populations' means. In general, therefore, H_a and H_0 are identical in that they must (1) deal with the same number of populations, (2) have the same statistical focus, and (3) involve the same variable(s).

The only difference between the null and alternative hypothesis is that the possible value of the population parameter included within H_a always differs from what is specified in H_0. If the null hypothesis is set up to say $H_0: \rho = .00$, then the alternative hypothesis might be set up to say $H_a: \rho \neq .00$, or, if a researcher specifies in Step 1 that $H_0: \mu_1 = \mu_2$, we might find that the alternative hypothesis is set up to say $H_a: \mu_1 \neq \mu_2$.

Excerpt 7.11 contains an alternative hypothesis, labeled H_1, as well as the null hypothesis with which it was paired. Notice that both H_0 and H_1 deal with the same two populations and have the same statistical focus (a percentage). The null

EXCERPT 7.11 • *The Alternative Hypothesis*

The null and alternative hypotheses are as follows:

$$H_0: p_1 - p_2 = 0$$

and

$$H_1: p_1 - p_2 \neq 0$$

where p_1 is the population proportion of administrators who select a certain outcome, and p_2 is the population proportion of school social workers who also select that outcome (for example, school social work services lead to increased attendance).

Source: Bye, L., Shepard, M., Patridge, J., & Alvarez, M. (2009). School social work outcomes: Perspectives of school social worker and school administrators. *Children & Schools, 31*(2), 97–108.

---

hypothesis states that the two populations—administrators and social workers—are identical in the proportion of the population choosing a particular outcome. The alternative hypothesis states the two populations are not identical.

As indicated in the previous section, the hypothesis testing procedure terminates (in Step 6) with a decision to either reject or fail to reject the null hypothesis. In the event that H_0 is rejected, H_a represents the state of affairs that the researcher considers to be probable. In other words, H_0 and H_a always represent two opposing statements as to the possible value of the parameter in the population(s) of interest. If, in Step 6, H_0 is rejected, then belief shifts *from H_0 to H_a*. Stated differently, if a reject decision is made at the end of the hypothesis testing procedure, the researcher will reject H_0 *in favor of* H_a.

Although researchers have flexibility in the way they set up alternative hypotheses, they normally will set up H_a either in a **directional** fashion or in a **nondirectional** fashion.[2] To clarify the distinction between these options for the alternative hypothesis, let's imagine that a researcher conducts a study to compare men and women in terms of intelligence. Further suppose that the statistical focus of this hypothetical study is on the mean, with the null hypothesis asserting that $H_0: \mu_{men} = \mu_{women}$. Now, if the alternative hypothesis is set up in a nondirectional fashion, the researcher simply states $H_a: \mu_{men} \neq \mu_{women}$. If, however, the alternative hypothesis is stated in a directional fashion, the researcher specifies a direction in H_a. This could be done by asserting $H_a: \mu_{men} > \mu_{women}$ *or* by asserting $H_a: \mu_{men} < \mu_{women}$.

The directional/nondirectional nature of H_a is highly important within the hypothesis testing procedure. The researcher must know whether H_a was set up in

---

[2]A directional H_a is occasionally referred to as a *one-sided* H_a; likewise, a nondirectional H_a is sometimes referred to as a *two-sided* H_a.

a directional or nondirectional manner in order to decide whether to reject (or to fail to reject) the null hypothesis. No decision can be made about H_0 unless the directional/nondirectional character of H_a is clarified.

In most empirical studies, the alternative hypothesis is set up in a nondirectional fashion. Thus, if I were to guess what H_a says in studies containing the null hypotheses presented as shown on the left, I would bet that the researchers had set up their alternative hypotheses as indicated on the right.

Possible H_0	Corresponding nondirectional H_a
$H_0: \mu = 100$	$H_a: \mu \neq 100$
$H_0: \rho = +.20$	$H_a: \rho \neq +.20$
$H_0: \sigma^2 = 4$	$H_a: \sigma^2 \neq 4$
$H_0: \mu_1 - \mu_2 = 0$	$H_a: \mu_1 - \mu_2 \neq 0$

Researchers typically set up H_a in a nondirectional fashion because they do not know whether the pinpoint number in H_0 is too large or too small. By specifying a nondirectional H_a, the researcher permits the data to point one way or the other in the event that H_0 is rejected. Hence, in our hypothetical study comparing men and women in terms of intelligence, a nondirectional alternative hypothesis allows us to argue that μ_{women} is probably higher than μ_{men} (in the event that we reject the H_0 because $M_{women} > M_{men}$); or, such an alternative hypothesis allows us to argue that μ_{men} is probably higher than μ_{women} (if we reject H_0 because $M_{men} > M_{women}$).

Occasionally, a researcher believes so strongly (based on theoretical consideration or previous research) that the true state of affairs falls on one side of H_0 's pinpoint number that H_a is set up in a directional fashion. So long as the researcher makes this decision prior to looking at the data, such a decision is fully legitimate. It is, however, totally inappropriate for the researcher to look at the data first and then subsequently decide to set up H_a in a directional manner. Although a decision to reject or fail to reject H_0 could still be made after first examining the data and then articulating a directional H_a, such a sequence of events would sabotage the fundamental logic and practice of hypothesis testing. Simply stated, decisions concerning how to state H_a (and how to state H_0) must be made without peeking at any data.

When the alternative hypothesis is set up in a nondirectional fashion, researchers sometimes use the phrase **two-tailed test** to describe their specific application of the hypothesis testing procedure. In contrast, directional H_as lead to what researchers sometimes refer to as **one-tailed tests.** Inasmuch as researchers rarely specify the alternative hypothesis in their technical write-ups, the terms *one-tailed* and *two-tailed* help us to know exactly how H_a was set up. For example, consider Excerpts 7.12 and 7.13. Here, we see how researchers sometimes use the term *two-tailed* or *one-tailed* to communicate their decisions to set up H_a in a nondirectional or directional fashion.

EXCERPTS 7.12–7.13 • *Two-Tailed and One-Tailed Tests*

All tests of significance were two-tailed.

Source: Miller, K. (2010). Using a computer-based risk assessment tool to identify risk for chemotherapy-induced febrile neutropenia. *Clinical Journal of Oncology Nursing, 14*(1), 87–91.

To investigate what variables might be important predictors of company support for fathers taking leave, [Pearson] correlations were calculated. . . . One-tailed tests of significance were used.

Source: Haas, L., & Hwang, P. C. (2010). Is fatherhood becoming more visible at work? Trends in corporate support for fathers taking parental leave in Sweden. *Fathering: A Journal of Theory, Research, & Practice about Men as Fathers, 7*(3), 303–321.

Step 4: Collection and Analysis of Sample Data

So far, we have covered Steps 1, 2, and 6 of the hypothesis testing procedure. In the first two steps, the researcher states the null and alternative hypotheses. In Step 6, the researcher either (1) rejects H_0 in favor of H_a or (2) fails to reject H_0. We now turn our attention to the principal "stepping stone" used to move from the beginning points of the hypothesis testing procedure to the final decision.

Inasmuch as the hypothesis testing procedure is, by its very nature, an empirical strategy, it should come as no surprise that the researcher's ultimate decision to reject or to retain H_0 is based on the collection and analysis of sample data. No crystal ball is used, no Ouija board is relied on, and no eloquent argumentation is permitted. Once H_0 and H_a are fixed, only scientific evidence is allowed to affect the disposition of H_0.

The fundamental logic of the hypothesis testing procedure can now be laid bare because the connections between H_0, the data, and the final decision are as straightforward as what exists between the speed of a car, a traffic light at a busy intersection, and a lawful driver's decision as the car approaches the intersection. Just as the driver's decision to stop or to pass through the intersection is made after observing the color of the traffic light, the researcher's decision to reject or to retain H_0 is made after observing the sample data. To carry this analogy one step further, the researcher looks at the data and asks, "Is the empirical evidence inconsistent with what one would expect if H_0 were true?" If the answer to this question is yes, then the researcher has a green light and rejects H_0. However, if the data turn out to be consistent with H_0, then the data set serves as a red light telling the researcher not to discard H_0.

Because the logic of hypothesis testing is so important, let us briefly consider a hypothetical example. Suppose a valid intelligence test is given to a random sample of 100 males and a random sample of 100 females attending the same university. If the null hypothesis was first set up to say H_0: $\mu_{male} = \mu_{female}$ and if the data reveal that the two sample means (of IQ scores) differ by only two-tenths of a point, the sample data would be consistent with what we expect to happen when two samples are selected from populations having identical means. Clearly, the notion of sampling error could fully explain why the two Ms might differ by two-tenths of an IQ point even if $\mu_{male} = \mu_{female}$. In this situation, there is no empirical justification for making the data-based claim that males at our hypothetical university have a different IQ, on average, than do their female classmates.

Now, let's consider what would happen if the difference between the two sample means turns out to be equal to 20 IQ points. If the empirical evidence turns out this way, we have a situation where the data are inconsistent with what one would expect if H_0 were true. Although the concept of sampling error strongly suggests that neither sample mean will turn out exactly equal to its population parameter, the difference of 20 IQ points between M_{males} and $M_{females}$ is quite improbable if, in fact, μ_{males} and $\mu_{females}$ are equal. With results such as this, the researcher would reject the arbitrarily selected null hypothesis.

To drive home the point I am trying to make about the way the sample data influence the researcher's decision concerning H_0, let's shift our attention to a real study that had Pearson's correlation as its statistical focus. In Excerpt 7.14, the hypothesis testing procedure was used to evaluate three bivariate correlations based on data that came from 90 men who had surgery after going to an infertility clinic. Each man was measured in terms of the number of left and right spermatic arteries as well the number of left and right lymphatic channels. Then, the left-right data were correlated for each of the two kinds of arteries and for the channels.

EXCERPT 7.14 • *Rejecting H_0 When the Sample Data Are Inconsistent with H_0*

An analysis of the relationship between the right and left spermatic cord anatomy in the bilateral varicocelectomy cases ($n = 90$) revealed a significant correlation between the number of right and left internal spermatic arteries ($r = 0.42, P < .05$). However, we did not identify a significant correlation between the number of right and left external spermatic arteries ($r = 0.13, P > .05$) or the number of right and left lymphatic channels ($r = 0.19, P > .05$).

Source: Libman, J. L., Segal, R., Baazeem, A., Boman, J., & Zini, A. (2010). Microanatomy of the left and right spermatic cords at subinguinal microsurgical varicocelectomy: Comparative study of primary and redo repairs. *Urology, 75*(6), 1324–1327.

In the study associated with Excerpt 7.14, the hypothesis testing procedure was used separately to evaluate each of the three sample rs. In each case, the null hypothesis stated H_0: $\rho = 0.00$. The sample data, once analyzed, yielded correlations of .42, .13, and .19. The first of these *r*s ended up being quite different from the null hypothesis number of 0.00. Statistically speaking, the *r* of .42 was so inconsistent with H_0 that sampling error alone was considered to be an inadequate explanation for why the observed correlation was so far away from the pinpoint number in the null hypothesis. Although we expect some discrepancy between 0.00 and the data-based value of *r* even if H_0 were true, we do *not* expect this big difference. Accordingly, the null hypothesis concerning the internal spermatic arteries—that there was no relationship between the number of left and right arteries—was rejected, as indicated by the phrase *significant correlation* and the notation $P < .05$.

The second and third correlations in Excerpt 7.14 turned out to be much closer to the pinpoint number in H_0, 0.00. The small differences between the null number and the *r*s of .13 and .19 could each be explained by sampling error. In other words, if the correlation in the population were truly equal to 0.00, it would not be surprising to have a sample *r* (with $n = 90$) be anywhere between $-.20$ and $+.20$. Accordingly, the null hypotheses concerning external spermatic arteries and the lymphatic channels were not rejected, as indicated by the notation $P > .05$.

In Step 4 of the hypothesis testing procedure, the summary of the sample data always leads to a single numerical value. Being based on the data, this number is technically referred to as the **calculated value** (or the **test statistic**). Occasionally, the researcher's task in obtaining the calculated value involves nothing more than computing a value that corresponds to the study's statistical focus. This was the case in Excerpt 7.14, where the statistical focus was Pearson's correlation coefficient and where the researcher needed to do nothing more than compute a value for *r*.

In most applications of the hypothesis testing procedure, the sample data are summarized in such a way that the statistical focus becomes hidden from view. For example, consider Excerpts 7.15 and 7.16. In the first of these excerpts, the calculated

EXCERPTS 7.15–7.16 • *The Calculated Value*

[A] *t*-test found that overall satisfaction levels of male students ($M = 4.14, SD = 1.10$) were significantly higher than those of female students ($M = 3.87, SD = 1.00$), $t(225) = 13.78, p < 0.05$ (two-tailed).

Source: Kim, H., Lee, S., Goh, B., & Yuan, J. (2010). Assessing College Students' Satisfaction with University Foodservice. Proceedings of the 15th Annual Graduate Student Research Conference in Hospitality and Tourism, Washington, DC, 34–46.

There was no difference in girls' ($M = 5$ years, 8 months; $SD = 1.52$) and boys' ($M = 5$ years, 10 months; $SD = 1.68$) ages, $F(1, 114) = 0.25, p > 05$.

Source: Tenenbaum, T. R., Hill, D. B., Joseph, N., & Roche, E. (2010). "It's a boy because he's painting a picture": Age differences in children's conventional and unconventional gender schemas. *British Journal of Psychology, 101*(1), 137–154.

value was labeled *t* and it turned out equal to 13.78. In Excerpt 7.16, the calculated value was *F*, and it was equal to 0.25. In each of these excerpts, the statistical focus was the mean.

In each of these excerpts, two sample means were compared. In Excerpt 7.15, the mean of 4.14 was compared against the mean of 3.87. In Excerpt 7.16, the means were 5 years, 8 months and 5 years, 10 months. Within each of these studies, the researchers put their sample data into a formula that produced the calculated value. The important thing to notice in these excerpts is that in neither case does the calculated value equal the difference between the two means being compared. In Chapter 10, we consider *t*-tests and *F*-tests in more detail, so you should not worry now if you do not currently comprehend everything that is presented in these excerpts. They are shown solely to illustrate the typical situation in which the statistical focus of a study is *not* reflected directly in the calculated value.

Before computers were invented, researchers always had a single goal in mind when they turned to Step 4 of the hypothesis testing procedure: the computation of the data-based calculated value. Now that computers are widely available, researchers still are interested in the magnitude of the calculated value derived from the data analysis. Contemporary researchers, however, are also interested in a second piece of information generated by the computer: the data-based *p*-value.

Whenever researchers use a computer to perform the data analysis, they either (1) tell the computer what the null hypothesis is going to be or (2) accept the computer's built-in default version of H_0. The researcher also specifies whether H_a is directional or nondirectional in nature. Once the computer knows what the researcher's H_0 and H_a are, it can easily analyze the sample data and compute the probability of having a data set that deviates as much or more from H_0 as does the data set being analyzed. The computer informs the researcher as to this probability by means of a statement that takes the form $p =$ ___, with the blank being filled by a single decimal value somewhere between 0 and 1.

Excerpt 7.17 illustrates nicely how a *p*-value is like a calculated value in that either one can be used as a single-number summary of the sample data. As you can see, three Pearson correlation coefficients are in this excerpt. The researchers associated

EXCERPT 7.17 • *Using p as the Calculated Value*

Correlation analyses and inspection of scatterplots between the PA composite and speech production variables showed that there was no significant relationship between PA and distortions ($r = .129, p = .429$), nor between PA and typical sound changes ($r = -.171, p = .273$). However, a significant relationship was found between PA and atypical sound changes ($r = -.362, p = .009$).

Source: Preston, J., & Louise Edwards, M. (2010). Phonological awareness and types of sound errors in preschoolers with speech sound disorders. *Journal of Speech, Language & Hearing Research, 53*(1), 44–60.

with this passage used a p-value to determine how likely it would be, assuming the null hypothesis to be true, to end up with a sample correlation as large or larger than each of their computed rs. Each p functioned as a measure of how inconsistent the sample data were compared with what would be expected to happen if H_0 were true.

Be sure to note in Excerpt 7.17 that there is an *inverse* relationship between the size of p and the degree to which the sample data deviate from the null hypothesis. The r that is furthest away from 0.00 (the pinpoint number in the null hypothesis) has the smallest p. In contrast, the smallest of the three rs has the largest p.

Step 5: The Criterion for Evaluating the Sample Evidence

After the researcher has summarized the study's data, the next task involves asking the question, "Are the sample data inconsistent with what would likely occur if the null hypothesis were true?" If the answer to this question is "Yes," then H_0 is rejected; however, a negative response to this query requires a fail-to-reject decision. Thus, as soon as the sample data can be tagged as consistent or inconsistent (with H_0), the decision in Step 6 is easily made. "But how," you might ask, "does the researcher decide which of these labels should be attached to the sample data?"

If the data from the sample(s) are in perfect agreement with the pinpoint numerical value specified in H_0, then it is obvious that the sample data are consistent with H_0. (This would be the case if the sample mean turned out equal to 100 when testing H_0: $\mu = 100$, if the sample correlation coefficient turned out equal to 0.00 when testing H_0: $\rho = 0.00$, etc.) Such a situation, however, is unlikely. There is almost always a discrepancy between H_0's parameter value and the corresponding sample statistic.

In light of the fact that the sample statistic (produced by Step 4) is almost certain to be different from H_0's pinpoint number (specified in Step 1), the concern over whether the sample data are inconsistent with H_0 actually boils down to the question, "Should the observed difference between the sample evidence and the null hypothesis be considered a big difference or a small difference?" If this difference (between the data and H_0) is judged to be large, then the sample data are looked on as being inconsistent with H_0 and, as a consequence, H_0 is rejected. If, however, this difference is judged to be small, the data and H_0 are looked on as consistent with each other and, therefore, H_0 is not rejected.

To answer the question about the sample data's being either consistent or inconsistent with what one would expect if H_0 were true, a researcher can use either of two simple procedures. Note that both of these procedures involve comparing a single-number summary of the sample evidence against a criterion number. The single-number summary of the data can be either the calculated value or the p-value. Our job now is to consider what each of these data-based indices is compared against, and what kind of result allows researchers to consider their samples to represent a large or a small deviation from H_0.

One available procedure for evaluating the sample data involves comparing the calculated value against something called the **critical value.** The critical value is nothing more than a number extracted from one of many statistical tables developed by mathematical statisticians. Applied researchers, of course, do not close their eyes and point to just any entry in a randomly selected table of critical values. Instead, they must learn which table of critical values is appropriate for their studies and also how to locate the single number within the table that constitutes the correct critical value.

As a reader of research reports, you do not have to learn how to locate the proper table that contains the critical value for any given statistical test, nor do you have to locate, within the table, the single number that allows the sample data to be labeled as being consistent or inconsistent with H_0. The researcher does these things. Occasionally, the critical value is included in the research report, as exemplified in Excerpts 7.18 and 7.19.

EXCERPTS 7.18–7.19 • *The Critical Value and the Decision Rule*

The observed value of 63.22 is greater than the critical *t* value of 1.96, and this is significant at a 0.05 level; hence the rejection of the null hypothesis.

Source: Oluwole, D. A. (2009). Spirituality, gender and age factors in cybergossip among Nigerian adolescents. *CyberPsychology & Behavior, 12*(3), 323–326.

[T]he calculated chi square (χ^2) value of 44.35 was greater than the critical χ^2 value of 9.49 at 0.05 level of significance with 4 degrees of freedom. This means that there is a significance relationship between the marital status of the nurses in Akwa-Ibom State and their being obese.

Source: Ogunjimi, L. O., Maria M. Ikorok, M. M., & Yusuf, O. O. (2010). Prevalence of obesity among Nigeria nurses: The Akwa Ibom State experience. *International NGO Journal, 5*(2), 045–049.

Once the critical value is located, the researcher compares the data-based summary of the sample data against the scientific dividing line that has been extracted from a statistical table. The simple question being asked at this point is whether the calculated value is larger or smaller than the critical value. With most tests (such as *t*, *F*, chi-square, and tests of correlation coefficients), the researcher follows a decision rule that says to reject H_0 if the calculated value is at least as large as the critical value. With a few tests (such as *U* or *W*), the decision rule tells the researcher to reject H_0 if the calculated value is smaller than the critical value. You do not need to worry about which way the decision rule works for any given test, because this is the responsibility of the individual who performs the data analysis.

The only things you must know about the comparison of calculated and critical values are that (1) this comparison allows the researcher to decide easily whether to reject or fail to reject H_0; and (2) some tests use a decision rule that says to reject H_0 if the calculated value is larger than the critical value, whereas other tests involve a decision rule that says to reject H_0 if the calculated value is smaller than the critical value.

The researchers associated with Excerpts 7.18 and 7.19 helped the readers of their research reports by specifying not only the critical value but also the nature of the decision rule that was used when the calculated value was compared against the critical value. In most research reports, you see neither of these things; instead, you are given only the calculated value. (On rare occasions, you do not even see the calculated value.) As indicated previously, however, you should not be concerned about this, because it is the researcher's responsibility to obtain the critical value and to know which way the decision rule operates. When reading most research reports, all you can do is trust that the researcher did these two things properly.

The second way a researcher can evaluate the sample evidence is to compare the data-based p-value against a preset point on the 0-to-1 scale on which the p must fall. This criterion is called the **level of significance,** and it functions much as does the critical value in the first procedure for evaluating sample evidence. Simply stated, the researcher compares his or her data-based p-value against the criterion point along the 0-to-1 continuum so as to decide whether the sample evidence ought to be considered consistent or inconsistent with H_0. The decision rule used in this second procedure is always the same: If the data-based p-value is equal to or smaller than the criterion, the sample is viewed as being *in*consistent with H_0; however, if p is larger than the criterion, the data are looked on as being consistent with H_0.

Excerpt 7.20 exemplifies the use of this second kind of criterion for evaluating sample evidence. Note that the data-based p-value of 0.006 was substantially smaller than the criterion number of 0.05. Accordingly, the null hypothesis (that the populations of men and women have equal levels of trust) was rejected.

EXCERPT 7.20 • *Comparing p to α*

A resultant p-value of 0.006 was compared to a significance level of 0.05 indicated that overall, there is a difference in trust levels for the males and females who participated in the study, with males generally having the higher trust in their organizations.

Source: Alston, F., & Tippett, D. (2009). Does a technology-driven organization's culture influence the trust employees have in their managers? *Engineering Management Journal, 21*(2), 1–10.

I discuss the level of significance in more depth in the next section, because it is a concept that must be dealt with by the researcher no matter which of the two procedures is used to evaluate the sample data. (With the second procedure, the level of significance *is* the criterion against which the data-based *p*-value is compared; with the first procedure, the level of significance influences the size of the critical value against which the calculated value is compared.) Before we leave this section, however, I must point out that the same decision is reached regarding H_0 no matter which of the two procedures is used in Step 5 of the hypothesis testing procedure. For example, suppose a researcher conducts an *F*-test and rejects H_0 because the calculated value is larger than the critical value. If that researcher were to compare the data-based *p* against the level of significance, it would be found that the former is smaller than the latter, and the same decision about H_0 would be made. Or, suppose a researcher conducts a *t*-test and fails to reject H_0 because the calculated value is smaller than the critical value. If that researcher were to compare the data-based *p* against the level of significance, it would be found that the former is larger than the latter, and the same fail-to-reject decision would be made.

Step 3: Selecting a Level of Significance

After the data of a study are collected and summarized, the six-step hypothesis testing procedure allows absolutely no subjectivity to influence, or bias, the ultimate decision that is made concerning the null hypothesis. This goal is accomplished by reliance on a scientific cutoff point to determine whether the sample data are consistent or inconsistent with H_0. By referring (in Step 5) to a numerical criterion, it becomes clear whether sampling error provides, by itself, a sufficient explanation for the observed difference between the single-number summary of the researcher's data (computed in Step 4) and H_0's pinpoint numerical value (articulated in Step 1). If the single-number summary of the data is found to lie on H_a's side of the criterion number (or if the data-based *p* lands on H_a's side of the level of significance), a decision (in Step 6) is made to reject H_0 in favor of H_a (set forth in Step 2); however, if the calculated value lands on H_0 's side of the critical value (or if the data-based *p* lands on H_0's side of the level of significance), a fail-to-reject decision is made.

Either the critical value or the level of significance serves as a scientific cutoff point that determines what decision will be made concerning the null hypothesis. The six-step hypothesis testing procedure not only allows the researcher to do something that affects the magnitude of this criterion—*it actually forces the researcher to become involved in determining how rigorous the criterion will be*. The researcher should not, as I have pointed out, do anything like this after the data have been collected and summarized. However, the researcher *must* do something prior to collecting data that has an impact on how large or small the criterion number will be.

After the null and alternative hypotheses have been set up, but before any data are collected, the researcher must select a level of significance. This third step of the hypothesis testing procedure simply asks the researcher to select a positive decimal value of the researcher's choosing. Although the researcher has the freedom to select any value between 0 and 1 for the level of significance, most researchers select a small number, such as .10, .05, or .01. The most frequently selected number is .05.

Before explaining how the researcher-selected level of significance influences the size of the critical value, I must alert you to the fact that not all researchers use the phrase *level of significance* to designate the decimal number that must be specified in Step 3. Instead of indicating, for example, that the level of significance is set equal to .05, some researchers state that "the **alpha level** (α) is set equal to .05," others assert that "$p = .05$," and still others indicate that "H_0 will be rejected if $p < .05$." Likewise, a decision to use the .01 level of significance might be expressed using statements such as "$\alpha = .01$," "$\alpha = .01$," or "results will be considered significant if $p < .01$."

In Excerpts 7.21 through 7.23, we see different ways in which researchers report what level of significance was selected within their studies.

EXCERPTS 7.21–7.23 • *The Level of Significance*

An alpha level of .05 was used for all statistical tests.

Source: Egan, P. M., & Giuliano, T. A. (2009). Unaccommodating attitudes: Perceptions of students as a function of academic accommodation use and test performance. *North American Journal of Psychology, 11*(3), 487–500.

An α level of .05 was selected for statistical significance.

Source: Fink, A. M., Sullivan, S. L., Zerwic, J. J., & Piano, M. R. (2009). Fatigue with systolic heart failure. *Journal of Cardiovascular Nursing, 24*(5), 410–417.

Statistical significance was set at $P < 0.05$.

Source: Langhammer, B., & Stanghelle, J. K. (2010). Exercise on a treadmill or walking outdoors? A randomized controlled trial comparing effectiveness of two walking exercise programmes late after stroke. *Clinical Rehabilitation, 24*(1), 46–54.

If the single-number summary of the sample data is a *p*-value, the pragmatic value of the level of significance is clear. In this situation, *p* is compared directly against α to determine whether H_0 should be rejected. However, even if the single-number summary of the sample data is a calculated value, the level of significance still performs a valuable and pragmatic function. This is because a critical value cannot be located (in Step 5) unless the level of significance has first been set. As indicated earlier, there are many tables of critical values. Once the proper table is located, the researcher still has the task of locating, inside the table, the single number that will serve as the critical value. The task of locating the critical value is easy, so long as the level of significance has been specified.[3]

[3]With certain tests, researchers cannot locate the critical value unless they also know how many *degrees of freedom* are connected with the sample data, a concept I discuss in several chapters, beginning with Chapter 10.

Although the level of significance plays an important pragmatic role within the six-step hypothesis testing procedure, it is even more important from a different perspective. When I introduce the concept of the null hypothesis and when I talk about the reject or fail-to-reject decision that researchers make regarding the null hypothesis, I am careful to use language that does *not* suggest that H_0 is ever *proved* to be true or false by means of hypothesis testing. Regardless of the decision made about H_0 after the p and α (or the calculated and critical values) are compared, it is possible that the wrong decision will be reached. If H_0 is rejected in Step 6, it is conceivable that this action represents a mistake, because H_0 may actually be true. Or, if H_0 is not rejected, it is conceivable that *this* action represents a mistake, because H_0 may actually be an inaccurate statement about the value of the parameter in the population(s).

In light of the fact that a mistake can conceivably occur regardless of what decision is made at the end of the hypothesis testing procedure, two technical terms have been coined to distinguish between these potentially wrong decisions. A **Type I error** designates the mistake of rejecting H_0 when the null hypothesis is actually true. A **Type II error,** however, designates the kind of mistake that is made if H_0 is not rejected when the null hypothesis is actually false. The following chart may help to clarify the meaning of these possible errors.

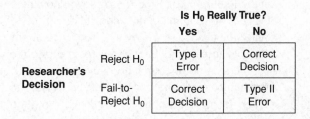

Beyond its pragmatic utility in helping the researcher locate the critical value (or in serving as the criterion against which the data-based p is compared), the level of significance is important because it establishes the probability of a Type I error. In other words, the selected alpha level determines the likelihood that a true null hypothesis will be rejected. If the researcher specifies, in Step 3, that $\alpha = .05$, then the chances of rejecting a true null hypothesis become equal to 5 out of 100. If, however, the alpha level is set equal to .01 (rather than .05), then the chances of rejecting a true null hypothesis become equal to 1 out of 100. The alpha level, therefore, directly determines the probability that a Type I error will be committed.[4]

[4]As discussed in Chapter 8, Chapter 9, and several other chapters, the alpha level defines the probability of a Type I error only if (1) important assumptions underlying the statistical test are valid and (2) the hypothesis testing procedure is used to evaluate only *one* null hypothesis.

After realizing that the researcher can fully control the likelihood of a Type I error, you may be wondering why the researcher does not select an alpha level that would dramatically reduce the possibility that a true H_0 will be rejected. To be more specific, you may be inclined to ask why the alpha level is not set equal to .001 (where the chance of a Type I error becomes equal to 1 out of 1,000), equal to .00001 (where the chance of Type I error becomes equal to 1 out of 100,000), or even equal to some smaller decimal value. To answer this legitimate question, we must consider the way in which a change in the alpha level has an effect on both Type I error risk *and* Type II error risk.

If the alpha level is changed, it is as if there is an apothecary scale in which the two pans hanging from opposite ends of the balance beam contain, respectively, Type I error risk and Type II error risk. The alpha level of a study could be changed so as to decrease the likelihood of a Type I error, but this change in alpha simultaneously has an opposite effect on the likelihood of a Type II error. Hence, researchers rarely move alpha from the more traditional level of .05 to levels that would greatly protect against Type I errors (such as .0001) because such a change in the alpha level serves to make the chances of a Type II error unacceptably high.

In light of the fact that the typical researcher likes to reject H_0 to gain empirical support for his or her hunch (that corresponds with the alternative hypothesis), and in light of the fact that a change in the level of significance has an impact on the likelihood of Type II errors, you now may be wondering why the researcher does not move alpha in the opposite direction. It is true that a researcher would decrease the chance of a Type II error by changing alpha—for example, from .05 to .40—because such a change makes it more likely that H_0 would be rejected. Researchers do not use such high levels of significance simply because the scientific community generally considers Type I errors to be more dangerous than Type II errors. In most disciplines, few people pay attention to researchers who reject null hypotheses at alpha levels higher than .20, because such levels of significance are considered to be too lenient (i.e., too likely to yield reject decisions that are Type I errors).

The most frequently seen level of significance, as illustrated in several of the excerpts we have just considered, is .05. This alpha level is considered to represent a happy medium between the two error possibilities associated with any application of the six-step hypothesis testing procedure. If, however, a researcher believes that it is more important to guard against the possibility of a Type I error, a lower alpha level (such as .01 or .001) is selected. On the other hand, if it is believed that a Type II error is more dangerous than a Type I error, then a higher alpha level (such as .10 or .15) is selected. Excerpts 7.24 and 7.25 illustrate how (and why) researchers sometimes set alpha equal to something other than .05. In Excerpt 7.24, the researchers wanted to guard against making Type II errors, so they set the level of significance equal to .10 rather than .05. In contrast, the researchers in Excerpt 7.24 wanted to guard against making Type I error, so they changed the level of significance from .05 to .01.

EXCERPTS 7.24–7.25 • *Reasons for Using Alpha Levels Other Than .05*

Significance was evaluated using $\alpha = 0.10$. . . . We used a larger than customary level of significance to reduce the likelihood of a Type II error (i.e., not detecting real differences), which we felt could pose more risk to murrelet management than the occurrence of a Type I error.

Source: Waterhouse, F. L., Burger, A. E., Lank, D. B., Ott, P. K., Krebs, E. A., & Parker, N. (2009). Using the low-level aerial survey method to identify Marbled Murrelet nesting habitat. *BC Journal of Ecosystems and Management, 10*(1), 80–96.

For the cross-sectional analysis, analyses of variance (ANOVAs) were used to examine group differences in continuous variables using a conservative threshold of $p < 0.01$ to control for Type I error.

Source: Drake, A. S., Weinstock-Guttman, B., Morrow, S. A., Hojnacki, D., Munschauer, F. E., & Benedict, R. H. B. (2010). Psychometrics and normative data for the Multiple Sclerosis Functional Composite: Replacing the PASAT with the Symbol Digit Modalities Test. *Multiple Sclerosis, 16*(2), 228–237.

Before concluding our discussion of the level of significance, I must clarify two points of potential confusion. To accomplish this goal, I want to raise and then answer two questions: (1) "Does the alpha level somehow determine the likelihood of a Type II error?" and (2) "If H_0 is rejected, does the alpha level indicate the probability that H_0 is true?"

The first point of potential confusion concerns the relationship between alpha and Type II error risk. Because alpha does, in fact, determine the likelihood that the researcher will end up rejecting a true H_0, and because it is true that a change in alpha affects the chance of a Type I error *and* the chance of a Type II error (with one increasing, the other decreasing), you may be tempted to expect the level of significance to dictate Type II error risk. Unfortunately, this is not the case. The alpha level specified in Step 3 does influence Type II error risk, but so do other features of a study such as sample size, population variability, and the reliability of the measuring instrument used to collect data.

The second point of potential confusion about the alpha level again concerns the decision reached at the end of the hypothesis testing procedure. If a study's H_0 is rejected in Step 6, it is *not* proper to look back to see what alpha level was specified in Step 3 and then interpret that alpha level as indicating the probability that H_0 is true. For example, if a researcher ends up rejecting H_0 after having set the level of significance equal to .05, you cannot legitimately conclude that the chances of H_0 being true are less than 5 out of 100. The alpha level in any study indicates only what the chances are that the forthcoming decision will be a Type I error. If alpha is set equal to .05, then the chances are 5 out of 100 that H_0 will be rejected *if H_0 is actually true*. Statisticians sometimes try to clarify this distinction by pointing out that the level of significance specifies "the probability of a reject decision, given a true H_0" and *not* "the probability of H_0 being true, given a reject decision."

Results That Are Highly Significant or Near Misses

As indicated earlier, the level of significance plays a highly important role in hypothesis testing. In a very real sense, it functions as a dividing line. Statistical significance is positioned on one side of that line, the lack of statistical significance on the other. That dividing line is clearly visible if the researcher decides to reject or fail to reject H_0 by comparing the data-based p against the level of significance. Yet even when the procedure for deciding H_0's fate involves comparing the data-based calculated value against a tabled critical value, the level of significance is still involved, because α influences the size of the critical value.

Because the level of significance plays such an important role—both pragmatically and conceptually—in hypothesis testing, it often is included when the decision about H_0 is declared. With the level of significance set at .05 (the most popular α-level), a decision to reject H_0 is often summarized by the notation $p < .05$, whereas a decision not to reject H_0 is summarized by the notation $p > .05$. Several of these notational summaries were presented earlier in this chapter.

Many researchers do not like to summarize their results by reporting simply that the null hypothesis either was or was not rejected. Instead, they want their readers to know how much of a discrepancy existed between the data-based p and the level of significance (or between the data-based calculated value and the critical value). In doing this, the researcher's goal is to provide evidence as to how strongly the data challenge H_0. In other words, these researchers want you to know if they beat the level of significance by a wide margin (presuming that H_0 was rejected) or if they just missed beating α (presuming that H_0 was retained).

Consider Excerpts 7.26 and 7.27. Notice the phrase **highly significant** that appears in Excerpt 7.27.

EXCERPTS 7.26–7.27 • *Rejecting the Null Hypothesis with Room to Spare*

The time that the paper disc was in the mouth/stomach of the fish significantly differed for the four conditions (Friedman ANOVA, $\chi^2[0.05, 3] = 67.955$, $n = 39 fish, p < 0.00001$).

Source: Wood, J. B., Maynard, A. E., Lawlor, A. G., Sawyer, E. K., Simmons, D. M., & Pennoyer, K. E., et al. (2010). Caribbean reef squid, *Sepioteuthis sepioidea*, use ink as a defense against predatory French grunts, *Haemulon flavolineatum*. *Journal of Experimental Marine Biology and Ecology, 388*(1–2), 20–27.

EXCERPTS 7.26–7.27 • (*continued*)

The correlation was strongly positive ($r = 0.95$) and highly significant ($p < 0.00000001$).

Source: Sripatil, A. P., & Olson, C. R. (2010). Global image dissimilarity in macaque inferotemporal cortex predicts human visual search efficiency. *Journal of Neuroscience, 30*(4), 1258–1269.

Although *p*-values like those shown in Excerpts 7.26 and 7.27 are not seen very often in research reports, I can assure you that you will frequently encounter *p*-less-than statements where the numerical value is smaller than .05. You will regularly see $p < .01$, you will come across $p < .001$ quite often, and you will see $p < .0001$ every now and then. Such statements do *not* indicate that the researcher initially set the level of significance equal to .01, .001, or .0001. Most likely, *p*-statements such as these come from studies in which α was set equal to .05. (This was the case in the studies from which Excerpts 7.26 and 7.27 were taken.) There is a simple reason why *p*-less-than statements often contain a number other than the level of significance.

Many researchers use an approach to hypothesis testing that involves reporting the most impressive *p*-statement that honestly describes their data. They first check to see if they have statistical significance at the .05 level. If they do, then they know they at least can say $p < .05$. They next check to see if the sample data would have been significant at the .01 level, had this been the selected alpha level. If the answer is yes, they then check again, this time to see if the data are significant at the .001 level. This process continues until either (1) the data cannot beat a more rigorous level of significance or (2) the researcher does not want to check further to see if *p* might beat an even more impressive α. It is clear that this approach to hypothesis testing was used in Excerpts 7.26 and 7.27.

Most researchers test more than one null hypothesis in the same study. In the research reports prepared by these investigators, you may see certain results summarized via the statement $p < .05$, other results summarized via the statement $p < .01$, and still other results summarized via the statement $p < .001$. (Recently, I read a research report in which four different *p*-statements—$p < .05$, $p < .01$, $p < .005$, and $p < .001$—were connected to the results presented in a single table.) In any one of these studies, it is highly unlikely that the researcher decided at the outset to use different alpha levels with the different null hypotheses being tested. Rather, it is far more probable that all H_0s were initially tested with α set equal to .05, with the researcher then revising α (as indicated in the previous paragraph) so that more impressive *p*-statements could be presented.

Now, let's shift gears and consider what happens if the data-based *p* is larger than the initially specified level of significance. If *p* is much larger than α, the situation is clear: the null hypothesis cannot be rejected. At times, however, *p* turns out to be just slightly larger than α. For example, p might turn out equal to .07 when α is set at .05. Many researchers consider this to be a near miss, and they communicate this observation via certain commonly seen phrases. When p fails to beat α by a small amount, researchers often say that they achieved *marginal significance,* that their findings *approached significance,* that there was a *trend toward significance,* or that the results indicate *borderline significance.* We see an example of this in Excerpt 7.28.

Other researchers deal with near misses in a different way. Believing that a miss is still a miss, they use an approach to hypothesis testing that has two clear rules: (1) choose the level of significance at the beginning of the study and then never change it, and (2) consider any result, summarized by $p,$ to lie on one side or the other side of α, with it making no difference whatsoever whether p is a smidgen or a mile away from α. According to this school of thought, the *only* thing that matters is whether p is larger or smaller than the level of significance. This approach was used in Excerpt 7.29. The null hypothesis was not rejected even though the data-based p-level was extremely close to being under the .05 level of significance.

EXCERPTS 7.28–7.29 • *Dealing with "Near Misses"*

When analyzed for all the veterans together, improvement approached significance pre- to posttreatment, $F(1, 8) = 5.26, p = .051$, and appeared to be maintained at follow-up.

Source: Ray, R. D., & Webster, R. (2010). Group Interpersonal Psychotherapy for Veterans with Posttraumatic Stress Disorder: A Pilot Study. *International Journal of Group Psychotherapy, 60*(1), 131–140.

Overall, arthroscopic evaluations were not statistically different between PRGF and control groups $(P = .051)$.

Source: Sánchez, M., Anitua, E., Azofra, J., Prado, R., Muruzabal, F., & Andia, I. (2010). Ligamentization of tendon grafts treated with an endogenous preparation rich in growth factors: Gross morphology and histology. *Arthroscopy: The Journal of Arthroscopic & Related Surgery, 26*(4), 470–480.

A Few Cautions

Now that you have considered the six-step hypothesis testing procedure from the standpoint of its various elements and its underlying rationale, you may be tempted to think that it will be easy to decipher and critique any research report in your field that has employed this particular approach to inferential statistics. I hope, of course, that this chapter has helped you become more confident about making sense out of statements such as these: "A two-tailed test was used," "A rigorous alpha level was employed to protect against the possibility of a Type I error," and "The results were significant ($p < .01$)." Before I conclude this chapter, however, it is important that I alert you to a few places where misinterpretations can easily be made by consumers of research reports (and by researchers themselves).

Alpha

The word *alpha* (or its symbol α) refers to two different concepts. Within the hypothesis testing procedure, alpha designates the level of significance selected by the researcher. In discussions of measuring instruments, *alpha* means something entirely different. In this latter context, *alpha* refers to the estimated internal consistency of data from the questionnaire, inventory, or test being discussed. Note that alpha must be a *small* decimal number in hypothesis testing in order to accomplish the task of protecting against Type I errors. In contrast, alpha must be a *large* decimal number in order to document high reliability.

The Importance of H_0

Earlier in this chapter, I presented excerpts from various journal articles wherein the null hypothesis was clearly specified. Unfortunately, most researchers do not take the time or space to indicate publicly the precise nature of H_0. Evidently, they presume that their readers will understand what their null hypothesis was in light of the number of samples involved in the study, the nature of the measurements collected, and the kind of statistical test used to analyze the data.

Right now, you may feel that you will never be able to discern H_0 unless it is specifically articulated. However, after becoming familiar with the various statistical tests used to analyze data, you will find that you can make accurate guesses as to the unstated null hypotheses you encounter. Many of the chapters in this book, beginning with Chapter 9, will help you acquire this skill.

This skill is important to have because the final decision of the hypothesis testing procedure always has reference to the point of departure. Researchers never end up by rejecting (or failing to reject) in the abstract; instead, they *always* terminate the hypothesis testing procedure by rejecting (or failing to reject) a *specific* H_0. Accordingly, no decision to reject should be viewed as important unless we consider what specifically has been rejected.

On occasion, the hypothesis testing procedure is used to evaluate a null hypothesis that could have been rejected, or not rejected, from the very beginning, strictly on the basis of common sense. Although it is statistically possible to test such an H_0, no real discovery is made by rejecting a null hypothesis that was known to be false from the outset, or by reaching a fail-to-reject decision when such an outcome was guaranteed from the start. To illustrate, consider Excerpt 7.30.

EXCERPT 7.30 • *Testing an Unimportant H₀*

Students were rank ordered and then matched on their scores on the Mathematical Problem Solving subtest of the Stanford Achievement Test–9 (SAT-9 MPS). Next, each member of a matched student pair was randomly assigned to either the intervention or comparison condition.

A one-way analysis of variance (ANOVA) indicated no statistically significant differences between the groups [regarding scores] on the SAT-9 MPS, $F(1, 58) = 0.00$, *ns*.

Source: Griffin, C. C., & Jitendra, A. K. (2009). Word problem-solving instruction in inclusive third-grade mathematics classrooms. *Journal of Educational Research, 102*(3), 187–202.

The material in Excerpt 7.30 is, perhaps, a truly classic case of the hypothesis testing procedure resulting in a decision about the null hypothesis that was fully guaranteed to be produced because of the way the two comparison groups were formed. In this excerpt, the nonsignificant result was no surprise whatsoever. Be on the lookout for other cases where either a null hypothesis did not need to be tested because the result was known a priori, or for cases where a trivial or silly null hypothesis was tested, even if no one knew whether the null hypothesis would be rejected or retained.

I cannot exaggerate the importance of the null hypothesis to the potential meaningfulness of results that come from someone using the hypothesis testing procedure. Remember that a reject or fail-to-reject decision, by itself, is not indicative of a useful finding. A reject decision could be easily brought about simply by setting up, in Step 1, an outrageous H_0; a fail-to-reject decision could be just as easily brought about by comparing two or more things that are known to be the same. Consequently, you should always be interested in not only the ultimate decision reached at the end of the hypothesis testing procedure but also the target of that decision—H_0.

The Ambiguity of the Word Hypothesis

In discussing the outcomes of their data analyses, researchers sometimes assert that their results support the hypothesis (or that the results do not support the hypothesis). However, which hypothesis is being referred to?

As you now know, the hypothesis testing procedure involves two formal hypotheses, H_0 and H_a. In addition, the person conducting the study may have a hunch (i.e., prediction) as to how things will turn out. Many researchers refer to such hunches as **research hypotheses**. Thus, in a single study, there can be three kinds of hypotheses![5] Usually, the full context of the research report help make clear which of these

---

[5]The researcher's hunch differs from *both* H_0 and H_a if the alternative hypothesis is set up to be nondirectional whereas the researcher's prediction is directional. This situation is not uncommon. Many researchers have been taught to conduct two-tailed tests—even though they have a directional hunch—in order to allow the data to suggest that reality, perhaps, is on the flip side of their hunch.

three hypotheses stands behind any statement about the hypothesis. At times, however, you must read very carefully to accurately understand what the researcher found.

To illustrate why I offer this caution, consider the short sentence in Excerpt 7.31 that comes from a study dealing with instant messaging among teenagers. As you can see, two hypotheses are referred to in this sentence, with one having been rejected whereas the other was accepted. Are these null hypotheses, alternative hypotheses, or investigators' research hypotheses?

EXCERPT 7.31 • *The Ambiguity of the Word Hypothesis*

H2a was rejected, while H2b was accepted.

Source: Hanyun H., & Leung, L. (2009). Instant messaging addiction among teenagers in China: Shyness, alienation, and academic performance decrement. *CyberPsychology & Behavior*, *12*(6), 675–679.

On first glance, the two hypotheses in Excerpt 7.31 might seem like null hypotheses (because null hypotheses get rejected or accepted). However, the sentence in this excerpt actually is referring to research hypotheses. (H2a predicted that teenagers scoring high in shyness would tend to be heavy users of instant messaging; H2b predicted that this same group would tend to rely on this form of communication.) Almost always, the materials contained in a research report clarify what kind(s) of hypotheses are being referred to. If you are not sure what the word *hypothesis* means, search around in the article until you find out. If you do not do this, you might think that the study showed one thing when it actually showed just the reverse.

When *p* Is Reported to Be Equal to or Less Than Zero

Whenever sample data are analyzed by a computer for the purpose of evaluating a null hypothesis, a *p*-value is produced. This *p* is a probability, and it can end up being any number between 0 and 1. As you now know, a small value of *p* causes H_0 to be rejected. The researcher takes that action because a small *p* signifies that a true H_0 population situation would not likely produce a randomly selected data set that, when summarized, is at least as far away from H_0's pinpoint number as is the researcher's actual data set. In most of the excerpts of this chapter, *p* turned out to be very low. In one case, *p* was equal to .051; in another, *p* was turned out equal to .006. We even saw one instance where *p* was shown to be less than .00000001.

Occasionally, you may encounter cases where the reported *p*-value is equal to or less than zero. Such *p*s are misleading, for they do not mean that an imaginary population defined by H_0 had no chance whatsoever (or less than no chance) to produce sample data like that obtained by the researcher. Rather, such *p*-statements are created when exceedingly small computer-generated *p*-values (e.g., $p = .00003$)

are rounded off to a smaller number of decimal places. It is important to know this to avoid falling into the trap of thinking that H_0 is proved to be wrong in those cases where p is reported to be zero or less than zero.

The Meaning of Significant

If the null hypothesis is rejected, the researcher may assert that the results are **significant.** Because the word *significant* means something different when used in casual everyday discussions than when it is used in conjunction with the hypothesis testing procedure, it is crucial that you recognize the statistical meaning of this frequently seen term. Simply stated, a statistically significant finding may not be very significant at all.

In our everyday language, the term *significant* means big, important, or noteworthy. In the context of hypothesis testing, however, the term *significant* has a totally different meaning. Within this inferential context, a significant finding is simply one that is not likely to have occurred if H_0 is true. So long as the sample data are inconsistent with what one would expect from a true null situation, the statistical claim can be made that the results are significant. Accordingly, a researcher's statement to the effect that the results are significant simply means that the null hypothesis being tested has been rejected. It does *not* necessarily mean that the results are *important* or that the absolute difference between the sample data and H_0 was found to be *large*.

Whether a statistically significant result constitutes an important result is influenced by (1) the quality of the research question that provides the impetus for the empirical investigation and (2) the quality of the research design that guides the collection of data. I have come across journal articles that summarized carefully conducted empirical investigations leading to statistically significant results, yet the studies seemed to be quite insignificant. Clearly, to yield important findings, a study must be dealing with an important issue.

Yet what if statistically significant results *are* produced by a study that focuses on an important question? Does this situation mean that the research findings are important and noteworthy? The answer, unfortunately, is no. In Chapter 8, we discuss how it is possible for a study to yield statistically significant results even though there is a tiny difference between the data and the null hypothesis. For example, in a recent study reported in the *Journal of Applied Psychology,* the researcher tested $H_0: \rho = 0$ within the context of a study dealing with correlation. After collecting and analyzing the sample data, this null hypothesis was rejected, with the report indicating that the result was "significant at the .001 level." The sample value that produced this finding was $-.03$!

Even if the issue being investigated is crucial, I cannot consider a correlation of $-.03$ to be very different in any meaningful way from the null value of 0. (With $r = -.03$, the proportion of explained variance is equal to .0009.) As you will soon learn, a large sample can sometimes cause a trivial difference to end up being statistically significant—and that is precisely what happened in the correlational study to which I am referring. In that investigation, there were 21,646 individuals in the sample. Because of the gigantic sample, a tiny correlation turned out to be statistically significant. Although significant in a statistical sense, the r of $-.03$ was clearly insignificant in terms of its importance.

You must be on guard as you come across research reports because many researchers, perhaps unconsciously, seem to exaggerate the importance of their findings, especially if they can say that their results are significant. Sometimes, but not often, you will come across a researcher who downplays a significant result that has been obtained. You can see an example of this in Excerpt 7.32. Give researchers such as these a heap of bonus points when you evaluate their work!

EXCERPT 7.32 • *Acknowledging That a Significant Finding Was Not Impressive*

All comparisons were significant at the $p < .0001$ level. However, Kendall's W for the total sample ($n = 46$) was .11. Kendall's W for the highly trained group ($n = 17$) was .14. Kendall's W for the less trained group ($n = 30$) was .13. These low values for the Kendall's W statistic indicate low levels of agreement despite the significant result. The null hypothesis for W is that concordance of ranks is not significantly different from 0 (i.e., random). Although the concordance of ranks was significantly different from 0, the level of agreement was not high.

Source: Miller, E. M. (2009). The effect of training in gifted education on elementary classroom teachers' theory-based reasoning about the concept of giftedness. *Journal for the Education of the Gifted*, *33*(1), 65–105.

Review Terms

Accept	Reject
Alpha level	Reliable difference
Alternative hypothesis	Research hypothesis
Calculated value	Significant
Critical value	Test statistic
Directional	Two-tailed test
Fail-to-reject	Type I error
Highly significant	Type II error
Hypothesis testing	α
Inexact H_0	H_0
Level of significance	H_a
Nondirectional	*ns*
Null hypothesis	*p*
One-tailed test	

The Best Items in the Companion Website

1. An email message sent from the author to his students entitled "Learning about Hypothesis Testing Is NOT Easy!"
2. An interactive online quiz (with immediate feedback provided) covering Chapter 7.
3. Ten misconceptions about the content of Chapter 7.
4. Chapter 7's best passage (selected by the author).
5. An interactive online resource called "Type I Errors."

To access the chapter outline, practice tests, weblinks, and flashcards, visit the companion website at http://www.ReadingStats.com.

Review Questions and Answers begin on page 531.

Effect Size, Power, CIs, and Bonferroni

In Chapter 7, we considered the basic six-step version of hypothesis testing. Although many researchers use that version of hypothesis testing, there is a definite trend toward using a seven step or nine-step procedure when testing null hypotheses. In this chapter, we consider the extra step(s) associated with these approaches to hypothesis testing. In addition, this chapter includes two related topics: the connection between hypothesis testing and confidence intervals, and the problem of an inflated Type I error rate brought about by multiple tests conducted simultaneously.

The Seven-Step Version of Hypothesis Testing: Estimating Effect Size

As you may recall from Chapter 7, the elements of the simplest version of hypothesis testing are as follows:

1. State the null hypothesis (H_0).
2. State the alternative hypothesis (H_a).
3. Select a level of significance (α).
4. Collect and analyze the sample data.
5. Refer to a criterion for evaluating the sample evidence.
6. Reject or fail to reject H_0.

To these six steps, a growing number of researchers add a seventh step. Instead of ending the hypothesis testing procedure with a statement about H_0, these researchers return to their sample data and perform an additional task. The purpose of the seventh step is to go beyond the decision made about H_0 in order to say something about the *degree* to which the sample data turned out to be incompatible with the null hypothesis.

161

Reason for the Seventh Step

Before discussing what researchers do in Step 7 of this (slightly expanded) version of hypothesis testing, I want to explain why competent researchers take the time to do this. Simply stated, they do this because a result that is deemed to be statistically significant can be, at the same time, completely devoid of *any* practical significance whatsoever. This is because there is a direct relationship between the size of the sample(s) and the probability of rejecting a false null hypothesis. If the pinpoint number in H_0 is wrong, large samples increase the likelihood that the result will be statistically significant—even if the sample data deviate from H_0 by a small amount. In such situations, a decision to reject H_0 in favor of H_a does not mean very much in a practical sense.

In Excerpts 8.1 and 8.2, this critically important distinction between **statistical significance** and **practical significance** is discussed.[1] In Excerpt 8.1, notice that the researchers argue that "beyond statistical significance," their finding appeared to be "of considerable practical significance as well." In Excerpt 8.2, the researchers first note that they had a giant sample size; then, they warn their readers not to automatically equate statistical significance with practical significance. It would be nice if all research reports contained statements like the ones in these two excerpts (or like the statement we saw earlier in the final excerpt of Chapter 7). Unfortunately, many researchers seem concerned with just one thing: statistical significance. This

EXCERPTS 8.1–8.2 • *Statistical Significance versus Practical Significance*

Beyond statistical significance, these associations appeared to be of considerable practical significance as well. Consider, for example, that youth with high levels of academic self-efficacy (i.e., one standard deviation above the mean) were 2.46 times as likely to endorse learning-oriented goals compared with youth with low levels of academic self-efficacy (i.e., one standard deviation below the mean).

Source: Baird, G. L., Scott, W. D., Dearing, E., & Hamill, S. K. (2009). Cognitive self-regulation in youth with and without learning disabilities: Academic self-efficacy, theories of intelligence, learning vs. performance goal preferences, and effort attributions. *Journal of Social & Clinical Psychology*, 28(7), 881–908.

We note that due to the size of the MYDAUS survey [$N = 80,428$], the power to detect differences is extremely high and the reader should be careful to not equate statistical significance with practical significance.

Source: O'Brien, L. M., Polacsek, M., MacDonald, P. B., Ellis, J., Berry, S., & Martin, M. (2010). Impact of a school health coordinator intervention on health-related school policies and student behavior. *Journal of School Health*, 80(4), 176–185.

[1]The term *clinical significance*, used frequently within medical research, means the same thing as *practical significance*.

is dangerous because it is quite possible for a study's results to be significant in a statistical sense without being important (i.e., significant) in a practical fashion.

We next look briefly at a procedure researchers use in an effort to measure the practical significance of their results. Doing this constitutes the seventh step of hypothesis testing. It involves computing an estimate of what's called *effect size*.

Estimating the Effect Size

Researchers who are sensitive to the distinction between statistical significance and practical significance often add a seventh step to the basic version of hypothesis testing by estimating the study's **effect size.** Once obtained, an effect size estimate can then be compared against a "yardstick" for assessing practical significance. Whereas the null hypothesis might be rejected in a study comparing two ways of treating a disease, an effect size estimate allows us to see whether the differential impact of the two treatments should be thought of as small or medium or large. Or, in a study involving the correlation of two variables that declares r to be significantly different from zero, an estimate of effect size allows the researcher to talk about the pure strength of the measured relationship, beyond saying simply that it is statistically significant.

When applying the concept of effect size to address the concept of practical significance, a researcher does two things. First, the researcher compares a simple summary the sample data to H_0 (and perhaps standardizes the result of this comparison). In doing this, the previously computed p-value is not considered at all; instead, attention is focused on the degree to which the sample data deviate from what is said in the null hypothesis. For example, if a study deals with correlation and if the null hypothesis says H_0: $\rho = 0.00$, the researcher looks to see how much the sample value of r deviates from 0.00. Or, in the two-group study in which the null hypothesis says H_0: $\mu_1 = \mu_2$, the researcher examines the sample data to see how different M_1 and M_2 are compared against the difference of zero contained in the null hypothesis.

Second, the researcher evaluates the size of the observed difference between the summary of the sample data and whatever was stated in H_0. This second part of an effect size analysis involves judging the size of the observed difference and declaring it to be tiny, small, medium, large, or gigantic. Most researchers use just the middle three of these labels. In Excerpt 8.3, we see an example where this was done

EXCERPT 8.3 • *Labeling the Effect Size*

A correlation between QB ratings and mean attractiveness ratings was conducted to determine whether athletic performance could in fact be assessed by simple examination of the QBs' faces. . . . Results demonstrated that attractiveness and QB ratings were positively correlated, $r = .31, p < .05$, exhibiting a small-to-medium effect size.

Source: Williams, K. M., Park, J. H., & Wieling, M. J. (2010). The face reveals athletic flair: Better National Football League quarterbacks are better looking. *Personality and Individual Differences, 48*(2), 112–116.

in a study dealing with NFL quarterbacks. Note the excerpt's final words: "a small-to-medium effect size."

To decide whether an observed discrepancy between the sample data and the null hypothesis should be described as small, medium, or large, many researchers refer to effect size criteria that have been recommended by statistical authorities. (Beginning in Chapter 9, these specific criteria, or *standards,* are considered when we examine different statistical procedures for testing null hypotheses.) For example, the researchers associated with Excerpt 8.3 may have used the phrase *small-to-medium effect size* by invoking the popular standards for evaluating the size of a sample-based correlation coefficient.

Instead of referring to the popular criteria for judging the magnitude of a study's estimated effect size, a researcher has the full right to make an evaluative judgment by considering what kinds of outcomes are thought to be small, medium, or large by practitioners or other researchers. In fact, certain statistical authorities recommend *avoiding* the popular criteria. Impressed by the argument that "one size does not fit all," I salute those researchers who use their knowledge of theory, previous research, and applied utility when trying to decide whether the practical significance of their finding is small, medium, or large.

In Excerpts 8.4 and 8.5, we see two studies that involved means rather than correlations. Although these excerpts are different in several ways, they are similar in that the researchers of each study first computed an estimated effect size and then made an evaluative judgment as to its magnitude. The final five words of each excerpt indicate the researchers' decision as to whether the estimated effect size should be considered small or medium or large. (This first of these excerpt shows nicely the value of adding this seventh step to the hypothesis testing procedure, as the statistically significant result was evaluated as having only small practical significance.)

EXCERPTS 8.4–8.5 • *Two Popular Estimates of Effect Size: d and Partial η^2*

The control group improved significantly in the Hostility measure ($t = 2.24$; $df = 15$; $P = 0.02$), while the effect size was small ($d = 0.21$).

Source: Thaut, M. H., Gardiner, J. C., Holmberg, D., Horwitz, J., Kent, L., Andrews, G., et al. (2009). Neurologic music therapy improves executive function and emotional adjustment in traumatic brain injury rehabilitation. *Annals of the New York Academy of Sciences, 1169*(1), 406–416.

The main effect of attention [$F(2, 45) = 55.248, p < .001$; partial $\eta^2 = .71$] was highly significant and represented a large effect size.

Source: Kushalnagar, P., Hannay, H. J., & Hernandez, A. E. (2010). Bilingualism and attention: A study of balanced and unbalanced bilingual deaf users of American Sign Language and English. *Journal of Deaf Studies and Deaf Education, 15*(3), 263–273.

In Excerpt 8.4, the effect was estimated by means of something called d, whereas the effect size in Excerpt 8.5 was estimated by something called *partial η^2*. (The notation η^2 is called *eta squared*.) These two indices turned out equal to .21 and .71, respectively, and each can be thought of as a measure of estimated effect size that has been standardized. Standardized indices such as these are needed by those researchers who opt to use the popular, more general criteria for deciding whether a result is small, medium, or large. Beginning in Chapter 9, we consider these and other standardized indices of effect size, and we see the criteria that brought forth the words *small* and *large* in Excerpts 8.4 and 8.5, respectively.

The Nine-Step Version of Hypothesis Testing: Power Analyses

Although many researchers utilize the six-step and seven-step versions of hypothesis testing, there is a definite trend toward using a nine-step approach. Six of the steps of this more elaborate version of hypothesis testing are identical to the six basic elements considered in Chapter 7, whereas the other three steps are related to the concepts of sample size, power, and a minimally important effect size. Listed in the order in which the researcher deals with them, the various elements of the nine-step version of hypothesis testing are as follows:

       1. State the null hypothesis, H_0.
       2. State the alternative hypothesis, H_a.
       3. Specify the desired level of significance, α.
(new) **4.** Specify the minimally important effect size.
(new) **5.** Specify the desired level of power.
(new) **6.** Determine the proper size of the sample(s).
       7. Collect and analyze the sample data.
       8. Refer to a criterion for assessing the sample evidence.
       9. Make a decision to discard/retain H_0.

The steps in the first third and final third of this nine-step version of hypothesis testing are identical to the six steps we discussed in Chapter 7. We focus here only on Steps 4, 5, and 6. Before we look at these three steps, it is important that you recognize two differences between the seven-step and nine-step versions of hypothesis testing. One difference concerns the timing that things are done relative to the basic six steps of hypothesis testing; the other concerns two very different notions of the term *effect size*.

First, there is the issue of timing. In the seven-step approach considered earlier in this chapter, the researcher executes the six basic steps and then adds a seventh step. That seventh step involves returning to the sample data *after* the decision has been made to reject/retain H_0. In contrast, the nine-step version of hypothesis testing requires that the researcher do the extra three steps *before* any data are collected.

Second, there is the meaning of *effect size*. In the seven-step version of hypothesis testing, the effect size is something that is estimated from the sample data. As discussed in the next section, the effect size in the nine-step version of hypothesis testing takes the form of an opinion, not a data-based fact.

With this preface to the nine-step version of hypothesis testing now behind us, let us turn our attention to Steps 4, 5, and 6.

Step 4: Specification of the Effect Size

As mentioned in the previous section, the term *effect size* has two meanings. We saw one of these earlier in this chapter when we considered the option researchers have for adding a seventh step to the basic bare-bones kind of hypothesis testing. Now, we must consider a different notion of effect size. In our present discussion, the effect size refers to an a priori specification of what constitutes the smallest study finding that the researcher considers to be worth talking about. Perhaps a picture will help you understand this new notion of effect size. You need a pen or pencil; this is going to be a picture that *you* draw.

Draw a line segment about 12 inches long. Mark the far left end of your line with this four-word sentence: "H_0 is totally true." At the far right end of your line, write these six words: "H_0 is false by a mile." Now, put a big dot somewhere on this line such that it divides the line into two parts. The portion of the line located to the left of the dot represents situations where the null hypothesis is false but false to only a trivial degree. It might help if you put the label "Trivial" on that side of the line. The portion of the line to the right of the dot represents situations where the null hypothesis is false by an amount that deserves to be thought of as "Big" or "Noteworthy." You might want to put the label "Important" on this segment of your line. As you may have guessed, the dot on this line represents the kind of effect size we now are considering.

To illustrate more specifically what this kind of effect size is and how it gets selected, suppose a researcher uses the hypothesis testing procedure in a study where there is one population, where the data are IQ scores, where the statistical focus is on the mean, and where the null and alternative hypotheses are H_0: $\mu = 100$ and H_a: $\mu > 100$, respectively. In this hypothetical study, the continuum of possible false null cases, as specified by H_a, extends from a value that is just slightly greater than 100 (e.g., 100.1) to whatever the maximum earnable IQ score is (e.g., 250). The researcher might decide to set 110 as the effect size. By so doing, the researcher would be declaring that (1) the true μ is judged to be only trivially different from 100 if it lies anywhere between 100 and 110, whereas (2) the difference between the true μ and 100 is considered to be important so long as the former is at least 10 points greater than H_0's pinpoint value of 100.

Researchers specify an effect size in one of two ways. On the one hand, the researcher can specify a **raw effect size.** On the other hand, he or she can specify a **standardized effect size.** Specifying a raw effect size is the better strategy, but it is often more difficult (or impossible) to do this. Standardized effect sizes are easy to specify, but they are not as good despite the fact that the word *standardized* gives the impression of scientific superiority.

The process of specifying a raw effect size is illustrated in the hypothetical IQ study we recently considered. In that example, the raw effect size is equal to an IQ of 110. Because the null hypothesis in that study was set equal to 100, the researcher could alternatively say that the effect size is equal to 10 IQ points (i.e., the difference between 110 and 100). Regardless of how the researcher might report what he or she has done, this process leads to a raw effect size because the researcher begins by specifying the line of demarcation between trivial and nontrivial outcomes directly on the score continuum of the study's dependent variable.

Many researchers who use the nine-step version of hypothesis testing choose to specify standardized effect sizes (rather than raw effect sizes). When using standardized effect sizes, researchers refer to established criteria just like the ones used in the seven-step version of hypothesis testing. For most statistical procedures, the standardized effect size criteria are numerical values that indicate what is considered to be a small effect, a medium effect, and a large effect. In choosing one of these standardized effect sizes, the researcher thinks about his or her study and then poses this three-part question:

> In the study I am going to conduct, do I want my statistical test to be sensitive to (and thus be able to detect) only a large effect, if that's what's truly "out there" in the real world? Or, do I want my study to have the added sensitivity that would allow it to detect either a large effect or a medium-size effect? Or, is it important for my study to have the high-level sensitivity that would allow it to detect not just large and medium effects, but small effects as well?

The criteria for **small, medium,** and **large standardized effect sizes** vary depending on the kind of statistical test the researcher is using. For example, if the researcher is going to compare two group means via a *t*-test, the criteria are .2, .5, and .8 for small, medium, and large effect sizes, respectively, but if the researcher is going to compute a correlation coefficient, the standardized effect sizes that define small, medium, and large *r*s are .1, .3, and .5, respectively. When we consider different test procedures in Chapters 9 through 18, I point out what the standardized effect size criteria are for each test procedure. At this point, all you must know is that (1) the nine-step version of hypothesis testing requires an a priori effect size specification, and (2) researchers have the option of specifying a raw effect size or a standardized effect size.

Excerpts 8.6, 8.7, and 8.8 illustrate the two kinds of effect sizes we have been considering. In Excerpts 8.6 and 8.7, the researchers used a standardized effect size by selecting .3 and .8, respectively. In Excerpt 8.8, we see a case where a raw effect size (equal to a 10% difference) was used. Despite the differences among these three excerpts, the common denominator is that in each instance the researchers specified the effect size. They did not compute it (or estimate it) from their data, they *set* it. Moreover, they set the effect size before the data were collected.

EXCERPTS 8.6–8.8 • *Selecting an Effect Size*

The effect size was set at 0.30 for medium size.

Source: Kato, G., Tamiya, N., Kashiwagi, M., Sato, M., & Takahashi, H. (2009). Relationship between home care service use and changes in the care needs level of Japanese elderly. *BMC Geriatrics, 9*(1), 1–9.

A power analysis was performed before the study [and] the effect size was set to 0.8. . . .

Source: Braun A., Jepsen S., Deimling D., & Ratka-Krüger, P. (2010). Subjective intensity of pain during supportive periodontal treatment using a sonic scaler or an Er:YAG laser. *Journal of Clinical Periodontology, 37*(4), 340–345.

We evaluated outcomes [following] surgery with the a priori clinically significant effect size set at 10% differences for all outcome measures.

Source: Cook, J. L., Luther, J. K., Beetem, J., Karnes, J., & Cook, C. R. (2010). Clinical comparison of a novel extracapsular stabilization procedure and tibial plateau leveling osteotomy for treatment of cranial cruciate ligament deficiency in dogs. *Veterinary Surgery, 39*(3), 315–323.

Step 5: Specification of the Desired Level of Power

The researcher's next task within the nine-step hypothesis testing procedure is to specify the level of power that is desired for rejecting H_0 if H_0 is off by an amount equal to the previously established effect size. Power is a probability value and can range from 0 to 1.0. Only high values are considered, however, because the complement of power is the probability of a Type II error.

The researcher does not know, of course, exactly how far off-target the null hypothesis is (or even if it is wrong at all). The specified effect size is simply the researcher's judgment as to what would or would not constitute a meaningful deviation from the null case. Note, however, that if the null hypothesis is wrong by an amount that is greater than the specified effect size, then the actual probability of rejecting H_0 is larger than the specified power level. Thus, the power level selected in Step 5 represents the lowest acceptable power for any of the potentially true H_a conditions that are considered to be meaningfully different from H_0.

To see illustrations of how researchers report desired levels level of power, review Excerpts 8.9 through 8.11. In the first of these excerpts, note that the researchers use the phase "a minimal clinically important difference of 5%." In saying that, the researchers specify their effect size. Then, they indicate that they want to have an 80 percent chance of detecting such a difference. That was the statistical power they wanted to have. In Excerpts 8.10 and 8.11, the researchers simply specify the power they desired: .80 and .90, respectively. In these two excerpts, the chosen level of significance appears in the same sentence with the chosen level of power. These two concepts—α and power—were put together in the same sentence because both are tied to potential inferential errors that might occur. Alpha specified the probability of making a Type I error; power specified the probability of not making a Type II error. These two probabilities are sometimes referred to as the **Type I error risk** and the **Type II error risk,** respectively.

EXCERPTS 8.9–8.11 • *Selecting a Desired Level of Power*

The test was two-tailed and the power was set at 80% to detect a minimal clinically important difference of 5% between the two groups (i.e., 10 vs. 15%). A difference of less than 5% would not be of clinical or substantive significance.

Source: Sword, W., Watt, S., Krueger, P., Thabane, L., Landy, C. K., Farine, D., et al. (2009). The Ontario Mother and Infant Study (TOMIS) III: A multi-site cohort study of the impact of delivery method on health, service use, and costs of care in the first postpartum year. *Pregnancy and Childbirth*, *9*(16), 1–12.

In this study, the power was set at .80, with $\alpha = .05$.

Source: Huijbregts, M. P. J., Teare, G. F., McCullough, C., Kay, T. M., Streiner, D., Wong, S. K. C., et al. (2009). Standardization of the Continuing Care Activity Measure: A multicenter study to assess reliability, validity, and ability to measure change. *Physical Therapy*, *89*(6), 546–555.

The 2-sided α was set at .05, and power was set at 0.90.

Source: Davidson, K. W., Rieckmann, N., Clemow, L., Schwartz, J. E., Shimbo, D., Medina, V., et al. (2010). Enhanced depression care for patients with acute coronary syndrome and persistent depressive symptoms: Coronary Psychosocial Evaluation Studies randomized controlled trial. *Archives of Internal Medicine*, *170*(7), 600–608.

In Excerpt 8.11, note that the power was set at .90, whereas it was set at .80 in Excerpts 8.9 and 8.10. On first glance, you might be inclined to think, "The higher the power, the better the study." However, there are two reasons why most researchers select a power of .80 and why you will probably never see a specified power greater than .90 in any applied study. First, extremely high power levels place unreasonable demands on researchers when they move to Step 6 and compute the sample size required to provide the desired power. Second, specified power level such as .95 or .99 increase the probability that trivial deviations from H_0 bring about a "reject" decision in the hypothesis testing procedure.

Step 6: Determination of the Needed Sample Size

After stating H_0 and H_a, after selecting a level of significance, and after specifying the effect size and the desired power, the researcher then uses a formula, a specially prepared table, or a computer program to determine the size of the sample(s) needed in the study. No judgment or decision-making comes into play at this **sample size determination** stage in the nine-step version of hypothesis testing, since the researcher simply calculates or looks up the answer to a very pragmatic question: How large should the sample be? At this point (and also in Steps 7 through 9), the researcher functions like a robot who performs a task, referred to as a **power analysis.**

Excerpts 8.12 and 8.13 illustrate the kinds of things researchers say when they talk about having computed their sample sizes in a power analysis. In the first of the excerpts, the power analysis told the researchers that they needed a sample size of 41. Note that the "ingredients" for this needed sample size include (among other things) a priori decisions regarding the level of significance ($\alpha = .05$), the effect size (.3), and power (.80). Note also that the effect size in this power analysis is of the standardized variety, with the researchers stating that a "medium" effect is considered big enough to be worth detecting.

EXCERPTS 8.12–8.13 • Determining the Needed Sample Size

A priori power analysis indicated that to achieve a power of .80 with $p < .05$ and a medium effect size of .30 and three predictors, a sample size of 41 would be required.

Source: Dahlbeck, D. T., & Lightsey, O. R. (2008). Generalized self-efficacy, coping, and self-esteem as predictors of psychological adjustment among children with disabilities or chronic illnesses. *Children's Health Care, 37*(4), 293–315.

The purpose of this study was to determine if giving 50 mg of meclizine the night before and on the day of surgery would effectively reduce postoperative nausea and vomiting (PONV) for the entire 24 hours after surgery in patients identified as being at high risk for PONV. . . . Before initiation of this study, a power analysis was [conducted] using an α of 0.05 and a β of 0.20 and revealed a need for approximately 40 subjects per group to achieve significance. Factoring in an attrition rate of 10%, this increased the necessary sample size to 44 subjects per group or 88 subjects for the total sample.

Source: Bopp, E. J., Estrada, J. L., Kilday, J. M., Spradling, J. C., Daniel, C., & Pellegrini, J. E. (2010). Biphasic dosing regimen of meclizine for prevention of postoperative nausea and vomiting in a high-risk population. *AANA Journal, 78*(1), 55–62.

In Excerpt 8.13, we see a power analysis from a medical study focused on preventing postsurgical nausea and vomiting. This excerpt is worth examining because of the symbol β that appears in the second sentence. Instead of saying that they wanted their study to have a power of .80, these researchers indicated that they wanted the probability of making a Type II error, or **beta error,** to be low, with that probability being .20. Because the sum of this kind of probability and the probability of correctly rejecting a false null hypothesis (i.e., power) must equal to 1.00, the statement that $\beta = .20$ is identical to the statement that power $= .80$.

As we finish our discussion of the nine-step version of hypothesis testing, I want to underscore the primary advantage of this approach to evaluating any null hypothesis. The eventual results of the statistical test become easier to interpret if the researcher has successfully wrestled with the issue of what ought to be viewed as a meaningful deviation from H_0, and if the sample size has been computed so as to create the desired level of power. In contrast, the six-step version of hypothesis testing can lead to a highly ambiguous finding.

If no consideration is given to the concepts of effect size and power, the researcher may end up very much in the dark as to whether (1) a fail-to-reject decision is attributable to a trivial (or zero) deviation from H_0 *or* is attributable to the test's insensitivity to detect important non-null cases due to a small sample size; or (2) a reject decision is attributable to H_0 being false by a nontrivial amount *or* is attributable to an unimportant non-null case being labeled *significant* simply because the sample size was so large. In Excerpts 8.14 and 8.15, we see examples of how murky results can be produced when the six-step approach to hypothesis testing is used. In Excerpt 8.14, the researchers tell us, in essence, that the statistically insignificant results may have been caused by insufficient power. In Excerpt 8.15, we see a research team that obtained statistically significant results but admits, in essence, that their findings may have been caused by an overly large sample size making the statistical tests too sensitive.

EXCERPTS 8.14–8.15 • *Problems Caused by Small and Large Sample Sizes*

The major limitation of the present study is the relatively small number of participants included for statistical analysis. A relatively small number of participants reduces power; thus, statistical outcomes are susceptible to Type II error.

Source: Fabiano-Smith, L., & Goldstein, B. A. (2010). Phonological acquisition in bilingual Spanish–English speaking children. *Journal of Speech, Language & Hearing Research*, *53*(1), 160–178.

The SF-36 vitality subscale was the only variable among the 18 variables related to emotional wellbeing that made an independent contribution in distinguishing between the two groups in the stepwise logistic regression. However, the effect for the SF-36 vitality subscale was very small (odds ratio of 0.99) and although it was statistically significant in this large sample [$N = 9,081$], the effect is not of practical significance.

Source: Rowlands, I., & Lee, C. (2009). Correlates of miscarriage among young women in the Australian Longitudinal Study on Women's Health. *Journal of Reproductive & Infant Psychology*, *27*(1), 40–53.

The advantage of the nine-step approach to hypothesis testing is *not* that the researcher is able to know whether the decision reached about H_0 is right or wrong. Regardless of the approach used, a reject decision might be correct or it might constitute a Type I error, and similarly a fail-to-reject decision might be correct or it might be a Type II error. The advantage of having effect size and power built into the hypothesis testing procedure is twofold: On the one hand, researchers know and control, on an a priori basis, the probability of making a Type II error, and on the other hand, they set up the study so that no critic can allege that a significant result, if found, was brought about by an overly sensitive test (or that a nonsignificant result, if found, was produced by an overly insensitive test).[2]

To summarize, the three extra components of the nine-step version of hypothesis testing deal with three concepts: a specified effect size, a chosen power level, and a power analysis that determines the needed sample size. The researchers who take the time to integrate these concepts into their studies deserve credit for being more careful and thoughtful in their empirical investigations. Be wary of the claims made by researchers who give no evidence of having considered these concepts.

Hypothesis Testing Using Confidence Intervals

Researchers can, if they wish, engage in hypothesis testing by means of using one or more confidence intervals, rather than by comparing a calculated value against a critical value or by comparing a *p*-level against α. Although this approach to hypothesis testing is not used as often as the approaches discussed in Chapter 7 and the earlier portion of this chapter, it is important for you to understand what is going on when a researcher uses confidence intervals within the context of hypothesis testing.

Whenever confidence intervals are used in this manner, everything about the hypothesis testing procedure remains the same except the way the sample data are analyzed and evaluated. To be more specific, this alternative approach to hypothesis testing involves the specification of H_0, H_a, and alpha, and the final step involves a reject or fail-to-reject decision regarding H_0. The concepts of Type I and Type II errors are still relevant, as are the opportunities to specify effect size and power and to compute the proper sample size if the nine-step version of hypothesis testing is being used.

As indicated in Chapter 7, calculated and critical values usually are numerical values that are metric-free. Such calculated and critical values have no meaningful connection to the measurement scale associated with the data. Although it is advantageous for the researcher to use metric-free calculated and critical values, such values provide little insight as to why H_0 ultimately is rejected or not rejected. The advantage of confidence intervals is that they help to provide that insight.

---

[2]In saying this, I assume that the hypothetical critic agrees with the researcher's decisions about H_0, H_a, α, and the effect size.

The way confidence intervals are used within hypothesis testing is easy to explain. If there is just a single sample involved in the study, the researcher takes the sample data and builds a confidence interval around the sample statistic. Instead of computing a calculated value, the researcher computes an interval, with the previously specified alpha level dictating the level of confidence associated with the interval (an α of .05 calls for a 95 percent interval, an α of .01 calls for a 99 percent interval, etc.). Instead of then turning to a critical value, the researcher turns to the null hypothesis and compares the confidence interval against the pinpoint number contained in H_0. The decision rule for the final step is straightforward: If the null number is outside the confidence interval, H_0 can be rejected; otherwise, H_0 must be retained.

Excerpt 8.16 illustrates the confidence interval approach to hypothesis testing. This excerpt comes from a study that looked at different regions of the country and examined the relationships between the number of fitness facilities, on the one hand, and six different measures of physical activity, on the other. In each case, the correlational null hypothesis said H_0: $\rho = 0$ where ρ was the correlation in the population. Four of these null hypotheses were rejected, as indicated by the very small p-values connected to the sample rs that appear in the excerpt.

EXCERPT 8.16 • *A Confidence Interval Approach to Hypothesis Testing*

Pearson correlations were performed between fitness facility density (number of facilities/100,000 people) and six summary measures of physical activity prevalence. . . . Direct correlations between fitness facility density and the percent of those physically active ($r = 0.27$, 95% CI 0.11, 0.42, $p = 0.0012$), those meeting moderate-intensity activity guidelines, ($r = 0.23$, 95% CI 0.07, 0.38, $p = 0.006$), and those meeting vigorous-intensity activity guidelines ($r = 0.30$, 95% CI 0.14, 0.44, $p = 0.003$) were found. An inverse correlation was found between fitness facility density and the percent of people physically inactive ($r = -0.45$, 95% CI -0.57, -0.31, $p < 0.0001$).

Source: Meissner, F. D. L. (2010). Physical activity levels and access to places to be physically active. Unpublished Master's thesis, University of Texas School of Public Health, Houston.

In Excerpt 8.16, notice that the 95 percent confidence interval for the first correlation ($r = 0.27$) extended from 0.11 to 0.42. Because the null number of 0 was not contained inside this interval, H_0 was rejected. Although last of the four correlations ($r = -0.45$) was negative, the same approach was taken to decide whether this r was statistically significant. The 95 percent confidence interval (extending from -0.57 to -0.31) did not contain the null number of 0, so this sample correlation between physical facility density and the percentage of physically inactive people caused H_0 to be rejected.

A confidence interval can also be used to determine whether two samples differ sufficiently to allow the researchers to reject a null hypothesis that says the corresponding populations have the same parameter value. Excerpt 8.17 illustrates how this is done. In the study associated with this excerpt, the mean birth weight of 16,464 babies born to fertile women was compared to the birth weight of 2,009 babies born to subfertile women. The null hypothesis was of the "no difference" variety: $\mu_{fertile} - \mu_{subfertile} = 0$. The mean difference in weight between these two groups was only 13 grams (compared to an average birth weight of about 3,550 grams). Because the 95 percent confidence interval built around this mean difference was found to overlap the null number of 0, the researchers stated that there was no statistically significant difference between the mean birth weights of the two groups of babies.

EXCERPT 8.17 • *Using a Confidence Interval to Compare Two Groups*

Our prospective cohort study [examined] pregnancies in subfertile women (who conceived spontaneously after 12+ months' waiting time to pregnancy) in comparison with the pregnancies of fertile women (who conceived spontaneously after less than 12 months' waiting time to pregnancy). . . . There was no statistically significant difference in the mean birth weight among infants of fertile and subfertile women (mean difference: 13 g; 95% CI, –9 to 36 g).

Source: Wisborg, K., Ingerslev, H. J., & Henriksen, T. B. (2010). In vitro fertilization and preterm delivery, low birth weight, and admission to the neonatal intensive care unit: A prospective follow-up study. *Fertility and Sterility*, *94*(6), 2102–2106.

Before completing our discussion of the confidence-interval approach to hypothesis testing, I must alert you (once again) to the difference between a *confidence interval* and a *standard error interval*.[3] Many researchers who compute calculated and critical values within one of the more traditional approaches to hypothesis testing summarize their sample data in terms of values of the statistic plus or minus the standard error of the statistic. Intervals formed by adding and subtracting the standard error to the sample statistic do *not* produce alpha-driven confidence intervals. Instead, the result is a 68 percent interval. (Alpha-driven confidence intervals are typically 95 percent intervals.)

Adjusting for an Inflated Type I Error Rate

In Chapter 7, I indicated that researchers have direct control over the probability that they will make a Type I error when making a judgment about H_0. (Recall that Type I errors occur when true null hypotheses are rejected.) This control is exerted

---

[3]The difference between confidence intervals and standard error intervals was first covered in Chapter 6.

when the researcher selects the level of significance. As long as the underlying assumptions of the researcher's statistical test are tenable, the alpha level selected in Step 3 of the hypothesis testing procedure instantly and accurately establishes the probability that a true H_0 will be rejected.

The fact that α dictates Type I error risk holds true *only* for situations where researchers use the hypothesis testing procedure just once within any given study. In many studies, however, more than one H_0 is tested. In Excerpt 8.18, we see an illustration of this common practice of applying the hypothesis testing procedure multiple times within the same study. As you can see, correlation coefficients appear in this excerpt. Because all possible bivariate correlations were computed among the four variables mentioned (victimization, shyness/withdrawal, aggression, and agreeableness), there actually were six correlations. Only three of these are presented, the ones that turned out to be statistically significant with $p < .05$.

EXCERPT 8.18 • *Hypothesis Testing Used More than Once*

Correlations examined associations between eighth-grade clustering variables ($p < .05$). Victimization and aggression were [significantly] correlated ($r = .14$), and shyness/withdrawal was [significantly] linked to both victimization ($r = .37$) and aggression ($r = -.16$). There were no statistically significant associations between agreeableness and any of the peer nomination variables.

Source: Laursen, B., Hafen, C. A., Rubin, K. H., Booth-LaForce, C., & Rose-Krasnor, L. (2010). The distinctive difficulties of disagreeable youth. *Merrill-Palmer Quarterly, 56*(1), 80–103.

When the hypothesis testing procedure is applied multiple times within the same study, the alpha level used within each of these separate tests specifies the Type I error risk that would exist if that particular test were the only one being conducted. However, with multiple tests being conducted in a study, the actual probability of making a Type I error somewhere within the set of tests *exceeds* the alpha level used within any given test. The term **inflated Type I error risk** is used to refer to this situation in which the alpha level used within each of two or more separate tests understates the likelihood that at least one of the tests will cause the researcher to reject a true H_0.

A simple example may help to illustrate the problem of an inflated Type I error rate. Suppose you are given a fair die and told to roll it on the table. Before you toss the die, also suppose that the person running this little game tells you that you will win $10 if your rolled die turns out to be anything but a six. If you get a six, however, you must pay $50. With an unloaded die, this is a fair bet, because your chances of winning are 5/6, whereas the chance of losing is 1/6.

However, what if you were handed a pair of fair dice and asked to roll both members of the pair simultaneously, with the rule being that you would win $10 if you can avoid throwing an evil six, but lose $50 if your roll of the dice causes either one or two "boxcars" to show up. This is not a fair bet for you, because the chances of avoiding a six are $5/6 \times 5/6 = 25/36$, a result that is lower than the 5/6 value needed to make the wager an even bet in light of the stakes ($10 versus $50). If you were handed five pairs of dice and were asked to roll them simultaneously, with the same payoff arrangement in operation (i.e., win $10 if you avoid a six, otherwise lose $50), you would be at a terrific disadvantage. With 10 of the six-sided cubes being rolled, the probability of your winning the bet by avoiding a six anywhere in the full set of results is equal to approximately .16. You would have a 16 percent chance of winning $10 versus an 84 percent chance of losing $50. That would be a very good arrangement for your opponent!

As should be obvious, the chances of having a six show up at least once increase as the number of dice being thrown increases. With multiple dice involved in our hypothetical game, there are two ways to adjust things to make the wager equally fair to both parties. One adjustment involves changing the stakes. For example, with two dice being rolled, the wager could be altered so you would win $11 if you avoid a six or lose $25 if you do not. The second adjustment involves tampering with the two little cubes so as to produce a pair of loaded dice. With this strategy, each die is weighted such that its chances of ending up as something other than a six are equal to a tad more than 10/11. This allows two dice to be used, in a fair manner, with the original stakes in operation ($10 versus $50).

When researchers use the hypothesis testing procedure multiple times, an adjustment must be made somewhere in the process to account for the fact that at least one Type I error somewhere in the set of results increases rapidly as the number of tests increases. Although there are different ways to effect such an adjustment, the most popular method is to change the level of significance used in conjunction with the statistical assessment of each H_0. If the researcher wants to have a certain level of protection against a Type I error anywhere within his or her full set of results, then he or she would make the alpha level more rigorous within each of the individual tests. By so doing, it is as if the researcher is setting up a fair wager in that the claimed alpha level truly matches the study's likelihood of yielding a Type I error.

The most frequently used procedure for adjusting the alpha level is called the **Bonferroni technique,** and it is quite simple for the researcher to apply or for consumers of research to understand. When there is a desire on the part of the researcher to hold the Type I error in the full study equal to a selected value, the alpha levels for the various tests being conducted must be chosen such that the sum of the individual alpha levels is equivalent to the full-study alpha criterion. This is usually accomplished by simply dividing the desired Type I error risk for the full study by the number of times the hypothesis testing procedure is going to be used. Excerpts 8.19 and 8.20 illustrate nicely how the Bonferroni technique works. In the first of these excerpts, 10 different tests were conducted, so the researchers divided .05 by 10; in Excerpt 8.20, the adjusted alpha level became .002, because 22 tests were conducted.

EXCERPTS 8.19–8.20 • *The Bonferroni Adjustment Procedure*

We used a Bonferroni adjusted significance criterion of .005 (.05/10) to correct for multiple tests.

Source: Prime, J. L., Carter, N. M., & Welbourne, T. M. (2009). Women "take care," men "take charge": Managers' stereotypic perceptions of women and men leaders. *Psychologist-Manager Journal, 12*(1), 25–49.

[A] total of 22 separate ANOVAs were performed. In order to hold the experiment-wise Type I error rate to less than 5%, the *p* values were adjusted for multiple comparisons (.05/22 = *p* < .002).

Source: Tasko, S. M., & Greilick, K. (2010). Acoustic and articulatory features of diphthong production: A speech clarity study. *Journal of Speech, Language & Hearing Research, 53*(1), 84–99.

In Excerpt 8.20, notice the term **experimentwise Type I error rate,** which refers to the probability that one or more of the multiple tests being conducted will end up rejecting a true null hypothesis. Thus, the numerical value of .05 that appears in Excerpts 8.19 and 8.20 can be thought of, and referred to as, the *desired experimentwise Type I error rate.* Most researchers—like the ones who gave us Excerpt 8.19—do not use this term, but instead say simply that they have applied the Bonferroni adjustment.

Because the Bonferroni technique leads to a more rigorous alpha level for each of the separate tests being conducted, each of those tests becomes more demanding. In other words, Bonferroni-adjusted alpha levels (as compared with an unadjusted level of significance) create a situation wherein the sample data must be even more inconsistent with null expectations before the researcher is permitted to reject H_0. If the researcher makes a decision about H_0 by comparing the data-based *p*-value against the adjusted alpha, that alpha criterion is smaller and therefore harder to beat. Or, if each test's calculated value is compared against a critical value, the researcher finds that the Bonferroni technique has again created a more stringent criterion. Thus, it does not make any difference which of these two paths the researcher takes in moving from the sample data to the ultimate decision about the null hypothesis; either way, more protection against Type I errors is brought about by making it harder for the researcher to reject H_0.

Excerpt 8.21 illustrates how the Bonferroni technique brings about a more demanding assessment of each study's set of statistical comparisons. Initially, the two groups of physicians were compared on each of the 66 survey items, with alpha set at .05. When the Bonferroni adjustment was made, the new alpha level became much more demanding, with the result being that half of the initial significant differences vanished. Take another look at Excerpt 8.18 to see a similar example. In

EXCERPT 8.21 • *Why Bonferroni Makes It Harder to Reject H_0*

[C]omparisons were conducted to detect statistical differences between the AAFP and AAFP NRN respondents on the 66 preselected survey items. . . . Overall, in 12% (8 of 66) of items across the 3 surveys, there was a significant difference between AAFP NRN and AAFP physicians. After correcting statistically for multiple comparisons using the Bonferroni technique, only 4 (6%) of these differences remained ($P < .001$).

Source: Galliher, J. M., Bonham, A. J., Dickinson, L. M., Staton, E. W., & Pace, W. D. (2009). Representativeness of PBRN physician practice patterns and related beliefs: The case of the AAFP National Research Network. *Annals of Family Medicine, 7*(6), 547–554.

that excerpt, Bonferroni was not used. If alpha had been adjusted (from .05 to .0083) because a total of six correlations were computed, the *r*s of −.16 and .14 would not have been significant.

It may seem odd that researchers who want to reject their null hypothesis choose to apply the Bonferroni technique and thereby make it more difficult to accomplish their goal. However, researchers who use the Bonferroni technique are not doing something stupid, self-defeating, or inconsistent with their own objectives. Although the Bonferroni technique does, in fact, create a more demanding situation for the researcher, it does not function to pull something legitimate out of reach. Instead, this technique serves the purpose of helping the researcher pull in the reins so he or she is less likely to reach out and grab something that, in reality, is nothing at all. The Bonferroni technique, of course, does not completely eliminate the chance that a Type I error will be made, but it does eliminate the problem of an *inflated* Type I error risk.

Although the Bonferroni procedure is the most frequently used technique for dealing with the inflated Type I error problem, other procedures have been developed to accomplish the same general procedure. One of these is formally called the Sidák modification of Dunn's procedure. Excerpt 8.22 shows how the **Dunn–Sidák modification** works. In this excerpt, 0.01695 is the adjusted level of significance

EXCERPT 8.22 • *The Dunn–Sidák Adjustment Procedure*

To take into account the greater probability of a type I error due to multiple comparisons, the level of significance for these Student's *t* tests was pre-set by 0.01695 ($\alpha = 1-[1-0.05]^{1/c}$); c = number of comparisons = 3; Dunn–Sidák correction).

Source: Mooij-van Malsen, J. G., van Lith, H. A., Oppelaar, H., Olivier, B., & Kas, M. J. H. (2009). Evidence for epigenetic interactions for loci on mouse chromosome 1 regulating open field activity. *Behavior Genetics, 39*(2), 176–182.

for the researchers' situation of wanting to conduct three tests. The information inside the parentheses is the adjustment formula, with α representing the modified alpha level, with 0.05 representing the desired study-wide Type I error risk, and with c representing the number of tests being conducted. To show yourself that the Dunn–Sidák and Bonferroni procedures are exceedingly similar, divide .05 by 3 and then compare the result to the modified alpha level in Excerpt 8.22.

A Few Cautions

As we come to the close of our two-chapter treatment of hypothesis testing, I want to offer a few more cautions that should assist you as you attempt to make sense out of technical summaries of empirical investigations. These tips (or warnings!) are different from the ones provided at the end of Chapter 7, so you may profit from a review of what I said there. In any event, here are four more things to keep in mind when you come across statistical inferences based on the hypothesis testing procedure.

Two Meanings of the Term Effect Size

When the seven-step version of hypothesis testing is used, the researcher computes an effect size. We saw three examples of this being done in Excerpts 8.3 through 8.5. The important thing to note is that the effect size involved in Step 7 of this version of hypothesis testing procedure is based on the sample data. This kind of effect size is *computed* using the evidence gathered in the researcher's study.

When the nine-step version of hypothesis testing is employed, a different kind of effect size comes into play. Within this strategy, researchers *specify* (rather than compute) the effect size, and this is done prior to the collection and examination of any data. When researchers specify the effect size in Step 4 of the nine-step version of hypothesis testing, they are not making a predictive statement as to the magnitude of the effect that will be found once the data are analyzed. Rather, they are indicating the minimum size of an effect that they consider to have practical significance. Most researchers hope that the magnitude of the true effect size exceeds the effect size specified prior to the collection of any data.

It is unfortunate that the same term—*effect size*—is used by researchers to refer to two different things. However, a careful consideration of context ought to clarify which kind of effect size is being discussed. If reference is made to the *effect size* within the research report's method section (and specifically when the sample size is being discussed), then it is likely that the nine-step version of hypothesis testing was used, with the effect size being a judgment call as to the dividing line between trivial and important findings. If, however, reference is made to the *effect size* during a presentation of the obtained results, this effect size is probably a data-based measure of how false the null hypothesis seems to be.

Small, Medium, and Large Effect Sizes

Regardless of whether a researcher's effect size is computed (in a post hoc sense) from the sample data or specified in the planning stages of the study (to help determine the needed sample size), it is not uncommon for the researcher to refer to the effect size as being *small, medium,* or *large.* As I indicated earlier in this chapter, criteria have been developed that help to define these standardized effect sizes. For example, the popular effect size standards for correlations indicate that .1, .3, and .5 represent small, medium, and large effect sizes, respectively.

Unfortunately, the criteria for small, medium, and large effect sizes vary depending on the study's statistical focus and the kind of effect size that is computed or specified. For example, if the effect size *d* is used in conjunction with a study that compares two sample means, the criteria for standardized effect sizes say .2 is small, .5 is medium, and .8 is large. Clearly, these criteria are different from the ones cited in the previous paragraph for a correlation coefficient.

In an effort to help you keep things straight when it comes to the criteria for standardized effect sizes, I have inserted a small chart into several of the following chapters. Each of these charts shows the names for the effect size measures associated with the statistical tests discussed in a given chapter, and then the criteria for small, medium, and large are presented. The information in these charts may prove useful to you, because it is not unusual to see a research report that contains a computed effect size (such as *d*) with absolutely no discussion about its meaning.

The Simplistic Nature of the Six-Step
Version of Hypothesis Testing

Many researchers test null hypotheses with the six-step version of the hypothesis testing procedure. This is unfortunate, because the important distinction between statistical and practical significance is not addressed in any way whatsoever by this simplistic approach to testing null hypotheses. Consequently, the outcome is ambiguous no matter what decision is reached about H_0. A reject decision may have been caused by a big difference between the single-number summary of the sample evidence and the pinpoint number in H_0; however, that same decision may have come about by a small difference being magnified by a giant sample size. Likewise, a fail-to-reject decision might be the result of a small difference between the sample evidence and the null hypothesis; however, the researcher's decision not to reject H_0 may have been the result of a big difference camouflaged by a small sample size.

To see examples of these two undesirable scenarios described in the previous paragraph, take another look at Excerpts 8.14 and 8.15. Before doing that, however, please formulate an answer to each of the two questions I'd like to ask you. In Excerpt 8.14, the researchers conceded that a limiting feature of their study reduced power and may have brought about Type II errors. Can you guess what that limitation was? In Excerpt 8.15, the researchers declared that one of their statistically significant findings "is not of practical significance." What feature of this study was primarily responsible for the caution that the statistically significant results were not very noteworthy? OK. Now see if your answers are correct.

It is relatively easy for a researcher to conduct a study using the basic six-step version of hypothesis testing. It is harder to use the nine-step version. Nevertheless, there is a giant payoff (for the researcher and for those who read or hear the research report) if the researcher (1) decides, in the planning stage of the study, what kind of results will represent trivial deviations from the null hypothesis versus results that are important, and (2) uses this informed opinion within the context of an a priori power analysis to determine how large the samples should be. When you come across a research report indicating that these two things were done, give that study's researcher(s) some big, big bonus points!

Despite its limitations, the basis six-step version of hypothesis testing is widely used. Whenever you encounter a researcher who has used this more simplistic version of hypothesis testing, *you* must be the one who applies the important seventh step. This is not an impossible task; it is something you *can* do!

If a correlation coefficient is reported to be statistically significant, look at the size of the r and ask yourself what kind of relationship (weak, moderate, or strong) was revealed by the researcher's data. Better yet, square the r and then convert the resulting coefficient of determination into a percentage; then make your own judgment as to whether a small or large amount of variability in one variable is being explained by variability in the other variable. If the study focuses on means rather than correlations, look carefully at the computed means. Ask yourself whether the observed difference between two means represents a finding that has practical significance.

I cannot overemphasize my warning that you can be (and will be) misled by many research claims if you look only at p-statements when trying to assess whether results are important. Many researchers use the simple six-step version of hypothesis testing, and the only thing revealed by this procedure is a yes or no answer to the question, "Do the sample data deviate from H_0 more than we would expect by chance?" Even if a result is statistically significant with $p < .0001$, it may be the case that the finding is completely devoid of *any* practical significance!

Inflated Type I Error Rates

My final caution is simply a reiteration of something I said earlier in this chapter. This has to do with the heightened chance of a Type I error when multiple tests are conducted simultaneously. This is a serious problem in scientific research, and this caution deserves to be reiterated.

Suppose a researcher measures each of several people on seven variables. Also suppose that the true correlation between each pair of these variables is exactly 0.00 in the population associated with the researcher's sample. Finally, suppose our researcher computes a value for r for each pair of variables, tests each r to see if it is significantly different from 0.00, and then puts the results into a correlation matrix.

If the .05 level of significance is used in conjunction with the evaluation of each r, the chances are about 66 percent that at least one of the rs will turn out to be significant. In other words, even though the alpha level is set equal to .05 for each separate test conducted, the collective Type I error risk has ballooned to about .66 because 21 separate tests are conducted.

My caution here is simple. Be wary of any researcher's conclusions if a big deal is made out of an unreplicated single finding of significance when the hypothesis testing procedure is used simultaneously to evaluate many null hypotheses. In contrast, give researchers extra credit when they apply the Bonferroni or Dunn–Sidák technique to hold down their study-wide Type I error risk.

Review Terms

Beta error	Power analysis
Bonferroni technique	Practical significance
Dunn–Sidák modification	Raw effect size
Effect size	Sample size determination
Experimentwise Type I	Small standardized effect size
error rate	Standardized effect size
Inflated Type I error risk	Statistical significance
Large standardized	Type I error risk
effect size	Type II error risk
Medium standardized	
effect size	

The Best Items in the Companion Website

1. An email message sent from the author to his students entitled "Binoculars and Significance."
2. An interactive online quiz (with immediate feedback provided) covering Chapter 8.
3. Ten misconceptions about the content of Chapter 8.
4. An email message sent by the author to his students concerning the seven-step and nine-step versions of hypothesis testing.
5. An interactive online resource called "Statistical Power."

To access the chapter outline, practice tests, weblinks, and flashcards, visit the companion website at http://www.readingstats.com.

Review Questions and Answers begin on page 531.

Statistical Inferences Concerning Bivariate Correlation Coefficients

In Chapter 3, we considered several descriptive techniques used by researchers to summarize the degree of relationship that exists between two sets of scores. In this chapter, we examine how researchers deal with their correlation coefficients inferentially. Stated differently, the techniques to be considered here are the ones used when researchers have access only to sample data, but wish to make educated guesses as to the nature of the population(s) associated with the sample(s). As you will see shortly, the techniques used most frequently to do this involve hypothesis testing. Occasionally, however, inferential guesses are made through the use of confidence intervals.

We begin this chapter with a consideration of the statistical tests applied to various bivariate correlation coefficients, along with an examination of the typical ways researchers communicate with the results of their analyses. I also point out how the Bonferroni technique is used in conjunction with tests on correlation coefficients, how researchers compare two (or more) correlation coefficients to see if they are significantly different, and how statistical tests can be applied to reliability and validity coefficients. Finally, I provide a few tips designed to help you become a more discerning consumer of research claims that emanate from studies wherein inferential statistics are applied to correlation coefficients.

Statistical Tests Involving a Single Correlation Coefficient

Later in this chapter, we consider the situation in which data are analyzed to see if a significant difference exists between two or more correlation coefficients. Before doing that, however, we consider the simpler situation where the researcher has a single sample and a single correlation coefficient. Although simple in nature because only one sample is involved, the inferential techniques focused on in the first part of this chapter are used far more frequently than the ones that involve comparisons between/among correlation coefficients.

The Inferential Purpose

Figure 9.1 has been constructed to help clarify what researchers are trying to do when they apply an inferential test to a correlation coefficient. I have set up this picture to make it consistent with a hypothetical study involving Pearson's product–moment correlation. However, by changing the symbols that are included, we could make our picture relevant to a study wherein any other bivariate correlation coefficient is tested.

As Figure 9.1 shows, a correlation coefficient is computed on the basis of data collected from a sample. Although the sample-based value of the correlation coefficient is easy to obtain, the researcher's primary interest lies in the corresponding value of the correlation in the population from which the sample has been drawn. However, the researcher cannot compute the value of the correlation coefficient in the population because only the objects (or persons) in the sample can be measured. Accordingly, an inference (i.e., educated guess) about the parameter value of the correlation is made on the basis of the known value of the statistic.

The nature of the inference that extends from the sample to the population could take one of two forms depending on whether the researcher wishes to use the techniques of estimation or to set up and evaluate a null hypothesis. Near the

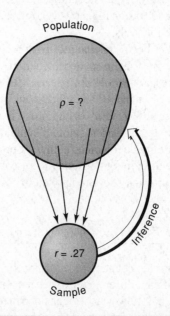

FIGURE 9.1   *The Inferential Purpose of a Test on a Correlation Coefficient*

end of the chapter, we examine the way confidence intervals are sometimes used to make inferences about correlation coefficients. We first turn our attention to the way researchers set up, evaluate, and report what happens to correlational null hypotheses.

The Null Hypothesis

When researchers are concerned about the relationship between two variables in a single population but can collect data only from a sample taken from that population, they are likely to attack their inferential question by means of hypothesis testing. In doing this, a null hypothesis serves as the hub around which the other involved statistical elements revolve.

In dealing with a single correlation, the null hypothesis is a pinpoint statement as to a possible value of the correlation in the population. Although researchers have the freedom to choose any value between -1.00 and $+1.00$ for inclusion in H_0, the correlational null hypothesis is almost always set up to say that there is, in the relevant population, a zero correlation between the two variables of interest. Stated symbolically, this null hypothesis takes the form $H_0: \rho = 0$.

In Excerpts 9.1 and 9.2, we see cases where researchers stated or alluded to the correlational null hypothesis. In the first of these excerpts, the null and alternative hypotheses are clearly articulated. These statements of H_0 and H_a appeared in the research report prior to any results being presented. The passage in Excerpt 9.2, taken from a different article, was located in a discussion of the study's results. In this excerpt, note the words "significantly different from zero."

EXCERPTS 9.1-9.2 • *The Null Hypothesis for Testing Correlations*

The purpose of this study is to examine differences in the relationships to GNP of the business cycles of various industrial sectors in Turkey. . . . The statistical significance of the correlation coefficients was evaluated by testing the null hypothesis that the unknown population correlation, ρ, is equal to zero, $H_o: \rho = 0$ against the two sided alternative $H_A: \rho \neq 0$, using the sample correlation coefficient r, obtained from our sample of industrial output.

Source: Bayar, G. (2009). Business cycles and foreign trade in the Turkish economy. *Middle Eastern Finance and Economics, 3,* 39–48.

The results indicated that the strength of association between the two variables was relatively high (r = .702), and that the correlation coefficient was significantly different from zero (P < 0.001).

Source: Wan, J., & Eastmond, N. (2008). A study on the use of cooperative learning strategies in a computer literacy course. *College & University Media Review, 14,* 21–63.

Because the correlational null hypothesis is usually set up to say that a zero correlation exists in the population, most researchers do not state (or allude to) the H_0 being tested. Instead, they take for granted that recipients of their research reports will know that the inferential conclusions refer to a null hypothesis of no relationship. Consider, for example, Excerpts 9.3 and 9.4. In each case, the sample r presented in the report was compared against the null value of zero—even though the tested H_0 never appeared in the technical write-ups.

EXCERPTS 9.3–9.4 • *Tests on r with No Reference to the Null Value of Zero*

The degree to which spirituality was used to cope was weakly related with the number of years respondents reported going to TASO ($r = .177, p = .037$).

Source: Hodge, D. R., & Roby, J. L. (2010). Sub-Saharan African women living with HIV/AIDS: An exploration of general and spiritual coping strategies. *Social Work*, *55*(1), 27–37.

Job alternatives were not significantly correlated with strategy fit ($r = .13$, *ns*), organizational commitment ($r = .06$, *ns*), or intention to stay ($r = -.05$, *ns*).

Source: Da Silva, N., Hutcheson, J., & Wahl, G. D. (2010). Organizational strategy and employee outcomes: A person–organization fit perspective. *Journal of Psychology*, *144*(2), 145–161.

In light of the fact that very few researchers either state the null hypothesis when applying a test to a sample correlation coefficient or refer to H_0's pinpoint number when discussing their results, you frequently will be forced into the position of having to guess what a researcher's H_0 was. In these situations, a safe bet is that H_0 was a statement of no relationship in the population. If researchers set up a null hypothesis that specifies a population correlation different from zero, I am confident that they will specify H_0's pinpoint number.

Deciding If r Is Statistically Significant

In conducting a statistical test on a single correlation coefficient, the value of r usually functions as the data-based calculated value. As you will soon see, statistical tests on means, variances, or percentages involve calculated values that are different from the means, variances, or percentages computed from the sample(s). However, with correlations, the sample-based correlation coefficient is, in its raw form, the calculated value.

When the sample value of r is considered to be the calculated value, there are two ways to determine whether it is statistically significant. If the data have been analyzed on a computer or Internet website, then the data-based value of p can be compared against the level of significance. If p is equal to or smaller than α, the null hypothesis will be rejected. In Excerpt 9.5, you can see this decision-making approach in operation.

Excerpt 9.6 illustrates the second method for determining whether a sample r is statistically significant. If the data have not been analyzed such that a data-based p is available, the researcher can compare the sample value of r against a tabled critical value. If the former equals or exceeds the latter, H_0 will be rejected. On occasion (but not often), you may come across a research report that includes the tabled critical value. An example of such a situation is presented in Excerpt 9.6. The single sentence in this excerpt was positioned beneath a correlation matrix that contained 91 correlation coefficients. To tell which of the correlation coefficients were significant (and at what level), the researchers as well as the readers of their research report had to compare each r against critical values shown in Excerpt 9.6.

EXCERPTS 9.5–9.6 • *Two Ways to Decide if r Is Statistically Significant*

The openness and honesty dimension correlated highest with culture having a correlation coefficient of 0.76. This dimension is believed to be of most importance in the trust-building process. . . . The resultant p-value of <0.001 was compared to a significance level of 0.05, indicating that there is a correlation between the trust dimensions and culture.

Source: Alston, F., & Tippett, D. (2009). Does a technology-driven organization's culture influence the trust employees have in their managers? *Engineering Management Journal*, *21*(2), 3–10.

For correlations greater than .20, $p < .05$; for correlations greater than .28, $p < .01$; and for correlations greater than .35, $p < .001$.

Source: Griffin, M. A., Parker, S. K., & Mason, C. M. (2010). Leader vision and the development of adaptive and proactive performance: A longitudinal study. *Journal of Applied Psychology*, *95*(1), 174–182.

One-Tailed and Two-Tailed Tests on r

Most researchers conduct their tests on r in a two-tailed fashion because they would like to know, as best they can, whether there is a positive or negative correlation in the population of interest. Sometimes, however, researchers use a one-tailed test to evaluate a sample r. In Excerpts 9.7 and 9.8, we see examples of these two options for testing any r.

EXCERPTS 9.7–9.8 • *Two-Tailed and One-Tailed Tests on r*

The correlation was found to be .256 ($p < .001$, two-tailed) for the first sample, and .279 ($p < .001$, two-tailed) for the second sample.

Source: Schermer, J. A., & Verno, P. A. (2010). The correlation between general intelligence (g), a general factor of personality (GFP), and social desirability. *Personality and Individual Differences, 48*(2), 187–189.

There was an expected negative correlation for ACS and STAI-t ($r = -.358$; $p = .031$; one-tailed).

Source: Putman, P., van Peer, J., Maimari, I., & van der Werff, S. (2010). EEG theta/beta ratio in relation to fear-modulated response-inhibition, attentional control, and affective traits. *Biological Psychology, 83*(2), 73–78

Unlike Excerpts 9.7 and 9.8, the typical research report dealing with tested correlations does not indicate whether r was evaluated in a one-tailed or two-tailed fashion. That is the case with most of the excerpts in this chapter, because there usually is not even a hint as to whether the alternative hypothesis was nondirectional (with H_a: $\rho \neq 0.00$) or directional (with H_a stating either $\rho > 0.00$ or $\rho < 0.00$). Why is this the case?

Because the vast majority of researchers conduct their tests on r in a two-tailed fashion, researchers presume that you understand this even if they do not say so directly. Therefore, you should guess that any test on r was conducted in a two-tailed manner unless the researcher says otherwise. When researchers perform a one-tailed test on r, they will be sure to point this out (as was the case in Excerpt 9.8).

Tests on Specific Kinds of Correlation

Until this point, we have been discussing tests of correlation coefficients in the generic sense. However, there is no such thing as a generic correlation. When correlating two sets of data, a specific correlational procedure must be used, with the choice usually being influenced by the nature of variables or the level of measurement of the researcher's instruments. As you may recall from Chapter 3, there are many different kinds of bivariate correlations: Pearson's, Spearman's, biserial, point biserial, phi, tetrachoric, and so on.

With any of the various correlation procedures, a researcher can apply the hypothesis testing procedure. When researchers report having tested r without specifying the type of correlation that was computed, you should presume that r represents Pearson's product–moment correlation. Thus, it is a good guess that the correlations presented or referred to in Excerpts 9.1 through 9.8 were all Pearson rs.

In Excerpts 9.9 through 9.11, we now see illustrations of other kinds of bivariate correlation coefficients being subjected to inferential testing.

EXCERPTS 9.9–9.11 • *Tests on Specific Kinds of Correlation*

The correlation between alcohol consumption and religiosity for the entire sample was $r_s = -.33$ $(p < .001)$.

Source: Wells, G. M. (2010). The effect of religiosity and campus alcohol culture on collegiate alcohol consumption. *Journal of American College Health*, *58*(4), 295–304.

Using this critical value [.07] for comparison, it is apparent that the point-biserial correlation for Q27 ($r = .06$) is not significant.

Source: Nadelson, L. S., & Southerland, S. A. (2010). Development and preliminary evaluation of the measure of understanding of macroevolution: Introducing the MUM. *Journal of Experimental Education*, *78*(2), 151–190.

Finally, Phi correlation was used to determine if articles discussed the MMR and thimerosal examples together. A negative correlation ($r_\Phi = -.234, p < .001$) suggested that when one topic appeared in an article, the other did not.

Source: Clarke, C. E. (2010). A case of conflicting norms? Mobilizing and accountability information in newspaper coverage of the autism–vaccine controversy. *Public Understanding of Science*, in press.

Tests on Many Correlation Coefficients (Each of Which Is Treated Separately)

In most of the excerpts presented so far in this chapter, inferential interest is focused on a single correlation coefficient. Although some researchers set up only one correlational null hypothesis (because each of their studies involves only one correlation coefficient), most researchers have two or more correlations that are inferentially tested in the same study. Our objective now is to consider the various ways in which such researchers present their results, to clarify the fact that a separate H_0 is associated with each correlation coefficient that is computed, and to consider the way in which the Bonferroni adjustment technique can help the researcher avoid the problem of an inflated Type I error risk.

Tests on the Entries of a Correlation Matrix

As we saw in Chapter 3, a **correlation matrix** is an efficient way to present the results of a correlational study in which there are three or more variables with a correlation coefficient computed between each possible pair of variables.[1] Typically, each of the entries within the correlation matrix will be subjected to an inferential test. In Excerpt 9.12, we see an illustration of this situation.

EXCERPT 9.12 • *Tests of Each r in a Correlation Matrix*

TABLE 4   *Correlation matrix showing Pearson correlations between locus of control on social life for all coping strategies*

	1	2	3	4	5	6
1. Locus of control social life	1					
2. Problem solving as coping	−.04	1				
3. Social support as coping	−.02	.57*	1			
4. Cognitive restructuring as coping	.03	.62*	.50*	1		
5. Wishful thinking as coping	−.19*	.25*	.28*	.14*	1	
6. Self-criticism as coping	−.10*	.31*	.27*	.17*	.59*	1

*Correlation is significant at the .001 level (2-tailed).

Source: Ljoså, C. H., & Lau, B. (2009). Shiftwork in the Norwegian petroleum industry: Overcoming difficulties with family and social life—a cross sectional study. *Journal of Occupational Medicine and Toxicology, 4*(22), 1–10.

The correlation matrix in Excerpt 9.12 contains 15 bivariate correlations that were computed among the study's six variables. Each of the resulting *r*s was subjected to a separate statistical test, and in each case the null hypothesis was a no-relationship statement about the population associated with the single sample of 1,697 Norwegian petroleum workers used in the investigation. As you can see, 12 of the 15 correlations turned out to be significant, with $p < .001$.

Tests on Several Correlation Coefficients Reported in the Text

Research write-ups often present the results of tests on many correlation coefficients in the text of the article rather than in a table. Excerpt 9.13 illustrates this approach to summarizing the results of inferential tests on multiple correlations.

---

[1]Whenever a correlation coefficient is computed, it is really not the variables per se that are being correlated. Rather, it is the measurements of one variable that are correlated with the measurements of the other variable. This distinction is not a trivial one, because it is possible for a low correlation coefficient to grossly underestimate a strong relationship that truly exists between the two variables of interest. Poor measuring instruments could create this anomaly.

EXCERPT 9.13 • *Tests on Several rs with Results in the Text*

Correlations between social distance scores and level of comfortability yielded results of $r = 0.34$ ($p = 0.036$) for white participants, and $r = 0.49$ ($p = 0.001$) for black African participants. Correlations between affective prejudice scores and level of comfortability yielded results of $r = -0.54$ ($p < 0.001$) for white participants, and $r = -0.44$ ($p = 0.005$) for black African participants.

Source: Schrieff, L. E., Tredoux, C. G., Finchilescu, G., & Dixon, J. A. (2010). Understanding the seating patterns in a residence-dining hall: A longitudinal study of intergroup contact. *South African Journal of Psychology*, *40*(1), 5–17.

In the study associated with Excerpt 9.13, each of the individuals in two groups—White and Black undergraduates attending residential colleges—was measured on several variables, three of which were social distance, prejudice, and level of comfort when eating in the dining hall at a table with members of other races. Excerpt 9.13 shows the obtained correlations, for each racial group, between "comfortability" and each of the other two variables. These significant correlations, the authors stated, suggest that more positive "outgroup" attitudes are associated with higher levels of comfort in being with members of other races.

The Bonferroni Adjustment Technique

In Chapter 8, I explained why researchers sometimes use the **Bonferroni technique** to adjust their level of significance. As you may recall, the purpose of doing this is to hold down the chances of a Type I error when multiple tests are conducted. I also hope that you remember the simple mechanics of the Bonferroni technique: Simply divide the desired study-wide Type I error risk by the number of tests being conducted.

In Excerpt 9.14, we see an example in which the Bonferroni technique was used in conjunction with correlation coefficients. This excerpt provides a nice review as to why the Bonferroni technique is used and how it works. In this case, the Bonferroni procedure reduced the level of significance from .05 to .005. This may

EXCERPT 9.14 • *Use of the Bonferroni Correction with Tests on Several Correlation Coefficients*

Using the Bonferroni approach to control for type I error across the 10 correlations, a *p* value of less than .005 (.05/10 = .005) was specified for significance.

Source: Chew, W., Osseck, J., Raygor, D., Eldridge-Houser, J., & Cox, C. (2010). Developmental assets: Profile of youth in a juvenile justice facility. *Journal of School Health*, *80*(2), 66–72.

strike you as an overly severe change in the alpha level. However, if a researcher uses the hypothesis testing procedure several times (as was the case in Excerpt 9.14), the actual Type I error risk become greatly inflated if the level of significance is not made more rigorous. The researchers associated with this excerpt would receive bonus points in our evaluation of their study because they recognized that they must "pay a price" for testing multiple correlations.

When you come across the report of a study that presents the results of inferential tests applied to several correlation coefficients, try to remember that the conclusions drawn can be radically different depending on whether some form of Bonferroni adjustment technique is used. For example, Excerpt 9.6 came from a study in which bivariate correlation coefficients were computed among 14 variables and presented in a large correlation matrix. Of these 91 rs, 48 were reported to be statistically significant. If the Bonferroni adjustment procedure had been used (because separate tests were conducted on so many correlations), the null hypothesis associated with 25 of these 48 correlations would have been retained rather than rejected.

Tests of Reliability and Validity Coefficients

As indicated in Chapter 4, many of the techniques for estimating reliability and validity rely totally or partially on one or more correlation coefficients. After computing these indices of data quality, researchers sometimes apply a statistical test to determine whether or not their reliability and validity coefficients are significant. Excerpts 9.15 and 9.16 illustrate such tests.

EXCERPTS 9.15–9.16 • *Tests of Reliability and Validity Coefficients*

Test–retest reliability was conducted by re-administering one session of the NVLA within 1 week of the first administration for 16 students. Test–retest correlation coefficient for the total test score of the NVLA was statistically significant ($p < .001$) at .970.

Source: Baker, J. N., Spooner, F., Shlgrim-Delzell, L., Flowers, C., & Browder, D. M. (2010). A measure of emergent literacy for students with severe developmental disabilities. *Psychology in the Schools, 47*(5), 501–513.

The predictive validation study of the GMAT was conducted . . . via Pearson correlation analysis procedure using a one-tailed test of significance. . . . The correlation between the GMAT score and the final MBA GPA was .60 ($p < .01$). . . . Results from the present study suggest that the GMAT is a strong predictor of academic performance in Central Europe and the Middle East.

Source: Koys, D. (2010). GMAT versus alternatives: Predictive validity evidence from Central Europe and the Middle East. *Journal of Education for Business, 85*(3), 180–185.

When you come across a research report in which reliability and validity co-efficients are tested for significance, be careful to focus your attention on the size of the coefficient (which should be large), not the reported *p*-level (no matter how small it may be). In Excerpt 9.15, for example, it is nice to know that the reported test–retest reliability coefficient turned out to be significantly different from zero (with $p < .001$); however, what is far more important is the impressive size of the stability coefficient: .97. Consider now both the *p* and the *r* in Excerpt 9.16. The *p* looks fairly good (because it is less than .01), but the validity coefficient of .60 is not so impressive (despite the author's claim that the GMAT was shown to be a "strong predictor"). In this study, the GMAT explained only 36 percent of the variability among the MBA students' GPAs.

Statistically Comparing Two Correlation Coefficients

At times, researchers have two correlation coefficients that they wish to compare. The purpose of such a comparison is to determine whether a significant difference exists between the two *r*s, with the null hypothesis being a statement of no difference between the two correlations in the population(s) associated with the study. For such tests, a no-difference H_0 is fully appropriate.

Figure 9.2 is designed to help you distinguish between two similar but different situations where a pair of correlation coefficients is compared. In Figure 9.2(a), we see that a sample is drawn from each of two populations, with a bivariate correlation coefficient computed, in each sample, between the same pair of variables. In this picture, these variables are labeled *X* and *Y;* the two variables might be height and weight, running speed and swimming speed, or any other pair of variables. The null hypothesis is that correlation between *X* and *Y* has the same value in each of the two populations. Notice that the single inference here is based on both groups of sample data and is directed toward the *set* of populations associated with the study.

In Figure 9.2(b), we see that a single sample is drawn from one population, but two correlation coefficients are computed on the basis of the sample data. One correlation addresses the relationship between variables *X* and *Y,* whereas the other correlation is concerned with the relationship between variables *X* and *Z*. The null hypothesis in this kind of study is that the parameter value of the correlation between *X* and *Y* is equal to the parameter value of the correlation between *X* and *Z*. Based on the sample's pair of correlation coefficients, a single inference is directed toward the unknown values of the pair of correlations in the one population associated with the study.

Excerpts 9.17 and 9.18 illustrate the two situations depicted in Figure 9.2. In the first of these excerpts, the correlation between two variables (endorsement of multiculturalism and ethnic identity) from one group of high school students ($n = 186$) in the Netherlands was compared with the correlation between these same two variables from a different group of students ($n = 140$) attending the same schools. In Excerpt 9.18, we again see that two correlation coefficients were statistically

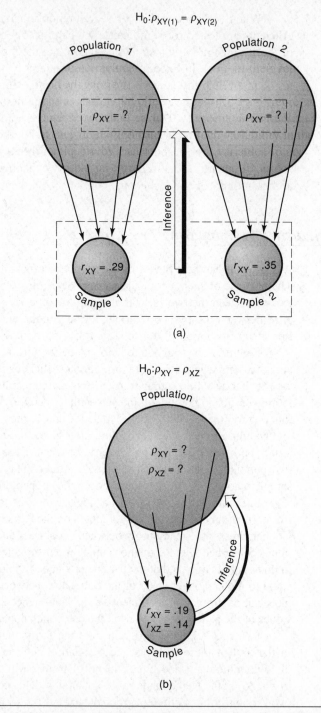

(a)

(b)

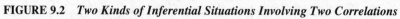

FIGURE 9.2   Two Kinds of Inferential Situations Involving Two Correlations

EXCERPTS 9.17–9.18 • *Statistical Comparison of Two*
Correlation Coefficients

There was a significant positive association between the endorsement of multiculturalism and ethnic identification for the ethnic minority group ($r = .26, p < .01$), and a negative association for the Dutch participants ($r = -.22, p < .01$). The difference between these two correlations is significant, $z = 4.25, p < .001$.

Source: Verkuyten, M. (2009). Self-esteem and multiculturalism: An examination among ethnic minority and majority groups in the Netherlands. *Journal of Research in Personality*, *43*(3), 419–427.

Fisher's Z-test indicated that, as predicted, the correlation between Internal Attributions and Guilt [$r = .52$] was significantly greater than the correlation between Guilt and External Attributions [$r = .21$], $Z = 2.56, p = .011$.

Source: Glenn, S. A., & Byers, E. S. (2009). The roles of situational factors, attributions, and guilt in the well-being of women who have experienced sexual coercion. *Canadian Journal of Human Sexuality*, *18*(4), 201–219.

---

compared. Here, however, the situation was different. In this case, the individuals in a single group—104 women who had been sexually coerced—were measured on three variables: internal attributions (i.e., blaming oneself), external attributions (blaming someone else), and guilt. Correlations were computed between guilt and each of the attribution variables, and then these two *r*s were statistically compared.

In Excerpts 9.17 and 9.18, the two correlations were compared by means of a statistical test called a *z*-test.[2] Sometimes, a *t*-test is used to make the correlational comparison. The only thing you need to know about these statistical tests is that in each case, the null hypothesis being evaluated is that the population values associated with the two sample correlations are identical. Stated symbolically, $H_0: \rho_1 = \rho_2$. Note that this null hypothesis says that the *difference* between the correlations is zero, not that the correlations themselves are zero. Thus, two large *r*s that would each be significantly different from zero if considered individually could end up causing H_0 to be *retained* when compared against each other. Although this did not happen in the two excerpts we have just considered, this outcome does occur.

The Use of Confidence Intervals around Correlation Coefficients

When researchers subject a data-based correlation coefficient to an inferential statistical procedure, they probably do so via hypothesis testing. All of the excerpts presented so far in this chapter have been taken from studies in which this testing

---

[2]The term **Fisher's *r*-to-*z* transformation** is often used to describe the test that's conducted to see if two correlations are significantly different from each other.

strategy was used. It is possible, however, for a researcher to deal inferentially with a correlation coefficient simply by placing a confidence interval around the sample value of r. Oddly, few researchers do this.

As was indicated previously, confidence intervals can be used *within* the context of hypothesis testing. In applying inferential tests to correlation coefficients, most researchers do not place confidence intervals around their sample values of r, but a few do. We see an illustration of this use of confidence intervals in Excerpt 9.19.

EXCERPT 9.19 • *Use of a Confidence Interval to Test r*

Pearson's correlation analysis (with 95% confidence intervals) was used [but] no significant correlation was found between treatment duration and BMI SDS ($r = 0.027$, 95% CI -0.29 to 0.34; $p = 0.867$). . . .

Source: Rauchenzauner, M., Griesmacher, A., Tatarczyk, T., Haberlandt, E., Strasak, A., Zimmerhackl, L., et al. (2010). Chronic antiepileptic monotherapy, bone metabolism, and body composition in non-institutionalized children. *Developmental Medicine & Child Neurology*, *52*(3), 283–288.

In Excerpt 9.19, notice that the confidence interval extends from -0.29 to 0.34. Thus, the CI overlaps 0.00. Because of this, the researcher failed to reject the null hypothesis, as indicated by the phrase *no significant correlation*. The null hypothesis under investigation was never mentioned in the research report, but it took the form H_0: $\rho = 0.00$, as is usually the case when a single correlation coefficient is tested.

Cautions

I feel obligated to end this chapter by suggesting a few cautions that you should keep in mind when trying to decipher (and critique) research reports based on correlation coefficients. As you will see, my comments here constitute a reiteration of some of the points presented at the end of Chapter 3 as well as some of the points offered at the conclusions of Chapters 7 and 8.

Relationship Strength, Effect Size, and Power

Many researchers seem to get carried away with the p-levels associated with their correlation coefficients and thus seem to forget that the estimated strength of a relationship is best assessed by squaring the sample value of r. Discovering that a correlation coefficient is significant may not really be very important—even if the results indicate $p < .01$ or $p < .001$—unless the value of r^2 is reasonably high. The result may be significant in a statistical sense (thus indicating that the sample data are not likely to have come from a population characterized by H_0), but it may be quite insignificant in a practical sense.

Consider Excerpt 9.20. The researchers are correct in saying that their two variables were "significantly and negatively correlated." (With a sample as large as theirs, $N = 1,164$, a two-tailed test of their r yields a p-value of .0166.) Nevertheless, I cannot help but think that an r of $-.07$ carries with it very little practical significance. Yet, one-third of the research report's abstract discussed the relationship between perceived teacher discrimination and student academic achievement without showing the size of the sample correlation.

EXCERPT 9.20 • *A Significant r with Questionable Usefulness*

The correlation results revealed that teacher discrimination is significantly and negatively correlated with academic achievement for total sample ($r = -.07, p \leq .05$).

Source: Thomas, O. N., Caldwell, C. H., Faison, N., & Jackson, J. S. (2009). Promoting academic achievement: The role of racial identity in buffering perceptions of teacher discrimination on academic achievement among African American and Caribbean black adolescents. *Journal of Educational Psychology, 101*(2), 420–431.

To see an example in which the important distinction between statistical significance and practical significance *was* kept in mind, take a look at Excerpt 9.21. In this excerpt, notice that the researchers seem to focus their attention on the size of their rs rather than on their ability to say that those rs were statistically significant. Perhaps they did this out of an awareness that an r of .24 describes a relationship in which less than that 6 percent of the variability in one variable is associated with variability in the other variable.

EXCERPT 9.21 • *Expressed Concern for the Strength of Statistically Significant Correlations*

Pearson correlation analyses showed generally low correlations among variables included in this study. Although some correlation coefficients were statistically significant, none was greater than .24.

Source: Molfese, V. J., Rudasill, K. M., Beswick, J. L., Jacobi-Vessels, J. L., Ferguson, M. C., & White, J. M. (2010). Infant temperament, maternal personality, and parenting stress as contributors to infant developmental outcomes. *Merrill-Palmer Quarterly, 56*(1), 49–79.

In Chapter 8, I pointed out how researchers can apply a seven-step version of hypothesis testing by discussing the concept of effect size. This can be done, of course, with correlation coefficients. In fact, we saw a case in Excerpt 8.3, where the correlation between the ratings of quarterback attractiveness and playing ability was judged to be "small-to-medium."

TABLE 9.1 *Effect Size Criteria for Correlations*

Small	Medium	Large
$r = .1$	$r = .3$	$r = .5$

Note: These standards for judging relationship strength are quite general and should be changed to fit the unique goals of any given research investigation.

In Excerpt 9.22, we see a case in which a group of researchers compared their obtained correlation coefficient against some common effect size criteria. (These criteria have been duplicated and put into Table 9.1, because these criteria apply to any of the correlational procedures we have considered.) Because the researcher's sample value of r ($-.506$) was larger than the criterion value of .50, the researchers felt justified in saying that their computed correlation was "of a large effect size." The small p-value had nothing to do with the researchers' decision to attach this effect size label to their computed r.

EXCERPT 9.22 • *Using Effect Size Criteria with Correlations*

The PB index correlated negatively with sexual inhibition due to fear of performance consequences [$r = -.506, p < .001$]. The correlation coefficient was of a large effect size.

Source: Winters, J., Christoff, K., & Gorzalka, B. B. (2010). Conscious regulation of sexual arousal in men. *Journal of Sex Research, 46*(4), 330–343.

As indicated by the note under Table 9.1, researchers should not blindly use the common effect size criteria for evaluating correlation coefficients. Depending on the specific context of a given study, it might be appropriate to think of a relationship as being strong when a correlation coefficient turns out to be .30 or .40 or to think of a correlation as weak even if it turns out to be .60 or .70. For example, if Pearson's r is being used to estimate test–retest reliability, a correlation coefficient of .50 would *not* be looked on as strong evidence of desirable stability over time. Also, in some fields of study where two variables have appeared independent, a new investigation (perhaps using better measuring instruments) yielding a correlation of .20 might well cause researchers to think that the new study's correlation is quite high. Although the effect size criteria for correlations are convenient and easy to use, give credit to those researchers who explain *why* these criteria either do or do not fit their particular studies.

Although researchers can demonstrate a concern for relationship strength by discussing effect size or by computing r^2, they can do an even better job if they use the nine-step version of hypothesis testing. As I hope you remember from Chapter 8, this involves setting up the study so that it has the desired power and the proper sample size. When a researcher's study is focused on one or more correlation coefficients, it is quite easy to add these extra tasks to the basic six-step version of hypothesis testing.

In Excerpt 9.23, we see an example of an a priori power analysis being conducted for a study dealing with Pearson's correlation. As indicated in this excerpt, the researchers' a priori power analysis indicated that they would need a sample size of 50 individuals in order to have the desired minimum power of .80 to detect a meaningful correlation at the .05 level of significance. In this study, the dividing line between trivial and meaningful rs was specified by the researchers when they selected, for the power analysis, an effect size of $r = .40$.

EXCERPT 9.23 • *An A Priori Power Analysis*

Fifty experienced hearing aid users (27 men and 23 women) between 46 and 89 years of age ($M = 75.36$, $SD = 9.33$) were recruited to participate in this study based on an a priori power analysis of a Pearson correlation ($r = .4$) using $p = .05$ and power = .8.

Source: Desjardins, J. L., & Doherty, K. A. (2009). Do experienced hearing aid users know how to use their hearing aids correctly? *American Journal of Audiology, 18*(1), 69–76.

When you come across a study that uses the seven-step approach to hypothesis testing, give the researcher some bonus points. When you come across a study in which the appropriate sample size was determined prior to the collection of any data, give the researcher even more bonus points for taking the time to set up the study with sensitivity to both Type I *and* Type II errors. And when you come across a study in which there is no mention whatsoever of effect size (of either type), award *yourself* some bonus points for detecting a study that could have been conducted better than it was.

Linearity, Homoscedasticity, and Normality

Tests on Pearson's r are conducted more frequently than tests on any other kind of correlation coefficient. Whenever tests on Pearson's r are conducted, three important assumptions about the population must hold true in order for the test to function as it was designed. One of these important prerequisite conditions is referred to as the *linearity assumption*. The second is referred to as the *equal variance assumption* (or, alternatively, as the *assumption of homoscedasticity*). The third is the *normality assumption*.

The assumption of **linearity** states that the relationship in the population between the two variables of interest must be such that the bivariate means fall on a straight line. The assumption of **homoscedasticity** states that (1) the variance of the Y variable around μ_y is the same regardless of the value of X being considered and (2) the variance of the X variable around μ_x is constant regardless of the value of Y being considered. The assumption of **normality** requires the population to have what is called *bivariate normality*. If a population is characterized by a curvilinear relationship between X and Y, by heteroscedasticity, or by non-normality, the inferential test on Pearson's r may provide misleading information when sample data are used to make an inferential guess regarding the direction and strength of the relationship in the population.

The easiest way for a researcher to check on these three assumptions is to look at a scatter diagram of the sample data. If the data in the sample appear to conform to the linearity, equal variance, and normality assumptions, then the researcher has good reason to suspect that the population is not characterized by curvilinearity, heteroscedasticity, or non-normality. In that situation, the test on r can then be performed. If a plot of the data suggests, however, that any of the assumptions is untenable, then the regular test on r should be bypassed in favor of one designed for different kinds of data sets.

As a reader of the research literature, my preference is to be able to look at scatter diagrams so I can judge for myself whether researchers' data appear to meet the assumptions that underlie tests on r. Because of space limitations, however, such visual displays of the data are rarely included in research reports. If scatter diagrams cannot be shown, then it is my feeling that researchers should communicate in words what *they* saw when they looked at their scatter diagrams.

Consider Excerpt 9.24. In the study associated with this excerpt, the researchers wanted to use Pearson's correlation to measure the relationship between two variables. Before computing r, however, the researchers checked their data to see if the linearity, equal variance, and normality assumptions were tenable. These investigators deserve high praise for paying attention to their statistical tool's assumptions. (They also deserve praise for reporting a coefficient of determination rather than the unsquared r.)

EXCERPT 9.24 • *Expressed Concern for Assumptions*

Because estimation of blood loss has repeatedly been found to be inaccurate in previous studies, the relationship between estimated blood loss and measured blood loss was investigated using Pearson product-moment correlation coefficient. Preliminary analyses were performed to ensure no violation of the assumptions of normality, linearity, and homoscedasticity. There was a strong positive correlation between the two variables [$r^2 = 0.54, n = 28, p < 0.001$].

Source: Schorn, M. N. (2009). The effect of guided imagery on the third stage of labor: A pilot study. *Journal of Alternative & Complementary Medicine, 15*(8), 863–870.

I believe that too many researchers move too quickly from collecting their data to testing their *r*s to drawing conclusions based on the results of their tests. Few take the time to look at a scatter diagram as a safety maneuver to avoid misinterpretations caused by violation of assumptions. I applaud the small number of researchers who take the time to perform this extra step.

Causality

When we initially looked at correlation from a descriptive standpoint in Chapter 3, I pointed out that a correlation coefficient usually should not be interpreted to mean that one variable has a causal impact on the other variable. Now that we have considered correlation from an inferential standpoint, I want to embellish that earlier point by saying that a correlation coefficient, even if found to be significant at an impressive alpha level, normally should not be viewed as addressing any cause-and-effect question.

In Excerpt 9.25, we see a situation in which a team of researchers warns their readers that correlation does not usually indicate causality. Take a close look at the final sentence of this excerpt. Not only do the researchers alert readers to the danger of drawing causal thoughts from a correlation, they also indicate why this can be problematic. As they point out, the causal force that brings about a relationship between two variables might not be one of those variables influencing the other variable; instead, there might be a third variable that has a causal impact on the first two.

EXCERPT 9.25 • *Correlation and Causality*

As hypothesized, the improvement in two crucial language functions, *naming* [$\rho = 0.7$; $p < 0.00001$] and *comprehension* [$\rho < 0.39$; $p = 0.007$], were associated with patients' baseline nonverbal visuo-spatial working memory. However, it should be emphasized that correlation between two variables does not necessarily indicate a direct causal relationship, as both variables might be correlated to a third variable or a group of variables.

Source: Seniów, J., Litwin, M., & Lesniak, M. (2009). The relationship between non-linguistic cognitive deficits and language recovery in patients with aphasia. *Journal of the Neurological Sciences, 283*(2–3), 91–94.

Attenuation

The inferential procedures covered in this chapter assume that the two variables being correlated are each measured without error. In other words, these procedures are designed for the case where each variable is measured with an instrument that has perfect reliability. Although this assumption may have full justification in a theoretical sense, it certainly does not match the reality of the world in which we live.

To the best of my knowledge, no researcher has ever measured two continuous variables and ended up with data that were perfectly reliable.

When two variables are measured such that the data have less than perfect reliability, the measured relationship in the sample data systematically underestimates the strength of the relationship in the population. In other words, the computed correlation coefficient is a **biased estimate** of the parameter if either or both of the variables are measured without error-free instruments. The term **attenuation** has been coined to describe this situation, where, using the product–moment correlation as an example, measurement error causes r to systematically underestimate ρ.

Once you come to understand the meaning (and likely occurrence) of attenuation, you should be able to see why statistical tests that yield fail-to-reject decisions are problematic in terms of interpretation. If, for example, a researcher computes Pearson's r and ends up not rejecting H_0: $\rho = 0.00$, this outcome *may* have come about because there is a very weak (or possibly no) relationship between the two variables in the population. However, the decision not to reject H_0 *may* have been caused by attenuation masking a strong relationship in the population.

In Chapter 4, we spent a great deal of time considering various techniques used by researchers to estimate the reliability of their measuring instruments. That discussion now becomes relevant to our consideration of inferential reports on correlation coefficients. If a researcher's data possess only trivial amounts of measurement error, then attenuation becomes only a small concern. However, reports of only moderate reliability coupled with correlational results that turn out nonsignificant leave us in a quandary as to knowing anything about the relationship in the population.

If researchers have information concerning the reliabilities associated with the measuring instruments used to collect data on the two variables being correlated, they can use a formula that adjusts the correlation coefficient to account for the suspected amount of unreliability. When applied, this **correction for attenuation** formula always yields an adjusted r that is higher than the uncorrected, raw r. In Excerpt 9.26, we see an example where a group of researchers conducted a correlational study and used the correction for attenuation formula.

EXCERPT 9.26 • *Correlation Coefficients and Attenuation*

Since the measures in this study had different levels of internal consistency, the correlations were corrected for attenuation due to unreliability. The FOCI Symptom Checklist correlated strongly and positively with the OCI-R ($r_c = .84$) and the STAI-A ($r_c = .66$), and moderately with STAI-D ($r_c = .41, p < .01$) and the BDI-II ($r_c = .48, p < .01$).

Source: Aldea, M. A., Rahman, O., & Storch, E. A. (2009). The psychometric properties of the Florida Obsessive Compulsive Inventory: Examination in a non-clinical sample. *Individual Differences Research*, 7(4), 228–238.

Attenuation, of course, is not the only thing to consider when trying to make sense out of a correlation-based research report. Several of these other relevant considerations have been addressed within our general discussion of cautions. Two points are worth reiterating, each now connected to the concept of reliability. First, it is possible that a correlation coefficient will turn out to be statistically significant even though H_0 is true and even though highly reliable instruments are used to collect the sample data; do not forget that Type I errors *do* occur. Second, it is possible that a correlation coefficient will turn out to be nonsignificant even when H_0 is false and even when highly reliable data have been collected; do not forget about the notion of Type II errors and power.

Review Terms

Attenuation	Fisher's *r*-to-*z*
Biased estimate	transformation
Bonferroni technique	Homoscedasticity
Correction for attenuation	Linearity
Correlation matrix	Normality

The Best Items in the Companion Website

1. An interactive online quiz (with immediate feedback provided) covering Chapter 9.
2. Nine misconceptions about the content of Chapter 9.
3. An email message sent from the author to his students entitled "Significant Correlations."
4. Four e-articles illustrating the use of hypothesis testing with correlations.
5. A delightful poem "A Word on Statistics."

To access the chapter outline, practice tests, weblinks, and flashcards, visit the companion website at http://www.ReadingStats.com.

Review Questions and Answers begin on page 531.

Inferences Concerning
One or Two Means

In Chapter 9, we saw how inferential statistical techniques can be used with correlation coefficients. Now, we turn our attention to the procedures used to make inferences with means. A variety of techniques are used by applied researchers to deal with their sample means, and we consider many of these inferential procedures here and in several of the following chapters. Multiple chapters are needed to deal with this broad topic because the inferential procedures used by researchers vary according to (1) how many groups of scores are involved, (2) whether underlying assumptions seem tenable, (3) how many independent variables come into play, (4) whether data on concomitant variables are used to increase power, and (5) whether people are measured under more than one condition of the investigation.

In this introductory chapter on inferences concerning means, we restrict our focus to the cases in which the researcher has computed either just one sample mean or two sample means. I illustrate how statistical tests are used in studies where interest lies in one or two means and the way interval estimation is sometimes used in such studies. Near the end of this chapter, we consider the assumptions that underlie the inferential procedures covered in this chapter, and we also examine the concept of *overlapping distributions*. With this overview now under your belt, let us turn to the simplest inferential situation involving means: the case where there is a single mean.

Inferences Concerning a Single Mean

If researchers have collected data from a single sample and wish to focus on *M* in an inferential manner, one (or both) of two statistical strategies are implemented. On the one hand, a confidence interval can be built around the sample mean. On the other hand, a null hypothesis can be set up and then evaluated by means of the hypothesis testing procedure.

The Inferential Purpose

Figure 10.1 has been constructed to help clarify what researchers are trying to do when they use the mean of a sample as the basis for building a confidence interval or for assessing a null hypothesis. As this figure shows, M is computed on the basis of data collected from the sample. Although the sample-based value of the mean is easy to obtain, primary interest lies in the corresponding value of μ, the population mean.[1] However, the researcher cannot compute the value of μ because only the objects in the sample can be measured. Accordingly, an inference (i.e., educated guess) about the unknown value of the population parameter, μ, is made on the basis of the known value of the sample statistic, M.

In summarizing their empirical investigations, many researchers discuss their findings in such a way that the exclusive focus seems to be on the sample data. The thick arrow in Figure 10.1 should help you remember that the different inferential techniques to which we now turn our attention are designed to allow a researcher to say something about the *population* involved in the study, not the sample. If concern rested with the sample, no inferential techniques would be necessary.

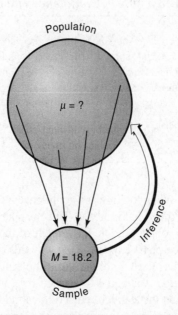

FIGURE 10.1   *The Inferential Purpose When One Sample's Mean Is Computed*

[1]If the researcher's data come from a probability sample, then M represents the mean of the study's *tangible* population. On the other hand, if the data come from a convenience or purposive sample (or some other form of nonprobability sample), then M represents the mean of the study's *abstract* population.

Interval Estimation

Of the two basic ways of applying inferential statistics to a sample mean, the **confidence interval** procedure is simpler. All the researcher does to implement this inferential strategy is (1) make a decision as to the level of confidence that will be associated with the interval to be built and (2) build the interval around M by using a formula that incorporates information from the sample (e.g., M, SD, and n) as well as a numerical value extracted from a statistical table. Of course, computer programs can quickly do the second of these tasks. The result is an interval that extends equally far above and below the sample value of M.

In Excerpt 10.1, we see a case in which a 95 percent confidence interval (CI) was placed around a sample mean. In a sense, this CI gives us a feel for how trustworthy the sample mean is. If this study were to be replicated, with another sample taken from the same population, we would expect sampling error to cause the mean age of people in the replicated study to be different from the mean age of this study's sample. But how much variation should we expect? The CI gives us a range within which we might expect to find that next sample mean.[2]

EXCERPT 10.1 • *Confidence Interval around a Single Sample Mean*

The scores obtained with the selected sample covered the whole possible metric space, and scores of both 0 and 21 points were recorded. Mean score was 8.75 points, with a standard deviation of 6.44 points. The 95% confidence interval for the mean ranged from 7.88 to 9.62.

Source: García-Campayo, J., Zamorano, E., Ruiz, M. A., Pardo, A., Pérez-Páramo, M., López-Gómez, V., et al. (2010). Cultural adaptation into Spanish of the Generalized Anxiety Disorder-7 (GAD-7) Scale as a screening tool. *Health and Quality of Life Outcomes, 8,* 1–11.

When looking at a confidence interval, many people make a big mistake in interpreting what it means. This mistake is thinking that a 95 percent CI indicates the range for the middle 95 percent of the scores used to generate the CI. By looking at Excerpt 10.1, you can see why this is an incorrect interpretation of this or any CI. Whereas the CI around the sample mean extends from 7.88 to 9.62, the standard deviation was equal to 6.44. If the middle 95 percent of the scores fell inside the CI, then the standard deviation would have been much smaller. Simply stated, a 95 percent CI does *not* indicate the range of scores for all but the highest and lowest 2½ percent of the scores. Instead, it gives us a feel for the likely difference between the sample mean and the population mean.

---

[2]Although it is technically wrong to think that a confidence interval indicates a "replication range," this notion is both simple to understand and not likely to be too inaccurate.

As you may recall from Chapter 8, it is technically wrong to interpret this (or other) CI by saying or thinking that there is a 95 percent chance that the population mean lies somewhere between the end points of the CI. Instead, you must imagine (1) that many samples are drawn randomly from the same population, (2) that a separate CI is built for each sample, and (3) that each CI is examined to see if it has captured the population mean. With these three things in mind, the correct way to interpret a 95 percent CI is to say or think that it is one of many (actually 95 percent) CIs that would, in fact, overlap μ rather than one of the few (actually 5 percent) that would not.

Researchers typically present more than just one CI in their research reports. If there is just one group in the study, the individuals in that group might be measured on several variables, with a CI built around the group mean on each variable. If there are two or more groups in the investigation, a separate CI can be built for each group around its mean on each variable for which data have been collected. In Excerpt 10.2, we see an example of the first of these situations.

EXCERPT 10.2 • *Confidence Intervals around Two Means from the Same Sample*

The outcome was assessed [via] the modified Harris Hip Score (MHHS) and the Non-Arthritic Hip Score (NAHS). Overall, at the last follow-up (mean 22 months, 12 to 72), the mean MHHS had improved by 15.3 points (95% confidence interval (CI), 8.9 to 21.7) and the mean NAHS by 15 points (95% CI, 9.4 to 20.5).

Source: Haviv, B., Singh, P. J., Takla, A., & O'Donnell, J. (2010). Arthroscopic femoral osteochondroplasty for cam lesions with isolated acetabular chondral damage. *Journal of Bone and Joint Surgery, 92-B*(5), 629–633.

There are two things to note about Excerpt 10.2. First, each CI is of the 95 percent variety. Such CIs are used far more often by researchers than 90 percent or 99 percent CIs. Second, the group's sample mean on each variable is located near the middle of each CI. If the means and interval end points had not been rounded off, you would have seen that each mean was positioned exactly at the midpoint of each CI. This happens whenever a CI is built around a sample mean because the interval's end points are computed by adding a certain amount to the mean and subtracting that *same* amount from the mean.

Because the number of patients involved in the study associated with Excerpt 10.2 was the same for each of the two CIs, you cannot see here the connection between interval width and sample size. However, there *is* a connection. If other things are held constant, larger sample sizes produce CIs that are narrower, whereas CIs based on smaller *n*s are wider. This relationship between *n* and CI width ought to seem reasonable to you. Simply stated, estimates based on more data are likely to be more precise than estimates based on small amounts of data.

Tests Concerning a Null Hypothesis

When researchers have a single sample (and thus a single population) and have inferential interest in the mean, they can approach the data by means of the hypothesis testing procedure. When this strategy is used, a null hypothesis must be articulated. In this kind of research situation, the null hypothesis takes the form

$$H_0 : \mu = a$$

where a stands for a pinpoint numerical value selected by the researcher.

After specifying H_0, researchers proceed to apply the various steps of the inferential testing strategy they have decided to follow. Regardless of whether the six-step, seven-step, or nine-step approach to hypothesis testing is used, researchers assess the discrepancy between the sample mean and H_0's pinpoint value. If the difference between M and H_0's μ-value is large enough, H_0 is rejected and viewed as not likely to be true because of the small value of p associated with the sample data.

There are several available test procedures that can be used to analyze the data of a one-sample study wherein the statistical focus is the mean. The two most popular of these test procedures are the *t*-test and the *z*-test. These two ways of testing the discrepancy between M and H_0's μ-value are identical in logic and have the same decision rule when comparing the calculated value against the critical value.[3] The only difference between the two tests is that the *z*-test yields a calculated value slightly larger than it ought to be (and a *p*-value slightly smaller than it ought to be). However, the amount of the bias is trivial when the sample size is at least 30.

Excerpts 10.3 and 10.4 illustrate how researchers will often present their results when they have a single sample and conduct a *z*-test or a *t*-test to evaluate a null hypothesis of the form $H_0: \mu = a$. In the first of these excerpts, the pinpoint number from the null hypothesis is not shown. However, it was equal to the mean

EXCERPTS 10.3–10.4 • *Use of z or t to Test the Mean of a Single Sample*

The one sample *z*-test was used to test the statistical significance of [mean] differences between the autistic children and the Japanese standard values. The standard values were obtained from the infant physical growth survey conducted by The Japanese Ministry of Health, Labour and Welfare [in order] to establish a standard of Japanese infant growth. . . . [Results indicated that] the head circumference at birth [$M = 33.27$] showed no significant difference from the standard value as determined by the Japanese Government study of 14,115 children [$z = 0.90, p = 0.366$].

Source: Fukumoto, A., Hashimoto, T., Ito, H., Nishimura, M., Tsuda, Y., Miyazaki, M., et al. (2008). Growth of head circumference in autistic infants during the first year of life. *Journal of Autism & Developmental Disorders*, *38*(3), 411–418.

(continued)

---

[3]This decision rule says to reject H_0 if (1) the calculated value is as large as or larger than the critical value or (2) the data-based p is equal to or smaller than the selected level of significance.

EXCERPTS 10.3–10.4 • (*continued*)

> Overall, the mean estimate of the number of drinks considered indicative of heavy drinking for "Kevin" was 6.09 (*SD* = 2.38). A one-sample *t* test . . . revealed a significant difference, $t(179) = 6.35, p = .001$ with participants' quantitative estimate of heavy drinking higher than the current definition of "binge" or "heavy episodic drinking" in the literature [5 drinks].
>
> *Source:* Segrist, D. J., & Pettibone, J. C. (2009). Where's the bar? Perceptions of heavy and problem drinking among college students. *Journal of Alcohol & Drug Education, 53*(1), 35–53.

score on the norms. In Excerpt 10.4, the null hypothesis took the form H_0: $\mu = 5.00$, with that level of drinking chosen because there was agreement in the literature that consuming five drinks at one sitting qualifies one to be called a "heavy drinker."

In the middle of Excerpt 10.4, notice that a number is positioned inside a set of parentheses located between the letter *t* and the calculated value of 6.35. This number, which in this particular excerpt is 179, is technically referred to as the **degrees of freedom** (which is often abbreviated *df*) for the *t*-test that was performed.[4] If you add 1 to the *df* number of a one-sample *t*-test, you get a number that equals the size of the sample. Thus, you can tell that 180 individuals provided scores that produced the mean of 6.35.

Inferences Concerning Two Means

If researchers want to compare two samples in terms of the mean scores using inferential statistics, they can utilize a confidence interval approach to the data or an approach that involves setting up and testing a null hypothesis. We consider the way in which estimation can be used with two means after we examine the way in which two means can be compared through a tested H_0. Before we do either of these things, however, I must draw a distinction between two 2-group situations: those that involve independent samples and those that involve correlated samples.

Independent versus Correlated Samples

Whether two samples are considered to be independent or correlated is tied to the issue of the nature of the groups *before* data are collected on the study's dependent variable. If the two groups have been assembled in such a way that a logical relationship exists between each member of the first sample and one and only one member of the second sample, then the two samples are **correlated samples.** However, if no such relationship exists, the two samples are **independent samples.**

[4]There is no *df* value in Excerpt 10.3 because *z*-tests do not utilize the *df* concept.

Correlated samples come into existence in one of three ways. If a single group of people is measured twice (e.g., to provide pretest and posttest data), then a relationship exists in the data because each of the pretest scores goes with one and only one of the posttest scores, because both come from measuring the same research participant. A second situation that produces correlated samples is *matching*. Here, each person in the second group is recruited for the study because he or she is a good match for a particular individual in the first group. The matching could be done in terms of height, IQ, running speed, or any of a multitude of possible matching variables. The matching variable, however, is never the same as the dependent variable to be measured and then used to compare the two samples. The third situation that produces correlated samples occurs when biological twins are split up, with one member of each pair going into the first sample and the other member going into the second group. Here, the obvious connection that ties together the two samples is genetic similarity.

When people, animals, or things are measured twice or when twin pairs are split up, it is fairly easy to sense which scores are paired together and why such pairing exists. When a study involves matching, however, things are slightly more complicated, because at least two data-based variables are involved. The data on one or more of these variables are used to create pairs of people such that the two members of any pair are as similar as possible on matching variables. Once the matched pairs are formed, then new data are examined on the dependent variable of interest to see if the two groups of individuals differ *on the dependent variable*. For example, a researcher might create matched pairs of students who have low academic self-concept, randomly split up the pairs to form an experimental group (which receives tutoring) and a control group (which does not), and then compare the two groups in terms of how they perform at the end of the term on a final course examination. In this hypothetical study, the matching variable is academic self-concept (with these scores discarded after being used to form matched pairs); the scores of primary interest—that is, the scores corresponding to the dependent variable—come from the final course examination.

If the two groups of scores being compared do not represent one of these three situations (pre/post, matched pairs, or twins), then they are considered to be independent samples. Such samples can come about in any number of ways. People might be assigned to one of two groups using the method of simple randomization, or possibly they end up in one or the other of two groups because they possess a characteristic that coincides with the thing that distinguishes the two groups. This second situation is exemplified by the multitude of studies that compare males against females, students who graduate against those who do not graduate, people who die of a heart attack versus those who do not, and so on. Or, maybe one of the two groups is formed by those who volunteer to undergo some form of treatment, whereas the other group is made up of folks who choose not to volunteer. A final example (of the ones to be mentioned) is created if the researchers simply designate one of two intact groups to be their first sample, which receives something that might help them, whereas the second intact group is provided with nothing at all or maybe a placebo.

In Excerpts 10.5 and 10.6, we see descriptions of data sets that represent independent and correlated samples. It is easy to tell that the data in the first of these studies should be thought of as independent samples because the sample sizes are different. (Whenever n_1 is different from n_2, it is impossible to have each score in one of the data sets paired logically with one of the scores in the second data set.) The two groups referred to in Excerpt 10.6 were correlated because the 20 individuals in the first group were individually matched with 20 individuals in the second group. It is not surprising that the two groups had nearly identical means on chronological age, as CA was the variable used to create the 20 matched pairs!

EXCERPTS 10.5–10.6 • *Independent and Correlated Samples*

Nine-hundred federal service dentists and 600 civilian dentists were [compared regarding] their knowledge, attitudes, and behaviors concerning MID (minimal intervention dentistry).

Source: Gaskin, E. B., Levy, S., Guzman-Armstrong, S., Dawson, D., & Chalmers, J. (2010). Knowledge, attitudes, and behaviors of Federal Service and civilian dentists concerning minimal intervention dentistry. *Military Medicine, 175*(2), 115–121.

Twenty-six participants with autism were recruited [but] to comply with the inclusion criteria [that had been established], the final sample comprised 20 individuals between 6 years 4 months and 18 years 4 months (mean 13 years 4 months). Participants with autism were individually matched to typically developing individuals of comparable CA [chronological age]. The CA-matched group ranged from 6 years 2 months to 18 years 8 months (mean 13 years 3 months). . . .

Source: Riby, D., & Hancock, P. J. B. (2009). Looking at movies and cartoons: Eye-tracking evidence from Williams syndrome and autism. *Journal of Intellectual Disability Research, 53*(2), 169–181.

Although this was not done in either of the two excerpts we have just considered, researchers sometimes indicate explicitly that their data came from independent samples or from correlated samples. When they do so, you will have no trouble knowing what kind of samples was used. However, they may use terms other than independent samples and correlated samples. Correlated samples are sometimes referred to as **paired samples, matched samples, dependent samples,** or **within samples,** whereas independent samples are sometimes called **unpaired samples, unmatched samples,** or **uncorrelated samples.**

To understand exactly what researchers did in comparing their two groups, you must develop the ability to distinguish between correlated samples and independent samples. The language used by the researchers can help to indicate the kind of samples involved in the study. If a descriptive adjective is not used, you must make a judgment based on the description of how the two samples were formed.

The Inferential Purpose

Before we turn our attention to the way researchers typically summarize studies that focus on two sample means, I want to underscore the fact that these comparisons of means are inferential in nature. Figure 10.2 is designed to help you visualize this important point.

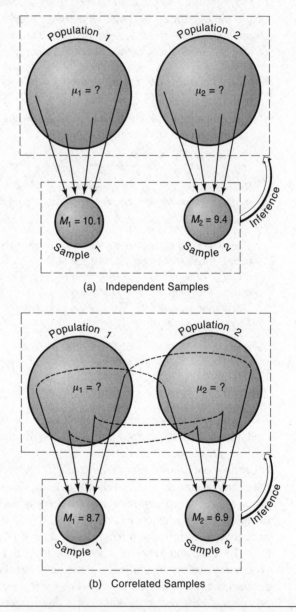

FIGURE 10.2   *Two Different Kinds of Inferential Situations Involving Two Means*

Panel A in Figure 10.2 represents the case where the means of two independent samples are compared. Panel B represents the case where two correlated samples of data are compared in terms of means. (In panel B, the dotted "chains" that extend from population 1 to population 2 are meant to denote the pairing or matching that is characteristic of correlated samples.)

Two points about Figure 10.2 must be highlighted. First, in both the independent-samples situation and in the correlated-samples situation, inferential statements are made about populations, not about samples. Unfortunately, researchers often discuss their results as if the samples were the total focus of their investigations. If you keep Figure 10.2 in mind when you are dealing with such research summaries, you can (and should) correct the discussion by having all conclusions apply to the study's populations.

My second point regarding Figure 10.2 concerns the fact that the statistical inference, in panel A or in panel B, extends from the full set of sample data to the study's *pair* of populations. Separate inferences are not made from each sample to its corresponding population because the purpose is to make a comparison between two things. The focus here is on how μ_1 compares with μ_2, and thus the inferential arrow in each picture points to the dotted box surrounding both populations involved in the study.

Setting Up and Testing a Null Hypothesis

The null hypothesis for the two-sample case having a focus on means can be expressed in the same form regardless of whether the samples are independent or correlated. The most general way to write the null hypothesis is to state

$$H_0 : \mu_1 - \mu_2 = a$$

where a represents any pinpoint number the researcher wishes to use in H_0. In most studies, researchers decide to set up a no-difference null hypothesis, and they accomplish this goal by saying $H_0 : \mu_1 - \mu_2 = 0$. Another way to express the notion of no difference is to say $H_0 : \mu_1 = \mu_2$. In Excerpt 10.7, we see an example of the null hypothesis (as well as the alternative hypotheses) being stated in a research report.

EXCERPT 10.7 • *Null Hypothesis for Comparing Two Means*

[In comparing] gym members and non-gym members . . . , the null and alternative hypotheses are:

H_0: $\mu_1 = \mu_2$ (population means are the same.)

H_1: $\mu_1 \neq \mu_2$ (population means are not the same.)

Source: Leeman, O., & Ong, J. S. (2008). Lost and found again: Subjective norm in gym membership. *DLSU Business & Economics Review, 18*(1), 13–28.

Unfortunately, the null and alternative hypotheses are rarely stated in studies where two means are inferentially compared. Most researchers assume that their readers will be able to discern the null hypothesis from the discussion of the way the sample data are summarized. A good rule of thumb to use when trying to decipher research reports is to presume that a test of two means revolved around a no-difference H_0 unless it is explicitly stated that some other kind of null hypothesis was set up. You should also presume that the alternative hypothesis was nondirectional unless you see the term *one-tailed* or *one-sided*.

After the sample data are collected, summarized, and analyzed, the results of the statistical comparison of the two sample Ms is presented within the text of the report or in a table. Excerpts 10.8 and 10.9 illustrate the way results are typically presented, with the studies associated with these excerpts involving independent and correlated samples, respectively.

EXCERPTS 10.8–10.9 • *Comparison of Two Sample Means Using a t-Test*

An independent samples *t*-test showed that girls were more accurate on no-go trials during the GNG task ($n = 73, M = 38.68, SD = 34.44$) than boys ($n = 62$, $M = 24.15, SD = 26.34$), $t(133) = 2.71, p < .01$.

Source: He, J., Degnan, K. A., McDermott, J. M., Henderson, H. A., Hane, A. A., Xu, Q., et al. (2010). Anger and approach motivation in infancy: Relations to early childhood inhibitory control and behavior problems. *Infancy*, *15*(3), 246–269.

A paired *t*-test showed a significant difference in the number of barriers [to engaging in exercise] endorsed by participants between pre-test and post-test ($t(57) = 5.01$; $p < .001$). The number of barriers endorsed by participants at pretest ($M \pm SD = 6.17 \pm 2.64$) was significantly reduced at post-test (4.48 ± 2.35).

Source: Brinthaupt, T. M., Kang, M., & Anshel, M. H. (2010). A delivery model for overcoming psycho-behavioral barriers to exercise. *Psychology of Sport and Exercise*, *11*(4), 259–266.

As you can see, a *t*-test is referred to in Excerpts 10.8 and 10.9. The *t*-test is a versatile statistical tool, because it can be used when a study's statistical focus is on a variety of other things (e.g., proportions and regression coefficients). Nevertheless, *t*-tests probably are used more often with means than anything else. In Excerpt 10.8, the *t*-test compared the two sample means (38.68 and 24.15) to see if they were further apart than we would expect them to be, if the null hypothesis were true. That *was* the case, as indicated by the notation $p < .01$. In Excerpt 10.9, the means of 6.17 and 4.48 were compared, and once again the null hypothesis was rejected.

Note that the authors of Excerpts 10.8 and 10.9, in reporting their *t*-test results, provide information as to the degrees of freedom associated with the tests that were conducted. These *df* values—located inside parentheses between the letter *t* and the calculated value—are useful because they allow us to know how much data each *t* was based on. When *t*-tests are conducted to compare the means of two independent samples, the total amount of data can be determined by adding 2 to the *t*-test's *df*. When *t*-tests are used to see if a significant difference exists between the means of two correlated samples of data, you can determine how many pairs of data were used by adding 1 to the *t*-test's *df*. Armed with this knowledge, we can verify that there were a total of 135 boys and girls involved in the study associated with Excerpt 10.8, and we can determine (on our own) that there were 58 individuals in the study associated with Excerpt 10.9.

Although a statistical test comparing the two means can be conducted using a *t*-test, it can also be accomplished by means of a *z*-test or an **F-test.** The *z*-test provides a result that is slightly biased in the sense that its probability of resulting in a Type I error is greater than the level of significance (with this bias being more pronounced when the sample sizes are small). In contrast, the *F*-test is not biased. The *F*-test's conclusion regarding H_0 is always identical to the conclusion reached by a *t*-test. Hence, it really does not matter whether researchers compare their two means using a *t*-test or an *F*-test.

In light of the fact that (1) some researchers opt to use an *F*-test when comparing two means and (2) the results of an *F*-test are typically presented in a way that requires an understanding of concepts not yet addressed, I am obliged to comment briefly about *F*-test results. Here, I focus attention exclusively on the use of *F*-tests to compare the means of two independent samples. In Chapter 14, I show how *F*-tests can be used with correlated samples.

To begin our discussion of *F*-tests applied to the means of two independent samples, consider the material in Excerpt 10.10. In this excerpt, note that two groups of teachers were being compared, that the focus was on the mean number of "professional development" hours required (21.1 for low-poverty districts and 29.9 for high-poverty districts), and that a statistically significant difference was found between these sample means, as indicated by the notation $p < .05$ at the end

EXCERPT 10.10 • *Comparison of Means from Two Independent Samples Using an F-Test*

High-poverty districts required on average 9 hours more to be spent [by teachers] on professional development than low-poverty districts (low-poverty districts $M = 21.1$, $SD = 31.1$; high-poverty districts $M = 29.9, SD = 42.3$), $F(1,596) = 8.510$, $p < .05$.

Source: Fall, A. M., & Billingsley, B. S. (2011). Disparities in work conditions among early career special educators in high- and low-poverty districts. *Remedial and Special Education*, *32*(1), 64–78.

of the excerpt. Also note that the calculated value turned out equal to 8.510, and that this value is referred to as F.

In Excerpt 10.10, also note that there are two degree of freedom values presented along with the calculated value. The dfs appear within a set of parentheses immediately to the right of the F, and they are separated by a comma. F-tests always have a pair of df values associated with them, and in this case the df values are equal to 1 and 596. Researchers always cite two df numbers in conjunction with the calculated value from their F-test.

The df values presented along with the results of an F-test can be used to discern the amount of data used to make the statistical comparison. When an F-test is used as in Excerpt 10.10 to compare the means of two independent samples, all you must do to determine the amount of data is add the two df values together and then add 1 to the resulting sum. Thus, in this study, the calculated value of 8.510 was based on a total of 598 pieces of data. Because each piece of data came from a different person, we know that there were 598 people involved in this study.

Sometimes, a table is used to present the results of the kind of F-test we have been discussing. An example of such a table is contained in Excerpt 10.11. In the study associated with this excerpt, 408 students taking a college algebra course filled out a survey that assessed each student's "locus of attribution" (i.e., the degree to which he or she felt responsible for doing well or poorly in the course). These attribution scores were then used to compare two subgroups of the students: those who passed the course and those who failed. The means of these groups (on a $1-9$ scale) were 6.76 and 5.94, respectively.

EXCERPT 10.11 • *F-Test Comparison of Two Sample Means*

The results for the locus of causality dimension are presented in Table 6. These findings indicate a statistically significant difference between the passing and failing students in the locus of causality dimension at $F(1,406) = 26.34, p < .0001$.

TABLE 6 *One-way ANOVA summary table for locus of causality by pass/fail*

Source	df	SS	MS	F
Between groups	1	66.447	66.447	26.340*
Within groups	406	1024.219	2.523	
Total	407	1090.666		

*$p < .0001$.

Source: Cortés-Suárez, G., & Sandiford, J. R. (2008). Causal attributions for success or failure of students in college algebra. *Community College Journal of Research & Practice, 32*(4–6), 325–346.

The first thing to note about Excerpt 10.11 is the researcher's use of the acronym ANOVA in the title of the table, which stands for the phrase **analysis of variance.** This phrase is a bit misleading, for it probably would lead an uninformed reader to think that the statistical focus was on variances. However, as you will see, the analysis of variance focuses on *means*. Also note that the term *one-way* appears in the table's title. There are several different kinds of ANOVAs, and the term *one-way* clarifies that the comparison groups were formed by considering just one thing when putting people into one or the other of the two groups. In this study, that one thing was whether a student passed or failed the algebra course.

The main thing going on in Excerpt 10.11 is a statistical comparison of the two sample means (6.76 and 5.94) that appeared in the text of the research report. The outcome of the inferential test comparing those two means is shown in the far right column of the ANOVA summary table. The number that appears in the column labeled *F* is the data-based calculated value, and it turned out to be significant, as indicated by the note beneath the table. Thus, the two sample means differed by an amount that was beyond the limits of chance sampling. Accordingly, the null hypothesis was rejected.

There are three *df* values presented in the analysis of variance table. On the row labeled "Between Groups," the *df* value is equal to 1; this is always the case when two sample means are being compared. The *df* value on the row labeled "Within Groups" is found first by subtracting 1 from each sample size and then by adding the resulting figures. The sum of the *df*s for the "Between Groups" and "Within Groups" rows is equal to 407, one less than the total number of people used in the analysis.

The column of numbers to the right of the *df* numbers is labeled with the abbreviation **SS,** which stands for "sum of squares." These numbers come from a statistical analysis of the sample data, and there is really no way to make sense out of this column of the analysis of variance table. The next column is labeled **MS,** the abbreviation for "mean square." The first of these values was found by dividing the first row's *SS* by that row's *df* ($66.447 \div 1 = 66.447$). In a similar fashion, the second row's *MS* of 2.523 was computed by dividing 1024.219 by 406. Finally, the calculated value for the *F* column was computed by dividing the "Between Groups" *MS* by the "Within Groups" *MS* ($66.447 \div 2.523 = 26.340$).

In one sense, all of the numbers in the *df, SS,* and *MS* columns of the analysis of variance table are used solely as stepping stones to obtain the calculated value. The top two *df* values are especially important, however, because the size of the appropriate critical value depends on these two *df* values (along with the selected level of significance). When the statistical analysis is being performed on a computer, the researcher's decision to reject or not reject H_0 is made by looking at the *p*-value provided by the computer (rather than by comparing the calculated *F* against the critical value). The computer's *p*-value, however, is influenced by the "between" and "within" *df* values (as well as by α and the computed *F*). Accordingly, the three most important numbers in the table are the first two values in the *df* column and the single number in the *F* column.

Interval Estimation with Two Means

As noted in Chapter 8, confidence intervals can be used to deal with a null hypothesis that a researcher wishes to test. Or, the confidence interval can be set up in studies where no test is being conducted on any H_0, with interest instead residing strictly in the process of interval estimation. Regardless of the researcher's objective, it is important to be able to decipher the results of a study in which the results are presented using a confidence interval around the difference between two means.

Consider Excerpt 10.12. Within this excerpt, notice how there is just one confidence interval, not two, even though each of the two groups was measured twice. Instead of building a separate CI around each group's pre-treatment and post-treatment means, and instead of building a separate CI around each group's mean change, a single CI was built around 12.9, the *difference* between one group's mean change and the other group's mean change. Because this CI did not overlap 0, the researchers were able to say that a significant difference existed between the two comparison groups. The null hypothesis—that $\mu_{1(pre-post)} = \mu_{2(pre-post)}$—was rejected.

EXCERPT 10.12 • *Using a Confidence Interval to Do Hypothesis Testing with Two Means*

The SPADI is a self-administered questionnaire consisting of [shoulder] pain and disability subscales, where the means of the 2 subscales are combined to produce a total score ranging from 0 (best) to 100 (worst). . . . . Differences in change scores for the SPADI for the success group were significantly better than for the nonsuccess group ($P < .001$), with a mean difference between groups of 12.9 (95% CI = 7.3, 18.5). The mean SPADI score for the success group decreased by more than 50% (from 38.1 to 18.4), whereas the mean SPADI score for the nonsuccess group decreased by 18% (from 37.9 to 30.4).

Source: Mintken, P. E., Cleland, J. A., Carpenter, K. J., Bieniek, M. L., Keirns, M., & Whitman, J. M. (2010). Some factors predict successful short-term outcomes in individuals with shoulder pain receiving cervicothoracic manipulation: A single-arm trial. *Physical Therapy*, *90*(1), 26–42.

Multiple Dependent Variables

If data are collected from one or two samples on two or more dependent variables, researchers with inferential interest in their data may build several confidence intervals or set up and test several null hypotheses, one for each dependent variable. A quick look at a few excerpts from recent studies illustrates how researchers often talk about such analyses.

Results Presented in the Text

In Excerpts 10.13 and 10.14, we see two examples of how researchers often discuss what they discovered when they compared two groups on multiple dependent variables. Although both studies involved two means per comparison, note how a *t*-test was used in Excerpt 10.13, whereas an *F*-test was used in Excerpt 10.14. Note also that you can use the degrees of freedom to determine how many individuals were involved in each study. In the study associated with the first excerpt, there were $89 + 1 = 90$ teachers who attended the workshop. In the study associated with Excerpt 10.14, data were gathered from a total of $1 + 642 + 1 = 644$ students.

EXCERPTS 10.13–10.14 • *Comparing Two Groups on Multiple Dependent Variables*

With regard to perceived confidence as a result of workshop participation, attendees felt more confident in their ability to incorporate technology into their lessons in a meaningful way, $t(89) = 5.28, p < .001$, to plan field trips that engage students in a meaningful learning experience, $t(89) = 7.80, p < .001$, and to use virtual field trips for student learning and various academic outcomes, $t(89) = 13.09, p < .001$, after participating in the workshop.

Source: Shriner, M., Clark, D. A., Nail, M., Schlee, B. M., & Libler, R. (2010). Social studies instruction: Changing teacher confidence in classrooms enhanced by technology. *Social Studies*, *101*(2), 37–45.

More academic responses $(M = .5196, SD = .30)$ were observed in the CM group than were observed in the No CM group $(M = .3030, SD = .46)$, $F(1, 642) = 32.479, p < .001$. Fewer competing responses were observed in the CM group $(M = .1423, SD = .35)$ than in the No CM group $(M = .2645, SD = 44)$, $F(1, 642) = 14.453, p < .001$.

Source: Lee, S., Wehmeyer, M. L., Soukup, J. H., & Palmer, S. B. (2010). Impact of curriculum modifications on access to the general education curriculum for students with disabilities. *Exceptional Children*, *76*(2), 213–233.

Results Presented in a Table

Excerpt 10.15 illustrates how a table can be used to convey the results of a two-sample comparison of means on several dependent variables. This table comes from a study in which 99 older residents of Crete completed a questionnaire about anxiety that had been translated into Greek. Three *t*-tests were conducted using data from the same people; the first compared men and women, the second compared

EXCERPT 10.15 • *Results of Several t-Tests in a Table*

TABLE 4 *Comparison of the Short Anxiety Screening Test (SAST) results for sex, age distribution and health centres*

	Frequency, N	SAST score, mean ($\pm$SD)	t Test
Sex			
Male	56	22.8 ($\pm$5.8)	$t = 3.105$,
Female	43	19.5 ($\pm$4.3)	$df = 97, P = 0.002$
Age distribution			
65 to 74	46	21.8 ($\pm$5.5)	$t = 0.837$,
≥ 75	53	20.9 ($\pm$5.5)	$df = 97, P = 0.404$
Health centre			
Spili	37	21.1 ($\pm$6.1)	$t = 0.382$,
Anogia	62	21.5 ($\pm$5.1)	$df = 97, P = 0.704$

Source: Grammatikopoulos, I. A., Sinoff, G., Alegakis, A., Kounalakis, D., Antonopoulou, M., & Lionis, C. (2010). The Short Anxiety Screening Test in Greek: Translation and validation. *Annals of General Psychiatry, 9*(1), 1–8.

those under 75 years old with those 75 or older, and the third compared the individuals located in two different areas of Crete.

In Excerpt 10.15, the *t*-test calculated values are presented in the right column. Hence, the first calculated value of 3.105 came from a comparison of SAST sample means of 22.8 and 19.5. As you can see, this first *t*-test revealed a significant difference between the gender subgroups. The other two *t*-tests were not significant. In each of the three cases, the null hypothesis stated that the population corresponding to the two samples being compared had equal means.

When reading research reports, try to remember that you can use the reported *df* numbers to help you understand how the study was structured, how many groups got compared, and how many participants were involved. In Excerpt 10.15, the sample sizes for each *t*-test were provided, so we know that there were 99 research participants and that each comparison was made via an independent-samples *t*-test. In some research reports, this information is not presented as clearly. In such cases, the *df* value(s) can help you figure out things about the study that are not stated in words.

Knowing how to use *df* numbers, of course, is not the most important skill to have when it comes to *t*- or *F*-tests. Clearly, it is more important for you to know what these tests compare, what the null hypothesis is, and what kind of inferential error might be made. Even though *df* numbers are *not* of critical importance, it is worth the effort to learn how to use them as an aid to interpreting what went on in the studies you read.

Use of the Bonferroni Adjustment Technique

When a researcher sets up and tests several null hypotheses, each corresponding to a different dependent variable, the probability of having at least one Type I error pop up somewhere in the set of tests is higher than indicated by the level of significance used in making the individual tests. As indicated in Chapter 8, this problem is referred to as the *inflated Type I error problem*. There are many ways to deal with this problem, but the most common strategy is the application of the **Bonferroni adjustment technique.**

In Excerpt 10.16, we see an example of the Bonferroni adjustment technique. This excerpt is worth considering for two reasons. First, it contains an explanation how the Bonferroni procedure works. Second, the adjusted alpha level is cited, near the end of the excerpt, as the more conservative criterion that was used in judging the wisdom, courage, and humanity *t*-tests to be nonsignificant.

EXCERPT 10.16 • *Use of the Bonferroni Adjustment Procedure*

Virtue scores were compared between students reporting alcohol use within the past month and those who had not consumed alcohol. Using a Bonferroni-adjusted alpha of .008 (.05/6), 3 of the 6 virtues were significantly higher in nondrinkers: justice $(t(412) = 3.43, p = .001)$, temperance $(t(412) = 4.77, p < .001)$, and transcendence $(t(412) = 3.36, p = .001)$. Wisdom, courage, and humanity did not differ significantly between the 2 groups $(p > .008)$.

Source: Logan, D. E., Kilmer, J. R., & Marlatt, G. A. (2010). The Virtuous drinker: Character virtues as correlates and moderators of college student drinking and consequences. *Journal of American College Health, 58*(4), 317–324.

As indicated in Chapter 8, application of the Bonferroni adjustment procedure makes it more difficult for a researcher to reject the null hypothesis because the modified level of significance becomes more rigorous. The adjusted level does not become overly conservative; it simply gets reset to its proper position. Stated differently, if the Bonferroni adjustment procedure is *not* used when it should be, the alpha level is more liberal than it seems.

Effect Size Assessment and Power Analyses

When dealing with one or two means using hypothesis testing, many researchers give no evidence that they are aware of the important distinction between statistical significance and practical significance. Those researchers seem content simply to reject or to fail to reject the null hypotheses they test. As indicated in Chapter 8, a growing number of researchers use the concept of **effect size** to address the notion of practical significance, either by computing an estimate of effect size as the last step of the seven-step version of hypothesis testing or by choosing an effect size during the nine-step version of hypothesis testing. Now, I want to illustrate how researchers actually do these things in studies where there is inferential interest in one or two means.

In Excerpts 10.17 and 10.18, we see how researchers use two popular indices to assess effect size. In the first of these excerpts, the index d accompanied the t-test results that compared 69 male athletes to 36 female athletes (all of whom were college undergraduates) in terms of their claimed willingness to take physical and psychological risks. In Excerpt 10.18, the researchers used **partial eta squared** (η_p^2) in conjunction with their F-test comparison of two groups of undergraduate students who individually stood on a two-story balcony and estimated its height off the ground. Before doing this, one group was shown 30 emotionally arousing pictures, while the other group saw 30 neutral pictures.

EXCERPTS 10.17–10.18 • *Effect Size Assessment with d and η_p^2*

Independent t-tests found that men [compared to women] have more positive attitudes towards taking physical ($t_{103} = 3.36, p < .01, d = 0.69$) and psychological risks ($t_{103} = 3.29, p < .01, d = 0.70$).

Source: Crust, L., & Keegan, R. (2010). Mental toughness and attitudes to risk-taking. *Personality and Individual Differences, 49*(3), 164–168.

We found that emotionally arousing stimuli influenced height perception, such that individuals who viewed arousing pictures ($M = 13.60\,\text{m}, SD = 2.64$) overestimated the height of the balcony more than did the individuals who viewed nonarousing pictures ($M = 11.40\,\text{m}, SD = 2.95$), $F(1, 27) = 4.35, p < .05, \eta_p^2 = .14.$

Source: Stefanucci, J. K., & Storbeck, J. (2009). Don't look down: Emotional arousal elevates height perception. *Journal of Experimental Psychology, 138*(1), 131–145.

The index d is computed as the difference between two sample means divided by the "average" of the two standard deviations. Thus, d is very much like z-score that "standardizes" any mean difference by eliminating the metric of the measurement. Partial eta squared is more difficult to compute, but it is fairly easy to interpret because it is analogous to the r^2 that is often paired with a bivariate correlation. After comparing two means via a t-test or an F-test, η_p^2 provides an index of the proportion of variability in the study's dependent variable that is associated with (i.e., explained by) the study's grouping variable. Just as r^2 is often converted into a percentage, partial eta squared is frequently interpreted as

TABLE 10.1   *Effect Size Criteria for Comparing Two Means*

Effect Size Measure	Small	Medium	Large
d	.20	.50	.80
Eta (η)	.10	.24	.37
Eta Squared (η^2)	.01	.06	.14
Omega Squared (ω^2)	.01	.06	.14
Partial Eta Squared (η_p^2)	.01	.06	.14
Partial Omega Squared (ω_p^2)	.01	.06	.14
Cohen's f	.10	.25	.40

Note: These standards for judging relationship strength are quite general and should be changed to fit the unique goals of any given research investigation.

indicating the percentage of variability in the dependent variable that is explained by the grouping variable.

In the research reports from which Excerpts 10.17 and 10.18 were taken, the values of d and η_p^2 were presented without any comment as to their size. Many other researchers like to attach one of three descriptive labels—small, medium, or large—to the effect size estimates they compute. Because you are likely to see these three labels used in research reports, it is important for you to know what criteria were used to determine if the practical significance of a finding should be considered large or medium or small.

Table 10.1 contains the popular criteria for small, medium, and large effect sizes for seven different measures that researchers use. It is recommended that each numerical value in this table be thought of as the lowest value needed in order to use the label at the top of that entry's column. Thus, the effect size estimates in Excerpt 10.17 are medium in size, whereas the effect size estimate in Excerpt 10.18 qualify as being large.

So far, we have seen and considered just two of the effect size indices: d and η_p^2. You are likely to encounter the others if you read many research reports. As you can see from Table 10.1, there are two versions of eta squared and two versions of omega squared. The two versions of each measure differ in terms of the way "explained variance" is conceptualized. However, for the situation where just two sample means are being compared, eta squared is identical to partial eta squared, and omega squared is identical to partial omega squared.

In terms of effect size measures, there are two additional things you must know. First, the computed value for any of these measures is based on sample data, which means that the actual effect size in the relevant populations is only estimated. Because of this, it is proper for researchers to place confidence intervals around their computed effect size so as to make clear that they are sample statistics and not population parameters. Second, neither researchers nor you should blindly use the values in Table 10.1 when interpreting effect size indices. Depending on the specific context of a given study, it might be fully appropriate to deviate from these general "rules of thumb" or to add supplemental information that supports the claim for any given estimated effect size.

In Excerpts 10.19 and 10.20, we see examples of researchers being careful with their effect-size assessments. In the first of these excerpts, a confidence interval was placed around the computed value of *d*. In Excerpt 10.20, the researchers provided supplemental information that helped clarify the "clinical meaningfulness" of the statistically significant finding.

EXCERPTS 10.19–10.20 • *Effect Size Assessment with Confidence Intervals & Supplemental Data*

However, the gains of the NBLI/FFW-L group did not differ significantly from the wait/NBLI controls, $t(40.00) = 0.46, p = .65, d = 0.22$ (90% CI $[-0.61, 1.05]$).

Source: Fey, M. E., Finestack, L. H., Gajewski, B. J., Popescu, M., & Lewine, J. D. (2010). A preliminary evaluation of Fast ForWord-Language as an adjuvant treatment in language intervention. *Journal of Speech, Language & Hearing Research, 53*(2), 430–449.

- -

Effect size estimates indicate that the reductions in school absences and pain ratings were in the medium range [$d = .59$ and $d = .55$, respectively], providing some support for the clinical significance of these changes. Further support for the clinical meaningfulness of these changes can be drawn from the finding that 39% of responding participants reported a decrease of two points or more on a 0–10 visual analogue scale assessing their worst pain following the intervention.

Source: Logan, D. E., & Simons, L. E. (2010). Development of a group intervention to improve school functioning in adolescents with chronic pain and depressive symptoms: A Study of feasibility and preliminary efficacy. *Journal of Pediatric Psychology, 35*(8), 823–836.

In the four excerpts we have just considered, the researchers' computation of effect size indices were all performed *after* the data in each study had been gathered. As pointed out in Chapter 8, however, it is possible to conduct a power analysis *before* any data are collected. The purpose of such an analysis is to determine how large the sample(s) should be so as to have a known probability of rejecting H_0 when H_0 is false by an amount at least as large as the researcher-specified effect size.

In Excerpt 10.21, we see what goes into, and what comes out of, an a priori power analysis. In doing this power analysis, the researchers first decided that their statistical focus would be the mean, that they would compare their sample means with an unpaired (i.e., independent-samples) *t*-test, that they would use the .05 level of significance, and that they wanted to have at least an 90 percent chance of

EXCERPT 10.21 • *An a Priori Power Analysis*

An unpaired *t*-test with an alpha level set at .05 was used to compare adherence [means] between the experimental group and comparison group. . . . An a priori power analysis was conducted to estimate the number of participants needed to obtain a statistical power of 0.90 at an alpha level of .05. The a priori power analysis estimated a sample size of 10 participants in the experimental group and 10 participants in the comparison group would detect a 2.0 standardized effect size with a statistical power of at least 0.90 at an alpha level of .05.

Source: Taylor, J. D., Fletcher, J. P., & Tiarks, J. (2009). Impact of physical therapist-directed exercise counseling combined with fitness center-based exercise training on muscular strength and exercise capacity in people with type 2 diabetes: A randomized clinical trial. *Physical Therapy, 89*(9), 884–889.

rejecting the null hypothesis (of equal population means) if the true and standardized $\mu_1 - \mu_2$ difference was as large as or larger than 2. After making these decisions, the power analysis indicated that the researchers would need 10 individuals in each comparison group.

In Excerpt 20.21, the computed sample sizes probably seem quite small, especially in light of the fact that power is set equal to .90. The reason why the a priori power analysis produced these small *n*s is the effect size specified by the researchers. When comparing two means with a *t*-test, a standardized effect size of 2.0 is enormous. (Recall that the widely used criteria for small, medium, and large effects are .2, .5, and .8, respectively.) In a power analysis, an *inverse* relationship exists between effect size and sample size. Small effects require big samples to detect them; big effects can be detected with small samples.

Underlying Assumptions

When a statistical inference concerning one or two means is made using a confidence interval or a *t*-, *F*-, or *z*-test, certain assumptions about the sample(s) and population(s) are typically associated with the statistical technique applied to the data. If one or more of these assumptions are violated, then the probability statements attached to the statistical results may be invalid. For this reason, well-trained researchers (1) become familiar with the assumptions associated with the techniques they use to analyze their data and (2) take the time to check out important assumptions before making inferences from the sample mean(s).

For the statistical techniques covered thus far in this chapter, there are four underlying assumptions. First, each sample should be a random subset of the population it represents. Second, there should be *independence of observations* (meaning that a particular person's score is not influenced by what happens to any other person during the study). Third, each population should be normally distributed in terms of the dependent variable being focused on in the study. And fourth, the two populations associated with studies involving two independent samples or two correlated samples should each have the same degree of variability relative to the dependent variable.

The assumptions dealing with the randomness and independence of observations are methodological concerns, and researchers rarely talk about either of these assumptions in their research reports. The other two assumptions, however, are often discussed by researchers. To be a discerning consumer of the research literature, you must know when the **normality assumption** and **equal variance assumption** should be considered, what is going on when these assumptions are tested, what researchers do if they find that their data violate these assumptions, and under what conditions a statistical test is robust to violations of the normality or equal variance assumptions. This section is intended to provide you with this knowledge.

Researchers should consider the normality and equal variance assumptions *before* they evaluate their study's primary H_0. Assumptions should be considered first, because the statistical test used to evaluate the study's H_0 may not function the way it is supposed to function if the assumptions are violated. In a sense, then, checking on the assumptions is like checking to see if there are holes in a canoe (or whether your companion has attached an outboard motor) before getting in and paddling out to the middle of a lake. Your canoe simply will not function the way it is supposed to if it has holes or has been turned into a motorboat.

When the normality or equal variance assumption is examined, the researcher uses the sample data to make an inference from the study's sample(s) to its population(s). This inference is similar to the one that the researcher wishes to make concerning the study's primary H_0, except that assumptions do not deal with the mean of the population(s). As their names suggest, the *normality assumption* deals with distributional shape, whereas the *equal variance assumption* is concerned with variability. Often, the sample data are used to test these assumptions. In such cases the researcher applies all of the steps of the hypothesis testing procedure, starting with the articulation of a null hypothesis and ending with a reject or fail-to-reject decision. In performing such tests, the researcher hopes that the null hypothesis of normality or of equal variance is *not* rejected, because then he or she is able to move ahead and test the study's main null hypothesis concerning the mean(s) of interest.

Excerpts 10.22 and 10.23 illustrate how the normality and equal variance assumptions are sometimes tested by applied researchers.[5] In each of these excerpts,

---

[5]In Excerpt 10.22, the researcher used the Kolmogorov–Smirnov test to check on the normality assumption. There are other available test procedures (e.g., the chi square goodness-of-fit test) that do the same thing. In Excerpt 10.23, Levene's test was used to check on the equal variance assumption. Other test procedures (e.g., Bartlett's chi square) can be employed to make the same data check.

EXCERPTS 10.22–10.23 • *Testing the Normality and Equal Variance Assumptions*

Normality of the data was tested using the One-Sample Kolmogorov-Smirnov test on the dependent variable which was the differences between pretest and posttest scores. The test statistics was 0.053 (not significant at the 0.05 level), and distribution of the data was normal. . . . The one-group t test was used for the pretest and posttest results within the same study group, and the two-group t test was used to compare scores between the two study groups.

Source: Lu, D. F., Lin, Z., & Li, Y. (2009). Effects of a web-based course on nursing skills and knowledge learning. *Journal of Nursing Education, 48*(2), 70–77.

Levene's Test for Equality of Variances was run before interpreting the results of *t* testing. Because the significance value was greater than .05 in Levene's test (.091), equal variances were assumed.

Source: Brock, S. E. (2010). Measuring the importance of precursor steps to transformative learning. *Adult Education Quarterly*, *60*(2), 122–142.

notice that the null hypothesis for the assumption was not rejected. That was the desired result, for the researchers were then permitted to move ahead and to what they were really interested in: a comparison of their two sample *means*.

The assumption of equal variances is often referred to as the **homogeneity of variance assumption.** This term is somewhat misleading, however, because it may cause you to think that the assumption specifies homogeneity *within* each population in terms of the dependent variable. That is not what the assumption means. The null hypothesis associated with the equal variance assumption says that $\sigma_1^2 = \sigma_2^2$. This assumption can be true even when each population has a large degree of variability. Homogeneity of variance exists if σ_1^2 is equal to σ_2^2, regardless of how large or small the common value of σ^2.

If a researcher conducts a test to see if the normality or equal variance assumption is tenable, it may turn out that the sample data do not argue against the desired characteristics of the study's populations. That was the case in Excerpts 10.22 and 10.23. But what happens if the test of an assumption suggests that the assumption is not tenable?

In the situation where the sample data suggest that the population data do not conform with the normality or equal variance assumptions, there are three options available to the researcher. These options include (1) using a special formula in the study's main test so as to compensate for the observed lack of normality or heterogeneity of variance; (2) changing each raw score by means of a data transformation designed to reduce the degree of nonnormality or heterogeneity of variance, thereby permitting the regular *t*-test, *F*-test, or *z*-test to be used when the study's main test focuses on the study's mean(s); or (3) using a test procedure other than *t, F,* or *z*—one that does not involve such rigorous assumptions about the populations. Excerpts 10.24, 10.25, and 10.26 illustrate these three options.

EXCERPTS 10.24–10.26 • *Options When Assumptions Seem Untenable*

For comparisons between two groups, a two-tailed *Student*'s *t*-test was conducted. If the *F*-test revealed that the group variances were significantly different, *Welch's* *t*-test was used in place of *Student*'s *t*-test.

Source: Kiya, T., & Kubo, T. (2010). Analysis of GABAergic and Non-GABAergic neuron activity in the optic lobes of the forager and re-orienting worker honeybee (*Apis mellifera* L.). *PLoS One.* 5(1), 1–8.

Data for PTP80 was transformed (square root transformation) to fulfill assumptions of normality. Paired t tests confirmed that baseline measures did not differ across the two experimental sessions.

Erickson, E., & Sivasankar, M. (2010). Evidence for adverse phonatory change following an inhaled combination treatment. *Journal of Speech, Language & Hearing Research, 53*(1), 75–83.

Student's *t* test was used to compare means of continuous variables. If assumptions of equality of variance and normality (assumed for the *t* test) were not met, the Mann-Whitney *U* test (a nonparametric equivalent of the *t* test) was performed as appropriate.

Source: Zyoud, A. H., Awang, R., Sulaiman, S. A. S., Sweileh, W. M., & Al-Jabi, S. W. (2010). Incidence of adverse drug reactions induced by *N*-acetylcysteine in patients with acetaminophen overdose. *Human & Experimental Toxicology, 29*(3), 153–160.

Excerpt 10.24 represents option 1, for a special version of the *t*-test (called Welch's *t*-test) has built-in protection against violations of the equal variance assumption. Excerpt 10.25 shows option 2, the strategy of transforming the data and then using the regular test procedure to compare the group means. Here, a square root transformation was used. In Excerpt 10.26, the researchers wanted to use a *t*-test to make comparisons of group means. However, in those cases where their data violated the normality assumption, the researchers chose option 3; instead of using the *t*-test, the researchers used the nonparametric Mann-Whitney *U* test that does not assume normality.

When researchers are interested in comparing the means of two groups, they often bypass testing the assumption of equal variances if the two samples are equally big. This is done because studies in theoretical statistics have shown that a test on means functions very much as it should even if the two populations have unequal amounts of variability, as long as $n_1 = n_2$. In other words, t-, F-, and z-tests are strong enough to withstand a violation of the equal variance assumption if the sample sizes are equal. Stated in statistical "jargoneze," equal ns make these tests **robust** to violations of the homogeneity of variance assumption.

Comments

Before concluding our consideration of inferences regarding one or two means, I want to offer five warnings that, if you heed them, will cause you to be a more informed recipient of research results. These warnings are concerned with (1) outcomes where the null hypothesis is not rejected, (2) outcomes where H_0 is rejected, (3) the typical use of t-tests, (4) practical significance, and (5) research claims that seem to neglect the possibility of a Type I or a Type II error.

A Nonsignificant Result Does Not Mean H_0 Is True

In Chapter 7, I indicated that a null hypothesis should *not* be considered to be true simply because it is not rejected. Researchers sometimes forget this important point, especially when they compare groups in terms of pretest means. In making this kind of comparison, researchers usually hope that the null hypothesis is *not* rejected, because they want to consider the comparison groups to have been the same at the beginning of the study. There are three reasons why it is dangerous to think that H_0 is true if it is not rejected.

The context for these three comments is a hypothetical study. Imagine that we have two groups, E and C (experimental and control), with pretest data available on each person in each group. Let's also imagine that the sample means, M_E and M_C, turn out equal to 16 and 14, respectively. Finally, imagine that a t-test or F-test is used to compare these two pretest Ms, with the result being that the null hypothesis ($H_0 : \mu_E = \mu_C$) is not rejected because $p_{\text{two-tailed}} > .05$.

The first reason for not accepting H_0 in this hypothetical study is purely logical in nature. If the null hypothesis had been set up to say that $H_0 : \mu_E - \mu_C = 1$, a fail-to-reject decision also would have been reached, which is also what would have happened if H_0's pinpoint number had been set equal to any other value between -2 and $+2.0$. Because the data support multiple null hypotheses that could have been set up (and that clearly are in conflict with each other), there is no scientific justification for believing that any one of them is right while the others are wrong.

The second reason for not accepting H_0 concerns data quality. In Chapter 9, I discuss attenuation and point out how measuring instruments that have less than perfect reliability can function to mask a true nonzero relationship that exists between two variables. The same principle applies to inferential tests that focus on things other than correlation coefficients, such as means. In our hypothetical study, data produced by a measuring instrument with low reliability could lead to a fail-to-reject decision; with a more reliable instrument, the sample means—even if they again turn out equal to 16 and 14—might end up producing a p that is lower than .05! Thus, our hypothetical study may have produced a nonsignificant finding because of unreliability in the data, not because $H_0 : \mu_E = \mu_C$.

A final consideration that mitigates against concluding that $\mu_E = \mu_C$ when H_0 is retained has to do with statistical power. As I have pointed out on several occasions, there is a direct relationship between sample size and the probability of detecting a situation in which H_0 is false. Thus, the failure to find a statistically significant finding in our hypothetical study may have been caused by ns that were too small. Perhaps μ_E and μ_C differ greatly, but our study simply lacked the statistical sensitivity to illuminate that situation.

For these three reasons (logic, reliability, and statistical power), be on guard for unjustified claims that H_0 is true following a decision not to reject H_0.

Overlapping Distributions

Suppose a researcher compares two groups of scores and finds that there is a statistically significant difference between M_1 and M_2. Notice that the significant difference exists between the *means* of the two groups. Be on guard for research reports in which the results are discussed without reference to the group means, thus creating the impression that every score in one group is higher than every score in the second group. Such a situation is *very* unlikely.

To illustrate what I mean by **overlapping distributions,** consider once again the information presented in Excerpt 10.9, in which we saw that the mean number of barriers (for not exercising) cited by the research participants dropped from 6.17 at pretest to 4.48 at posttest. These two sample means were compared with a paired t-test, and it turned out that there was a statistically significant difference between the group means. The null hypothesis of equal pretest and posttest population means was rejected with $p < .001$.

Did all research participants decrease the number of barriers they cited? The evidence contained in Excerpt 10.9 allows us to answer this question with a resounding "No." Return to this excerpt and take a look at the standard deviations for each group of scores. These standard deviations (along with the means) strongly suggest that the two distributions of scores overlapped because the two group means were 1.69 points apart, whereas the two standard deviations were 2.64 and 2.35.

Be on guard for researchers who make a comparison between two different groups (or between a single group that is measured twice), who reject the null hypothesis, and who then summarize their findings by saying something like "Girls outperformed boys" or "The treatment produced higher scores than did the control" or "Participants improved between pretest and posttest." Such statements are often seen in the abstracts of research reports. When you see these phrases, be sure to insert the three words "On the average" at the beginning of the researcher's summary.

Also, keep in mind that overlapping distributions are the rule, not the exception, in research investigations.

The Typical Use of t-Tests

In this chapter, you have seen how a *t*-test can be used to evaluate a null hypothesis dealing with one or two means. You will discover that *t*-tests can also be used when the researcher's statistical focus is on things other than means. For example, a *t*-test can be used to see if a correlation coefficient is significantly different from zero, or if there is a significant difference between two correlations. For this reason, it is best to consider a *t*-test to be a general tool that can be used to accomplish a variety of inferential goals.

Although a *t*-test can focus on many things, it is used most often when the researcher is concerned with one or two means. In fact, *t*-tests are used so frequently to deal with means that many researchers equate the term *t-test* with the notion of a test focusing on the mean(s). These researchers use a modifying phrase to clarify how many means are involved and the nature of the samples, thus leading to the terms *one-sample t-test, independent-samples t-test, correlated-samples t-test, matched t-test, dependent-samples t-test,* and *paired t-test.* When any of these terms is used, a safe bet is that the *t*-test being referred to had the concept of *mean* as its statistical focus.

Practical Significance versus Statistical Significance

Earlier in this chapter, you saw how researchers can do certain things in an effort to see whether a statistically significant finding is also meaningful in a practical sense. Unfortunately, many researchers do not rely on computed effect size indices or power analyses to help them avoid the mistake of "Making a mountain out of a molehill." They simply use the six-step version of hypothesis testing and then get excited if the results are statistically significant.

Having results turn out to be statistically significant can cause researchers to go into a trance in which they willingly allow the tail to wag the dog. This is what happened, I think, to the researchers who conducted a study a few years ago comparing the attitudes of two groups of women. In the researchers' technical report I examined, they first indicated that the means turned out equal to 67.88 and 71.24 (on a scale that ranged from 17 to 85) and then stated, "Despite the small difference in means, there was a significant difference."

To me, the final 11 words of the previous paragraph conjure up the image of statistical procedures functioning as some kind of magic powder that can be sprinkled on one's data and transform a molehill of a mean difference into a mountain that deserves others' attention. However, statistical analyses lack that kind of magical power. Had the researchers who obtained those means of 67.88 and 71.24 not been blinded by the allure of statistical significance, they would have focused their attention on the small difference and not the significant difference. And had they done this, their final words would have been, "Although there was a significant difference, the difference in means was small."

Estimates of effect size and power analyses can help keep researchers (and you) alert to the important distinction between practical significance and statistical significance. However, do not be reluctant to use your own knowledge (and common sense) when it comes to judging the "meaningfulness" of statistical results. In some cases, you will be able to make confident decisions on your own as to whether a "big difference" exists between two sample means. You ought to be able to do that when you examine Excerpt 10.27. (The two means reported in this excerpt were on a scale that extended from 1.0 to 5.0.)

EXCERPT 10.27 • *Practical Significance: Is This Mean Difference "Big"?*

Male students (Mean = 3.08, SD = 0.42) showed significantly higher adoption of scientific attitudes than did female students (Mean = 3.01, SD = 0.41), $F(1,550) = 4.13, p < .05$.

Source: Chen, C., & Howard, B. (2010). Effect of live simulation on middle school students' attitudes and learning toward science. *Journal of Educational Technology & Society, 13*(1), 133–139.

Type I and Type II Errors

My final comment concerns the conclusion reached whenever the hypothesis testing procedure is used. Because the decision to reject or fail to reject H_0 is fully inferential in nature (being based on sample data), there is *always* the possibility that a Type I or Type II error will be committed. You must keep this in mind as you read technical research reports, as most researchers do not allude to the possibility of inferential error as they present their results or discuss their findings. In certain cases, the researcher simply presumes that you know that a Type I or Type II error may occur whenever a null hypothesis is tested. In other cases, the researcher unfortunately may have overlooked this possibility in the excitement of seeing that the statistical results were congruent with his or her research hypothesis.

When reading research reports, you will encounter many articles in which the researchers talk as if they have discovered something definitive. The researchers' assertions typically reduce to the claim that "The data confirm our expectations, so now we have proof that our research hypotheses were correct." Resist the temptation to bow down in front of such researchers and accept everything and anything they might say, simply because they have used fancy statistical techniques when analyzing their data. Remember that *inferences* are *always* involved whenever (1) confidence intervals are placed around means or differences between means and (2) null hypotheses involving one or two means are evaluated. Nothing is *proved* by any of these techniques, regardless of how bold the researchers' claims might be.

Review Terms

Analysis of variance	*MS*
Bonferroni adjustment technique	Normality assumption
	Omega squared
Confidence interval	Overlapping distributions
Correlated samples	Paired samples
Dependent samples	Partial eta squared
Degrees of freedom (*df*)	Power analysis
Effect size	Pseudo-Bonferroni adjustment procedure
Equal variance assumption	Robust
	SS
Eta squared	*t*-test
F-test	Uncorrelated samples
Homogeneity of variance assumption	Unmatched samples
	Unpaired samples
Independent samples	Within samples
Matched samples	*z*-test

The Best Items in the Companion Website

1. An interactive online quiz (with immediate feedback provided) covering Chapter 10.
2. Nine misconceptions about the content of Chapter 10.
3. An email message sent from the author to his students entitled "A Little *t*-Test Puzzle."
4. One of Chapter 10's best passages: "Inference and Proof."
5. Two good jokes about statistics.

To access the chapter outline, practice tests, weblinks, and flashcards, visit the companion website at http://www.ReadingStats.com.

Review Questions and Answers begin on page 531.

11

Tests on Three or More Means Using a One-Way ANOVA

In Chapter 10, we considered various techniques used by researchers when they apply inferential statistics within studies focusing on one or two means. I now wish to extend that discussion by considering the main inferential technique used by researchers when their studies involve three or more means. The popular technique used in these situations is called analysis of variance and it is abbreviated as **ANOVA.**

As I pointed out in Chapter 10, the analysis of variance can be used to see if there is a significant difference between two sample means. Hence, this particular statistical technique is quite versatile. It can be used when a researcher wants to compare two means, three means, or any number of means. It is also versatile in ways that will become apparent in Chapters 13, 14, and 19.

The analysis of variance is an inferential tool that is widely used in many disciplines. Although a variety of statistical techniques have been developed to help applied researchers deal with three or more means, ANOVA ranks first in popularity. Moreover, there is a big gap between ANOVA and whatever ranks second!

In the current chapter, we will focus our attention on the simplest version of ANOVA, something called a one-way analysis of variance. I begin with a discussion of the statistical purpose of a one-way ANOVA, followed by a clarification of how a one-way ANOVA differs from other kinds of ANOVA. Then, we turn our attention to the way researchers present the results of their one-way ANOVAs, with examples to show how the Bonferroni adjustment technique is used in conjunction with one-way ANOVAs, how the assumptions underlying a one-way ANOVA are occasionally tested, and how researchers sometimes concern themselves with power analyses and effect size. Finally, I offer a few tips that should serve to make you better able to decipher and critique the results of one-way ANOVAs.

The Purpose of a One-Way ANOVA

When a study has been conducted in which the focus is centered on three or more groups of scores, a **one-way ANOVA** permits the researcher to use the data in the samples for the purpose of making a single inferential statement concerning the means of the study's populations. Regardless of how many samples are involved, there is just one inference that extends from the set of samples to the set of populations. This single inference deals with the question, "Are the means of the various populations equal to one another?"

Figure 11.1 illustrates what is going on in a one-way ANOVA. There are three things to notice about this picture. First, our picture illustrates the specific situation where there are three comparison groups in the study; additional samples and populations can be added to parallel studies that have four, five, or more comparison groups. Second, there is a single inference made from the full set of sample data to the group of populations. Finally, the focus of the inference is on the population means, even though each sample is described in terms of M, SD, and n.

Although you may never come across a journal article that contains a picture like that presented in Figure 11.1, I hope my picture helps you understand what is going on when researchers talk about having applied a one-way ANOVA to their data. Consider, for example, Excerpt 11.1, which comes from a study focused on the use of multimedia technology to assist students as they read an electronic essay. All 69 of the research participants (college freshmen) individually read the same essay. However, the technology-based learning aids were different. Students in one group ($n = 23$) could click on any of 42 highlighted vocabulary words and see a

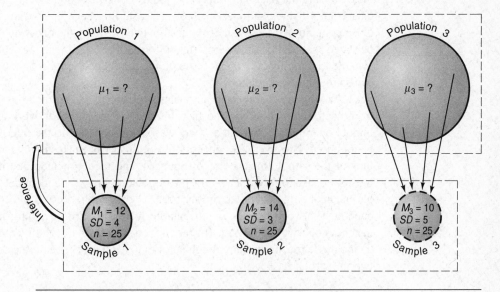

FIGURE 11.1 *Illustration of a One-Way ANOVA's Inferential Objective*

EXCERPT 11.1 • *Typical Data Used in a One-Way ANOVA*

TABLE 4   *Means and standard deviations for reading comprehension test*

Group	Mean	SD
Definition (D)	15.30	3.57
Definition & picture (DP)	17.43	3.69
Definition & movie	16.09	3.55

Source: Akbulut, Y. (2007). Effects of multimedia annotations on incidental vocabulary learning and reading comprehension of advanced learners of English as a foreign language. *Instructional Science, 35*(6), 499–517.

definition. Members of the second group ($n = 23$) had the same words highlighted, but a click on any of them brought forth a picture along with the definition. Those in the third group ($n = 23$), if they clicked a highlighted word, saw the definition plus a brief 10–15 second video. After reading the essay, all students were given a 34-item reading comprehension test covering the material in the essay.[1]

With Figure 11.1 fresh in your mind, you should be able to look at Excerpt 11.1 and discern what the researcher was trying to accomplish by using a one-way ANOVA. Each row of data in this excerpt, of course, corresponds to one of the study's three samples. Connected to each sample was an abstract population (i.e., a larger group of students, like the ones in the study, who theoretically could be given the essay and the reading comprehension test). The researchers' goal was to use the data from all three samples to make a single inference concerning the means of those populations. The statistical question dealt with by the one-way ANOVA could be stated as: "In light of the empirical information available in the samples, is it reasonable to think that the mean score on the reading comprehension test is the same across the three populations?"

As you can see, the sample means in Excerpt 11.1 turned out to be different from each other. Based on the fact that the *M*s in this study were dissimilar, you might be tempted to think that there was an easy answer to the inferential question being posed. However, the concept of sampling error makes it *impossible* to simply look at the sample means, see differences, and then conclude that the population means are also different. Possibly, the population means are identical, with the sample means being dissimilar simply because of sampling error. Or, maybe the discrepancy between the *M*s *is* attributable to dissimilarities among the population means. A one-way ANOVA helps researchers decide, in a scientific manner, whether their sample means are far enough apart to place their eggs into the second of these two proverbial baskets.

[1]This study actually involved more data, hypotheses, and statistical analyses than described here.

The Distinction between a One-Way ANOVA and Other Kinds of ANOVA

In this chapter, we are focusing our attention on the simplest kind of ANOVA; the kind that is referred to as a one-way ANOVA, as a **one-factor ANOVA,** or as a **simple ANOVA.** Because there are many different kinds of analysis variance, it is important to clarify the difference between the kind that we are considering in this chapter and the more complex kinds of ANOVA that are discussed in Chapters 13, 14, and 19. (Some of the more complex kinds of analysis of variance have the labels *two-way ANOVA, repeated measures ANOVA,* and *multivariate ANOVA.*)

Although all ANOVAs are alike in that they focus on means, they differ in three main respects: the number of independent variables, the number of dependent variables, and whether the samples are independent or correlated. In terms of these distinguishing characteristics, a one-way ANOVA has *one* independent variable, it focuses on *one* dependent variable, and it involves samples that are *independent.* It is worthwhile to consider each of these defining elements of a one-way ANOVA because researchers sometimes use the term *ANOVA* by itself without the clarifying adjective *one-way.*

When we say that there is just one independent variable, this means that the comparison groups differ from one another, prior to the collection and analysis of any data, in *one* manner that is important to the researcher. The comparison groups can differ in terms of a qualitative variable (e.g., favorite TV show) or in terms of a quantitative variable (e.g., number of siblings), but there can be only one characteristic that defines how the comparison groups differ. Because the terms **factor** and **independent variable** mean the same thing within the context of analysis of variance, this first way in which a one-way ANOVA differs from other ANOVAs can be summed up in this manner: A one-way ANOVA has a single factor (i.e., one independent variable).

Excerpt 11.2 comes from a study in which 904 school children in grades 3 through 6 completed a personality inventory called the SEARS-C. This instrument had been developed to measure a child's opinion of his or her social–emotional strengths,

EXCERPT 11.2 • *The Independent and Dependent Variables in a One-Way ANOVA*

Data on the SEARS-C were analyzed using a one-way, between-subjects analysis of variance to estimate the degree of difference across grade levels. . . . Grade was the independent variable with four levels: (a) 3rd grade, (b) 4th grade, (c) 5th grade, and (d) 6th grade. Total sum score on the SEARS-C was the dependent variable.

Source: Cohn, B., Merrell, K. W., Felver-Grant, J., Tom, K., & Endrulat, N. R. (2009). Strength-based assessment of social and emotional functions. Paper presented at the meeting of the National Association of School Psychologists, Boston.

with Likert items such as "I make friends easily" and "I stay calm when there is an argument or a problem." A one-way ANOVA was used to see if differences existed among the four grade levels in terms of the children's total score on the SEARS-C.

As illustrated by Excerpt 11.2, some researchers identify explicitly the independent variable associated with their one-way ANOVA. However, many researchers choose not to do this and instead presume that their readers can figure out what the independent variable was based on a description of samples used in the study. By the end of this chapter, I believe that you will have little difficulty identifying the independent variable in any one-way ANOVA you encounter.

As you might suspect, a two-way ANOVA has two independent variables, a three-way ANOVA has three independent variables, and so on. Beginning in Chapter 13, we consider some of these more complex ANOVAs. In this chapter, however, we restrict our focus to the kind of ANOVA that has a single independent variable.

Even if there is just one independent variable within a study in which the analysis of variance is applied, the ANOVA may or may not be a one-way ANOVA. The second criterion that distinguishes one-way ANOVAs from many other kinds of ANOVAs has to do with the number of dependent variables involved in the analysis. With a one-way ANOVA, there is always just one dependent variable. (If there are two or more dependent variables involved in the same analysis, then you are likely to see the analysis described as a multivariate ANOVA, or MANOVA.)

The **dependent variable** corresponds to the measured characteristic of people, animals, or things from whom or from which data are gathered. For example, in the study from which Excerpt 11.1 was taken, the dependent variable was the students' reading comprehension as measured by the 34-item test administered after the students read the multimedia essay. In that excerpt, the table's title lets us know what the dependent variable was. In Excerpt 11.2, the researchers came right out and told us what their dependent variable was.

The third distinguishing feature of a one-way ANOVA concerns the fact that the comparison groups are independent (rather than correlated) in nature. As you may recall from the discussion in Chapter 10 of independent versus correlated samples, this means that (1) the people or animals who provide the scores in any given group are different from those who provide data in any other comparison group, and (2) there is no connection across comparison groups because of matching or because several triplets or litters were split up (with one member of each family being put into each of the comparison groups). It is possible for an ANOVA to be applied to the data that come from correlated samples, but I delay my discussion of that form of analysis until Chapter 14.

As indicated in Excerpt 11.2, researchers sometimes refer to their study's independent variable as being the **between-subjects variable.** The adjective *between-subjects* is used to clarify that comparisons are being made with data that have come from independent samples. (Chapter 14 discusses one-way ANOVAs used in studies where the data come from correlated samples, and you will see that the independent variables in those studies are considered to be *within-subjects* in nature.) Because each of the one-way ANOVAs discussed in this chapter involves data collected from separate groups of individuals who have not been matched in any way, every independent variable we encounter here is a between-subjects independent variable.

Now, we turn our attention to the specific components of a one-way ANOVA, and we begin with a consideration of the one-way ANOVA's null and alternative hypotheses.

The One-Way ANOVA's Null and Alternative Hypotheses

The null hypothesis of a one-way ANOVA is always set up to say that the mean score on the dependent variable is the same in each of the populations associated with the study. The null hypothesis is usually written by putting equal signs between a set of μs, with each μ representing the mean score within one of the populations. For example, if there were four comparison groups in the study, the null hypothesis would be $H_0: \mu_1 = \mu_2 = \mu_3 = \mu_4$.

If you recall my claim (Chapter 7) that every null hypothesis must contain a pinpoint parameter, you may now be wondering how the symbolic statement at the end of the preceding paragraph qualifies as a legitimate null hypothesis because it does not contain a pinpoint number. In reality, there *is* a pinpoint number contained in that H_0, but it is simply hidden from view. If the population means are all equal to one another, then there is no variability among those means. Therefore, we can bring H_0's pinpoint number into plain view by rewriting the null hypothesis as $H_0: \sigma_\mu^2 = 0$. As we said earlier, however, you are more likely to see H_0 written with Greek mus and equal signs and no pinpoint number (e.g., $H_0: \mu_1 = \mu_2 = \mu_3$) rather than with a sigma squared (with μ as a subscript) set equal to zero.

In Excerpts 11.3 and 11.4, we see examples of one-way ANOVA null hypotheses that have appeared in research summaries. Notice that these two excerpts

EXCERPTS 11.3–11.4 • *The Null Hypothesis in a One-Way ANOVA*

The null hypothesis is constructed as $H_0: \mu_1 = \mu_2 = \mu_3$ (the improvement are equal), which essentially implies that the three feedback strategies are identically effective.

Source: Shen, Y. (2010). Evaluation of an eye tracking device to increase error recovery by nursing students using human patient simulation. Unpublished Master's theses, University of Massachusetts, Amherst.

ANOVA test was used to identify the relationship between BMI categories and age. The null-hypothesis is $(H_0: \mu_1 = \mu_2 = \mu_3 = \mu_4)$ was tested.

Source: Rosnah, M.Y., Mohd R. H., & Sharifah_Norazizan, S.A.R. (2009*)*. Anthropometry dimensions of older Malaysians: Comparison of age, gender and ethnicity. *Asian Social Science*, *5*(6), 133–140.

are similar in that each null hypothesis deals with its study's population means. Moreover, both null hypotheses have been set up to say that there are no differences among the population means. These excerpts differ, however, in that the first study involved three populations, whereas there were four in the second study.

The researchers associated with Excerpts 11.3 and 11.4 deserve high praise for taking the time to articulate the null hypothesis associated with their one-way ANOVAs. The vast majority of researchers do not do this. They tell us about the data they collected and what happened in terms of results, but they skip over the important first step of hypothesis testing. Perhaps they assume that readers will know what the null hypothesis was.

In hypothesis testing, of course, the null hypothesis must be accompanied by an alternative hypothesis. This H_a always says that at least two of the population means differ. Using symbols to express this thought, we get $H_a : \sigma_\mu^2 \neq 0$. Unfortunately, the alternative hypothesis is rarely included in technical discussions of research studies. Again, researchers evidently presume that their readers are familiar enough with the testing procedure being applied and familiar enough with what goes on in a one-way ANOVA to know what H_a is without being told.

Presentation of Results

The outcome of a one-way ANOVA is presented in one of two ways. Researchers may elect to talk about the results within the text of their report and to present an ANOVA **summary table.** However, they may opt to exclude the table from the report and simply describe the outcome in a sentence or two of the text. (At times, a researcher wants to include the table in the report but is told by the journal editor to delete it due to limitations of space.)

Once you become skilled at deciphering the way results are presented within an ANOVA summary table, I am confident that you will have no difficulty interpreting results presented within a "tableless" report. For this reason, I begin the next section with a consideration of how the results of one-way ANOVAs are typically presented in tables. I divide this discussion into two sections because some reports contain the results of a single one-way ANOVA, whereas other reports present the results of many one-way ANOVAs.

Results of a Single One-Way ANOVA

In Excerpt 11.5, we see the ANOVA summary table for the data presented earlier in Excerpt 11.1. As you may recall, that earlier excerpt, and now this new one, come from a study wherein the researchers wanted to know if different kinds of multimedia "learning aids" differ in their ability to help students with their reading comprehension when confronted with an electronic essay. You might want to take a quick look at Excerpt 11.1 before proceeding.

EXCERPT 11.5 • *Results from a One-Way ANOVA Presented in a Table*

The ANOVA summary table is provided in Table 5. As the table suggests, the reading comprehension scores of the groups did not differ significantly ($F_{2,66} = 2.054$, $p = .136$).

TABLE 5 *ANOVA summary table for the reading comprehension test*

Source	SS	df	MS	F
Between groups	53.420	2	26.710	2.054
Within groups	858.348	66	13.005	
Total	911.768	68		

Source: Akbulut, Y. (2007). Effects of multimedia annotations on incidental vocabulary learning and reading comprehension of advanced learners of English as a foreign language. *Instructional Science, 35*(6), 499–517.

In Excerpt 11.5, the number 2.054 is the calculated value, and it is positioned in the column labeled **F**. That calculated value was obtained by dividing the **mean square (MS)** on the "**Between groups**" row of the table (26.710) by the mean square on the "**Within groups**" row of the table (13.005). Each row's MS value was derived by dividing that row's **sum of squares (SS)** value by its *df* value.[2] Those *SS* values came from an analysis of the sample data. The *df* values, however, came from simply counting the number of groups, the number of people within each group, and the total number of participants—with 1 subtracted from each number to obtain the *df* values presented in the table.[3]

The first two *df* values determined the size of the critical value against which 2.054 was compared. (That critical value, at the .05 level of significance, was equal to 3.14.) I am confident that a computer used the *df* values of 2 and 66 to determine the critical value, and then the computer said that *p* was equal to .136. This is the standard way of getting a precise *p*-value like the one reported in Excerpt 11.5. (In a few cases, the researcher uses the two *df* values to look up the size of the critical value in a statistical table located in the back of a statistics book. Such tables allow researchers to see how big the critical values are at the .05, .01, and .001 levels of significance.)

Whereas the *df* numbers in a one-way ANOVA have a technical and theoretical meaning (dealing with things called *central and noncentral F distributions*), those *df* numbers can be useful to *you* in a very practical fashion. To be more specific, you can use the first and the third *df* to help you understand the structure of a completed study. To

---

[2]A mean square is never computed for the total row of a one-way ANOVA or for the total row of any other kind of ANOVA.

[3]The within *df* was computed first by subtracting 1 from each of the three sample sizes, and then by adding the three *n* – 1 values.

show you how this is done, let's focus on Excerpts 11.5 and 11.1. By adding 1 to the **between groups** *df*, you can determine, or verify, there were $2 + 1 = 3$ groups in this study. By adding 1 to the total *df*, you can figure out that there were $68 + 1 = 69$ students who read the multimedia essay and then took the reading comprehension test.

In the ANOVA summary table displayed in Excerpt 11.5, the second row of numbers was labeled "Within groups." I would be remiss if I did not warn you that a variety of terms are used by different researchers to label this row of a one-way ANOVA summary table. On occasion, you are likely to see this row referred to as *Within, Error, Residual,* or *Subjects within groups*. Do not let these alternative labels throw you for a loop. If everything else about the table is similar to the table we have just examined, then you should presume that the table you are looking at is a one-way ANOVA summary table.

Some of the one-way ANOVA summary tables you see are likely to be modified in two ways beyond the label given to the middle row of numbers. Some researchers delete entirely the bottom row (for Total), leaving just the rows for "Between groups" and "Within groups." Other researchers either switch the location of the *df* and *SS* columns or delete altogether the *SS* column. These variations do not affect the main things being communicated by the ANOVA summary table, nor do they hamper your ability to use such tables to as you try to make sense out of a study's structure and its results.

Because the calculated *F*-value and the *p*-value are considered to be the two most important numbers in a one-way ANOVA summary table, those values are sometimes pulled out of the summary table and included in a table containing the comparison group means and standard deviations. By doing this, space is saved in the research report because only one table is needed rather than two. Had this been done in Excerpt 11.1, a note might have appeared beneath Table 4 saying "$F(2, 66) = 2.054, p = .136$." Because this kind of note indicates the *df* for between groups and within groups, you can use these numbers to determine how many people were involved in the study if the sample sizes are not included in the table. A note like this does not contain *SS* or *MS* values, but you really do not need them.

Although the results of a one-way ANOVA are sometimes presented in a table similar to the one we have just considered, more often the outcome of the statistical analysis is discussed in the text of the report, with no table included. In Excerpt 11.6, we see an illustration of a one-way ANOVA being summarized in five sentences.

EXCERPT 11.6 • *Results from a One-Way ANOVA Presented Without a Table*

The purpose of our experiment was to study how different output modalities of a GPS navigation guide affect drivers and driving performance during real traffic driving. We included three output modalities namely audio, visual, and audio-visual [and used]

(continued)

EXCERPTS 11.6 • (*continued*)

30 people ranging between 21–38 years of age. . . . We utilized a between-subject experimental design [and] assigned five GPS system users and five non-users to each of the three configurations (which constitute three groups of ten). . . . Our experiment showed that participants using the audio configuration on average had 8.8 [speeding] violations during the trails whereas visual participants on average had 17.9 violations and audio-visual participants had 19.3 violations. An ANOVA test showed significant difference among the three configurations, $F(2,27) = 6.67, p < 0.01$.

Source: Jensen, B. S., Skov, M. B., & Thiruravichandran, N. (2010). Studying driver attention and behaviour for three configurations of GPS navigation in real traffic driving. *Proceedings of the 28th International Conference on Human Factors in Computing Systems,* 1271–1280.

In the study associated with Excerpt 11.6, the independent variable was the kind of output produced by a car's GPS navigation system. Ten of the study's drivers had a NAV system that just talked, 10 others had a system that provided only visual output, and a third group of 10 had a system that provided both auditory and visual output. The dependent variable was the number of speeding violations recorded on videotape as the drivers drove 16 kilometers through both rural and urban sections of Denmark. The excerpt's final sentence contains the ANOVA's results, with the calculated value (6.67) being shown. The *p*-statement at the end of the excerpt indicates that the three samples had mean scores that were further apart than would be expected by chance. Accordingly, the null hypothesis was rejected.

Excerpt 11.6 also contains two numbers in parentheses next to the letter *F*. These are the *df* values taken from the between groups and within groups rows of the one-way ANOVA summary table. By adding 1 to the first of these *df* values, you can verify or determine how many groups were compared. To figure out or verify how many people were involved in the study, you must add the two *df* values together and then add 1 to the sum. Thus, it is possible from the excerpt to determine that 30 drivers provided data for the ANOVA.

As you can see, Excerpt 11.6 ends with the statement "$p < 0.01$." This small decimal number (0.01) was *not* the researchers' level of significance. In their research report, the investigators did not indicate their alpha level; however, I can tell from their full set of results that it was set equal to .05. By reporting *p* as being less than .01 (rather than by saying $p < .05$), the researchers wanted others to realize that the three population means, if identical, would have been *very* unlikely to yield sample means as dissimilar as those actually associated with this study's three comparison groups.

Results of Two or More One-Way ANOVAs

Data are often collected on two or more dependent variables in studies characterized by at least three comparison groups, a concern for means, and a single independent variable. Although such data sets can be analyzed in various ways, many researchers choose to conduct a separate one-way ANOVA for each of the multiple dependent variables. The ANOVA results are sometimes presented in tables; however, more often than not the findings are presented only in the text of the research report. Excerpt 11.7 contains such a presentation.

In the study from which Excerpt 11.7 was taken, each respondent to the online survey was put into one of three groups based on his or her engineering experience. Those in the "high" group had engineering degrees and first-hand engineering experience; those in the "intermediate" group were learners of engineering; those in the "low" group were non-engineers who had little or no interest in engineering. The survey instrument assessed one's self-concept for conducting an engineering design task, and it had four scales: self-efficacy, motivation, expectancy, and anxiety. A separate one-way ANOVA was used to compare the means of the three groups on each scale of the self-confidence instrument.

As you read the passage in Excerpt 11.7, you ought to be able to determine what the independent variable was, how many individuals were in the study, how many null hypotheses were tested, and what decision was made regarding each H_0. You also should be able to indicate what level of significance was used and which numbers are the calculated values. Most important, you should be able to explain the meaning of each null hypothesis that was tested.

EXCERPT 11.7 • *Results from Several Separate One-Way ANOVAs*

A 36-item online instrument was developed and administered to [several] individuals with different levels of engineering experience. . . . A one-way ANOVA was conducted to compare mean scores on self-efficacy, motivation, outcome expectancy, and anxiety toward engineering design for the three groups. There were statistically significant effects on all four task-specific self-concepts at the $p < 0.05$ level for the three groups $[F_{\text{self-efficacy}}(2,199) = 79.16, p < 0.001); F_{\text{motivation}}(2,199) = 71.73, p < 0.001); F_{\text{expectancy}}(2,199) = 77.91, p < 0.001); F_{\text{anxiety}}(2,199) = 8.76, p < 0.001)]$. . . . The results of this study demonstrate [that] engineering design self-concept is highly dependent on engineering experiences. This is evident in significant differences in task-specific self-concepts among high, intermediate, and low engineering experience groups.

Source: Carberry, A. R., Lee, S., & Ohland, M. W. (2010). Measuring engineering design self-efficacy. *Journal of Engineering Education, 99*(1), 71–79.

If they had been presented, each null hypothesis in the engineering study would have looked the same: $H_0 : \mu_{High} = \mu_{Intermediate} = \mu_{Low}$. They would differ, however, with respect to the data represented by the μs. For the first one-way ANOVA, the three population means would be dealing with self-efficacy; in the second null hypothesis, the three means would be dealing with motivation; and so on.

Thus, the four null hypotheses connected to the passage in Excerpt 11.7 were identical in terms of the number of μs and the group represented by each μ. Those population means differed, however, based on which of the four dependent variables was under consideration.

Excerpt 11.7 is worth considering for one other reason. Note that the researchers indicate that the level of significance was set at .05, yet the result of each F-test is reported to be $p < 0.001$. This excerpt is like Excerpt 11.6 in the sense that both sets of researchers wanted to show that they rejected their null hypotheses "with room to spare." This is a common practice, but it is not universal. A small number of researchers want their alpha level and their p-statements to be fully congruent. They do this by reporting $p < .05$ if the null hypothesis is rejected at the .05 level of significance, even if the computer analysis reveals that the data-based p is smaller than .01 (or even smaller than .001).

The Bonferroni Adjustment Technique

In the preceding section, we looked at an example where separate one-way ANOVAs were used to assess the data from multiple dependent variables. In a situation such as this, there is an inflated Type I **error** risk unless something is done to compensate for the fact that multiple tests are being conducted. In other words, if the data associated with each of several dependent variables are analyzed separately by means of a one-way ANOVA, the probability of incorrectly rejecting at least one of the null hypotheses is greater than the common alpha level used across the set of tests. (If you have forgotten what an inflated Type I error risk is, return to Chapter 8 and read again the little story about the two gamblers, dice, and the bet about rolling, or not rolling, a six.)

Several statistical techniques are available for dealing with the problem of an inflated Type I error risk. Among these, the **Bonferroni adjustment procedure** appears to be the most popular choice among applied researchers. As you may recall from our earlier consideration of this procedure, the researcher compensates for the fact that multiple tests are being conducted by making the alpha level more rigorous on each of the separate tests.

In Excerpts 11.8 and 11.9, we see two cases where the Bonferroni technique was used. In the first of these excerpts, the researchers conducted two one-way ANOVAs, whereas five one-way ANOVAs were used in the second excerpt. In each of these studies, the researchers wanted the Type I error risk for their *set* of one-way ANOVAs to be no greater than .05. Therefore, they simply divided .05 by the numbers of tests being conducted. This caused alpha to change from .05 to .025 in Excerpt 11.8 and to .01 in Excerpt 11.9. Each modified α in each study became the criterion against which its ANOVAs' p-values were compared.

There is no law, of course, that directs all researchers to deal with the problem of an inflated Type I error risk when multiple one-way ANOVAs are used. Furthermore, there are circumstances where it would be unwise to take any form of

EXCERPTS 11.8–11.9 • *Use of the Bonferroni Adjustment*

Finally, we conducted one-way ANOVAs using a Bonferroni-corrected α of .025 (.05/02) to test for the two hypothesized rater differences [across the three groups] in mean fundamental and advanced MET adherence.

Source: Martino, S., Ball, S., Nich, C., Frankforter, T. L. & Carroll, K. M. (2009). Correspondence of motivational enhancement treatment integrity ratings among therapists, supervisors, and observers. *Psychotherapy Research, 19*(2): 181–193.

- -

One-way ANOVA was used to detect differences [among the three gang groups]. The Bonferroni adjustment technique (alpha/number of statistical tests performed) was used to control for type I error due to alpha inflation. This was done for each individual question. Hence, the adjusted level of significance is .05/5 = .01.

Source: Lanier, M. M., Pack, R. P., & Akers, T. A. (2010). Epidemiological criminology: Drug use among African American gang members. *Journal of Correctional Health Care, 16*(1), 6–16.

corrective action. Nevertheless, I believe that you should value more highly those reports wherein the researcher either (1) does something (e.g., uses the Bonferroni procedure) to hold down the chances of a Type I error when multiple tests are conducted, or (2) explains why nothing was done to deal with the inflated Type I error risk. If neither of these things is done, you have a right to downgrade your evaluation of the study.

Assumptions of a One-Way ANOVA

In Chapter 10, we considered the four main assumptions associated with *t*-tests, *F*-tests, and *z*-tests: independence, randomness, **normality,** and **homogeneity of variance.** My earlier comments apply as much now to cases in which a one-way ANOVA is used to compare three or more means as they did to cases in which two means are compared. In particular, I hope you recall the meaning of these four assumptions and my point about how these tests tends to be robust to the equal variance assumption when the sample sizes of the various comparison groups are equal.

Many researchers who use a one-way ANOVA seem to pay little or no attention to the assumptions that underlie the *F*-test comparison of their sample means. Consequently, I encourage you to feel better about research reports that (1) contain discussions of the assumptions, (2) present results of tests that were conducted to check on the testable assumptions, (3) explain what efforts were made to get the data in line with the assumptions, or (4) point out that an alternative test having fewer assumptions than a regular one-way ANOVA *F*-test was used. Conversely, I encourage you to lower your evaluation of research reports that do none of these things.

Consider Excerpts 11.10 and 11.11 (that deal with boars and bullfrogs, respectively). In both studies, the researchers used a one-way ANOVA. Before doing so, however, they screened their data regarding the normality and equal variances assumption. In the first excerpt, the Kolmogorov–Smirnov and Levene tests were applied to the sample data to see if the normality and homoscedasticity (i.e., equal variance) assumptions, regarding the populations, were untenable. Excerpt 11.11 illustrates how these same two assumptions can be dealt with by means of alternative procedures: the Shapiro–Wilk test and the Bartlett test. Whenever researchers use one of these tests to check an assumption, they hope that the test procedure will cause the null hypothesis (concerning the shape of and variability in the study's populations) to be retained, not rejected. Assumptions should be met, not violated.

EXCERPTS 11.10–11.11 • *Attending to the Normality and Equal Variance Assumptions*

These results were analysed using a one-way analysis of variance (ANOVA) with an independent factor (the treatment) and a dependent variable (a sperm quality parameter). . . . Before the ANOVA test was applied, the data were tested for normality and homoscedasticity using the Kolmogorov–Smirnov and Levene tests.

Source: Yeste, Briz, M., Pinart, E., Sancho, S., Bussalleu, E., & Bonet, S. (2010). The osmotic tolerance of boar spermatozoa and its usefulness as sperm quality parameter. *Animal Reproduction Science, 119*(3–4), 265–274.

Normality and homogeneity of variance of the data were tested using Shapiro–Wilk normality test and Bartlett's test, respectively. Differences of target gene (StAR) expression were analyzed by one way analysis of variance (ANOVA).

Source: Paden, N. E., Carr, J. A., Kendall, R. J., Wages, M., & Smith, E. E. (2010). Expression of steroidogenic acute regulatory protein (StAR) in male American bullfrog (*Rana catesbeiana*) and preliminary evaluation of the response to TNT. *Chemosphere, 80*(1), 41–45.

The researchers associated with Excerpts 11.10 and 11.11 set a good example by demonstrating a concern for the normality and equal variance assumptions of a one-way ANOVA. Unfortunately, many researchers give no indication that they thought about either of these assumptions. Perhaps they are under the mistaken belief that the *F*-test is always robust to violations of these assumptions. Or, perhaps they simply are unaware of the assumptions. In any event, I salute the researchers associated with Excerpts 11.10 and 11.11 for checking their data before conducting a one-way ANOVA.

Sometimes, preliminary checks on normality and the equal variance assumption suggest that the populations are not normal or have unequal variances. When this happens, researchers have four options: They can (1) identify and eliminate any existing outliers, (2) transform their sample data in an effort to reduce nonnormality or stabilize the variances, (3) use a special test that has a built-in correction for violations of assumptions, or (4) switch from the one-way ANOVA F-test to some other test that does not have such rigorous assumptions. In Excerpts 11.12, 11.13, and 11.14, we see cases in which researchers took the 2nd, 3rd, and 4th of these courses of action.

EXCERPTS 11.12–11.14 • *Options When Assumptions Seem Violated*

Assumptions of nonnormality were investigated graphically with the Kolmogorov-Smirnov test, and when significant, the distribution of log-transformed variables was also checked in the same way. The data were analyzed by one-way ANOVA. . . .

Source: Kim, M-K., Tanaka, K, Kim, M-J., Matsuo, T., Tomita, T., Ohkubo, H., et al. (2010). Epicardial fat tissue: Relationship with cardiorespiratory fitness in men. *Medicine and Science in Sports and Exercise, 42*(3), 463–469.

Homogeneity of variances was tested by Levene's test and Welch's ANOVA was used to compare group means when the group variances were unequal.

Source: Shrestha, S., Ehlers, S. J., Lee, J., Fernandez, M., & Koo, S. I. (2009). Dietary Green Tea Extract Lowers Plasma and Hepatic Triglycerides and Decreases the Expression of Sterol Regulatory Element-Binding Protein-1c mRNA and Its Responsive Genes in Fructose-Fed, Ovariectomized Rats. *Journal of Nutrition, 139*(4), 640–645.

[T]he assumption of homogeneity of variance failed for all variables, as the Levene's tests turned out to be significant. Because parametric testing was not justified, non-parametric tests (Kruskal-Wallis) were conducted.

Source: Terband, H., Maassen, B., Guenther, F. H., & Brumberg, J. (2009). Computational neural modeling of speech motor control in childhood apraxia of speech (CAS). *Journal of Speech, Language & Hearing Research, 52*(6), 1595–1609.

Of the four assumptions associated with a one-way ANOVA, the one that is neglected most often is the **independence assumption.** In essence, this assumption says that a particular person's (or animal's) score should not be influenced by the measurement of any other people or by what happens to others while the study is conducted. This assumption would be violated if different groups of students (perhaps different intact classrooms) are taught differently, with each student's exam score being used in the analysis.

In studies where groups are a necessary feature of the investigation, the recommended way to adhere to the independence assumption is to have the **unit of analysis** (i.e., the scores that are analyzed) be each group's mean rather than the scores from the individuals in the group. Many researchers shy away from using the group mean as the unit of analysis because doing this usually causes the sample size to be reduced dramatically. The solution, of course, is to use lots of groups!

Statistical Significance versus Practical Significance

Researchers who use a one-way ANOVA can do either of two things to make their studies more statistically sophisticated than would be the case if they use the crude six-step version of hypothesis testing. The first option involves doing something after the sample data have been collected and analyzed. Here, the researcher can compute an estimate of the **effect size.** The other option involves doing something on the front end of the study, not the tail end. With this option, the researcher can conduct an **a priori power analysis.**

Unfortunately, not all of the researchers who use a one-way ANOVA take the time to perform any form of analysis designed to address the issue of **practical significance versus statistical significance.** In far too many cases, researchers simply use the simplest version of hypothesis testing to test their one-way ANOVA's H_0. They collect the amount of data that time, money, or energy allows, and then they anxiously await the outcome of the analysis. If their F-ratios turn out significant, these researchers quickly summarize their studies, with emphasis put on the fact that "significant findings" have been obtained.

I encourage you to upgrade your evaluation of those one-way ANOVA research reports in which the researchers demonstrate that they were concerned about practical significance as well as statistical significance. Examples of such concern appear in Excerpts 11.15 and 11.16. In each case, the effect size associated with the ANOVA was estimated. As you can see, the effect size measures used in these studies were eta squared and Cohen's f.

EXCERPTS 11.15–11.16 • *Estimating the Effect Size*

A one-way between groups ANOVA [indicated] that there were significant differences among marital status groups in the change score for sexual feelings and behavior toward men, $F(2,107) = 4.43, p < .05$. The effect size calculated using eta squared was .08, indicating a moderate effect.

Source: Karten, E. Y., & Wade, J. C. (2010). Sexual orientation change efforts in men: A client perspective. *Journal of Men's Studies, 18*(1), 84–102.

(continued)

EXCERPTS 11.15–11.16 • (continued)

In order to examine the conversational differences among these three groups, a One-way Analysis of variance was used [and revealed] significant overall differences among the groups on all three of these scales, with very large effect sizes . . . (Pragmatic Behaviors: $F = 34.2, p < .0001, f = 5.8$; Speech/Prosody Behaviors: $F = 20.2$, $p < .0001$, $f = 4.5$; Paralinguistic Behaviors: $F = 14.8, p < .0001, f = 3.9$, very large).

Source: Paul, R., Orlovski, S. M., Marcinko, H. C., & Volkmar, F. (2009). Conversational behaviors in youth with high-functioning ASD and Asperger Syndrome. *Journal of Autism & Developmental Disorders, 39*(1), 115–125.

In Excerpts 11.15 and 11.16, notice the terms *moderate effect* and *very large effect sizes*. The researchers used these terms in an effort to interpret their effect size estimates. Most likely, they used the relevant information in Table 11.1 as a guide when trying to judge whether the estimated effect size number deserved the label *small, medium,* or *large*. (This table does not include any information about *d*, because this measure of effect size cannot be used when three or more means are compared.)

In Excerpts 11.17 and 11.18, we see two cases in which teams of researchers performed a power analysis to determine the needed sample size for their one-way ANOVA. In the first of these excerpts, notice that the researchers specify an unstandardized effect size (10 degrees), a desired power level (80%), the standard deviation (15 degrees), and a level of significance (.05) as the ingredients of their power analysis. Doing this, they determined the needed sample size (35). The information in Excerpt 11.18 is less specific, but it is clear that those researchers also conducted an a priori power analysis to determine the needed sample size for each of the three comparison groups.

TABLE 11.1   *Effect Size Criteria for a One-Way ANOVA*

Effect Size Measure	Small	Medium	Large
Eta (η)	.10	.24	.37
Eta Squared (η^2)	.01	.06	.14
Omega Squared (ω^2)	.01	.06	.14
Partial Eta Squared (η_p^2)	.01	.06	.14
Partial Omega Squared (ω_p^2)	.01	.06	.14
Cohen's f	.10	.25	.40

Note: These standards for judging relationship strength are quite general and should be changed to fit the unique goals of any given research investigation.

EXCERPTS 11.17–11.18 • *An A Priori Power Analysis*

A priori power analysis showed that in order to have a power of 80% to detect a difference of as little as 10 degrees at the 0.05 level of significance assuming a standard deviation of 15 degrees, 35 women would be needed in each group. The increased enrolment improved the power of the study. Statistical analysis was performed using the one factor ANOVA.

Source: Papadakis, M., Papadokostakis, G., Kampanis, N., Sapkas, G., Papadakis, S. A., & Katonis, P. (2010). The association of spinal osteoarthritis with lumbar lordosis. *BMC Musculoskeletal Disorders, 11*, 1–6.

We carried out an a priori power analysis for one-way analysis of variance with 3 groups of participants, to calculate number of subjects needed to have sufficient power to detect effect sizes similar to those found in previous positive studies (ES = 0.6 to 0.9 for amygdala and ES = 0.9 for hippocampal volume decreases in pediatric or adolescent patients with BD) as statistically significant.

Source: Hajek, T., Gunde, E., Slaney, C., Propper, L., MacQueen, G., Duffy, A., et al. (2009). Amygdala and hippocampal volumes in relatives of patients with bipolar disorder: A high-risk study. *Canadian Journal of Psychiatry, 54*(11), 726–733.

Cautions

Before concluding this chapter, I want to offer a few tips that will increase your skills at deciphering and critiquing research reports based on one-way ANOVAs.

Significant and Nonsignificant Results from One-Way ANOVAs

When researchers say that they have obtained a statistically significant result from a one-way ANOVA, this means that they have rejected the null hypothesis. Because you are unlikely to see the researchers articulate the study's H_0 (in words or symbols) or even see them use the term *null hypothesis* in discussing the results, it is especially important for you to remember (1) what the one-way ANOVA H_0 stipulates, (2) why H_0 is rejected, and (3) how to interpret correctly the decision to reject H_0.

Although a one-way ANOVA can be used to compare the means of two groups, this chapter focuses on the use of one-way ANOVAs to compare three or more means. If the data lead to a significant finding when more than two means have been contrasted, it means that the sample data are not likely to have come from populations having the same μ. This one-way ANOVA result does not provide any information as to how many of the μ values are likely to be dissimilar, nor does it provide any information as to whether any specific pair of populations are likely to have different μ values. The only thing that a significant result indicates is that the variability among the full set of sample means is larger than would be expected if all population means were identical.

Usually, a researcher wants to know more about the likely state of the population means than is revealed by a one-way ANOVA. To be more specific, the typical researcher wants to be able to make comparative statements about pairs of population means, such as "μ_1 is likely to be larger than μ_2, but μ_2 and μ_3 cannot be looked on as different based on the sample data." To address these concerns, the researcher must move past the significant ANOVA F and apply a subsequent analysis. In Chapter 12 we consider such analyses, which are called, understandably, *post hoc* or *follow-up tests*.

In any research report discussing the results of a one-way ANOVA, you are given information about the sample means. If a significant result has been obtained, you may be tempted to conclude that each population mean is different from every other population mean. Do not fall prey to this temptation!

To gain an understanding of this very important point, consider once again Excerpt 11.7. In that excerpt, the results of four one-way ANOVAs are presented from a study in which three groups differing in engineering experience responded to a survey that measured four components of self-confidence in performing an engineering design task. The three means compared on *anxiety* were 38.77, 49.46, and 62.16, and the F-test yielded statistical significance with $p < .001$. This result indicates that the null hypothesis of equal population means was rejected. Hence, it is legitimate to think that $H_0: \mu_1 = \mu_2 = \mu_3$ is probably not true. However, it is *not* legitimate to think that all three of the population means are different from each other. Perhaps two of the μs are the same, but different from the third μ.[4]

You must also be on guard when it comes to one-way ANOVAs that yield nonsignificant Fs. As I have pointed out on several occasions, a fail-to-reject decision should not be interpreted to mean that H_0 is true. Unfortunately, many researchers make this inferential mistake when different groups are compared in terms of mean scores on a pretest. The researchers' goal is to see whether the comparison groups are equal to one another at the beginning of their studies, and they mistakenly interpret a nonsignificant F-test to mean that no group began with an advantage or a disadvantage.

One-way ANOVAs, of course, can produce a nonsignificant F-value when groups are compared on things other than a pretest. Consider once again Excerpt 11.5. This ANOVA (comparing the mean reading comprehension scores of three multimedia groups) yielded a nonsignificant F of 2.054. This results should *not* be interpreted to mean that individuals like those involved in this study—computer savvy college freshmen from Turkey taking a course in English from the foreign language department—have the same mean reading comprehension scores regard-

---

[4]If I tell you that the three coins in my pocket are not all identical, you might quickly guess that all three coins are different. That guess would be wrong if two of my coins are the same.

less of whether their multimedia learning aid is just definitions, definitions plus pictures, or definitions plus a brief movie when reading an online essay. In other words, neither you nor the researchers who conducted this study should draw the conclusion that $\mu_1 = \mu_2 = \mu_3$ because the *p*-value associated with the ANOVA *F* turned out to be larger than the selected level of significance. A one-way ANOVA, if nonsignificant, cannot be used to justify such an inference.

Confidence Intervals

In Chapter 10, you saw how a confidence interval can be placed around the difference between two sample means. You also saw how such confidence intervals can be used to test a null hypothesis, with H_0 rejected if the null's pinpoint numerical value lies beyond the limits of the interval. As we now conclude our consideration of how researchers compare three or more means with a one-way ANOVA, you may be wondering why I have not said anything in this chapter about the techniques of estimation.

When a study's focus is on three or more means, researchers occasionally build a confidence interval around each of the separate sample means. This is done in situations where (1) there is no interest in comparing all the means together at one time or (2) there is a desire to probe the data in a more specific fashion after the null hypothesis of equal population means has been tested. Whereas researchers sometimes use interval estimation (on individual means) in lieu of or as a complement to a test of the hypothesis that all μs are equal, interval estimation is not used as an alternative strategy for testing the one-way ANOVA null hypothesis. Stated differently, you are not likely to come across any research studies where a confidence interval is put around the variance of the sample means in order to test $H_0: \sigma_\mu^2 = 0$.

Other Things to Keep in Mind

If we momentarily lump together this chapter with the ones that preceded it, it is clear that you have been given a slew of tips or warnings designed to help you become a more discerning recipient of research-based claims. Several of the points are important enough to repeat here.

1. The mean is focused on in research studies more than any other statistical concept. In many studies, however, a focus on means does not allow the research question to be answered because the question deals with something other than central tendency.
2. If the researcher's interest resides exclusively in the group(s) from which data are collected, only descriptive statistics should be used in analyzing the data.
3. The reliability and validity of the researcher's data are worth considering. To the extent that reliability is lacking, it is difficult to reject H_0 even when H_0 is false. To the extent that validity is lacking, the conclusions drawn will be

unwarranted because of a mismatch between what is truly being measured and what the researcher thinks is being measured with a one-way ANOVA.

4. With a one-way ANOVA, nothing is proved regardless of what the researcher concludes after analyzing the data. Either a Type I error or a Type II error always will be possible, no matter what decision is made about H_0.

5. The purpose of a one-way ANOVA is to gain an insight into the population means, not the sample means.

6. Those researchers who talk about (and possibly test) the assumptions under-lying a one-way ANOVA deserve credit for being careful in their utilization of this inferential technique.

7. A decision not to reject the one-way ANOVA's H_0 does not mean that all pop-ulation means should be considered equal.

8. Those researchers who perform an a priori power analysis or compute effect size indices (following the application of a one-way ANOVA) are doing a more conscientious job than are those researchers who fail to do anything to help distinguish between statistical significance and practical significance.

9. The Bonferroni procedure helps to control the risk of Type I errors in studies where one-way ANOVAs are conducted on two or more dependent variables.

10. The *df* values associated with a one-way ANOVA (whether presented in an ANOVA summary table or positioned next to the calculated F-value in the text of the research report) can be used to determine the number of groups and the total number of participants involved in the study.

A Final Comment

We have covered a lot of ground in this chapter. We have looked at the basic ingre-dients that go into any one-way ANOVA, seen different formats for showing what pops out of this kind of statistical analysis, considered underlying assumptions, and observed how conscientious researchers make an effort to discuss practical signifi-cance as well as statistical significance. You may have assimilated everything pre-sented in this chapter, you may have assimilated only the highlights (with a review perhaps in order), or you may be somewhere between these two extremes. Regard-less of how well you now can decipher research reports based on one-way ANOVAs, it is exceedingly important that you leave this chapter with a crystal clear understanding of one essential point, and unless you heed the advice embodied in this final comment, you are likely to lose sight of the forest for all the trees.

A one-way ANOVA (like any other statistical analysis) cannot magically transform a flawed study into a sound one. And where can a study be flawed the most? The answer to this question is unrelated to F-tests, equal variance assump-tions, effect size indices, or Bonferroni adjustments, because the potential worth of any study is connected, first and foremost, to the research question that sets the data collection and analysis wheels in motion. If the research question is silly or irrelevant, a one-way ANOVA cannot make the study worthwhile. Hence, do not be impressed by any result from a one-way ANOVAs *until* you have first considered the merits of the research question being addressed.

If you would like to see an example of a silly one-way ANOVA, consider Excerpt 11.19. The full research report from which this excerpt was taken discussed many analyses that dealt with important issues. However, this particular one-way ANOVA does not deserve that kind of positive evaluation. For this analysis, the 644 children in the study were put into three groups based on their scores on the Diagnostic Evaluation of Language Variation—Screening Test (DELV-S), a test designed to measure dialect variation in language (i.e., deviation from "mainstream American English"). These DELV-S groups were then compared on how the children scored on the DVAR, a measure of "dialect variation" computed as "the percentage of scored items that were observed to vary from MAE." Because the DELV-S and the DVAR instruments were so similar in what they were measuring, it is not surprising at all that the groups formed on the basis of one of these tests had extremely differing means on the other test. The three sample means were miles apart, as indicated by the gigantic calculated value generated by the one-way ANOVA. You likely will never see an *F*-value as large as this one ever again!

EXCERPT 11.19 • *An Unnecessary One-Way ANOVA*

In the sample, 28% [of the children] were classified on the DELV–S as speaking with strong variation, 9% with some variation, and 62% with no variation from MAE [Mainstream American English]. . . . Students in the MAE group achieved significantly lower DVAR scores ($M = 12.4$) compared with students whose dialect varied somewhat ($M = 50.3$) or strongly ($M = 78.7$) from MAE, $F(2, 641) = 1,854.2, p < .001$.

Source: Terry, N. P., Connor, C. M., Thomas-Tate, S., & Love, M. (2010). Examining relationships among dialect variation, literacy skills, and school context in first grade. *Journal of Speech, Language & Hearing Research, 53*(1), 126–145.

Review Terms

A priori power analysis	Error
ANOVA	Factor
Between groups	*f*
Between subjects variable	*F*
Bonferroni adjustment procedure	Homogeneity of variance assumption
	Independence assumption
Dependent variable	Independent variable
df	Mean square (*MS*)
Effect size	Normality assumption

One-factor ANOVA
One-way ANOVA
Practical significance
   versus statistical
   significance
Simple ANOVA

Source
Sum of squares (*SS*)
Summary table
Unit of analysis
Within groups

The Best Items in the Companion Website

1. An interactive online quiz (with immediate feedback provided) covering Chapter 11.
2. Eight misconceptions about the content of Chapter 11.
3. An email message sent from the author to his students entitled "A Closed Hydraulic System."
4. The author-selected best passage from Chapter 11: "One-Way ANOVAs and What's Really Important."
5. An interactive online resource entitled "One-Way ANOVA (a)."

To access the chapter outline, practice tests, weblinks, and flashcards, visit the companion website at http://www.ReadingStats.com.

Review Questions and Answers begin on page 531.

12

Post Hoc and Planned Comparisons

In Chapter 11, we examined the setting, purpose, assumptions, and outcome of a one-way analysis of variance that compares three or more groups. In this chapter, we turn our attention to two categories of inferential procedures closely related to the one-way ANOVA. As with a one-way ANOVA, the procedures looked at in this chapter involve one independent variable, one dependent variable, no repeated measures, and a focus on means.

The two classes of procedures considered here are called **post hoc comparisons** and **planned comparisons.** Post hoc comparisons were developed because a one-way ANOVA F, if significant, does not provide any specific insight into what caused the null hypothesis to be rejected. To know that all population means are probably not equal to one another is helpful, but differing scenarios fit the general statement that not all μs are identical. For example, with three comparison groups, it might be that two μs are equal, but the third is higher; or maybe two μs are equal, but the third is lower; or it could be that all three μs are different. By using a post hoc procedure, the researcher attempts to probe the data to find out which of the possible non-null scenarios is most likely to be true.

Planned comparisons were developed because researchers sometimes pose questions that cannot be answered by rejecting or failing to reject the null hypothesis of the more general one-way ANOVA H_0. For example, a researcher might wonder whether a specific pair of μs is different, or whether the average of two μs is different from a third μ. In addition to allowing researchers to answer specific questions about the population means, planned comparisons have another desirable characteristic. Simply stated, the statistical power of the tests used to answer specific, preplanned questions is higher than is the power of the more generic F-test from a one-way ANOVA. In other words, planned comparisons allow a researcher to deal with specific, a priori questions with less risk of a Type II error than does a two-step approach involving an ANOVA F-test followed by post hoc comparisons.

Researchers use post hoc comparisons more often than they do planned comparisons. For this reason, we first consider the different test procedures and reporting schemes used when a one-way ANOVA yields a significant F and is followed by a post hoc investigation. We then turn our attention to what researchers do when they initially set up and test planned comparisons instead of following the two-step strategy of conducting a one-way ANOVA followed by a post hoc analysis. Finally, we look at a set of related issues, including some special terminology, the importance of assumptions to planned and post hoc comparisons, and the distinction between statistical versus practical significance.

Post Hoc Comparisons

Definition and Purpose

There is confusion among researchers as to what is or is not a post hoc test. I have come across examples where researchers conducted a post hoc investigation but used the term *planned comparisons* to describe what they did. I have also come across research reports where planned comparisons were conducted by means of a test procedure that many researchers consider to be post hoc in nature. To help you avoid getting confused when you read research reports, I want to clarify what does and does not qualify as a post hoc investigation.

If a researcher conducts a one-way ANOVA and uses the outcome of the F-test to determine whether additional specific tests should be conducted, then I refer to the additional tests as being *post hoc* in nature. As this definition makes clear, the defining criterion of a post hoc investigation has nothing to do with the name of the test procedure employed, with the number of tests conducted, or with the nature of the comparisons made. The only thing that matters is whether the ANOVA F-test must first be checked to see if further analysis of the data set is needed.

In turning to a post hoc investigation, the researcher's objective is to better understand why the ANOVA yielded a significant F. Stated differently, a post hoc investigation helps the researcher understand why the ANOVA H_0 was rejected. Because the H_0 specifies equality among all population means, you might say that a set of post hoc comparisons is designed to help the researcher gain insight into the pattern of μs. As we indicated at the outset of this chapter, the ANOVA F can turn out to be significant for different reasons—that is, because of different possible patterns of μs. The post hoc analysis helps researchers in their efforts to understand the true pattern of the population means.

In light of the fact that empirical studies are usually driven by research hypotheses, it is not surprising to find that post hoc investigations are typically conducted to find out whether such hypotheses are likely to be true. Furthermore, it should not be surprising that differences in research hypotheses lead researchers to do different things in their post hoc investigations. Sometimes, for example, researchers set up their post hoc investigations to compare each sample mean against every other sample mean. On other occasions they use their post hoc tests to compare the mean associated with each of several experimental groups against a control group's mean, with no comparisons made among the experimental groups. On rare occasions, a post hoc investigation is implemented to compare the mean of one of the comparison groups against the average of the means of two or more of the remaining groups. I illustrate each of these post hoc comparisons later in the chapter.

Terminology

Various terms are used in a synonymous fashion to mean the same thing as the term **post hoc test.** The three synonyms that show up most often in the published literature are **follow-up test, multiple comparison test,** and **a posteriori test.** Excerpts 12.1 through 12.3 show how three of these four terms have been used.

EXCERPTS 12.1–12.3 • *The Term Post Hoc and Its Synonyms*

Statistical analysis of electrophysiology data was performed by a one-way ANOVA and *post hoc* tests with significance assessed as $p < 0.05$.

Source: Sharma, A., Hoeffer, C. A., Takayasu, Y., Miyawaki, T., McBride, S. M., Klann, E., et al. (2010). Dysregulation of mTOR signaling in Fragile X Syndrome. *Journal of Neuroscience, 30*(2), 694–702.

One-way ANOVA was used to analyse potential differences between the climate change adaptation options of the skier types. Multiple comparisons were made between the skier types [in order] to find out which groups differ and whether the differences are statistically significant at $p \leq 0.05$.

Source: Lanmdauer, M., Sievänen, T., & Neuvonen, M. (2009). Adaptation of Finnish cross-country skiers to climate change. *Fennia, 187*(2), 99–113.

[W]e conducted a one-way analysis of variance [and then] follow-up tests were performed to evaluate pairwise differences among the means.

Source: Zingerevich, C., Greiss-Hess, L., Lemons-Chitwood, K., Harris, S. W., Hessl, D., Cook, K., et al. (2009). Motor abilities of children diagnosed with fragile X syndrome with and without autism. *Journal of Intellectual Disability Research, 53*(1), 11–18.

You may come across a research report in which the term *contrast* appears. The word **contrast** is synonymous with the term **comparison.** Hence, post hoc contrasts are nothing more than post hoc comparisons. Follow-up contrasts are nothing more than follow-up comparisons. A posteriori contrasts are nothing more than a posteriori comparisons.

It is also worth noting that the F-test used in the preliminary ANOVA is sometimes referred to as the **omnibus F-test.** This term seems appropriate because the ANOVA's H_0 involves *all* of the population means. Because post hoc (and planned) investigations often use F-tests to accomplish their objectives, it is helpful when researchers use the term *omnibus* (when referring to the ANOVA F) to clarify which F is being discussed. Excerpt 12.4 illustrates the use of this term.

EXCERPT 12.4 • *The Omnibus F-Test*

The only statistically significant difference in means [among] conditions as tested by a one way ANOVA omnibus F test was for the Go-Signal Respond RT difference.

Source: Liddle, E. B., Scerif, G., Hollis, C. P., Batty, M. J., Groom, M. J., Liotti, M., et al. (2009). Looking before you leap: A theory of motivated control of action. *Cognition, 112*(1), 141–158.

Finally, the terms *pairwise* and *nonpairwise* often pop up in discussions of post hoc (and planned) comparisons. The term *pairwise* simply means that groups are being compared two at a time. For example, **pairwise comparisons** among three groups labeled A, B, and C would involve comparisons of A versus B, A versus C, and B versus C. With four groups in the study, a total of six pairwise comparisons would be possible.

A **nonpairwise** (or **complex**) **comparison** involves three or more groups, with these comparison groups divided into two subsets. The mean score for the data in each subset is then computed and compared. For example, suppose there are four comparison groups in a study: A, B, C, and D. The researcher might be interested in comparing the average of groups A and B against the average of groups C and D. This would be a nonpairwise comparison, as would a comparison between the first group and the average of the final two groups (with the second group omitted from the comparison).

In Excerpts 12.5 and 12.6, reference is made to the pairwise and nonpairwise (complex) comparisons the researchers set up in their studies. In the first of these excerpts, pairwise comparisons were made following a one-way ANOVA.

EXCERPTS 12.5–12.6 • *Pairwise and Nonpairwise Comparisons*

Data were compared in the three groups using one way ANOVA and then post hoc pairwise comparisons between groups were performed.

Source: Kim, C., Marcus, C. L., Bradford, R., Gallagher, P. R., Torigian, D., Victor, U., et al. (2010). Upper airway soft tissue differences in apneic adolescent children compared to BMI matched controls. Paper presented at the American Thoracic Society International Conference, New Orleans.

(continued)

EXCERPTS 12.5–12.6 • (*continued*)

This study involved one independent variable (thinking aloud) with three treatment conditions and a control. The three TA conditions were the traditional technique, the speech-communication technique, and coaching. The control condition was silence: Participants in this condition did not think aloud. . . . The one-way analysis of variance (ANOVA) with alpha = 0.05 shows that condition has a significant effect on Accuracy ($F_{3,636} = 13.48$, $p < 0.0001$). To understand the result of the study, we determined which condition had the biggest effect on Accuracy, and whether all the conditions were significantly different from the control condition. The first [nonpairwise] comparison compared the control with all the other conditions. The next [nonpairwise] comparison was the coaching condition against the first two conditions, traditional and speech communication, ignoring the control in this contrast.

Source: Olmsted-Hawala, E. L., Murphy, E. D., Hawala, S., & Ashenfelter, K. T. (2010). Think-aloud protocols: A comparison of three think-aloud protocols for use in testing data-dissemination Web sites for usability. *Proceedings of the ACM Conference on Human Factors in Computer Systems, 28*, 2381–2390.

In Excerpt 12.6, we see two nonpairwise comparisons conducted in a post hoc fashion. One of these tests compared the mean of the control group against the single mean of the three treatment groups combined together. In the second nonpairwise comparison, the mean of one of the treatment groups was compared against the single mean of the other two treatment groups combined. Of the two kinds of comparisons illustrated in these two excerpts, you are likely to see the pairwise kind far more often than the nonpairwise kind.

Test Procedures Frequently Used in Post Hoc Analyses

A wide array of statistical procedures is available for making post hoc comparisons. Many of these you are unlikely to see, simply because they are not used very often. Three procedures are used by a few researchers, and thus you may come across **Fisher's LSD test, Duncan's multiple range test,** or the **Newman-Keuls test.** The three most frequently used procedures are called the **Bonferroni test,** the **Tukey test,** and **Scheffé test.** Excerpts 12.7 through 12.9 show how researchers indicate that they have chosen to use these three popular tests.

EXCERPTS 12.7–12.9 • *Test Procedures Frequently Used in Post Hoc Investigations*

Tukey's post hoc test was used to compare pairwise differences between mean values.

Source: Graham, J. B., & Vandewalle, K. S. (2010). Effect of long-term storage temperatures on the bond strength of self-etch adhesives. *Military Medicine, 175*(1), 68–71.

(*continued*)

EXCERPTS 12.7–12.9 • (*continued*)

We compared changes in knowledge, attitudes, and beliefs across the 3 randomized groups using [one-way] analyses of variance (ANOVA), and used a Bonferroni test to assess pairwise comparisons among the 3 ad types.

Source: Murphy-Hoefer, R., Hyland, A., & Rivard, C. (2010). The influence of tobacco counter-marketing ads on college students' knowledge, attitudes, and beliefs. *Journal of American College Health, 58*(4), 373–381.

One-way analysis of variance (ANOVA) was performed [and then] Scheffe's *post-hoc* test was used to compare the disability scores across the subgroups.

Source: Thirthalli, J., Venkatesh, B. K., Naveen, M. N., Venkatasubramanian, G., Arunachala, U., Kishore Kumar, K.V., & Gangadhar, B. N. (2010). Do antipsychotics limit disability in schizophrenia? A naturalistic comparative study in the community. *Indian Journal of Psychiatry, 52*(1), 37–41.

You may have been surprised to see, in Excerpt 12.8, that the Bonferroni procedure can be used to conduct post hoc tests, but it can. Suppose there are four comparison groups (A, B, C, and D) in a study, suppose that a one-way ANOVA has yielded a significant F, and finally suppose that the researcher uses $\alpha = .05$. A post hoc investigation involving the Bonferroni test involves a set of six independent-samples t-tests within which each group's mean is compared with every other group's mean, two at a time (M_A vs. M_B, M_A vs. M_C, etc.), with these post hoc tests conducted at a reduced alpha level of .0083 (i.e., .05/6). Using the Bonferroni procedure is logically equivalent to using it in a two-group study where six t-tests are conducted because there are six dependent variables.

Instead of dealing with the problem of an inflated Type I error risk by adjusting the level of significance (as is done when the Bonferroni technique is applied), the Duncan, Newman-Keuls, Tukey, and Scheffé procedures make an adjustment in the size of the critical value used to determine whether an observed difference between two means is significant. To compensate for the fact that more than one comparison is made, larger (and more rigorous) critical values are used. However, the degree to which the critical value is adjusted upward varies according to which test procedure is used.

When the critical value is increased only slightly (as compared with what would have been the critical value in the situation of a two-group study), the test procedure is considered to be **liberal**. In contrast, when the critical value is increased greatly, the test procedure is referred to as being **conservative**. Liberal procedures provide less control over Type I errors, but this disadvantage is offset by increased power (i.e., more control over Type II errors). Conservative procedures do just the opposite; they provide greater control over Type I error risk, but do so at the expense of lower power (i.e., higher risk of Type II errors).

The Fisher LSD test procedure is the most liberal of the test procedures, because it makes no adjustment for the multiple tests being conducted. In a very real sense, it is just like comparing every pair of means with a t-test. On the other end of the liberal–conservative continuum is the Scheffé test. It has enormous protection against Type I errors, because it was designed for the situation where the researcher wishes to make all possible pairwise comparisons *plus* all possible nonpairwise comparisons. Few researchers need or want that level of protection! The other test procedures (such as Bonferroni and Tukey) lie between these two liberal–conservative extremes.

You are likely to come across two additional test procedures that are used in post hoc investigations. These tests have special purposes compared with the ones we have considered so far. One of these procedures is Dunnett's test; the other one is called Tamhane's test.

Dunnett's test makes pairwise comparisons, as do the other test procedures we have considered. However, Dunnett's test does not pair every mean with every other mean. Instead, the Dunnett test compares the mean of a particular group in the study against each of the remaining group means. This procedure might be used, for example, if a researcher cares only about how each of several versions of an experimental treatment affects the dependent variable, compared with a controlled (or placebo) condition. In Excerpt 12.10, we see a case where the Dunnett test was used in just this manner.

The second of our two special case test procedures is **Tamhane's post hoc test.** This test—referred to by many researchers as *Tamhane's T2 test*—was created so that researchers could make pairwise comparisons among group means in the situation where the equal variance assumption seems untenable. Excerpt 12.11 illustrates the use of the Tamhane test.

EXCERPTS 12.10–12.11 • *The Dunnett and Tamhane Tests*

One-way ANOVA test was used to make comparisons among three groups, and the Dunnett's test was further used to compare each treated group with the control group.

Source: Maa, P., Wang, Z., Pflugfelder, S. C., & Li, D. (2010). Toll-like receptors mediate induction of peptidoglycan recognition proteins in human corneal epithelial cells. *Experimental Eye Research, 90*(1), 130–136.

In checking the assumptions of [the one-way ANOVA] it was noted that the only constructs to pass the homogeneity of variance assumption were those of Centrality and Sign in Study 1. As such, the relatively stringent Tamhane's post hoc analysis was relied on to assess if significant differences were actually present in all tests other than for Centrality and Sign in Study I.

Source: Beaton, A. A., Funk, D., C., & Alexandris, K. (2009). Operationalizing a theory of participation in physically active leisure. *Journal of Leisure Research, 41*(2), 177–203.

The Null Hypotheses of a Post Hoc Investigation

In the next section, we look at the different ways researchers present the results of their post hoc analyses. In such presentations, you rarely see reference made, through symbols or words, to the null hypotheses that are associated with the test results. Consequently, you must remember that all of the post hoc procedures are inferential in nature and are concerned with null hypotheses.

In most post hoc analyses, two or more contrasts are investigated, each involving a null hypothesis. For example, in a study involving three groups (A, B, and C) and pairwise comparisons used to probe a significant result from a one-way ANOVA, three null hypotheses would be tested: H_0: $\mu_A = \mu_B$, H_0: $\mu_A = \mu_C$, and H_0: $\mu_B = \mu_C$.[1] With a similar analysis involving four groups, there are six null hypotheses. With Dunnett's test, there is one fewer null hypothesis than there are comparison groups.

The purpose of a post hoc analysis is to evaluate the null hypothesis associated with each contrast that is investigated. As I have pointed out several times, many applied researchers seem to forget this exceedingly important point. They often talk about their findings with reference only to their sample means, and they discuss their results in such a way as to suggest that they have proved something in a definitive manner. When they do so, they are forgetting that their "discoveries" are nothing more than inferences regarding unseen population means, with every inference potentially being nothing more than a Type I or Type II error.

Presentation of Results

Researchers often summarize the results of their post hoc investigations through the text of the technical report. Usually it is not difficulty to figure out what the researcher has concluded when results are presented in this fashion. Sometimes, however, you must read carefully. Consider, for example, Excerpts 12.12 and 12.13. The

EXCERPTS 12.12–12.13 • Results of Post Hoc Investigations

A one-way ANOVA indicated a statistically significant difference in post-test moods scores based on music condition, $F(2, 51) = 18.79$, $p < 0.001$. Post-hoc Tukey tests indicated that the students nested in the heavy metal music ($M = 46.28$) condition had significantly higher scores (indicating higher anxious moods) than those in both the classical ($M = 31.52$) and pop music ($M = 33.53$) conditions.

Source: Rea, C., MacDonald, P., & Carnes, G. (2010). Listening to classical, pop, and metal music: An investigation of mood. *Emporia State Research Studies, 46*(1), 1–3.

(*continued*)

---

[1]Although the null hypotheses of a post hoc investigation theoretically can be set up with something other than zero as H_0's pinpoint number, you are unlikely to ever see a researcher test anything except no-difference null hypotheses in a post hoc analysis.

EXCERPTS 12.12–12.13 • (*continued*)

Results of the one-way ANOVA reveals that there was a statistically significant difference among the four groups, $F[3, 160] = 7.401, p < .0001$. . . . Post-hoc analyses [Bonferroni] were conducted to explore differences among the four CLBP groups. There was a significant difference between Groups 4 and 1 [mean difference $= -12.40$, $p < .001$, 95 percent confidence interval $(-19.63, -5.18)$]. Furthermore, there was a significant difference between Group 4 and Group 3 [mean difference $= -8.23$, $p = .0011$, 95 percent confidence interval $(1.27, 15.19)$].

Source: DeCarvalho, L. T. (2010). Important missing links in the treatment of chronic low back pain patients. *Journal of Musculoskeletal Pain, 18*(1), 11–22.

first of these excerpts conveys the results of four separate tests that were conducted. One involved the omnibus ANOVA comparison of all three comparison groups. The other three were pairwise comparisons using the Tukey test, with reference made to only the two of these that turned out to be significant. Turning our attention to Excerpt 12.13, can you tell how many null hypotheses were set up and tested?[2]

Excerpts 12.12 and 12.13 nicely illustrate the fact that different researchers not only use different test procedures in their post hoc investigation but also use different techniques while reporting the results of such investigations. In Excerpt 12.12, we are given the sample means for all groups involved in the study. In contrast, Excerpt 12.13 contains a confidence interval positioned around the mean difference for only the pairwise comparisons that turned out to be significant.

Take one final look at Excerpt 12.13 and notice that the researchers say that they used "post hoc Tukey tests" after the ANOVA turned out to be significant. Actually, there are several versions of the Tukey post hoc test procedure, and researchers often specify which particular Tukey test they used. The three most popular version are referred to as *Tukey HSD, Tukey–Kramer,* and *Tukey-B*. For all practical purposes, these different versions of Tukey's test are nearly equivalent.[3] They all compare means in a pairwise fashion.

When the results of a post hoc analysis are presented graphically, one of three formats is typically used. These formats involve (1) a table of means with attached letters, (2) a table of means with one or more notes using group labels and less-than or greater-than symbols to indicate which groups were found to be significantly

---

[2]A total of seven null hypotheses were tested. Of these, one was connected to the omnibus F-test that compared all four sample means at once; the other six were connected to various pairwise tests needed to compare each sample mean with every other sample mean.

[3]The Tukey-Kramer test was designed to handle situations where the sample sizes of the comparison groups differ; the Tukey HSD (honestly significantly different) test was designed for the situation where the researcher desires to make all possible pairwise comparison; and the Tukey-B test (also called the Tukey WSD test) was created for situations where interest lies in fewer than all possible pairwise comparisons or where there is a desire to use a test that is a bit more liberal than the Tukey HSD test.

different from one another, and (3) a figure containing lines drawn above vertical bars. These three formats are identical in that they reveal where significant differences were found among the comparison groups.

Consider the table Excerpt 12.14. In the study from which this table was taken, there were four groups of pharmacy workers who differed in terms of job title. Separate one-way ANOVAs compared these groups in terms of age and experience, and in each case the F-test was significant. Two post hoc investigations were then conducted, one involving all possible pairwise comparisons of the four age means and the other making the same six group-versus-group comparisons of the four experience means. The subscripts attached to each row of means indicate the post hoc results. To interpret those subscripts, we must read the second and third sentences in the note beneath the table.

EXCERPT 12.14 • *Results of a Post Hoc Investigation Presented in a Table with Letters Attached to Group Means*

TABLE 7   *Comparison of Staff Characteristics between Job Titles*

	Job Title					
	PharmD	BS Pharm	CPhT	Pharmacy Technician	df	F
Age (years)	33.96_a (8.73)	54.45_b (11.65)	36.30_a (11.06)	33.75_a (16.14)	3, 95	24.12^{***}
Experience (years)	6.00_a (4.83)	20.63_b (14.65)	5.65_a (6.10)	3.09_a (2.44)	3,135	24.81^{***}

Note: $^{***} = p \leq .001$. Standard deviations appear in parentheses below means. Results of LSD post hoc paired comparisons are shown using subscripts (a, b). Means with the same subscript are not significantly different while means with different subscripts are significantly different from one another at the $p \leq .05$.

Source: Wilkerson, T. W. (2009). An exploratory study of the perceived use of workarounds utilized during the prescription preparation process of pharmacies in Alabama. Unpublished doctoral dissertation, Auburn University, Auburn, Alabama.

When looking at a table in which the results of a post hoc investigation are shown via letters attached to the means, you must *carefully* read the note that explains what the letters mean. It is important to do this because all authors do not set up their tables in the same way. In some tables, the same letter attached to any two means indicates that the tested null hypothesis was *not* rejected; in other tables, common letters indicate that the tested null hypothesis *was* rejected.

Excerpt 12.14 provides an example of the first (and more typical) situation. The second (less typical) situation can be illustrated best by these words that I recently saw in a note positioned beneath a table: "means sharing the same subscript were all significant at $p < .05$."

The second method for summarizing the results of a post hoc investigation also involves a table of group means, as did the first method. Instead of attaching letters to those means, however, the second method involves (1) an ordering of abbreviations or numbers that represent the group means and (2) the use of the symbols $>$, $<$, and $=$ (positioned within those group abbreviations or numbers) to indicate the findings of the pairwise comparisons. If this method had been used in the study from which Excerpt 12.14 was taken, perhaps the four groups would have been abbreviated as PD (for PharmD), BS (for BS Pharm), CT (for CPhT), and PT (for Pharmacy Technician). Using these abbreviation, the researcher could have put the results of the post hoc investigation concerning age in a note beneath the table saying either "BS>CT=PD=PT" or "BS>CT,PD,PT."

Sometimes researchers use a graph of some type to help others see what was discovered in a post hoc investigation. Excerpt 12.15 contains an example of this helpful strategy. In this case, each of the three means was plotted as a bar, and the brackets above the bars indicate which pairwise comparisons were significant.

EXCERPT 12.15 • *Results of a Post Hoc Investigation Displayed in a Bar Graph*

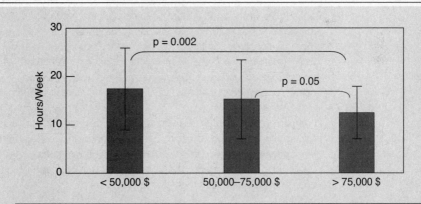

FIGURE 5 *Differences in weekly hours watching TV by household income (Study 2).*

Source: Drenowatz, C., Eisenmann, J. C., Pfeiffer, K. A., Welk, G., Heelan, K., Gentile, D., et al. (2010). Influence of socio-economic status on habitual physical activity and sedentary behavior in 8- to 11-year old children. *BMC Public Health, 10*(214), 1–11.

Planned Comparisons

So far we have considered the comparison of group means using a two-step strategy that involves conducting a one-way ANOVA followed by a post hoc investigation. Researchers can, if they wish, bypass the ANOVA F-test and move directly to one or more specific comparisons of particular interest. Such comparisons among means (without reliance on a green light from a significant omnibus F-test) are called *planned comparisons*.[4] Although planned comparisons are used less frequently than post hoc comparisons, they show up in the research literature often enough to make it important for you to recognize and understand this kind of statistical test on means.

Excerpt 12.16 illustrates the use of planned comparisons. As you will see, the researchers bypassed the more typical one-way ANOVA because they considered their planned comparisons to be better aligned with the research hypotheses they wanted to test. One of their three planned comparisons was pairwise in nature; the other two were of the nonpairwise variety.

EXCERPT 12.16 • Planned Pairwise and Nonpairwise Comparisons

One hundred thirty-seven students . . . were randomly assigned to one of the following five conditions: (a) deciding for themselves (self), (b) deciding for a close same-sex friend (friend–decide), (c) predicting the decisions of a close same-sex friend (friend–predict), (d) deciding for a typical same-sex student in the United States (typical student–decide), and (e) predicting the decisions of a typical same-sex student in the United States (typical student–predict). . . . We had three primary questions we were interested in addressing in these analyses [and] because each of these issues corresponds to a specific contrast or set of contrasts, we analyzed these issues via planned comparisons rather than by a full ANOVA. . . . Aggregating across impact level, a comparison of the self condition ($M = 9.72$) to the average of the friend–decide ($M = 10.59$) and typical student–decide ($M = 10.33$) conditions was not significant, $t(127) = 1.11, p = .27$, partial $\eta^2 = .010$. . . . A planned comparison between the two deciding-for-others conditions found no difference between the friend–decide and typical student–decide conditions, $t(127) = 0.88$, $p = .38$, partial $\eta^2 = .006$. . . . A planned contrast comparing the self condition to the average of the predict conditions showed that our participants did not predict that others would decide differently than they themselves decide, $t(127) = 0.44, p = .66$, partial $\eta^2 = .002$.

Source: Stone, E. R., & Allgaier, L. (2008). A social values analysis of self-other differences in decision making involving risk. *Basic & Applied Social Psychology, 30*(2), 114–129.

---

[4]The term *a priori comparison* means the same thing as *planned comparison*.

In the study from which Excerpt 12.16 was taken, there were five groups. As you can see, each of those groups was labeled with a letter. The single nonpairwise contrast involved a comparison of the sample means from groups "b" and "d." One of the two nonpairwise contrasts involved comparing the mean from group "a" against the mean created by combining groups "b" and "d." The other nonpairwise contrast involved comparing the mean from group "a" against the mean created by combining groups "c" and "e."

In Excerpt 12.16, each of the planned comparisons was evaluated statistically by means of an independent-samples *t*-test. That was the case for the single pairwise comparison that was tested as well as for each of the two cases in which nonpairwise comparisons were tested. The use of this kind of *t*-test in these situation seems logical when you realize that two means were involved in each of the three comparisons. Granted, the nonpairwise comparisons each involved three of the original groups. However, each of these comparisons boiled down to a statistical look at just two means after two of the three original groups involved in each comparison were combined.

In general, planned comparisons are good for two distinct reasons. First of all, this kind of comparison usually is more powerful than is the omnibus *F*-test. In other words, Type II error risk is typically lower when planned comparisons are examined instead of the generic one-way ANOVA *F*-test. Second, well-trained researchers usually have justifiable hypotheses (gleaned from their own earlier studies and their knowledge of others' research findings), and this expertise shows through better in a set of planned comparisons than it does in the two-step strategy of doing a one-way ANOVA followed by a post hoc investigation. Therefore, give researchers a double set of bonus points when you see them discuss planned comparisons in their technical written reports or in their oral presentations. That was what I did when I came across the passage that appears in Excerpt 12.17.

EXCERPT 12.17 • *Decision to Bypass Using a One-Way ANOVA*

Given that we hypothesised that there would be specific SES group differences in the places that children undertook physical activities, we employed the use of a priori comparisons rather than using ANOVA procedures because a priori comparisons have greater power and improve the clarity of comparisons over the omnibus *F*-test.

Source: Ziviani, J., Wadley, D., Ward, H., Macdonald, D., Jenkins, D., & Rodger, S. (2008). A place to play: Socioeconomic and spatial factors in children's physical activity. *Australian Occupational Therapy Journal, 55*(1), 2–11.

Comments

As we come to the end of this chapter on planned and post hoc comparisons, there are a few final things to consider. If you take the time to consider these end-of-chapter issues, you will be better able to decipher and critique research reports. These issues are positioned here at the end of the chapter because each of them has relevance to both post hoc and planned comparisons.

Terminology

Earlier in this chapter, you encountered six technical terms: post hoc, planned, comparison, contrast, pairwise, and nonpairwise. (A seventh term, **a priori,** appears in a footnote.) We now must consider two additional terms: **1 *df* F-test** and **orthogonal**. After you add these two terms to your working vocabulary, you will be able to understand just about any research report wherein the researchers discuss their post hoc and planned comparisons.

In reading discussions of post hoc and planned comparisons, you may come across the term **one-degree-of-freedom *F*-test.** This term pops up every so often when nonpairwise contrasts are conducted via *F*-tests, and it simply refers to the fact that the first of the two *df* values of such an *F*-test is always 1, no matter how many groups are involved in the comparison being made. Thus, if a study involves the personality trait of "hot-temperedness" among groups of blonds, brunettes, and redheads, the researcher might want to pool together the blond-haired men and the brown-haired men and compare that combined group's mean against the mean of the red-haired men. An *F*-test used to do this would have a between-groups *df* equal to 1.

My hair color example, of course, is quite artificial. To see an example of this special kind of *F*-test, take another look at Excerpt 12.16. Three *t*-tests appear in that excerpt, two of which involved nonpairwise comparisons. All three of those tests could have been conducted via *F*-tests rather than *t*-test, with identical results. Had that been done, each *F*-test's *df*s would have been 1 and 127. The first of those two *df* number would have been 1, even for the two nonpairwise comparisons that involved three of the five groups, because such comparisons really were comparing just two means, not three.[5]

The second new term for us to consider is **orthogonal.** In a researcher's planned or post hoc investigation, two contrasts are said to be orthogonal to one another if the information yielded by one contrast is new and different (i.e., independent) from what is revealed by the other contrast. For example, with three groups in a study (A, B, and C), a contrast comparing group A against the average of groups B and C would be orthogonal to a contrast comparing group B against group C because knowing how the first contrast turned out would give you no clue as to how the second contrast will turn out.

In Excerpt 12.18, we see case where a group of researchers set up and tested a pair of orthogonal contrasts just like those discussed in the preceding paragraph. As you can see, both of these contrasts turned out to be significant. However, the orthogonal

---

[5]The second of the two *df* values in Excerpt 12.16 is correct despite the fact that there were 137 research participants in the study. The researchers used a 2 × 5 research design, with 9 *df* associated with the two main effects and the interaction (even though none of these three effects was actually tested).

EXCERPT 12.18 • *Orthogonal Comparisons*

We conducted two a priori orthogonal comparison tests, the adults compared to the average of the 2 child groups and the second test compared the mean of the typical child group to the mean of the group of children with SPD. The a priori tests revealed that adults had a significantly smaller mean P50 T/C ratio than the average of both child groups, $t(68) = -2.44, p = .017, d = .67$. Typical children also demonstrated a statistically significant smaller mean P50 T/C ratio than children with SPD, $t(68) = -2.01, p < .049, d = .49$.

Source: Davies, P. L., Chang, W., & Gavin, W. J. (2009). Maturation of sensory gating performance in children with and without sensory processing disorders. *International Journal of Psychophysiology, 72*(2), 187–197.

nature of these two *t*-tests meant that the first test could have rejected its null hypothesis, whereas the second *t*-test retained its H_0. Or, the results of the two *t*-tests could have been just the opposite in their outcomes. With orthogonal comparisons, any combination of results is possible, because the tests are independent of each other.

Assumptions

The various planned and post hoc test procedures mentioned earlier in this chapter will function as they are supposed to function only if four underlying assumptions hold true for the populations and samples involved in the study. These assumptions are the same ones that underlie a one-way ANOVA *F*-test, and they are referred to by the terms *randomness, independence, normality,* and *homogeneity of variance.* I hope you remember the main points that I made in Chapter 11 about these assumptions.

Although the various test procedures covered so far in this chapter generally are robust to the normality assumption, the same point cannot be made regarding the equal variance assumption—especially in situations where the sample sizes are dissimilar. If researchers conduct planned comparisons, they ought to talk about the issue of assumptions. If the study's sample sizes vary, a test should be applied to assess the homogeneity of variance assumption. With a post hoc investigation, the assumptions should have been discussed in conjunction with the omnibus *F*-test; those assumptions do not have to be discussed or tested a second time when the researcher moves from the one-way ANOVA to the post hoc comparisons.

If the equal variance assumption is tested and shown to be untenable (in connection with planned comparisons or with the one-way ANOVA), the researcher will likely make some form of adjustment when a priori or post hoc contrasts are tested. This adjustment might take the form of a data transformation, a change in the level of significance employed, or a change in the test procedure used to compare means. If the latter approach is taken, you are likely to see the Welch test applied to the data (because the Welch model does not assume equal population variances).

Many of the test procedures for making planned or post hoc comparisons were developed for the situation in which the various samples are the same size. When used with samples that vary in size, researchers may indicate that they used a variation of one of the main techniques. Thus, Kramer's extension of Tukey test or of Duncan's multiple range test simply involves a modification of the regular Tukey or Duncan test procedure to make it usable in studies in which the ns vary. Do not let such extensions or modifications cause you to shy away from deciphering research reports in the same way you would if the regular planned or post hoc test had been used.

The Researcher's Choice of Test Procedure

As I pointed out near the outset of this chapter, the various post hoc procedures differ in terms of how liberal or conservative they are. Ideally, a researcher ought to choose among these procedures after considering the way they differ in terms of power and control of Type I errors. Realistically, however, the decision to use a particular test procedure is probably influenced most by what computer programs are available for doing the data analysis, by what procedure was emphasized in a specific textbook or by a specific instructor, or by habit.

In Excerpt 12.19, a team of researchers did the right thing; they explained *why* they chose the test procedure they used to make multiple comparisons among means. As indicated in this excerpt, they used the Tukey test because they wanted their post hoc investigation to be based on a test procedure that was "middle of the road" in terms of the liberal/conservative continuum. In other words, they wanted to use a test that balanced the risks of making Type I and Type II errors.

EXCERPT 12.19 • Explaining Why a Test Procedure Was Used

Finally, a post hoc analysis was conducted, whose results allowed [us] to evaluate the statistical significance of the units' autonomy and verify them. From among several methods at hand the Tukey's test (HSD) was chosen due to its medium position on the scale of conservativeness in comparison with a more conservative Scheffe's test, giving statistically less significant results between the averages, and a less conservative Newman-Keuls's [procedure].

Source: Zubel, P., Gugnacka-Fiedor, W., Rusinek, A., & Barcikowski, A. (2008). Different faces of polar habitats extremity observed from the angle of arctic tundra plant communities (West Spitsbergen). *Ecological Question, 9*(1), 25–36.

Regardless of the stated (or unstated) reasons why a researcher chooses to use a particular test procedure, *you* are in full control of how you interpret the results presented in the research report. If a researcher uses a test procedure that is too liberal or too conservative for *your* taste, remember that you have the undisputed right to accept only a portion of the researcher's full set of conclusions. In the extreme case, you can, if you wish, reject the totality of what is "discovered" in a research study because the test procedure employed to make statistical inferences was far too liberal (or far too conservative) for your taste.

Statistical Significance versus Practical Significance

We have considered the distinction between statistical significance and practical significance in earlier chapters. My simple suggestion at this point is to keep this distinction in mind when you come into contact with the results of planned and post hoc comparisons.

In Excerpt 12.20, we see a case in which the measure *d* was used in a post hoc investigation to assess the practical significance of the two pairwise comparisons that turned out to be statistically significant. In this excerpt, notice that different measures of effect size were computed for the ANOVA *F*-test and the post hoc pairwise comparisons. Also note that the researchers labeled the *d* estimates of .71 and .49 with the terms *medium* and *small-to-medium,* respectively. When you come across passages like the one in Excerpt 12.20, upgrade your evaluation of the researchers' work.

EXCERPT 12.20 • *Concern for Practical Significance in a Post Hoc Investigation*

A one-way ANOVA revealed a statistically significant effect, $F(3, 203) = 5.15$, $p < .01$, eta-squared $= 0.07$. Post-hoc analysis, Tukey's HSD, of all possible paired comparisons revealed two differences among the conditions. Students in the GO-only group answered more target-material items correctly $(M = 9.75)$ than did students in the SD-only condition $(M = 7.80)$, $d = 0.71$, with a medium effect size [while those] in the Control group answered more target-material questions correctly than did students in the SD-only condition $(M = 9.20$ compared to $M = 7.80$, respectively), $d = 0.49$, with a small-to-medium effect size [being estimated].

Source: Rowland-Bryant, E., Skinner, C. H., Skinner, A. L., Saudargas, R., Robinson, D. H., & Kirk, E. R. (2009). Investigating the interaction of graphic organizers and seductive details: Can a graphic organizer mitigate the seductive-details effect? *Research in the Schools, 16*(2), 29–40.

As exemplified by Excerpt 12.20, researchers often use terms such as *small, medium,* or *large* to describe the estimated effect sizes associated with pairwise comparisons. When they do this, they are using a set of criteria for interpreting *d* that are widely used by researchers in many different disciplines. Those criteria for comparing two means were presented earlier in Chapter 10. If you refer to Table 10.1, you will be able to tell whether the use of the effect size labels in Excerpt 12.20 was justified.

The distinction between statistical significance and practical significance is just as important to planned comparison as it is to post hoc comparisons. To see examples where researchers attended to this distinction with their planned comparisons, look again at Excerpts 20.16 and 20.18. They illustrate the use of partial η^2 in conjunction with both pairwise and nonpairwise comparisons, as well as the use of d with orthogonal comparisons.

Other Test Procedures

In this chapter, we considered several test procedures that researchers use when comparing means within planned and post hoc investigations. The excerpts we have considered demonstrate the popularity of the Tukey and Bonferroni test procedures. However, we have seen additional excerpts that illustrate the use of other test procedures, such as Tamhane's test, Scheffé's test, and Dunnett's test. All these procedures help hold down the chances of a Type I error when two or more contrasts are evaluated.

Although we have seen examples of a variety of test procedures in this chapter, there are additional test procedures that we have not considered. The tests mentioned in the preceding paragraph are the ones I believe you will encounter most often when you read research reports. However, you may come across one or more techniques not discussed in this text. If this happens, I hope you will not be thrown by the utilization of a specific test procedure different from those considered here. If you understand the general purpose served by the planned and post hoc tests we *have* considered, I think you will have little difficulty understanding the purpose and results of similar test procedures that we have *not* considered.

Review Terms

A posteriori test	Multiple comparison test
A priori	Newman-Keuls test
Bonferroni test	Nonpairwise (or complex) comparison
Comparison	Omnibus F-test
Complex comparison	One-degree-of-freedom F-test
Conservative	Orthogonal
Contrast	Pairwise comparison
Duncan's multiple	Planned comparisons
range test	Post hoc comparisons
Dunnett test	Post hoc test
Fisher's LSD test	Scheffé test
Follow-up test	Tamhanes' post hoc test
Liberal	Tukey test

The Best Items in the Companion Website

1. An interactive online quiz (with immediate feedback provided) covering Chapter 12.
2. Eight misconceptions about the content of Chapter 12.
3. An email message sent from the author to his students entitled "Seemingly Illogical Results."
4. Two funny jokes about statisticians.
5. One of the best passages from Chapter 12: "Your Right to Be Liberal or Conservative."

To access the chapter outline, practice tests, weblinks, and flashcards, visit the companion website at http://www.ReadingStats.com.

Review Questions and Answers begin on page 531.

Two-Way Analyses
of Variance

In Chapters 10 and 11, we saw how one-way ANOVAs can be used to compare two or more sample means in studies involving a single independent variable. In this chapter, I want to extend our discussion of analysis of variance to consider how this extremely popular statistical tool is used in studies characterized by two independent variables. It should come as no surprise that the kind of ANOVA to be considered here is referred to as a two-way ANOVA. Because you may have come across the term *multivariate analysis of variance* or the abbreviation *MANOVA*, it is important to clarify that this chapter does not deal with multivariate analyses of variance. The first letter of the acronym *MANOVA* stands for the word multivariate, but the letter *M* indicates that multiple dependent variables are involved in the same unitary analysis. Within the confines of this chapter, we will look at ANOVAs that involve multiple independent variables but only one dependent variable. Accordingly, the topics in this chapter (along with those of earlier chapters) fall under the general heading **univariate analyses.**

Similarities between One-Way and Two-Way ANOVAs

A two-way ANOVA is similar to a one-way ANOVA in several respects. Like any one-way ANOVA, a two-way ANOVA focuses on group means. (As you will soon see, a minimum of four Ms are involved in any two-way ANOVA.) As with a one-way AVOVA, any two-way ANOVA is actually concerned with the set of μ values that correspond to the sample means that are computed from the study's data. With both kinds of ANOVAs, the inference from the samples to the populations is made through the six-, seven-, or nine-step version of hypothesis testing. Statistical assumptions may need to be tested with each kind of ANOVA, and the research questions dictate whether planned or post hoc comparisons are used in conjunction with (or in lieu of) the two-way ANOVA, as is the case with a one-way ANOVA. Despite these similarities between one-way and two-way ANOVAs, the kind of ANOVA to which we now turn is substantially different from the kind we examined in Chapter 11.

The Structure of a Two-Way ANOVA

Before we discuss what kinds of research questions can be answered by a two-way ANOVA, it is essential that you understand how a two-way ANOVA is structured. Therefore, now I explain (1) how factors and levels come together to form cells; (2) how randomization is used to fill the ANOVA's cells with the people, animals, or things from which data are eventually collected; and (3) why this chapter deals exclusively with two-way ANOVAs having "between-subjects" factors.

Factors, Levels, and Cells

A two-way ANOVA always involves two independent variables. Each independent variable, or **factor,** is made up of, or defined by, two or more elements called **levels.** When looked at simultaneously, the levels of the first factor and the levels of the second factor create the conditions of the study to be compared. Each of these conditions is referred to as a **cell.**

To help you see how factors, levels, and cells form the basic structure of any two-way ANOVA, let's consider a recent study that involved 18 male and 20 female college undergraduates who participated in a simulated job-hiring task. Each of these students initially examined the résumés of two fictitious applicants and had to decide which one to hire. Half of each gender group then wrote a rejection letter to the applicant who was not hired, whereas the other half of each gender group wrote a private critique of either résumé.

After participating in some additional superfluous activities, all 38 students read a description of a new club on campus and then indicated how likely they were to join the new club to make new friends. Each person's response (on the 1–12 scale) became his or her score that went into the two-way ANOVA.

I have pulled a few of the summary statistics from the research report of the simulated job-hiring study and put them into a kind of picture. This picture was not included in the actual research report, but it is useful here because it permits us to see the factors, levels, and cells of this particular two-way ANOVA.

In my picture, the term *Gender* labels the two main rows, whereas the term *Treatment* labels the two main columns. These are the two independent variables, or factors, involved in this study. The specific rows and columns indicate the levels that went together to make up the two factors. Thus the factor of Gender was made up of two levels, Male and Female, whereas the Treatment factor was made up of two levels, Rejection Letter and Résumé Critique. If you take either row of my picture

	Treatment	
	Write Rejection Letter	Write Résumé Critique
Male	M = 6.72 SD = 2.38 n = 9	M = 7.38 SD = 1.08 n = 9
Female	M = 5.55 SD = 2.36 n = 10	M = 8.15 SD = 1.36 n = 10

Gender

Source: Data from Zhou, X., Zheng, L., Zhou, L., & Guo, N. (2009). The act of rejecting reduces the desire to reconnect: Evidence for a cognitive dissonance account. *Journal of Experimental Social Psychology, 45*(1), 44–50.

and combine it with either of the columns, you end up with one of the four cells associated with this particular two-way ANOVA. Each of these cells represents the home of one of the four subgroups of the college students.

Within each cell of my picture, there is a mean, a standard deviation, and a sample size. These three numerical values constitute a summary of the scores on the dependent variable—desire to join the new campus club—collected from the college students who were in each cell. As you will soon see, these data were very important to the two-way ANOVA that was conducted in conjunction with this study. The factors, levels, and cells provided the structure for the two-way ANOVA; without data on the dependent variable, however, there would have been no way to probe any of the research questions of interest.

As indicated earlier, all two-way ANOVAs involve two factors. Researchers tell you what factors were involved in their studies, but they are not consistent in their descriptions. Sometimes factors are called *independent variables,* sometimes they are called *main effects,* and sometimes they are not called anything. Two of the variations in the way researchers label the factors of their two-way ANOVAs are illustrated in Excerpts 13.1 and 13.2.

EXCERPTS 13.1–13.2 • *Alternative Names for a Two-Way ANOVA's Factors*

The data analysis (relating to the ability to apply reading strategies) was based on two-way ANOVA. The independent variables are: group (experimental or control) and Chinese reading ability (high or low).

Source: Chang, K., Lan, Y., Chang, C., & Sung, Y. (2010). Mobile-device-supported strategy for Chinese reading comprehension. *Innovations in Education & Teaching International, 47*(1), 69–84.

(*continued*)

EXCERPTS 13.1–13.2 • (*continued*)

> Statistical analyses were performed using a two-way ANOVA for non-repeated measures using training and MTP inhibition as the main effects.
>
> *Source:* Chapados, N. A., & Lavoie, J. (2010). Exercise training increases hepatic endoplasmic reticulum (er) stress protein expression in MTP-inhibited high-fat fed rats. *Cell Biochemistry and Function, 28*(3), 202–210.

When describing their two-way ANOVAs, most researchers indicate how many levels were in each factor. They do this by using terms such as 2×2 ANOVA, 2×4 ANOVA, 3×5 ANOVA, and 2×3 ANOVA. When such notation is used, the first of the two numbers that precede the acronym ANOVA specifies how many levels went together to make up the first factor, while the second number indicates how many levels composed the second factor. Excerpts 13.3 and 13.4 illustrate the use of this kind of notation.

EXCERPTS 13.3–13.4 • *Delineating a Two-Way ANOVA's Dimensions*

> A 2×2 analysis of variance design (averaging method $\times$ testing frequency) is used to explore the effects of the two independent variables on student achievement.
>
> *Source:* Vaden-Goad, R. E. Leveraging summative assessment for formative purposes. *College Teaching, 57*(3), 153–155.
>
> --
>
> A 2×3 ANOVA was performed, with frame and occupational status as independent variables and bonus importance perceptions as a dependent variable.
>
> *Source:* Lozza, E., Carrera, S., & Bosio, A. C. (2010). Perceptions and outcomes of a fiscal bonus: Framing effects on evaluations and usage intentions. *Journal of Economic Psychology, 31*(3), 400–404.

The researchers who are most helpful in describing their two-way ANOVAs are the ones who indicate not only the names of the factors and the number of levels in each factor but also the names of the levels. An example of this kind of description appears in Excerpt 13.5. Based on the information contained in this excerpt, you can and should create a picture (either on paper or in your mind) like the one I created in my picture for the simulated job-hiring study. Here, however, the picture has five rows, two columns, and 10 cells. Collectively, the five rows are labeled *Faculty* (i.e., the equivalent of a college within a university), with the specific rows being *Technical studies, Natural sciences, Social sciences, Arts,* and

EXCERPT 13.5 • *Naming Factors and Levels*

We performed a univariate factorial ANOVA with [students'] global trait EI as the dependent variable, and faculty (technical studies/natural sciences/social sciences/arts/humanities) and gender (male/female) as the independent variables.

Source: Sánchez-Ruiz, M. J., Pérez-González, J. C., & Petrides, K. V. (2010). Trait emotional intelligence profiles of students from different university faculties. *Australian Journal of Psychology, 62*(1), 51–57.

Humanities. Collectively, the two columns would be labeled with the word *Gender,* with the specific columns corresponding to *Males* and *Females.*[1]

Later in this chapter we consider the null hypotheses typically tested when researchers use a two-way ANOVA, and we also look at the different reporting schemes used by researchers to report the results of such tests. I cannot overemphasize how important it is to understand the concepts of factor, level, and cell *before* considering what a two-way ANOVA tries to accomplish. Stated differently, if you cannot create a picture that shows the structure of a researcher's two-way ANOVA, there is no way you can understand the results or evaluate whether the researcher's claimed discoveries are supported by the empirical data.

Active versus Assigned Factors and the Formation of Comparison Groups

All two-way ANOVAs are the same in that the levels of the two factors jointly define the cells. However, there are different ways to fill each cell with the things (people, animals, or objects) from which measurements will be taken. In any given study, one of three possible procedures for forming the comparison groups is used depending on the nature of the two factors. Because any factor can be classified as being *assigned* or *active* in nature, a two-way ANOVA could be set up to involve two assigned factors, two active factors, or one factor of each type.

An **assigned factor** deals with a characteristic of the things being studied that they "bring with them" to the investigation. In situations where the study focuses on people, for example, such a factor might be gender, handedness, birth order, intellectual capability, color preference, grade point average (GPA), or personality type. If the study focuses on dogs, an assigned factor might be breed, size, or age. The defining element of an assigned factor is that a person's (or animal's) status for this kind of independent variable is determined by the nature of that person (or animal) on entry into the study.

---

[1]The picture created for Excerpt 13.5 could be set up with rows corresponding to Gender and columns corresponding to Faculty. If we set it up that way, there would be two rows and five columns. With two-way ANOVAs, it makes no difference whether a particular factor is used to define the picture's rows or its columns.

The second kind of factor is called an **active factor.** Here, a participant's status on the factor is determined within the investigation, because active factors deal with conditions of the study that are under the control of the researcher. Simply put, this means that the researcher can decide, for any participant, which level of the factor that participant will experience. Examples of active factors include type of diet, time allowed to practice a task, gender of the counselor to whom the participant is assigned, and kind of reward received following the occurrence of desirable behavior. The hallmark of these and all other active factors is the researcher's ability to decide which level of the factor any participant experiences during the investigation.

If a two-way ANOVA involves two assigned factors, the researcher simply puts the available participants into the various cells of the ANOVA design based on the characteristics of the participants. If the factors are both active, the researcher forms the comparison groups by randomly assigning participants to the various cells of the design. If there is one active factor and one assigned factor, the researcher forms the comparison groups by taking the participants who share a particular level on the assigned factor and randomly assigning them to the various levels of the active factor. Examples of these three ways of getting people into the cells of a two-way ANOVA can be found in Excerpts 13.5, 13.2, and 13.1, respectively.

Between-Subjects and Within-Subjects Factors

Each of the factors in a two-way ANOVA can be described as being either *between subjects* or *within subjects* in nature. The distinction between these two kinds of factors revolves around the simple question, "Are the study's participants measured under (i.e., have exposure to) just one level of the factor, or are they measured repeatedly across (and thus exposed to) all levels of the factor?" If the former situation exists, the factor is a between-subjects factor; otherwise, it is a within-subjects factor.

To help clarify the difference between these two kinds of factors, let's consider a simple (yet hypothetical) study. Imagine that a researcher wants to see if a golfer's ability to putt accurately is influenced by whether the flag is standing in the hole and whether the golf ball's color is white or orange. Further imagine that 20 golfers agree to participate in our study, that all putts are made on the same green from the same starting spot 25 feet from the hole, and that putting accuracy (our dependent variable) is measured by how many inches each putted ball ends up away from the hole.

In our putting investigation, we might design the study so both of our independent variables (flag status and ball color) are between-subjects factors. If we do that, we create four comparison groups (each with $n = 5$) and have the golfers in each group putt under just one of the four conditions of our study (e.g., putting an orange ball toward a flagless hole). Or, we might want to have both factors be within-subjects in nature. If that were our choice, we would have all 20 golfers putt under all four conditions of the study. There's also a third possibility. We could have one between-subjects factor and one within-subjects factor. For example, we could have all golfers putt both white and orange balls, with half of the golfers putting toward the hole with the flag in and the other half putting toward the hole with the flag out. In this third example, the flag's status would be a between-subjects factor, whereas ball color would be a within-subjects factor.

In this chapter, we consider only two-way ANOVAs involving two between-subjects factors. If a researcher indicates that he or she used a two-way ANOVA (without any specification as to the type of factors involved), you should presume that it is the kind of two-way ANOVA being discussed in this chapter. You can feel relatively confident doing this because most researchers use the generic phrase *two-way ANOVA* when both factors are of the between-subjects variety.[2] Occasionally, as illustrated in Excerpt 13.6, you will see a clear indication that two between-subjects factors were involved in the ANOVA being discussed.

EXCERPT 13.6 • *Between-Subjects Factors*

ANOVAs were conducted for each trial with sex (male, female) and age (9–13, 14–17, 18 and older) as between-subjects factors.

Source: Lawson, R. (2010). People cannot locate the projection of an object on the surface of a mirror. *Cognition*, *115*(2), 336–342.

Samples and Populations

The samples associated with any two-way ANOVA are always easy to identify. There are as many samples as there are cells, with the research participants who share a common cell creating each of the samples. Thus there are four distinct samples in any 2 × 2 ANOVA, six distinct samples in any 2 × 3 ANOVA, 12 distinct samples in any 3 × 4 ANOVA, and so on.

As is always the case in inferential statistics, a distinct population is associated with each sample in the study. Hence, the number of cells designates not only the number of samples (i.e., comparison groups) in the study, but also the number of populations involved in the investigation. Although it is easy to tell how many populations are involved in any two-way ANOVA, care must be exercised in thinking about the nature of these populations, especially when one or both factors are active.

Simply put, each population in a two-way ANOVA should be conceptualized as being made up of a large group of people, animals, or objects that are similar to those in the corresponding sample represented by one of the cells. Suppose, for example, that the dart-throwing ability of college students is measured via a two-way ANOVA in which the factors are gender (male and female) and handedness (right

---

[2]If one or both of the factors are of the within-subjects variety, researchers will normally draw this to your attention by labeling the factor(s) in that way or by saying that they used a repeated-measures ANOVA, a mixed ANOVA, or a split-plot ANOVA. We will consider such ANOVAs in Chapter 14.

and left). One of the four populations in this study is right-handed male college students. Each of the other three populations in this study is defined in a similar fashion by the combination of one level from each factor. If the four samples in this study were extracted from a larger group of potential participants, then each population is considered tangible in nature. If, however, all available dart-throwers were used, then the populations are abstract in nature.[3]

If a two-way ANOVA involves one or two active factors, the populations associated with the study are definitely abstract in nature. To understand why this is true, let's consider a study that produced results we will see later (in Excerpt 13.9). In this study concerning the vicarious effects of weight-related teasing, 88 college women participated, 44 of whom were binge eaters. Each woman read one of two short vignettes about a female shopping at a mall. These vignettes were identical except that the main character in one of the vignettes was teased about being overweight (with phrases such as, "Hey, Fatty") by people she encountered at the mall. Half of the binge-eating and half of the non-binge-eating research participants were given each vignette. After reading the story about the mall, all women filled out a questionnaire designed to measure a personality trait called *negative affect*.

In this study, the populations were abstract because they were characterized by people like the study's research participants *who read the vignettes*. There undoubtedly were (and are) lots of college women similar to the ones who served as participants in this investigation. However, there probably is no one outside the study who has read one of the two vignettes and then has taken the personality inventory that measures negative effect. Even though the women in this study were drawn randomly from a larger, tangible group of potential participants, the study's four populations (defined by binge-eating status and which vignette is read) became fully unique and abstract because of the experimental conditions created by the researchers in their investigation.

Three Research Questions

To gain an understanding of the three research questions that are focused on by a two-way ANOVA, let's continue our examination of the simulated job-hiring study discussed earlier in the chapter. As you may recall, male and female college students in this study first read two résumés and had to reject one of the two applicants. Then, half of the males and half of the females had to write a rejection letter to the person not hired. After some irrelevant tasks (designed to hide the study's true purpose), all research participants were asked to rate their interest in joining a new campus club. To facilitate our current discussion, I reproduce here a picture containing the original four cells means and sample sizes, plus I added four additional numbers that we must consider. As you may recall, the means in this study reflect the college students' stated desire to join the new club on a 1 to 12 scale.

[3]I discussed the difference between tangible and abstract populations in Chapter 5.

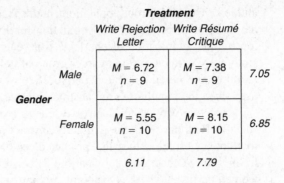

When the researchers applied a two-way ANOVA to the data provided by the 38 college students, they obtained answers to three research questions. Although these three research questions were tied to the specific independent and dependent variables involved in this study, the nature of their questions is identical to the nature of the three research questions that are posed and answered in any two-way ANOVA. These three questions, in their generic form, can be stated as follows: (1) Is there a statistically significant main effect for the first factor? (2) Is there a statistically significant main effect for the second factor? (3) Is there a statistically significant interaction between the two factors?

The first research question asked whether there was a statistically significant main effect for Gender. To get a feel for what this first research question was asking, you must focus your attention on the **main effect means** for the Gender factor. These means, located on the right side of the box containing the four cells, turned out equal to 7.05 and 6.85. The first of these means is simply the overall mean for the 18 males who were asked to rate their interest in joining the new club. (Because there are 9 males in each of the top two cells, the top row's main effect mean is equal to the arithmetic average of 6.72 and 7.38.) The second main effect mean for the Gender factor is the overall mean for the 20 female students who rated their interest in joining the club. Those 20 students are located, so to speak, within the two cells on the bottom row of the box.

In any two-way ANOVA, the first research question asks whether there is a statistically significant **main effect** for the factor that corresponds to the rows of the two-dimensional picture of the study. Stated differently, the first research question is asking whether the main effect means associated with the first factor are further apart from each other than would be expected by chance. There are as many such means as there are levels of the first factor. In the study we are considering, there were two levels (Male and Female) of the first factor (Gender), with the first research question asking whether the difference between 7.05 and 6.85 was larger than could be accounted for by chance. In other words, the first research question asks, "Is there a statistically significant difference between the means from the 18 males and the 20 males when they rated their interest in joining the new club?"

The second research question in any two-way ANOVA asks whether there is a significant main effect for the factor that corresponds to the columns of the two-dimensional picture of the study. The answer to this question is yes if the main effect means for the second factor turn out to be further apart from each other than would be expected by chance. In the study we are considering, there are two main effect means for Treatment, one for those who wrote the rejection letter and one for those who wrote a private critique of one of the job applicant's résumés. These means turned out equal to 6.11 and 7.79, respectively. Simply put, the second research question in this study asks, "Is there a statistically significant difference, regarding interest in joining the new club, between the mean of the 19 research participants who had to write a rejection letter and the mean of the 19 others who simple wrote a private critique of a résumé?"

The third research question in any two-way ANOVA asks whether there is a statistically significant **interaction** between the two factors involved in the study. As you will soon see, interaction deals with cell means, not main effect means. Therefore, there are always four means involved in the interaction of any 2×2 ANOVA, six means involved in any 2×3 ANOVA, and so on.

Interaction exists to the extent that the difference between the levels of the first factor changes when we move from level to level of the second factor. To illustrate, consider again the simulated job-hiring study. The difference between the means for the males who wrote a rejection letter and the males who critiqued a résumé was .66, with the mean in the right cell being larger than the mean in the left cell. If this difference of .66 were to show up again in the bottom row (with the same ordering of those means in terms of their magnitude), there would be absolutely no interaction. However, if this difference in the bottom row is either smaller or larger than .66 (or if it is .66 with a reverse ordering of the means), then interaction is present in the data.

When we look at the cell means in the bottom row, we can see that the difference between them does not mirror what we saw in the top row. The females who critiqued a résumé had a mean that was 2.60 higher than the mean for the females who wrote a rejection letter. Hence, there is some interaction between the Gender and Treatment factors. But is the amount of interaction contained in the four cell means more than what one would expect by chance? If so, then it can be said that a statistically significant interaction exists between the study's two factors, Gender and Treatment.

It should be noted that the interaction of a two-way ANOVA can be thought of as dealing with the difference between the levels of the column's factor as one moves from one level to another level of the row's factor, or it can be thought of as dealing with the difference between the levels of the row's factor as one moves from one level to another level of the column's factor. For example, in the study we are considering, the difference between the cell means in the left column is 1.17, with the top mean being larger, whereas the difference between the means in the right column is −.77. (The negative sign is needed because the top mean is now smaller than the bottom mean.) Although these two differences (1.17 and −.77) are not the same as the differences discussed in the preceding two paragraphs (.66 and 2.60), note that in both cases the difference between the differences is exactly the same: 1.94. My point here is simply that there is only one interaction in a two-way ANOVA; the order in which the factors are named (or used to compare cell means) makes no difference whatsoever.

The Three Null Hypotheses (and Three Alternative Hypotheses)

There are three null hypotheses examined within a two-way ANOVA. One of these null hypotheses is concerned with the main effect of the row's factor, the second is concerned with the main effect of the column's factor, and the third is concerned with the interaction between the two factors. Rarely are these null hypotheses referred to in a research report. In Excerpt 13.7, however, we see a case where a team of researchers enumerated the main effect and interaction null hypotheses associated with their two-way ANOVA.[4]

EXCERPT 13.7 • The Three Null Hypotheses of a Two-Way ANOVA

In order to determine the effects of REBT [rational–emotive behavior therapy] on the test anxiety level of groups at the end of treatment, the following hypotheses were tested using test anxiety level as dependent variable.

H_{01}: There is no significant difference in the test anxiety level of groups subjected to REBT therapy and control after treatment.

H_{02}: There is no significant difference in the test anxiety level of groups with moderate and high entry anxiety level at the end of treatment.

H_{03}: There is no significant interaction effect of treatment by entry test anxiety level on test anxiety level at the end of treatment.

Source: Egbochuku, E. O., Obodo, B., & Obada, N. O. (2009). Efficacy of rational-emotive behaviour therapy on the reduction of test anxiety among adolescents in secondary schools. *European Journal of Social Sciences, 6*(4), 152–164.

To explain how each of these null hypotheses should be conceptualized, I want to reiterate that the group of participants that supplies data for any cell of the two-way ANOVA is only a sample. As was pointed out earlier in this chapter, a population is connected to each cell's sample. Sometimes each of these populations is concrete in nature, with participants randomly selected from a finite pool of potential

---

[4]It would have been better if the word *significant* had not appeared in these three null hypotheses. Whether results are or are not statistically significant is dependent upon sample data. Null hypotheses, in contrast, are statements about a study's populations and any H_0 is not influenced whatsoever by sample data.

participants. In many studies, each population is abstract, with the nature of the population tailored to fit the nature of the group within each cell and the condition under which data are collected from the participants in that group.

In the 2×2 ANOVA from the simulated job-hiring experiment, four populations were involved. As indicated earlier, each of these populations was abstract (rather than tangible). One of them should be conceptualized as being made up of male college students (like those used in the study) who first review two résumés, then write a rejection letter to the person not selected, and finally (following some irrelevant activities designed simply to conceal the study's true purpose) rate their interest in joining a new campus club. The other three populations should be conceptualized in this same way, with changes made in the gender of the students (substituting females for males) and the writing task (substituting a critique of the résumé for the rejection letter). Each of these four populations is created, in our minds, to match the gender of the students and treatment condition associated with each of the ANOVA's cells.

The first null hypothesis in any two-way ANOVA deals with the main effect means associated with the rows factor of the study. This null hypothesis asserts that the population counterparts of these sample-based main effect means are equal to each other. Stated in symbols for the general case, this null hypothesis is as follows: $H_0: \mu_{row1} = \mu_{row2} = \ldots = \mu_{bottom\ row}$. For the study dealing with the hiring decision, writing task, and club rating, this null hypothesis took the form $H_0: \mu_{males} = \mu_{females}$.

The second null hypothesis in any two-way ANOVA deals with the main effect means associated with the columns factor. This null hypothesis asserts that the population counterparts of these sample-based main effect means are equal to each other. For the general case, the null hypothesis says $H_0: \mu_{column1} = \mu_{column2} = \ldots = \mu_{last\ column}$. For the study dealing with the simulated hiring decision, this null hypothesis took the form $H_0: \mu_{rejection\ letter} = \mu_{résumé\ critique}$.

Before we turn our attention to the third null hypothesis of a two-way ANOVA, I must clarify the meaning of the μs that appear in the null hypothesis for the main effects. Each of these μs, like the data-based sample mean to which it is tied, actually represents the average of cell means. For example, μ_{row1} is the average of the μs associated with the cells on row 1, while $\mu_{column1}$ is the average of the μs associated with the cells in column 1. Each of the other main effect μs similarly represents the average of the μs associated with the cells that lie in a common row or in a common column. This point about the main effect μs is important to note because (1) populations are always tied conceptually to samples and (2) the samples in a two-way ANOVA are located in the cells. Unless you realize that the main effect μs are conceptually derived from averaging cell μs, you might find yourself being misled into thinking that the number of populations associated with any two-way ANOVA can be determined by adding the number of main effect means to the number of cells. Hopefully, my earlier and current comments help you see that a two-way ANOVA has only as many populations as there are cells.

The third null hypothesis in a two-way ANOVA specifies that there is no interaction between the two factors. This null hypothesis deals with the cell means, not the main effect means. This null hypothesis asserts that whatever differences exist among the population means associated with the cells in any given column of the two-way layout are equal to the differences among the population means associated with the cells in each of the other columns. Stated differently, this null hypothesis says that the relationship among the population means associated with the full set of cells is such that a single pattern of differences, in any specific column, accurately describes what exists within every other column.[5]

To express the interaction null hypothesis using symbols, we must first agree to let j and j' stand for any two different rows in the two-way layout, and to let k and k' stand for any two different columns. Thus the intersection of row j and column k designates cell jk, with the population mean associated with this cell being referred to as μ_{jk}. The population mean associated with a different cell in the same column would be symbolized as $\mu_{j'k}$. The population means associated with two cells on these same rows, j and j', but in a different column, k', could be symbolized as $\mu_{jk'}$ and $\mu_{j'k'}$, respectively. Using this notational scheme, we can express the interaction null hypothesis of any two-way ANOVA as follows:

$$H_0: \mu_{jk} - \mu_{j'k} = \mu_{jk'} - \mu_{j'k'}, \text{ for all rows and columns}$$
$$\text{(i.e., for all combinations of both } j \text{ and } j', k \text{ and } k')$$

To help you understand the meaning of the interaction null hypothesis, I have constructed sets of hypothetical population means corresponding to a 2×2 ANOVA, a 2×3 ANOVA, and a 2×4 ANOVA. In each of the hypothetical ANOVAs, the interaction null hypothesis is completely true.

$\mu = 20$	$\mu = 40$
$\mu = 10$	$\mu = 30$

$\mu = 10$	$\mu = 30$	$\mu = 29$
$\mu = 5$	$\mu = 25$	$\mu = 24$

$\mu = 2$	$\mu = 12$	$\mu = 6$	$\mu = 24$
$\mu = 4$	$\mu = 14$	$\mu = 8$	$\mu = 26$

Before turning our attention to the alternative hypotheses associated with a two-way ANOVA, it is important to note that each H_0 we have considered is independent from the other two. In other words, any combination of the three null hypotheses can be true (or false). To illustrate, I have constructed three sets of hypothetical population means for a 2×2 layout. Moving from left to right, we see a case in which only the row means differ, a case in which only the interaction null hypothesis is false, and a case in which the null hypotheses for both the row's main effect and the interaction are false.

---

[5]The interaction null hypothesis can be stated with references to parameter differences among the cell means within the various rows (rather than within the various columns). Thus, the interaction null hypothesis asserts that whatever differences exist among the population means associated with the cells in any given row of the two-way layout are equal to the differences among the population means associated with the cells in each of the other rows.

$\mu = 10$	$\mu = 10$
$\mu = 5$	$\mu = 5$

$\mu = 20$	$\mu = 10$
$\mu = 10$	$\mu = 20$

$\mu = 10$	$\mu = 30$
$\mu = 20$	$\mu = 0$

Because the three null hypotheses are independent of each other, a conclusion drawn (from sample data) concerning one of the null hypotheses is specific to that particular H_0. The same data set can be used to evaluate all three null statements, but the data must be looked at from different angles in order to address all three null hypotheses. This is accomplished by computing a separate F-test to see if each H_0 is likely to be false.

If the researcher who conducts the two-way ANOVA evaluates each H_0 by means of hypothesis testing there will usually be three alternative hypotheses. Each H_a is set up in a nondirectional fashion, and they assert that:

1. The row μs are *not* all equal to each other;
2. The column μs are *not* all equal to each other;
3. The pattern of differences among the cell μs in the first column (or the first row) *fails* to describe accurately the pattern of differences among the cell μs in at least one other column (row).

Presentation of Results

The results of a two-way ANOVA can be communicated through a table or within the text of the research report. We begin our consideration of how researchers present the findings gleaned from their two-way ANOVAs by looking at the results of the study dealing with job applicants, writing tasks, and ratings of a new campus club. We then consider how the results of other two-way ANOVAs were presented. Near the end of this section, we look at the various ways researchers organize their findings when two or more two-way ANOVAs have been conducted within the same study.

Results of the Two-Way ANOVA from the Simulated Job-Hiring Study

In the research report of the study we considered at the beginning of this chapter, the findings were not presented in a two-way ANOVA summary table. If such a table had been prepared, it probably would have looked like Table 13.1.

In Table 13.1, notice that this summary table is similar to the summary table for a one-way ANOVA in terms of (1) the number and names of columns included in the table, (2) each row's *MS* being computed by dividing the row's *df* into its *SS*, (3) the total *df* being equal to one less than the number of participants used in the investigation, and (4) calculated values being presented in the *F* column. Despite these similarities, one-way and two-way ANOVA summary tables differ in that the

TABLE 13.1   *ANOVA Summary Table for Ratings for Joining the New Club*

Source	SS	df	MS	F	p
Gender	.38	1	.38	.11	.747
Treatment	25.21	1	25.21	7.06	.012
Gender × Treatment	8.82	1	8.82	2.47	.124
Within Groups	121.42	34	3.57		
Total	155.83	37			

latter contain five rows (rather than three) and three F-ratios (rather than one). Note that in the two-way summary table, the *MS* for the next-to-bottom row (which is usually labeled *error* or *within groups*) was used as the denominator in computing each of the three F-ratios: $.38/3.57 = .11, 25.21/3.57 = 7.06, 8.82/3.57 = 2.47$.

There are three values in the F column of a two-way ANOVA summary table because there are three null hypotheses associated with this kind of ANOVA. Each of the three Fs addresses a different null hypothesis. The first two Fs are concerned with the study's main effects; in other words, the first two Fs deal with the two sets of main effect means. The third F deals with the interaction between the two factors, with the focus of this F being on cell means. In the ANOVA summary table, look at the results for the two main effects. The first F was small and not significant. If you look again at the table I created to display the cell, row, and column means, you see that the main effect means for the Gender factor were quite similar: 7.05 and 6.85. With F-tests in ANOVAs, a homogeneous set of means causes the F to be small and nonsignificant. Statistically speaking, the observed difference between the row main effect means was not large enough to cast doubt on the Gender null hypothesis: $H_0: \mu_{males} = \mu_{females}$. The two main effect means for Treatment (6.11 and 7.79) were further apart, and this larger difference caused the F to be larger: 7.06. At the .05 level of significance, this F is statistically significant.

In the ANOVA summary table, you can see that the interaction F was not significant because its p was larger than .05. As we noted earlier, there was some interaction in the sample data because the difference between the means for the two groups of males (.66) was not the same as the difference between the means for the two groups of females (2.60). The two-way ANOVA considered the difference between these differences and declared that it was "within the limits of chance sampling." In other words, this degree of observed interaction in the sample data was not unexpected if the four samples had come from populations in which there was no interaction. Accordingly, the interaction null hypothesis was not rejected, as indicated by the p of .124 next to the F of 2.47.

Results from Additional Two-Way ANOVA Studies

In Excerpt 13.8, we see the summary table from a study in which a 2×3 ANOVA was used. It is worth the effort to compare this table to Table 13.1 so as to see how

EXCERPT 13.8 • *Two-Way ANOVA Summary Table*

TABLE 5   *Summary of ANOVA for marking frequency* ($N = 236$)

Source	SS	df	MS	F
MEF	718.21	1	718.21	13.99*
Examinees' EA	169.97	2	84.98	1.66
MEF × Examinees' EA	417.23	2	208.62	4.06*
w. cell (error)	11810.15	230	51.35	

*$p < .05$

Source: Chen, L. J., Ho, R. G., & Yen, Y. C. (2010). Marking strategies in metacognition-evaluated computer-based testing. *Journal of Educational Technology & Society, 13*(1), 246–259.

different authors vary the way they present the results of their two-way ANOVAs, even when such results are put into a table. Can you determine the three main difference between this ANOVA summary table and the one we examined earlier?

First, notice that there are only four rows of numbers, not five. As is the case with many ANOVA summary tables, the row for Total has been discarded. (You can still figure out how many scores went into the analysis from a table like this; simply add up all of the *df* numbers and then add 1 to the total.) The second thing to notice is that the fourth row of numbers has a different label here: "w. cell (error)." As you might guess, the *w* stands for *within.* (The word in parentheses, *error,* is often used by itself to label this row of the ANOVA summary table.) Finally, notice that there is no column of actual *p*-levels provided to the right of the computed *F*-values; instead, an asterisk has been attached to the *F*-values that were statistically significant, with a note beneath the table explaining the statistical meaning of the asterisk.

I have seen summary tables for two-way ANOVAs that differ in other ways from the two we have just considered. Sometimes the position of the *SS* and *df* columns is reversed. In a few instances, the row that's usually called "Within Groups" or "Error" is called "Residual." On occasion, there will be no column of *SS* numbers. (In this last situation, you are not at a disadvantage by not having *SS* values, because you could, if you so desired, compute all five missing *SS* values by multiplying each row's *df* by its *MS.*)

Despite the differences between Table 13.1 and Excerpt 13.8, these two ANOVA summary tables are similar in several respects. In each case, the title of the table reveals what the dependent variable was. In each case, the names of the first two rows reveal the names of the two factors. In each case, the *df* values for the main effect rows of the table allow us to know that there there how many levels were in each factor. In each case, the three *F*-values were computed by dividing the *MS* values on the first three rows by the *MS* value located on the fourth row (i.e., within groups or error). And in each case, three calculated F-values are presented, with each one addressing a different null hypotheses associated with the two-way ANOVA.

In reporting the outcome of their two-way ANOVAs, researchers often talk about the results within the text of the research report without including a summary table. In Excerpt 13.9, we see how this was done for a study we considered earlier in the chapter. In this investigation, an equal number of binge-eating and non-binge-eating females first read one of two vignettes concerning a girl shopping at a mall (who, in one vignette, was teased about her weight); then, all of the research participants filled out a personality inventory designed to measure a component of mood called negative effect.

EXCERPT 13.9 • *Results of a Two-Way ANOVA Presented in the Text*

A 2 × 2 between subjects ANOVA with binge eating status and vignette as independent variables and negative mood as the dependent variable was conducted. This analysis revealed a significant main effect of vignette, $F(1, 84) = 4.57, p = .04$, $eta^2 = .05$, such that participants who read the weight-related teasing vignette reported significantly more negative affect ($M = 15.75, SD = 4.99$) than those who read the neutral vignette ($M = 13.87, SD = 3.42$). There was no main effect of binge status ($p = .12$), nor was there an interaction between binge status and vignette ($p = .39$).

Source: Aubie, C. D., & Jarry, J. L. (2009). Weight-related teasing increases eating in binge eaters. *Journal of Social & Clinical Psychology*, *28*(7), 909–936.

Based on the information in Excerpt 13.9, you should be able to discern what the three null hypotheses were, what decision was reached regarding each H_0, and what level of significance was probably used by the researchers. In addition, you should be able to figure out how many college women supplied data for this two-way ANOVA.[6] Finally, you should also be able to determine whether the two means contained in this passage are cell means or main effect means.

Follow-Up Tests

If none of the two-way ANOVA Fs turns out to be significant, no follow-up test will be conducted. However, if at least one of the main effects is found to be significant, or if the interaction null hypothesis is rejected, you may find that a follow-up investigation is undertaken in an effort to probe the data. We consider first the follow-up tests used in conjunction with significant main effect F-ratios. Then, we examine post hoc strategies typically employed when the interaction F turns out to be significant.

[6]If you are tempted to think that there were 86 research participants in this study, make another guess.

Follow-Up Tests to Probe Significant Main Effects

If the F-test for one of the factors turns out to be significant and if there are only two levels associated with that factor, no post hoc test is applied. In this situation, the outcome of the F-test indicates that a significant difference exists between that factor's two main effect means, and the only thing the researcher must do to determine where the significant difference lies is to look at the two row (or two column) means. Whichever of the two means is larger is significantly larger than the other mean. If you take another look at Excerpt 13.9, you see an example in which just one of the main effect Fs was significant in a 2×2 ANOVA. Because there were only two main effect means associated with the significant F, the researchers interpreted these results directly. That interpretation is presented in the second portion of the second sentence in Excerpt 13.9, where the focus is on the teasing and non-teasing main effect means.

If the two-way ANOVA yields a significant F for one of the two factors, and if that factor involves three or more levels, the researcher is likely to conduct a **post hoc** investigation in order to compare the main effect means associated with the significant F. This is done for the same reasons that a post hoc investigation is typically conducted in conjunction with a one-way ANOVA that yields a significant result when three or more means are compared. In both cases, the omnibus F that turns out to significant must be probed to allow the researcher (and others) to gain insight into the likely pattern of population means.

Excerpt 13.10 shows how a post hoc investigation can help to clarify the meaning of a significant main effect in a two-way ANOVA. In the study associated

EXCERPT 13.10 • *Post Hoc Investigation Following a Significant Main Effect*

To test hypothesis two, a two-way ANOVA was run. The interaction between PA level and HR intensity level was not significant. The main effect of PA level was not significant. However, ANOVA results indicated a significant main effect [of] heart rate intensity level [on] attention change. The $F(3, 112) = 3.24$ was statistically significant ($p = .025$). . . . The high intensity group had the highest average inattention change improvement ($M = 3.36, SD = 4.16$), followed by the self-regulated group ($M = 1.81, SD = 3.40$), the low intensity group ($M = 1.27, SD = 3.96$), and the control group ($M = .83, SD = 3.80$). A Tukey post hoc multiple comparison of means revealed significant ($p = .044$) mean differences between high running intensity and the control group (2.531), but not between high intensity and self-regulated or low intensity groups. There were not statistically significant mean differences between the self-regulated group mean and low, high, or the control group means. There were not statistically significant mean differences between the low group mean and the self-regulated, high, or the control group means.

Source: Norling, J. C., Sibthorp, J., Suchy, Y., Hannon, J. C., & Ruddell, E. (2010). The benefit of recreational physical activity to restore attentional fatigue: The effects of running intensity level on attention scores. *Journal of Leisure Research, 42*(1), 135–152.

with this excerpt, 120 recreational runners were initially classified according to three levels of physical activity (PA) and four levels of running intensity. These individuals then took a test on a computer that measured their ability to keep their attention focused on a visual task. In this passage from the research report, notice how Tukey's test was used to compare, in a pairwise fashion, the four main effect means of the intensity factor.

If each of the factors in a two-way ANOVA turns out significant and if each of those factors contains three or more levels, then it is likely that a separate post hoc investigation will be conducted on each set of main effect means. The purpose of the two post hoc investigations would be the same: to identify the main effect means associated with each factor that are far enough apart to suggest that the corresponding population means are dissimilar. When both sets of main effect means are probed by means of post hoc investigations, the same test procedure (e.g., Tukey's) is used to make comparisons among each set of main effect means.

Follow-Up Tests to Probe a Significant Interaction

When confronted with a statistically significant interaction, researchers typically do two things. First, they *refrain* from interpreting the F-ratios associated with the two main effects. Second, post hoc tests are conducted or a **graph of the interaction** is prepared to help explain the specific nature of the interaction within the context of the study that has been conducted. Before turning our attention to the most frequently used follow-up strategies employed by researchers after observing a statistically significant interaction, let's consider what they typically *do not do*.

Once the results of the two-way ANOVA become available, researchers usually first look to see what happened relative to the interaction F. If the interaction turns out to be nonsignificant, they move their attention to the two main effect Fs and interpret them in accordance with the principles outlined in the previous section. If, however, the interaction turns out to be significant, little or no attention is devoted to the main effect F-ratios, because conclusions based on main effects can be quite misleading in the presence of significant interactions.

To illustrate how the presence of interaction renders the interpretation of main effects problematic, consider the three hypothetical situations presented in Figure 13.1. The number within each cell of each diagram is meant to be a sample mean, the numbers to the right of and below each diagram are meant to be main effect means (assuming equal cell sizes), and the abbreviated summary table provides the results that would be obtained if the samples were large enough or if the within-cell variability were small enough.

In situation 1 (on the left), both main effect Fs turn out nonsignificant. These results, coupled with the fact that there is no variability within either set of main effect means, might well lead one to think that the two levels of factor A are equivalent and to think that the three levels of factor B are equivalent. An inspection of the cell means, however, shows that those conclusions based on main effect means

	B			
A	10	15	20	15
	20	15	10	15
	15	15	15	

Source	F
A	ns
B	ns
A × B	*

(1)

	B			
A	5	10	15	10
	25	20	15	20
	15	15	15	

Source	F
A	*
B	ns
A × B	*

(2)

	B			
A	10	10	10	10
	10	10	40	20
	10	10	25	

Source	F
A	*
B	*
A × B	*

(3)

FIGURE 13.1 *Hypothetical Results from Three Two-Way ANOVAs*

would cause one to overlook potentially important findings. The two levels of factor A produced different means at the first and third levels of factor B, and the three levels of factor B were dissimilar at each level of factor A.

To drive home this point about how main effect Fs can be misleading when the interaction is significant, pretend that factor A is gender (males on the top row, females on the bottom row), that factor B is a type of headache medicine given to relieve pain (brands X, Y, and Z corresponding to the first, second, and third columns, respectively), with each participant asked to rate the effectiveness of his or her medication on a 0 to 40 scale (0 = no relief; 40 = total relief) 60 minutes after being given a single dose of one brand of medication. If one were to pay attention to the main effect Fs, one might be tempted to conclude that men and women experienced equal relief and that the three brands of medication were equally effective. Such conclusions would be unfortunate because the cell means suggest strongly (1) that males and females differed, on the average, in their reactions to headache medications X and Z, and (2) that the three medications differed in the relief produced (with brand X being superior for females, brand Z being superior for males).

In the second of our hypothetical situations (located in the center of Figure 13.1), notice again how the main effect Fs are misleading because of the interaction. Now, the main effect of factor A is significant, and one might be tempted to look at the main effect means and draw the conclusion that males experienced less relief from their headache medicines than did females. However, inspection of the cell means clearly shows that no difference exists between males and females when given brand Z. Again, the main effect F for factor B would be misleading for the same reason as it would be in the first set of results.

In the final hypothetical situation (on the right in Figure 13.1), both main effect Fs are significant. Inspection of the cell means reveals, however, that the levels of factor A do not differ at the first or at the second levels of factor B, and that the levels of factor B do not differ at the first level of factor A. Within the context of our hypothetical headache study, the main effect Fs, if interpreted, would lead one to suspect that females experienced more relief than males and that the three medicines differed in their effectiveness. Such conclusions would be misleading, for males and females experienced differential relief only when given brand Z, and the brands seem to be differentially effective only in the case of females.

When the interaction F turns out to be significant, the main effect Fs must be interpreted with extreme caution—or not interpreted directly at all. This is why most researchers are encouraged to look at the interaction F first when trying to make sense out of the results provided by a two-way ANOVA. The interaction F serves as a guidepost that tells the researchers what to do next. If the interaction F turns out to be *non*significant, this means that they have a green light and may proceed to interpret the F-ratios associated with the two main effects. If, however, the interaction F *is* significant, this is tantamount to a red light that says, "Don't pay attention to the main effect means but instead focus your attention on the cell means."

One of the strategies used to help gain insight into a statistically significant interaction is a **graph of the cell means.** We will look at such a graph that was included in a recent research report; first, however, you must become acquainted with the study.

In this study's experiment, half of the research participants (who were college students) were asked to think of someone about their age with whom they had an especially close relationship, whereas the others were asked to think of an age-level peer who was more of an acquaintance rather than a close friend. Next, each research participant was asked to read narrative accounts of two situations where one person betrayed someone else. When reading these accounts, each research participant was asked to imagine that he or she was the one who had been betrayed, with the betrayer being the individual they had been asked to think of. Not all research participants read the same betrayal narratives, however. Half of the students were given narrative accounts that involved "high severity" betrayal; the other received accounts that were "low severity" instances of betrayal.

After the college students had finished reading their pair of betrayal narratives, they filled out a questionnaire that assessed their opinions of how a variety of "forgiveness concepts" applied to the instances of betrayal they had just considered. The students' total scores from the forgiveness instrument became the data that went into in a 2 × 2 ANOVA. The independent variables were defined by the status of the imagined betrayer ("close" or "not close") and by the severity level of the betrayal ("high" or "low").

In Excerpt 13.11, we see a graph of the cell means from the study dealing with betrayal and forgiveness. Most researchers set up their graphs like the one in Excerpt 13.11 in that (1) the ordinate represents the dependent variable, (2) the points on the abscissa represent the levels of one of the independent variables, and

EXCERPT 13.11 • *The Graph of a Significant Interaction*

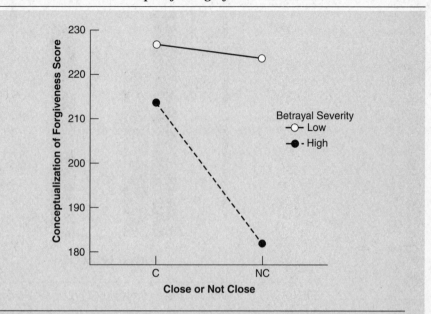

FIGURE 1   *The Effects of the Interaction between Betrayal Severity and Relational Closeness on the Conceptualization of Forgiveness Score*

Source: Dixon, L. J. (2009). The effects of betrayal characteristics on laypeople's ratings of betrayal severity and conceptualization of forgiveness. Unpublished doctoral dissertation, University of Tennessee, Knoxville.

(3) the lines in the graph represent the levels of the other independent variable. In setting up such a graph, either of the two independent variables can be used to label the abscissa. For example, the researcher associated with Excerpt 13.11 could have put the factor Betrayal Severity and its two levels (Low and High) on the baseline with the lines in the graph representing the two levels (Close and Not Close) of the Relational Closeness factor.

Notice how the graph in Excerpt 13.11 allows us to see whether an interaction exists. First recall that the absence of interaction means that the difference between the levels of one factor remains constant as we move from one level to another level of the other factor. Now look at Excerpt 13.11. In the graph, there is about a 13-point difference between the mean scores which define the left ends of the two lines. However, things change when we consider the right end points of these lines. Here, there is more than a 41-point difference between the two means. Considered as a whole, this graph suggests that people like this study's research participants extend the notion of forgiveness about as much to "betrayers" who are close or not close, so long as the betrayal issue is low in severity. However, if the

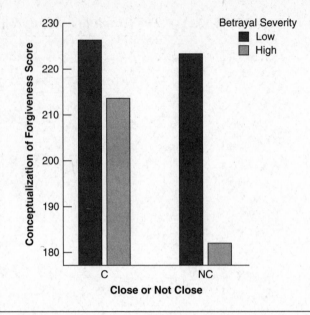

FIGURE 13.2 *Interaction from Betrayal and Forgiveness Study Displayed as a Bar Graph*

issue is high in severity, forgiveness is extended much more readily to those who are close friends rather than mere acquaintances.[7]

The graph of an interaction can be and often is set up such that vertical bars (rather than dots above the baseline) represent the cell means. I have created such a graph using the data from the betrayal and forgiveness study, and it appears in Figure 13.2. The amount of interaction shown in this figure, of course, is exactly the same as that shown in Excerpt 13.11. Both graphs do a nice job of showing that the variable of "closeness" makes a greater difference to the amount of forgiveness that is extended when the nature of the betrayal is severe.

Another strategy used by researchers to help understand the nature of a significant interaction is a statistical comparison of cell means. Such comparisons are usually performed in one of two ways. Sometimes, all of the cell means will be compared simultaneously in a pairwise fashion using one of the test procedures discussed in Chapter 12. In other studies, cell means are compared in a pairwise fashion one row or one column at a time using a post hoc strategy referred to as a simple main effects analysis. In the following five paragraphs, excerpts from actual studies are used to illustrate each of these post hoc strategies.

When a researcher probes a statistically significant interaction via tests of **simple main effects,** the various levels of one factor are compared in such a way

[7]Notice that the two lines in Excerpt 13.16 are not parallel. If the interaction F turns out to be significant, this is because the lines in the graph are nonparallel to a degree that is greater than what one expects by chance. For this reason, some authors define interaction as a departure from parallelism.

that the other factor is held constant. This is accomplished by comparing the cell means that reside in individual rows or in individual columns of the two-dimensional arrangement of cell means. This strategy of making tests of simple main effects is illustrated in Excerpt 13.12.

EXCERPT 13.12 • *Tests of Simple Main Effects*

The mean proportions of correct prospective memory responses [went] into a 3 (age) $\times$ 3 (task) between-participants ANOVA. This analysis revealed a significant main effect of task [but] this main effect was qualified by a significant age by condition interaction. . . . Tests of simple main effects showed that age effects were significant in the activity-based, $F(2, 214) = 4.87, MSE = .112, p = .009, \eta^2 = .04$, and the time-based conditions, $F(2, 214) = 16.14, MSE = .112, p < .0001, \eta^2 = .13$, but not in the event-based condition, $F(2, 214) = 2.12, MSE = .112, p = .12$, $\eta^2 = .019$. Follow-up [Tukey] post hoc tests indicated that in the activity-based condition performance of young and young-old participants did not reliably differ from each other $(p = .24)$ but young were reliably better than old-old participants $(p = .002)$. Although young-old participants were numerically better $(M = .85)$ than old-old participants $(M = .66)$, this difference was marginally significant $(p = .05)$. In contrast, in the time-based condition young participants were reliably better than both young-old and old-old participants (both $ps < .0001$) who did not differ from each other $(p = .24)$.

Source: Kvavilashvili, L., Kornbrot, D. E., Mash, V., Cockburn, J., & Milne, A. (2009). Differential effects of age on prospective and retrospective memory tasks in young, young-old, and old-old adults. *Memory, 17*(2), 180–196.

The first thing to note about Excerpt 13.12 is that we're being given the results of a 3 $\times$ 3 ANOVA. One of the factors was called *age,* and its levels were called *young, young-old,* and *old-old.* The other factor is initially referred to as *task,* but thereafter this same factor is referred to as *condition.* The levels of this second factor were the *activity-based* condition, the *time-based* condition, and the *event-based* condition. You must use these "discoveries" to create a picture of the study. In my picture, I have a square with three rows (for age) and three columns (for condition). Finally, in looking at your picture, you must realize that there was a group of individuals inside each of the nine cells, with each person providing a memory response score.

Because the interaction between age and condition was significant, the researchers conducted tests of simple main effects. As you can see, they did this by comparing, within each level of condition, the cell means for the three age groups. In my picture, this amounts to comparing the three means in the left column, then comparing the three means in the middle column, and finally comparing the three means in the right column. My picture helps me understand what was being compared in each of the three tests of simple main effects.

Each of the three tests of simple main effects conducted in Excerpt 13.12 is highly analogous to a one-way ANOVA *F*-test. The results of these three tests are reported in the third sentence of the excerpt, two of which yielded a significant result. Because each of these significant simple effect comparisons involved three means, the researchers conducted a post hoc investigation to probe each significant simple effect (just as a post hoc investigation typically is applied following a significant finding from a one-way ANOVA involving three or more means). The results of the Tukey tests that compared, in a pairwise fashion, the means of the three age groups in the activity-based condition are presented first. Then, in the last sentence, we see what was discovered when the time-based means were compared.

In Excerpt 13.12, the tests of simple main effects (and subsequent Tukey tests) compared the three age groups within each condition. If the researchers had wanted to, they could have conducted their tests of simple main effects to compare the three conditions within each age group. Researchers usually choose to run their tests of simple main effects in the direction that makes most sense to them after considering their research questions. Occasionally (but not often), researchers conduct tests of simple main effects going *both* directions.

A third strategy exists for statistically comparing cell means after the interaction null hypothesis has been rejected. Instead of comparing the means that reside within individual rows or columns of the two-dimensional layout of a two-way ANOVA, some researchers conduct all possible pairwise comparisons among the cell means. In Excerpt 13.13, we see an example in which this approach was taken.

EXCERPT 13.13 • *Pairwise Comparison of All Cell Means*

[Cell means were compared] to identify whether the differences of the pairs were statistically significant. The analyses involved [six comparisons]: the various unique combinations of field dependent/independent and type of preset learning goal. The greatest difference was for field dependence, specific preset learning goal versus field dependence general preset learning goal, which was statistically significant: $t(88) = 4.06, p = .00, d = .86$. There were statistically significant results for field dependence general preset learning goal versus field independence specific preset learning goal: $t(88) = -3.25, p = .00, d = .69$ and for field dependence, general preset learning goal versus field independence, general preset learning goal: $t(88) = -2.72, p = .00, d = .57$. Three statistically insignificant values were also reported. These were field dependence, specific preset learning goal versus field independence, specific preset learning goal: $t(88) = .52, p = .60$; field dependence, specific preset learning goal, versus field independence, general preset learning goal: $t(88) = .81, p = .42$: and field independence, specific preset learning goal, versus field independence, general preset learning goal: $t(88) = .30, p = .76$.

Source: Ku, D. T., & Soulier, J. S. (2009). The effects of learning goals on learning performance of field-dependent and field-independent late adolescents in a hypertext environment. *Adolescence, 44,* 651–664.

There were four cells involved in this study (because it was a 2 × 2 ANOVA), created by the combination of two levels of the learning-style factor (field-dependence and field-independence) and two levels of the preset learning goal factor (general and specific). Each of these cell means was compared in a pairwise fashion with each of the other three cell means, producing six tests.

Planned Comparisons

In Chapter 12, we saw several cases in which planned comparisons were used instead of or in addition to a one-way ANOVA. (See Excerpts 12.16, 12.17, and 12.18.) It should come as no surprise that such comparisons can also be used in conjunction with (or instead of) two-way ANOVAs. Not many researchers do this, but there are some who do. Consider, for example, a recent study involving spiders.

In the study associated with Excerpt 13.14, the researchers had two factors, each of which had two levels. Thus, they could have conducted a standard 2 × 2 ANOVA that would have produced two main effect *F*s and one interaction *F*. Instead of conducting those three tests, the researchers performed a single planned comparison. If you read the excerpt closely, you see that this planned comparison involved all four cells. The mean from one of the four cells was compared against the mean of the other three cells combined. In the excerpt's final sentence, the researchers provide all four cell means. However, the final three of these cell means

EXCERPT 13.14 • *A Planned Comparison Instead of a Two-Way ANOVA*

The study employed a 2 × 2 design with spider fear (low versus high) and spider prime (spider present in testing room versus no spider) as the between-subject variables, creating four groups (low fear – spider: $N = 28$; low fear + spider: $N = 29$; high fear – spider: $N = 26$; high fear + spider: $N = 28$). . . . [This] design can best be thought of as including one experimental group (high fear + spider) and three control groups (low fear – spider, low fear + spider, and high fear – spider) that vary the presence of the spider prime and level of predisposing spider fear, the two elements hypothesized to be necessary for expression of the memory bias. . . . To test the central hypothesis of preferential recall of spider words, a [preplanned] contrast was run to examine the mean number of spider words recalled by the high fear + spider group (contrast weight = + 3) when compared to the other three groups (contrast weights = −1, −1, −1). As predicted, the high fear + spider group recalled significantly more spider words than the other groups (high fear + spider: $M = 6.39 \pm 1.66$; high fear − spider: $M = 5.30 \pm 1.66$; low fear + spider: $M = 5.32 \pm 1.28$; low fear − spider: $M = 5.32 \pm 1.87$), $F(1, 100) = 8.81, P = .004, f = .30$.

Source: Smith-Janik, S. B., & Teachman, B. A. (2008). Impact of priming on explicit memory in spider fear. *Cognitive Therapy & Research, 32*(1), 291–302.

were combined into a single mean. Note the first of the *F*-test's two *df* values. If all four of the separate cell means had been compared, that *df* would have been 3. It is 1 because only two things were being compared: the mean of high-fear + spider cell versus the mean of the data in the other three cells.

The researchers associated with Excerpt 13.14 deserve enormous credit for having specific plans in mind when they designed their study *and* when they analyzed their data. Their research question guided their statistical analysis, and their analysis did not follow the conventional rules for making statistical comparisons in a two-way ANOVA. Far too many applied researchers erroneously think that (1) *F*-tests for main and interaction effects must be computed and (2) comparisons of cell means can be made only if the interaction is significant. This is unfortunate for several reasons, the main one being that planned comparisons of cell means can sometimes produce interesting findings that would remain undetected by the standard *F*-tests of a two-way ANOVA.

Assumptions Associated with a Two-Way ANOVA

The assumptions associated with a two-way ANOVA are the same as those associated with a one-way ANOVA: randomness, independence, normality, and homogeneity of variance. As I hope you recall from the discussion of assumptions contained in Chapter 11, randomness and independence are methodological concerns; they are dealt with (or *should* be dealt with) when a study is set up, when data are collected, and when results are generalized beyond the participants and conditions of the researcher's investigation. Although the randomness and independence assumptions can ruin a study if they are violated, there is no way to use the study's sample data to test the validity of these prerequisite conditions.

The assumptions of normality and homogeneity of variance *can* be tested and in certain circumstances *should* be tested. The procedures used to conduct such tests are the same as those used by researchers to check on the normality and equal variance assumptions when conducting *t*-tests or one-way ANOVAs. Two-way ANOVAs are also similar to *t*-tests and one-way ANOVAs in that (1) violations of the normality assumption usually do not reduce the validity of the results, and (2) violations of the equal variance assumption are more problematic when the sample sizes differ.

Because violations of the normality and equal variance assumptions are less disruptive to the *F*-tests of a two-way ANOVA when the *n*s are large and equal, many researchers make an effort to set up the studies with equal cell sizes. Frequently, however, it is impossible to achieve this goal. On occasion, a researcher begins with equal cell sizes but ends up with cell *n*s that vary because of equipment failure, subject dropout, or unusable answer sheets. On other occasions, the researcher has varying sample sizes at the beginning of the study but does not want to discard any data so as to create equal *n*s because such a strategy would lead to a loss of statistical power. For either of these reasons, a researcher may end up with cell sizes that vary. In such situations, researchers frequently concern themselves with the normality and homogeneity of variance assumptions.

In Excerpts 13.15 and 13.16, we see examples where the **normality assumption** and the **equal variance assumption** were tested. In the first of these excerpts, the Kolmogorov and Levene tests were utilized to check on these assumptions. In Excerpt 13.16, we see two other procedures—Lilliefors test and Bartlett's test—being used. In these and other cases where assumptions are tested, researchers usually hope that the assumption's null hypothesis will not be rejected. When this is the case, they can proceed directly from these preliminary tests to their study's main tests.

EXCERPTS 13.15–13.16 • *The Normality and Equal Variance Assumptions*

Each set of data was analyzed first with the Kolmogorov-Smirnov to test for normality and then with Levene's test for homogeneity of variances. Differences among means were then analyzed by two-way analysis of variance.

Source: Zalups, R. K., & Bridges, C. C. (2010). Seventy-five percent nephrectomy and the disposition of inorganic mercury in 2, 3-dimercaptopropanesulfonic acid-treated rats lacking functional multidrug-resistance protein 2. *Journal of Pharmacology and Experimental Therapeutics*, *332*(3), 866–875.

Normality and homogeneity of the variance were studied with the Lilliefors and Bartlett tests, respectively, at 5% probability. [Acceptable] data were submitted to two-way analysis of variance and the means were compared by the Tukey test at 5% probability.

Source: Manfroi, L., Silva, P. H. A., Rizzon, L. A., Sabaini, P. S., & Glória, M. B. A. (2009). Influence of alcoholic and malolactic starter cultures on bioactive amines in Merlot wines. *Food Chemistry*, *116*(1), 208–213.

As indicated in Chapter 11, researchers have several options when it becomes apparent that their data sets are characterized by extreme nonnormality or heterogeneity of variance. One option is to apply a **data transformation** before testing any null hypotheses involving the main effect or cell means. In Excerpt 13.17, we see a case where this was done using a log transformation after the Cochran test indicated heterogeneity of variance. In other research reports, you are likely to see different kinds of transformations used (e.g., the square root transformation, the arcsine transformation). Different kinds of transformations are available because nonnormality or heterogeneity of variance can exist in different forms. It is the researcher's job, of course, to choose an appropriate transformation that accomplishes the desired objective of bringing the data into greater agreement with the normality and equal variance assumptions.

EXCERPT 13.17 • *Using a Data Transformation*

Significant differences in number of fish detected by the riverwatcher were investigated using a two-way ANOVA using turbidity and number of fish as factors. Cochran's test identified heterogeneous variances within the data and a subsequent $\log(x + 1)$ transformation was undertaken.

Source: Baumgartner, L., Bettanin, M., McPherson, J., Jones, M., Zampatti, B., & Beyer, K. (2010). *Assessment of an infrared fish counter (Vaki Riverwatcher) to quantify fish migrations in the Murray-Darling Basin* (Fisheries Final Report Series No. 116). Industry & Investment NSW, Narrandera, Australia.

Instead of using data transformations, researchers can deal with untenable assumptions in other ways. One procedure is to have equal ns in the various cells, either by setting up the study like that or by discarding data randomly from those cells with larger ns. A two-way ANOVA with equal cells sizes is **robust,** meaning that the main effect and interaction F-tests operate as intended even if the assumptions are violated. A second strategy is to switch from the regular two-way ANOVA to a test procedure that has less rigorous assumptions. These alternative procedures are often referred to as *nonparametric tests,* and we consider some of these procedures in Chapter 18. A third procedure is to change the way the dependent variable is measured, or to change the dependent variable itself. Regardless of which procedure is used, well-trained researchers pay attention to the assumptions underlying the F-tests of a two-way ANOVA.

If a researcher conducts a two-way ANOVA but does not say anything at all about the normality and equal variance assumptions, then you have a right—even an obligation—to receive the researcher's end-of-study claims with a big grain of salt. You have a right to do this, because F-tests can be biased if the assumptions are violated. That **bias** can be positive or negative in nature, thus causing the ANOVA's F-tests to turn out either too large or too small, respectively. If the former problem exists, the computed p-value associated with a calculated F-value will be too small, thereby exaggerating how much the sample data deviate from null expectations. In this situation, the nominal alpha level understates the probability of a Type I error. If the bias is negative, the p-values associated with computed F-values will be too large. This may cause the researcher to not reject one or more null hypotheses that would have been rejected if evaluated with unbiased ps.

Estimating Effect Size and Conducting Power Analyses in Two-Way ANOVAs

As indicated in Chapter 8, various techniques have been developed to help researchers assess the extent to which their results are significant in a practical sense. It is worth repeating that such techniques serve a valuable role in quantitative studies wherein null hypotheses are tested; it is possible for a result to end up being declared statistically significant even though it is totally unimportant from a practical standpoint. Earlier, we saw how these techniques have been used in conjunction with *t*-tests and one-way ANOVAs. We now consider their relevance to two-way ANOVAs.

When researchers use a two-way ANOVA, they can estimate the practical significance of the main and interaction effects, of post hoc comparisons, or of planned contrasts. Four of the excerpts we have already considered illustrate the use of *d*, eta squared, and *f*. Take another look at Excerpts 13.9, 13.12, 13.13, and 13.14 to see how these effect size indices were used. As you will discover, *d* is relevant only when two means are being compared, whereas eta squared and *f* can be used when two or more means are involved. In Excerpt 13.18, we see a case in which partial eta squared was used in conjunction with a two-way ANOVA.

EXCERPT 13.18 • *Assessing Practical Significance*

Data were analysed using a two-way between subject ANOVA with a between subject factor of car (Fiesta and Bentley) and a between subject factor of sex (female and male). Analysis revealed no main effect of sex [$F(1, 56) = 1.3, p > .05$, partial $\eta^2 = .001$], a main effect of car status [$F(1, 56) = 276.5, p < .01$, partial $\eta^2 = .49$], and importantly no significant car $\times$ sex interaction [$F(1, 56) = 0.1, p > .05$, partial $\eta^2 = .001$].

Source: Dunn, M. J.; Searle, R. (2010). Effect of manipulated prestige-car ownership on both sex attractiveness ratings. *British Journal of Psychology, 101*(1), 69–80.

Both partial eta squared and eta squared are used frequently by applied researchers. In a one-way ANOVA, they are identical in size, so it does not matter which one is used. These two measures of effect size, however, usually turn out to be dissimilar in size in a two-way ANOVA, because **eta squared** is computed as the variability associated with an effect (such as a main effect) divided by the total variability in the ANOVA. In contrast, **partial eta squared** is computed as the variability associated with an effect divided by the sum of that variability and within-group (i.e., error) variability. There are differences of opinion as to whether eta squared or partial eta squared should be used. In a real sense, either one is better than nothing!

When estimating the effect size of an *F*-ratio or the difference between two sample means, researchers ought to do more than simply report a numerical value. To be more specific, researchers ought to *discuss* and *interpret* the magnitude of the estimated effect. Unfortunately, you will come across many research reports where this is not done. Because certain journals now require estimates of effect size to be presented, some researchers simply "jump through the hoop" to include such estimates when summarizing their studies. It is sad but true that in some cases these researchers do not say anything because they have no idea how to interpret their effect size estimates. They just toss them into their report in an effort to make things look good.

TABLE 13.2 *Effect Size Criteria for Two-Way ANOVAs*

Effect Size Measure	Small	Medium	Large
d	.20	.50	.80
Eta (η)	.10	.24	.37
Eta Squared (η^2)	.01	.06	.14
Omega Squared (ω^2)	.01	.06	.14
Partial Eta Squared (η_p^2)	.01	.06	.14
Partial Omega Squared (ω_p^2)	.01	.06	.14
Cohen's f	.10	.25	.40

Note. These standards for judging effect size are quite general and should be changed to fit the unique goals of any given research investigation.

To make sure that *you* are not in the dark when it comes to interpreting measures of effect size computed for two-way ANOVA results, I have put the common standards for small, medium, and large effect sizes in Table 13.2. The information contained in this table is not new, because it was presented earlier in Table 10.1. It is helpful to have these standards handy now, however, because they help when interpreting the five excerpts of this chapter already seen that contained statements about effect size.

To demonstrate a concern for practical significance, researchers are advised to conduct an **a priori power analysis** in conjunction with their two-way ANOVAs. When this occurs, it is done in the design phase of the study to determine the needed sample sizes for the two-way ANOVA to function as desired. In Excerpt 13.19, we see an example of this kind of power analysis. The researchers who conducted this study deserve to be commended for taking the time to determine how large their sample sizes needed to be.

EXCERPT 13.19 • A Power Analysis to Determine the Needed Sample Size

In order to more systematically determine the sample size needed for Study 2, a power analysis using G-POWER was conducted. Estimating an effect size of .75, and a .05 significance level, the analysis indicated that 108 participants were needed to achieve statistical power of .80. . . . Therefore, a minimum of 18 participants were needed in each of the six conditions [of the 2 × 3 ANOVA].

Source: Aubie, C. D., & Jarry, J. L. (2009). Weight-related teasing increases eating in binge eaters. *Journal of Social & Clinical Psychology, 28*(7), 909–936.

In your reading of the research literature, you are likely to encounter many studies in which a two-way ANOVA functions as the primary data analytic tool. Unfortunately, many of the researchers who use this tool formally address only the concept of statistical significance, with the notion of practical significance automatically (and incorrectly) superimposed on each and every result that turns out to be statistically significant. Consequently, you must remain vigilant for instances of this unjustified and dangerous misinterpretation of results.

The Inflated Type I Error Rate in Factorial ANOVAs

When data are subjected to a standard two-way ANOVA, three F-values are computed—one for each main effect and one for the interaction. If the same level of significance (e.g., .05) is used in assessing each F-value, you may have been thinking that the probability of a Type I error occurring somewhere among the three F-tests is greater than the alpha level used to evaluate each F-value. Accordingly, you may have been expecting me to point out how conscientious researchers make some form of adjustment to avoid having an inflated Type I error rate associated with their two-way ANOVAs.

Although it is clear that the computation of three F-values in a two-way ANOVA leads to a greater-than-alpha chance that one or more of the three null hypotheses will be incorrectly rejected, the vast majority of applied researchers do not adjust anything in an effort to deal with this problem. This is because most applied researchers consider each F-test separately rather than look at the three F-tests collectively as a set. When the F-tests are viewed in that manner, the Type I error risk is *not* inflated, because the researcher's alpha level correctly specifies the probability that any given F-test causes a true H_0 to be rejected.

When a given level of significance is used to evaluate each of the three F-values, it can be said that the **familywise error rate** is set equal to the alpha level. Each *family* is defined as the set of contrasts represented by an F-test and any post hoc tests that might be conducted if the F turns out to be statistically significant. The familywise error rate is equal to the common alpha level employed to evaluate all three F-tests because the chances of a Type I error, *within each family,* are equal to the alpha level.

If a researcher analyzes the data from two or more dependent variables with separate two-way ANOVAs, you may find that the Bonferroni procedure is used to adjust the alpha level. In Excerpt 13.20, we see where this was done in a study dealing with two adhesives used by dentists. Because five two-way ANOVAs were conducted

EXCERPT 13.20 • *The Bonferroni Adjustment Used with Multiple Two-Way ANOVAs*

Two different types of self-etching adhesive bonding agents were evaluated. . . . The shear bond strength data were analyzed with a two-way ANOVA per adhesive to evaluate the effect of storage temperature (2 levels) or storage time (3 levels) on shear bond strength. Additionally, a two-way ANOVA per storage time was used to analyze the effect of storage temperature (2 levels) or adhesive type (2 levels) on

(continued)

EXCERPTS 13.20 • (*continued*)

shear bond strength. Tukey's post hoc test was used to compare pairwise differences between mean values. . . . A Bonferroni correction with an α level of 0.01 was applied as a multiple-comparison correction because [five two-way ANOVAs] were performed simultaneously.

Source: Graham, J. B., & Vandewalle, K. S. (2010). Effect of long-term storage temperatures on the bond strength of self-etch adhesives. *Military Medicine*, *175*(1), 68–71.

in this study, the researchers lowered their alpha level from .05 to .01 via the standard Bonferroni procedure. This adjusted level of significance was then used in evaluating the two main effect and interaction *F*-ratios generated by each of the five ANOVAs.

A Few Warnings Concerning Two-Way ANOVAs

Before concluding this chapter, I want to offer a few cautionary comments that I hope you will tuck away in your memory bank and then bring up to consciousness whenever you encounter a research report based on a two-way ANOVA. Although I have touched on some of these issues in previous chapters, your ability to decipher *and* critique research summaries may well improve if I deliberately reiterate a few of those earlier concerns.

Evaluate the Worth of the Hypotheses Being Tested

I cannot overemphasize how important it is to critically assess the worth of the hypotheses being tested within any study based on a two-way ANOVA. No matter how good the study may be from a statistical perspective and no matter how clear the research report is, the study cannot possibly make a contribution unless the questions being dealt with are interesting. In other words, the research questions that prompt the investigator to select the factors and levels of the two-way ANOVA must be worth answering and must have no clear answer before the study is conducted. If these things do not hold true, then the study has a fatal flaw in its foundation that cannot be overcome by appropriate sample sizes, rigorous alpha levels, high reliability and validity estimates, tests of assumptions, Bonferroni corrections, *F*-ratios that are statistically significant, and elaborate post hoc analyses. The old saying that "You can't make a silk purse out of a sow's ear" is as relevant here as anywhere else.

Remember That Two-Way ANOVAs Focus on Means

As with most *t*-tests and all one-way ANOVAs, the focus of a two-way ANOVA is on means. The main effect means and the cell means serve as the focal points of the three research questions associated with any two-way ANOVA. When the main effect and interaction *F*-tests are discussed, it is essential for you to keep in mind that conclusions should be tied to means.

I recently read a study (utilizing a two-way ANOVA) that evaluated the impact of an outdoor rock-climbing program on at-risk adolescents. There were two main dependent variables: alienation and personal control. The researchers asserted, in the abstract of the research report, that "after experiencing the climbing program, the experimental group was less alienated than its control counterparts" and "demonstrated a stronger sense of personal control than did the control group." Many people reading those statements would think that *everyone* in the experimental group scored lower on alienation and higher on personal power than *anyone* in the control group. However, the group means and standard deviations included in the research report—on both measures (alienation and personal power)—make it absolutely clear that some of the members of the control group had better scores than did some of the members in the experimental group.

Many researchers fail to note that their statistically significant findings deal with means, and the literal interpretation of the researchers' words says that all of the folks in one group outperformed those in the comparison group(s). If the phrase *on average* or some similar wording does not appear in the research report, make certain that you insert it as you attempt to decipher and understand the statistical findings. If you do not, you will end up thinking that comparison groups were far more different from one another than was actually the case.

Remember the Possibility of Type I and Type II Errors

The third warning that I offer is not new. You have encountered it earlier in our consideration of tests on correlations, *t*-tests, and one-way ANOVAs. Simply stated, I encourage you to remember that regardless of how the results of a two-way ANOVA turn out, there is always the possibility of either a Type I or Type II error whenever a decision is made to reject or fail to reject a null hypothesis.

Based on the words used by many researchers in discussing their results, it seems to me that the notion of *statistical significance* is quite often amplified (incorrectly) into something on the order of a firm discovery—or even proof. Far too seldom do I see the word *inference* or the phrase *null hypothesis* in the technical write-ups of research investigations wherein the hypothesis testing procedure has been used. Although you do not have the ability to control what researchers say when they summarize their investigations, you most certainly *do* have the freedom to adjust, in your mind, the research summary to make it more accurate.

Sooner or later, you are bound to encounter a research report wherein a statement is made on the order of (1) "Treatment A works better than Treatment B" or (2) "Folks who possess characteristic X outperform those who possess characteristic Y." Such statements come from researchers who temporarily forgot not only the difference between sample statistics and population parameters but also the ever-present possibility of an inferential error when a finding is declared either significant or nonsignificant. You can avoid making the mistake of accepting such statements as points of fact by remembering that no *F*-test *ever* proves anything.

Be Careful When Interpreting Nonsignificant F-tests

In Chapter 7, I pointed out that it is wrong to consider a null hypothesis to be true simply because the hypothesis testing procedure results in a fail-to-reject decision. Any of several factors (e.g., small sample sizes, unreliable measuring instruments, too much within-group variability) can cause the result to be nonsignificant, even if the null hypothesis being tested is actually false. This is especially true when the null hypothesis is false by a small amount.

Almost all researchers who engage in hypothesis testing have been taught that it is improper to conclude that a null hypothesis is true simply because the hypothesis testing procedure leads to a fail-to-reject decision. Nevertheless, many of these same researchers use language in their research reports suggesting that they have completely forgotten that a fail-to-reject decision does not logically permit one to leave a study believing that the tested H_0 is true. In your review of studies that utilize two-way ANOVAs (or, for that matter, any procedure for testing null hypotheses), remain vigilant for erroneous statements as to what a nonsignificant finding means.

Review Terms

A priori power analysis	Interaction
Active factor	Level
Assigned factor	Main effect *F*
Biased *F*-test	Main effect mean
Cell	Normality assumption
Cohen's *f*	Omega squared
Data transformation	Partial eta squared
Equal variance assumption	Post hoc tests
Eta squared	Power analysis
Factor	Robust
Familywise error rate	Simple main effect
Graph of an interaction	Univariate analysis
Graph of cell means	

The Best Items in the Companion Website

1. An interactive online quiz (with immediate feedback provided) covering Chapter 13.
2. Nine misconceptions about the content of Chapter 13.

3. An email message sent from the author to his students entitled "Can One Cell Create an Interaction?"
4. An interactive online resource entitled "Two-Way ANOVA (a)."
5. One of the best passages from Chapter 13: "You Can't Make a Silk Purse Out of a Sow's Ear."

To access chapter outlines, practice tests, weblinks, and flashcards, visit the companion website at http://www.ReadingStats.com.

Review Questions and Answers begin on page 531.

Analyses of Variance with Repeated Measures

In this chapter, we consider three different ANOVAs that are characterized by repeated measures. In particular, the focus here is on one-way ANOVAs with repeated measures, two-way ANOVAs with repeated measures on both factors, and two-way ANOVAs with repeated measures on just one factor. Although there are other kinds of ANOVAs that involve repeated measures (e.g., a four-way ANOVA with repeated measures on all or some of the factors), the three types considered here are the ones used most often by applied researchers.

The one-way and two-way ANOVAs examined in this chapter are similar in many respects to the ANOVAs considered in Chapters 10, 11, and 13. The primary difference between the ANOVAs of this chapter and those looked at in earlier chapters is that the ANOVAs to which we now turn our attention involve repeated measures on at least one factor. This means that the research participants are measured once under each level (or combination of levels) of the factor(s) involved in the ANOVA.

Perhaps an example will help to distinguish between the ANOVAs considered in earlier chapters and their repeated measures counterparts examined in this chapter. If a researcher has a 2×3 design characterized by no repeated measures, each participant in the study can be thought of (1) as being located inside *one* of the six cells of the factorial design and (2) as contributing *one* score to the data set. In contrast, if a researcher has a 2×3 design characterized by repeated measures across both factors, each participant can be thought of (1) as being in *each of the six* cells and (2) as contributing *six* scores to the data set.

Before we turn our attention to the specific ANOVAs of this chapter, three introductory points are worth noting. First, each of the ANOVAs to be considered here is univariate in nature. Even though participants are measured more than once within the same ANOVA, these statistical procedures are univariate in nature—not multivariate—because each participant provides only one score to the data set for each level or combination of levels of the factor(s) involved in the study. The ANOVAs of this chapter could be turned into multivariate ANOVAs if each participant were measured repeatedly within each cell of the design, with these within-cell repeated measurements corresponding to different dependent variables. Such multivariate repeated measures ANOVAs, however, are not considered in this chapter.

Second, it is important to understand the distinction between (1) two or more separate ANOVAs, each conducted on the data corresponding to a different dependent variable, with all data coming from the same participants; and (2) a single unified ANOVA in which there are repeated measures across levels of the factor(s) of the study. In Chapters 10, 11, and 13, you have seen many examples of multiple but separate ANOVAs being applied to different sets of data, each corresponding to a unique dependent variable. The ANOVAs to which we now turn our attention are different from those referred to in the preceding sentence in that the ones considered here always involve a single, consolidated analysis.

My final introductory point concerns different kinds of repeated measures factors. To be more specific, I want to distinguish between some of the different circumstances in a study that can create a **within-subjects factor.**[1] You will likely encounter three such circumstances as you read technical research reports.

One obvious way for a factor to involve repeated measures is for participants to be measured at different points in time. For example, a researcher might measure people before and after an intervention, with the factor being called *time* and its levels being called *pre* and *post.* Or, in a study focused on the acquisition of a new skill, the factor might be called *trials* (or *trial blocks*), with levels simply numbered 1, 2, 3, and so on. A second way for a factor to involve repeated measures is for participants to be measured once under each of two or more different treatment conditions. In such studies, the factor might well be called *treatments* or *conditions,* with the factor's levels labeled to correspond to the specific treatments involved in the study. A third kind of repeated measures factor shows up when the study's participants are asked to rate different things or are measured on different characteristics. Here, the factor and level names would be chosen to correspond to the different kinds of data gathered in the study.

In Excerpts 14.1 through 14.3, we see how different kinds of situations can lead to data being collected from each participant across levels of the repeated measures factor. Although the factors referred to in these excerpts are all within-subjects factors, they involve repeated measures for different reasons. In Excerpt 14.1, the data were collected at different *times;* in Excerpt 14.2, the data were collected under different *treatment conditions;* and in Excerpt 14.3, the data were collected on different *variables.*

---

[1]The terms *repeated-measures factor, within-subjects factor,* and *within-participants factor* are synonymous.

EXCERPTS 14.1–14.3 • *Different Kinds of Repeated Measures Factors*

[T]o assess senior high school students' mood changes during their preparation for a very important academic examination, the Brazilian vestibular . . ., 231 students were asked to answer the PANAS-X (Positive and Negative Affect Schedule–Expanded Form) three times: in March (start of the academic year) . . ., in August . . ., and in late October (15 days before the vestibular).

Source: Peluso, M. A. M., Savalli, C., Cúri, M., Gorenstein, C., & Andrade, L. H. (2010). Mood changes in the course of preparation for the Brazilian university admission exam—A longitudinal study. *Revista Brasileira de Psiquiatria, 32*(1), 30–36.

The aim of this study was to determine the differential effects of three commonly used [bicycle] crank lengths (170, 172.5 and 175 mm) on performance measures relevant to female cross-country mountain bike athletes ($n = 7$) of similar stature. All trials were performed in a single blind and balanced order with a 5- to 7-day period between trials.

Source: Macdermid, P. W., & Edwards, A. M. (2010). Influence of crank length on cycle ergometry performance of well-trained female cross-country mountain bike athletes. *European Journal of Applied Physiology, 108*(1), 177–182.

The DAP [questionnaire] yields quantitative scores on internal, external, and social context areas of developmental assets for youth aged 11 to 18. . . . [This study] was conducted to evaluate differences in means among the internal, external, and social context areas.

Source: Chew, W., Osseck, J., Raygor, D., Eldridge-Houser, J., & Cox, C. (2010). Developmental assets: Profile of youth in a juvenile justice facility. *Journal of School Health, 80*(2), 66–72.

One-Way Repeated Measures ANOVAs

When researchers use a one-way repeated measures ANOVA, they usually tip you off that their ANOVA is different from the kind we considered in Chapters 10 and 11 (where no repeated measures are involved). They do this by including the phrase *repeated measures* or *within-subjects* or *within-participants* as a descriptor of their ANOVA or of their ANOVA's single factor. Examples of this practice appear in Excerpts 14.4 and 14.5.

EXCERPTS 14.4–14.5 • *Different Labels for a One-Way Repeated Measures ANOVA*

To examine if participants predicted differences between racial groups on AGIB–Math scores, we conducted a one-way within-subjects ANOVA. In this analysis, the

(continued)

EXCERPTS 14.4–14.5 • (*continued*)

within-subjects factor was racial group (Asian Americans, African Americans, Caucasians/Whites, and Hispanics/Latinos) and the dependent variable was predicted AGIB–Math score

Source: Unzueta, M. M., & Lowery, B. S. (2010). The impact of race-based performance differences on perceptions of test legitimacy. *Journal of Applied Social Psychology, 40*(8), 1948–1968.

To evaluate differences between target toy touch in the neutral attention and negative emotion trials, a repeated measures ANOVA with trial (neutral attention, negative emotion) as the within-subjects factor was conducted separately for children at 12, 18, and 24 months.

Source: Nichols, S. R., Svetlova, M., & Brownell, C. A. (2010). Toddlers' understanding of peers' emotions. *Journal of Genetic Psychology, 171*(1), 35–53.

Purpose

The purpose of a one-way repeated measures ANOVA is identical to the purpose of a one-way ANOVA not having repeated measures. In each case, the researcher is interested in seeing whether (or the degree to which) the sample data cast doubt on the null hypothesis of the ANOVA. That null hypothesis, for the within-subjects case as well as the between-subjects case, states that the μs associated with the different levels of the factor do not differ. Because the researcher who uses a one-way within-subjects ANOVA is probably interested in gaining an insight into how the μs differ, post hoc tests are normally used (as in a between-subjects ANOVA) if the overall null hypothesis is rejected and if three or more levels compose the ANOVA's factor.

To illustrate, suppose a researcher collects reaction-time data from six people on three occasions: immediately upon awakening in the morning, one hour after awakening, and two hours after awakening. Each of the six people is measured three times, with a total of 18 pieces of data available for analysis. In subjecting the data to a one-way repeated measures ANOVA, the researcher is asking whether the three sample means, each based on six scores collected at the same time during the day, are far enough apart to call into question the null hypothesis that says all three population means are equivalent. In other words, the purpose of the one-way repeated measures ANOVA in this study is to see if the average reaction time of folks similar to the six people used in the study varies depending on whether they are tested 0, 60, or 120 minutes after awakening.

It is usually helpful to think of any one-way repeated measures ANOVA in terms of a two-dimensional matrix. Within this matrix, each row corresponds to a different person and each column corresponds to a different level of the study's factor.

Hours Since Awakening

	Zero	One	Two
Person 1	1.6	1.0	1.1
Person 2	2.0	1.5	1.8
Person 3	1.7	0.6	1.5
Person 4	2.9	0.9	1.3
Person 5	1.8	1.5	1.9
Person 6	2.0	0.5	1.4
	$M = 2.0$	$M = 1.0$	$M = 1.5$

FIGURE 14.1 *Data Setup for the One-Way Repeated Measures ANOVA in the Hypothetical Reaction-Time Study*

A single score is entered into each cell of this matrix, with the scores on any row coming from the same person. Such a matrix, created for our hypothetical reaction-time study, is presented in Figure 14.1. Such illustrations usually do not appear in research reports. Therefore, you must create such a picture (in your mind or on a piece of scratch paper) when trying to decipher the results of a one-way repeated measures ANOVA. This is usually easy to do because you are given information as to the number of people involved in the study, the nature of the repeated measures factor, and the sample means that correspond to the levels of the repeated measures factor.

Presentation of Results

The results of a one-way repeated measures ANOVA are occasionally presented in an ANOVA summary table. In Table 14.1, I have prepared such a table for our hypothetical study on reaction time. This summary table is similar in some ways to the one-way ANOVA summary tables contained in Chapter 11, yet it is similar, in other respects, to the two-way ANOVA summary tables included in Chapter 13. Table 14.1 is like a one-way ANOVA summary table in that a single *F*-ratio is contained in the right column of the table. (Note that this *F*-ratio is computed by dividing the *MS* for the study's factor by the *MS* for residual.) It is like a two-way ANOVA summary table in that (1) the row for people functions, in some respects, as a second factor of the study, and (2) the numerical values on the row for residual are computed in the same way as if this were the interaction from a two-way ANOVA. (Note that the *df* for residual is computed by multiplying together the first two *df* values.) In fact, we could have used the term *Hours* $\times$ *People* to label this row instead of the term *Residual*.

TABLE 14.1   *ANOVA Summary Table for the Reaction-Time Data Contained in Figure 14.1*

Source	SS	df	MS	F
Hours since awakening	3.00	2	1.50	10.0*
People	0.99	5		
Residual	1.49	10	0.15	
Total	5.48	17		

*$p < .001$.

Regardless of whether Table 14.1 resembles more closely the summary table for a one-way ANOVA or a two-way ANOVA, it contains useful information for anyone trying to understand the structure and the results of the investigation. First, the title of the table indicates what kind of data were collected. Second, we can tell from the Source column that the study's factor was Hours since awakening. Third, the top two numbers in the *df* column inform us that the study involved three levels $(2 + 1 = 3)$ of the factor and six people $(5 + 1 = 6)$. Fourth, the bottom number in the *df* column indicates that a total of 18 pieces of data were analyzed $(17 + 1 = 18)$. Finally, the note beneath the table reveals that the study's null hypothesis was rejected, with .01 being one of three things: (1) the original level of significance used by the researcher, (2) the revised alpha level that resulted from a Bonferroni adjustment, or (3) the most rigorous of the three standard alpha levels (i.e., .05, .01, .001) that could be beaten by the data.

In our hypothetical study on reaction time, Table 14.1 indicates that the null hypothesis of equal population means is not likely to be true. To gain an insight into the pattern of the population means, a post hoc investigation would probably be conducted. Most researchers would set up this follow-up investigation such that three pairwise contrasts are tested, each involving a different pair of means (M_0 versus M_1, M_0 versus M_2, and M_1 versus M_2).

In Excerpt 14.6, we see the results of a real study that used a one-way repeated-measures ANOVA. The data summarized by this table came from 20 individuals who each walked four different routes using different navigation devices (referred to as *modes*) to help them reach their destinations. The study's dependent variable was a "navigation score" based on each walker's accuracy and confidence during each of the four tasks. Although the ANOVA summary table contains two *F*-values, only the one on the row labeled Mode was of interest to the researchers. Note that this table contains no SS or MS values; also, the next-to-the-bottom row is called Error rather than Residual. Despite these differences from the summary table I created for the hypothetical reaction-time study, you should be able to look at the table in Excerpt 14.6 and see that the null hypothesis concerning the four navigation devices was rejected.

EXCERPT 14.6 • *Results of a One-Way Repeated Measures ANOVA Presented in a Summary Table*

TABLE II   *One-way, repeated measures analysis of variance summary table*

Source	df	F	η^2	p
Mode	3	4.78*	0.11	0.0049
Subject	19	3.25*	0.46	0.0003
Error	57			
Total	79			

*p < 0.05.

Source: Sohlberg, M. M., Fickas, S., Hung, P.-F., & Fortier, A. (2007). A comparison of four prompt modes for route finding for community travellers with severe cognitive impairments. *Brain Injury, 21*(5), 531–538.

Although it is helpful to be able to look at the ANOVA summary table, researchers often are unable to include such tables in their reports because of space considerations. In Excerpt 14.7, we see how the results of a one-way repeated measures ANOVA were presented wholly within the text of the report. Note how a post hoc investigation was conducted by the researchers because (1) the omnibus *F*-test yielded a statistically significant result and (2) more than two means were being compared. Note also the inclusion of an estimate of the effect size.

EXCERPT 14.7 • *Results of a One-Way Repeated Measures ANOVA Presented Without a Summary Table*

Based on the results of the factor analysis, the 15 homework purpose items were reduced to three scales [learning-oriented reasons, peer-related reasons, adult-oriented reasons] for use in subsequent analyses. . . . A one-way, within-subjects ANOVA revealed a statistically significant difference among these three scales, with a large effect size, $F(2, 1798) = 624.76, p < .001, \eta^2 = .410$. An adjusted Bonferroni post hoc comparison detected specific differences among these factors. The mean score for learning-oriented reasons ($M = 2.82, SD = 0.61$) was statistically significantly higher than adult-oriented reasons ($M = 2.70, SD = 0.73$), which was, in turn, statistically significantly higher than was peer-oriented reasons ($M = 2.31, SD = 0.73$).

Source: Xu, J. (2010). Homework purposes reported by secondary school students: A multilevel analysis. *Journal of Educational Research, 103*(3), 171–182.

Sometimes researchers apply a one-way repeated measures ANOVA more than once within the same study. They do this for one of two reasons. On one hand, each participant in the study may have provided two or more pieces of data at each level of the repeated measures factor, with each of these scores corresponding to a different dependent variable. Given this situation, the researcher may utilize a separate one-way repeated measures ANOVA to analyze the data corresponding to each dependent variable. On the other hand, the researcher may have two or more groups of participants, with just one score collected from each of them at each level of the within-subjects factor. Here, the researcher may decide to subject the data provided by each group to a separate one-way repeated measures ANOVA.

The Presentation Order of Levels of the Within-Subjects Factor

As indicated earlier, the factor in a one-way repeated measures ANOVA can take one of three basic forms. In some studies, the within-subjects factor corresponds to time, with the levels of the factor indicating the different points in time at which data are collected. The second kind of within-subjects factor corresponds to different treatments or conditions given to or created for the participants, with a measurement taken on each person under each treatment or condition. The third kind of within-subjects factor is found in studies where each participant is asked to rate different things, take different tests, or in some other way provide scores on different variables.

If the one-way repeated measures ANOVA involves data collected at different points in time, there is only one order in which the data can be collected. If, however, the within-subjects factor corresponds to treatment conditions or different variables, then there are different ways in which the data can be collected. When an option exists regarding the order in which the levels of the factor are presented, the researcher's decision regarding this aspect of the study should be taken into consideration when *you* make a decision as to whether to accept the researcher's findings.

If the various treatment conditions, things to be rated, or tests to be taken are presented in the same order, then a systematic bias may creep into the study and function to make it extremely difficult—or impossible—to draw clear conclusions from the statistical results. The systematic bias might take the form of a **practice effect,** with participants performing better as they warm up or learn from what they have already done; a **fatigue effect,** with participants performing less well on subsequent tasks simply because they get bored or tired; or **confounding,** with things that the participants do or learn outside the study's setting between the points at which the study's data are collected. Whatever its form, such bias can cause different treatment conditions (or the different versions of a measuring device) to look different when they are really alike or to look alike when they are really dissimilar.

To prevent the study from being wrecked by practice effects, fatigue effects, and confounding due to order effects, a researcher should alter the order in which the treatment conditions, tasks, questionnaires, rating forms, or whatever are presented to participants. This can be done in one of three ways. One design strategy is to randomize the order in which the levels of the within-subjects factor are presented. A second strategy is to utilize all possible presentation orders, with an equal proportion of the participants assigned randomly to each possible order. The third strategy involves counterbalancing the order in which the levels of the repeated measures factor are presented; here, the researchers make sure that each level of the factor appears equally often in any of the ordered positions.

In Excerpt 14.8, we see an example in which a set of researchers **counterbalanced** the order in which this study's participants played a table soccer game using three different control devices. The **Latin square** referred to in this excerpt is simply an ordering arrangement that made sure that each of the three treatment conditions occurs equally often in the first, second, and final positions. For example, if we let the letters A, B, and C represent the three control devices used in this study, 10 of the study's 30 participants would receive them in the order A-B-C. A different third of the participants would get the treatments in the order B-C-A. The order for the final third of the participants would be C-A-B.

EXCERPT 14.8 • *Counterbalancing the Order of Treatments*

A total of 30 undergraduate students participated in this evaluation. . . . The experiment followed a within-subject design. . . . For each control method [head tracking control, eye tracking control and traditional joystick control], participants had 5 minutes to play the table soccer game. . . . The order effect was counterbalanced by using a Latin square.

Source: Zhu, D., Gedeon, T., & Taylor, K. (2010). Natural interaction enhanced remote camera control for teleoperation. *Proceedings of the 28th Conference on Human Factors in Computing Systems, ACM,* 3229–3234.

Carryover Effects

In studies where the repeated-measures factor is related to different kinds of treatments, the influence of one treatment might interfere with the assessment of how the next treatment works. If so, such a **carryover effect** interferes with the comparative evaluation of the various treatments. Even if the order of the treatments is varied, the disruptive influence of carryover effects can make certain treatments appear to be more or less potent than they really are.

One way researchers can reduce or eliminate the problem of carryover effects is to delay presenting the second treatment until after the first treatment has "run its course." In studies comparing the effectiveness of different pharmaceutical drugs, there usually is a 48-hour time period between the days on which the different drugs are administered. These so-called "washout intervals" are designed to allow the effects of each drug to dissipate totally before another drug is introduced.

The Sphericity Assumption

In order for a one-way repeated measures ANOVA to yield an *F*-test that is valid, an important assumption must hold true. This is called the **sphericity assumption,** and it should be considered by *every* researcher who uses this form of ANOVA. Even though the same amount of data is collected for each level of the repeated measures factor, the *F*-test of a one-way repeated measures ANOVA is *not* robust to violations of the sphericity assumption. If this assumption is violated, the *F*-value from this ANOVA is positively biased; this means the calculated value will be larger than it should be, thus increasing the probability of a Type I error above the nominal alpha level.

The sphericity assumption says that the population variances associated with the levels of the repeated measures factor, in combination with the population correlations between pairs of levels, must represent one of a set of acceptable patterns. One of the acceptable patterns is for all the population variances to be identical and for all the bivariate correlations to be identical. There are, however, other patterns of variances and correlations that adhere to the requirements of sphericity.

The sample data collected in any one-factor repeated measures investigation can be used to test the sphericity assumption. This test was developed by J. W. Mauchly, and researchers now refer to it as the **Mauchly sphericity test.** If the application of Mauchly's test yields a statistically significant result (thus suggesting that the condition of sphericity does not exist), there are various things the researcher can do in an effort to help avoid making a Type I error when the one-way repeated measures ANOVA is used to test the null hypothesis of equal population means across the levels of the repeated measures factor. The two most popular strategies both involve using a smaller pair of *df* values to determine the critical *F*-value used to evaluate the calculated *F*-value. This adjustment results in a larger critical value and a greater likelihood that a fail-to-reject decision will be made when the null hypothesis of the one-way repeated measures ANOVA is evaluated.

One of the two ways to adjust the *df* values is to use a simple procedure developed by two statisticians, S. Geisser and S. W. Greenhouse. Their procedure involves basing the critical *F*-value on the *df* values that would have been appropriate if there had been just two levels of the repeated measures factor. This creates a drastic reduction in the critical value's *df*s, because it presumes that the sphericity assumption is violated to the maximum extent. Thus the **Geisser–Greenhouse** approach to dealing with significant departures from sphericity creates a **conservative *F*-test** (because the true Type I error rate is smaller than that suggested by the level of significance).

The second procedure for adjusting the degrees of freedom involves first using the sample data to estimate how extensively the sphericity assumption is violated. This step leads to ϵ, a fraction that turns out to be smaller than 1.0 to the extent that the sample data suggest that the sphericity assumption is violated. Then, the "regular" *df* values associated with the *F*-test are multiplied by ϵ, thus producing adjusted *df* values and a critical value that are tailor-made for the study being conducted. When researchers use this second procedure, they often report that they have used the **Huynh–Feldt correction.**

In Excerpt 14.9, we see a case in which a team of researchers took corrective action after the sphericity assumption was tested and found not to be tenable. These researchers applied the Huynh–Feldt correction. This caused the F-test's degrees of freedom (which were originally 2 and 126) to get smaller, thereby making the critical value larger. This change in the critical value eliminated the positive bias in the F-test that would have existed (due to a lack of sphericity) if the regular critical value had been used.

EXCERPT 14.9 • *Dealing with the Sphericity Assumption*

A [one-way] repeated measures ANOVA was conducted on composite self-reported scores for each of the three disposition categories. In an examination of the composite mean scores for disposition 1: professional commitment, Mauchly's test for sphericity was found to be significant $\chi^2(2) = 15.125, p < .05$]. As such, the degrees of freedom were corrected using the Huynh-Feldt correction $\epsilon = .841$. The results show significant changes in candidates' self-perceptions of professional commitment in each phase of the program $F(1.682, 105.988) = 40.611 p < .01$.

Source: Rinaldo, V. J., Denig, S. J., Sheeran, T. J., Cramer-Benjamin, R., Vermette, P. J., Foote, C. J., et al. (2009). Developing the intangible qualities of good teaching: A self-study. *Education, 130*(1), 42–52.

Regardless of which strategy is used to deal with the sphericity assumption, I want to reiterate my earlier statement that this is an important assumption for the ANOVAs being discussed in this chapter. If a researcher conducts a repeated measures ANOVA and does not say anything at all about the sphericity assumption, the conclusions drawn from that investigation probably ought to be considered with a *big* grain of salt. If the data analysis produces a statistically significant finding when no test of sphericity is conducted or no adjustment is made to the critical value's *df,* you have the full right to disregard the inferential claims made by the researcher.

Two-Way Repeated Measures ANOVAs

We now turn our attention to ANOVAs that contain two repeated measures factors. As you will see, there are many similarities between this kind of ANOVA and the kind we examined in Chapter 13. However, there are important differences between two-way ANOVAs that do or do not involve repeated measures. For this reason, you must be able to distinguish between these two kinds of analysis.

If researchers state that they have used a two-way ANOVA but say nothing about repeated measures, then you should presume that it is the kind of ANOVA we considered in Chapter 13. However, if researchers use the phrase *repeated measures, within-subjects,* or *within-participants* when describing each of their ANOVA's two factors, then you must remember the things you learn in this section of the book. Excerpts 14.10 and 14.11 illustrate how researchers usually provide a tip-off that they used a two-way ANOVA with repeated measures.

EXCERPTS 14.10–14.11 • *Different Labels for a Two-Way Repeated Measures ANOVA*

On the basis of the preliminary analysis, the mean RT data were collapsed and analyzed in a two-way [2 × 12] within-subjects ANOVA, with character facing (2) and orientation (12) as factors.

Source: Kung, E., & Hamm, J. P. (2010). A model of rotated mirror/normal letter discriminations. *Memory & Cognition, 38*(2), 206–220.

A repeated measures ANOVA with trial (neutral attention, negative emotion) and toy (target, distracter) as within-subjects factors was conducted separately for children at 12, 18, and 24 months.

Source: Nichols, S. R., Svetlova, M., & Brownell, C. A. (2010). Toddlers' understanding of peers' emotions. *Journal of Genetic Psychology, 171*(1), 35–53.

In Excerpt 14.10, it is clear that both of the ANOVA's factors were of the *within* (i.e., repeated measures) variety. The ANOVA referred to in Excerpt 14.11 also had repeated measures on both factors. Sometimes, however, things are not so clear. I say that because some researchers report that they used a two-way within-subjects ANOVA when just one of the two factors involved repeated measures. We consider such ANOVAs later in the chapter. Our focus now is on the type of ANOVA having two within factors.

Purpose

The purpose of a two-way repeated measures ANOVA is identical to the purpose of a two-way ANOVA not having repeated measures. In each case, the researcher uses inferential statistics to help assess three null hypotheses. The first of these null hypotheses deals with the main effect of one of the two factors. The second null hypothesis deals with the main effect of the second factor. The third null hypothesis deals with the interaction between the two factors.

Although two-way ANOVAs with and without repeated measures are identical in the number and nature of null hypotheses that are evaluated, they differ in two main respects. In terms of the way data are collected, the kind of two-way ANOVA considered in Chapter 13 requires that each participant be positioned in a single cell, with only one score per person going into the data analysis. In contrast, a two-way repeated measures ANOVA requires that each participant travel across all cells created by the two factors, with each person being measured once within *each* cell. For example, in a recent study, eight well-trained male athletes rode a bicycle on two occasions, each time for 100 miles. During one ride, the outside temperature was cold (0°C); during the other ride, it was much warmer (19°C). One of the study's dependent variables was heart rate, with measures taken just before each ride, immediately after each ride, and then again 24 hours after each ride. This study's two factors were temperature and time of measurement, and you should be able to imagine how each of the athletes was measured across the study's six cells.

The second main difference between two-way ANOVAs with and without repeated measures involves the ANOVA summary table. We return to this second difference in the next section when we consider how researchers present the results of their two-way repeated measures ANOVAs. Now, we must concentrate on the three null hypotheses dealt with by this kind of ANOVA and the way the necessary data must be collected.

To help you gain insight into the three null hypotheses of any two-way repeated measures ANOVA, let's consider an imaginary study. This study involves the game "Simon," which is a battery-operated device with four colored buttons. After the player flips the start switch, a sequence of Simon's buttons lights up, with each light accompanied by a unique tone. The task of the person playing this game involves (1) watching and listening to what Simon does and then, after Simon stops, (2) pushing the same sequence of buttons that Simon just illuminated.

Suppose now that you are the player. If the red button on Simon lights up, you must press the red button. Then, if the red button lights up first followed by the green button, you must press these same two buttons in this same order to stay in the game. Every time you successfully mimic what Simon does, you receive a new string of lighted buttons that is like the previous one, except that the new sequence is one light longer. At first, it is easy for you to duplicate Simon's varied but short sequences, but after the sequences lengthen, it becomes increasingly difficult to mimic Simon.

For my study, imagine that I have each of six people play Simon. The dependent variable is the length of the longest sequence that the player successfully duplicates. (For example, if the player works up to the point where he or she correctly mimics an eight-light sequence but fails on the ninth light, that person's score is 8.) After three practice rounds, I then ask each person to play Simon four times, each under a different condition defined by my study's two factors: tones and words. The two levels of the tones factor are *on* and *off*, meaning that the Simon device is set up either to provide the auditory cues or to be silent while the player plays. The two levels of the word factor are *color names* or *people names*. With color names, the player is instructed to say out loud the color of the lights as Simon presents them in a sequence (i.e., red, blue, green, and yellow). With people names, the player is instructed to say out loud one of these names for each color: Ron for red, Barb for blue, Greg for green, and Yoko for yellow. Finally, imagine that I randomly arrange the order in which the four conditions of my study are ordered for each of the six Simon players.

	Color Names		People Names	
	Tones on	Tones off	Tones on	Tones off
Player 1	6	8	6	3
Player 2	8	3	2	4
Player 3	7	5	6	6
Player 4	9	6	3	5
Player 5	8	6	6	3
Player 6	10	8	7	3
$M =$	8	6	5	4

Names (for Simon's Lights)

		Colors	People	
Tones	On	8	5	6.5
	Off	6	4	5.0
		7.0	4.5	

FIGURE 14.2   *How the Data from the Simon Study Would Be Arranged*

Figure 14.2 contains the scores from my hypothetical study, with the order of each person's four scores arranged so as to fit accurately under the column headings. This figure also contains a 2 × 2 matrix of the four cell means, with the main effect means positioned on the right and bottom sides of the cell means.

As indicated earlier, there are three null hypotheses associated with any two-way repeated measures ANOVA. In our hypothetical study, the null hypothesis for the main effect of tones states that there is no difference, in the populations associated with our samples, between the mean performance on the Simon game when players hear the tone cues as compared to when they do not hear the tone cues. In a similar fashion, the null hypothesis for the main effect of words states that there is no difference, in the populations associated with our samples, between the mean performance on the Simon game when players say color words when trying to memorize each sequence as compared to when they say people's names. Finally, the interaction null hypothesis states the positive (or negative) impact on players of having the tone cues is the same regardless of whether they must say color names or people names as Simon's buttons light up.

The lower portion of Figure 14.2 shows the four cell means and the two main effect means for each of the two factors. The null hypothesis for tones will be rejected if the means of 6.5 and 5.0 are found to be further apart from each other than we would expect by chance. Likewise, the null hypothesis for words will be rejected if the means of 7.0 and 4.5 are found to be further apart from each other than would be expected by chance. The interaction null hypothesis is rejected if the difference between the cell means on the top row $(8 - 5 = 3)$ varies more from the difference between the cell means on the bottom row $(6 - 4 = 2)$ than would be expected by chance.

Illustrations such as that presented in the upper portion of Figure 14.2 rarely appear in research reports. However, it is usually quite easy to construct pictures of cell means and main effect means. This picture-constructing task is easy because you are almost always given information as to (1) the factors and levels involved in the study, (2) the nature of the dependent variable, and (3) the sample means. Having such a picture is highly important, because a study's results are inextricably tied to its table of cell and main effect means.

Presentation of Results

Occasionally, the results of a two-way repeated measures ANOVA are presented using an ANOVA summary table. Table 14.2 shows such a table for the Simon study.

The summary table shown in Table 14.2 is similar, in some very important ways, to the two-way ANOVA summary tables contained in Chapter 13. Most important, it contains three calculated F-values, one for the main effect of words, one for the main effect of tones, and one for the words-by-tones interaction. These three F-values speak directly to the null hypotheses discussed in the previous section.

There are two main differences between the summary table shown in Table 14.2 and the summary tables we examined in Chapter 13. First, there are three error rows in Table 14.2, whereas there is just one such row in the summary table for a two-way ANOVA without repeated measures. If you look closely at the workings inside

TABLE 14.2   ANOVA Summary Table for Performance Scores on Simon

Source	SS	df	MS	F
Players	15.5	5	3.1	
Words	37.5	1	37.5	19.74*
Error 1	9.5	5	1.9	
Tones	13.5	1	13.5	12.27*
Error 2	5.5	5	1.1	
Words × Tones	1.5	1	1.5	.29
Error 3	25.5	5	5.1	
Total	108.5	23		

*$p < 0.05$.

Table 14.2, you see that the *MS* for error 1 is used to obtain the calculated *F*-value for words, that the *MS* for error 2 is used to obtain the calculated *F*-value for tones, and that the *MS* for error 3 is used to obtain the calculated *F*-value for the interaction.[2]

The second difference between Table 14.2 and the ANOVA summary tables contained in Chapter 13 concerns the meaning of the *df* for total. As you can see, this *df* number is equal to 23. If this were a regular two-way ANOVA, you could add 1 to df_{Total} in order to figure out how many people were in the study. You cannot do that here, obviously, because there were only six players in our Simon study, yet df_{Total} is much larger than this. The problem gets solved completely when you realize that in all ANOVA summary tables, adding 1 to df_{Total} gives you the total number of pieces of data that were analyzed. That is true for ANOVAs with and without repeated measures. If there are no repeated measures, then the number of pieces of data is the same as the number of people (because each person provides, in those cases, just one score). When there *are* repeated measures, however, you must remember that adding 1 to the *df* for the top row of the summary table (not the bottom row) allows you to know how many people were involved.

If you ever encounter a summary table like that presented in Table 14.2, do not overlook the valuable information sitting in front of you. From such a table, you can determine how many people were involved in the study $(5 + 1 = 6)$, what the dependent variable was (performance on the Simon game), what the two factors were (tones and words) and how many levels made up each factor $(1 + 1 = 2$ in each case), how many total pieces of data were involved in the analysis $(23 + 1 = 24)$, and which null hypotheses were rejected.

In Excerpt 14.12, we see how a team of researchers summarized the results of their two-way repeated measures ANOVA in the text of their report. Note that this

EXCERPT 14.12 • *Results of a Two-Way Repeated Measures ANOVA Presented without a Summary Table*

[The 14 college students] did not know the task condition until they actually moved the mouse. Participants moved the mouse with right, left, or both hands. That is, we used six conditions (two visual feedback conditions $\times$ three hand conditions), and participants engaged in 30 trials (six conditions $\times$ five repetitions) arranged in random order. ... Analysis of overshooting with a two-way ANOVA revealed significant main effects for "feedback" $(F_{1,13} = 11.7, P < 0.01)$ and "hand condition" $(F_{2,26} = 5.71, P < 0.01)$ and for the interaction of "feedback $\times$ hand condition" $(F_{2,26} = 4.18, P < 0.05)$. The simple main effect of "feedback" was significant under the right-hand condition $(P < 0.001)$, and the simple main effect of "hand

(continued)

[2]The *df* numbers for these error rows are all equal to 5, but they were computed differently. Each was found by multiplying together the *df* for players and the *df* for the row immediately above the error being considered. For example, the *df* for error 2 was found by multiplying $df_{Players}$ by df_{Tones}.

EXCERPT 14.12 • (*continued*)

condition" was significant under the no-visual-feedback condition ($P < 0.001$). A post-hoc multiple comparison using Ryan's method revealed significant differences between the right-hand and bimanual conditions and between the right- and left-hand conditions ($P < 0.01$) for the no-visual-feedback condition.

Source: Asai, T., Sugimori, E., & Tanno, Y. (2010). Two agents in the brain: Motor control of unimanual and bimanual reaching movements. *PLoS ONE, 5*(4), 1–7.

passage is extremely similar to the textual summary of a two-way ANOVA without repeated measures. There are two factors (feedback and hand conditions), separate *F*-tests of the two main effects and the interaction (all of which turned out to be significant), and a *p*-value associated with each of these three *F*-tests.

Because the interaction between feedback and hand condition was significant, the researchers conducted a post hoc investigation that involved two sets of tests of simple main effects. First, the researchers compared participants' performance under the two visual feedback conditions, and they did this separately for each hand condition. Then, they ran their tests of simple main effects the other direction, this time comparing the three hand conditions separately under each visual feedback conditions. One of these latter tests was significant, so the researchers then used Ryan's multiple comparison procedure to make pairwise comparisons of participants' performance under the three hand conditions when there was no visual feedback.

Although the ANOVA discussed in Excerpt 14.12 resembles, in many ways, the kind of two-way ANOVA considered in Chapter 13, there is a subtle yet important difference. The *df* values for the three *F*-tests are not the same. The first of the two *df* numbers next to each *F*, of course, are not the same simply because there were two levels in one factor and three levels in the other factor. However, look at the second *df* next to each *F*. These *df*s vary because there were three different values for MS_{Error} involved in this analysis, each of which was used as the denominator for one of three *F*-ratios. Had this been a two-way ANOVA without repeated measures, just one MS_{Error} would have been used to get all three of the *F*s, thus causing each *F*'s second *df* to be equal to the same value.

The Presentation Order of Different Tasks

Earlier in this chapter, I indicated how a repeated measures factor can take one of three basic forms: different points in time, different treatment conditions, or different measures, tests, or variables. With a two-way repeated measures ANOVA, any combination of these three kinds of factors is possible. The most popular combinations, however, involve either (1) two factors, each of which is defined by different versions (i.e., levels) of a treatment, or (2) one treatment factor and one factor that involves measurements taken at different points in time.

When the levels of one or both factors in a two-way repeated measures ANOVA correspond with different treatment conditions or different variables, those levels should not be presented to the research participants in the same order. If they are, certain problems (e.g., practice effect, fatigue effect) might develop, and if that happens, the meaning of the F-tests involving the repeated measures factor(s) will be muddied.

The Sphericity Assumption

The calculated F-values computed in a two-way repeated measure ANOVA will turn out to be too large unless the population variances and correlations that correspond to the sample data conform to one or more acceptable patterns. This is the case even though the sample variances and correlations in a two-way repeated measures ANOVA are based on sample sizes that are equal (a condition brought about by measuring the same people, animals, or things repeatedly). Therefore, it is important for researchers, when using this kind of ANOVA, to attend to the assumption concerning population variances and correlations. This assumption is popularly referred to as the **sphericity assumption.**

Any of three strategies can be used when dealing with sphericity assumption. As is the case with a one-way repeated measures ANOVA, the researcher can (1) subject the sample data to Mauchly's test for sphericity, (2) bypass Mauchly's test and instead use the Geisser–Greenhouse conservative *df*s for locating the critical value needed to evaluate the calculated F-values, or (3) utilize the sample data to compute the index that estimates how badly the sphericity assumption is violated and then reduce the critical value(s) *df*s to the extent indicated by the ϵ index.

In the research report from which Excerpt 14.12 was taken, there was no mention of the sphericity assumption. If the researchers *had* attended to this important assumption, would it have made a difference? Let's consider what would have happened if the researchers had used Geisser–Greenhouse conservative *df*s in their analysis. With this change, the calculated F of 5.71 still would have been significant, but with $p < .05$ instead of $p < .01$. More importantly, application of the Geisser–Greenhouse adjustment would have caused the feedback $\times$ hand condition interaction to change from being significant (with $p < .05$) to being *non*significant (with $p > .05$).

Practical versus Statistical Significance

Throughout this book, I emphasize repeatedly the important point that statistical significance may or may not signify practical significance. Stated differently, a small p does not necessarily indicate that a research discovery is big and important. In this section, I try to show that well-trained researchers who use two-way repeated measures ANOVAs do *not* use the simple six-step version of hypothesis testing.

There are several options available to the researcher who is concerned about the meaningfulness of his or her findings. These options can be classified into two categories: a priori and post hoc. In the first category, one option involves conducting an a priori power analysis for the purpose of determining the proper sample size. A second option in that category involves checking to see if there is adequate power for a fixed sample size that already exists. In the post hoc category, the researcher can use the study's sample data to estimate effect size.

In Excerpt 14.13, we see a case in which a pair of researchers attended to the issue of practical significance of the statistically significant results that emerged from their two-way repeated measures ANOVA. The specific technique used to do this was the computation of eta squared. In this excerpt, note the *p*-level associated with the two main effects that were statistically significant. Also note the way the researchers evaluated those findings in terms of practical significance.

EXCERPT 14.13 • *Assessing Practical Significance*

A two-way repeated measures ANOVA was performed. . . . First, the analysis on the free recall scores revealed no statistically significant interaction between text type and stuttering frequency, however, a significant main effect of text type was found, $F(1, 56) = 75.77, p < .001$, which shows that the percentage of free recall units was greater for narrative than for expository texts. The effect size (η^2) for the significant main effect of text type [indicated] a moderate practical significance. The main effect of stuttering frequency associated with free recall was also statistically significant, $F(3, 56) = 7.41, p < .001$. The effect size for the significant main effect of stuttering frequency [indicated] a small practical significance.

Source: Panico, J., & Healey, E. C. (2009). Influence of text type, topic familiarity, and stuttering frequency on listener recall, comprehension, and mental effort. *Journal of Speech, Language, and Hearing Research, 52*(2), 534–546.

The criteria for judging effect size indices in a two-way repeated measures ANOVA are the same as the standards for these indices when they are computed for a two-way ANOVA without any repeated measures. As indicated in Chapter 13, the criteria for judging η^2 equate .01 with a small effect, .06 with a medium effect, and .14 with a large effect. To see the criteria for judging the magnitude of other ways of estimating effect size, refer to Table 13.2. Give researchers some bonus points when they use any of these standardized criteria for assessing practical significance, or when they discuss the notion of impact in terms of raw (i.e., unstandardized) effect sizes.

Two-Way Mixed ANOVAs

We now turn our attention to the third and final kind of ANOVA to be considered in this chapter. It is called a **two-way mixed ANOVA.** The word *mixed* is included in its label because one of the two factors is between subjects in nature whereas the other factor is within subjects.

Labels for This Kind of ANOVA

Unfortunately, all researchers do not use the same label to describe the kind of ANOVA that has one **between-subjects** factor and one **within-subjects** factor. Therefore, the first thing you must do relative to two-way mixed ANOVAs is familiarize yourself with the different ways researchers indicate that they have used this kind of ANOVA. If you do not do this, you might be looking at a study that involves a two-way mixed ANOVA and not even realize it.

When researchers use a two-way mixed ANOVA, some of them refer to it as a *two-way ANOVA with repeated measures on one factor.* Others call it an *ANOVA with one between-subjects factor and one within-subjects factor.* Occasionally it is called a **split-plot** ANOVA or a *two-factor between-within ANOVA.* In Excerpts 14.14 through 14.16, we see three different ways researchers chose to describe the two factor mixed ANOVA that they used.

EXCERPTS 14.14–14.16 • *Different Labels for a Two-Way Mixed ANOVA*

A 2 × 2 (Gender of Participant × Recipient of Tease: self vs. other) ANOVA, with the second variable as a repeated measure, was conducted on the children's hurt feelings ratings following the eight hypothetical teasing scenarios.

Source: Barnett, M. A., Barlett, N. D., Livengood, J. L., Murphy, D. L., & Brewton, K. E. (2010). Factors associated with children's anticipated responses to ambiguous teases. *Journal of Genetic Psychology, 171*(1), 54–72.

A job type × patient classification mixed ANOVA was performed on the participants' ratings.

Source: Hughes, A., & Gilmour, N. (2010). Attitudes and perceptions of work safety among community mental health workers. *North American Journal of Psychology, 12*(1), 129–144.

The research design was a 3 (evaluation type: teacher, self and peer) × 2 (lesson-plan version: draft and final) factorial design. Evaluation type was a between-subjects variable and lesson-plan version was a within-subjects variable. The primary data analysis comparing student performance across the three evaluation conditions on the two versions of the lesson plans was a 3 × 2 repeated measures analysis of variance (ANOVA).

Source: Ozogul, G., & Sullivan, H. (2009). Student performance and attitudes under formative evaluation by teacher, self and peer evaluators. *Educational Technology Research & Development, 57*(3), 393–410.

Data Layout and Purpose

To understand the results of a two-way mixed ANOVA, you must be able to conceptualize the way the data were arranged prior to being analyzed. Whenever you

		Time of Day		
Gender		8 A.M.	2 P.M.	8 P.M.
Male	Participant 1	6	3	8
	Participant 2	7	6	8
	Participant 3	4	2	10
	Participant 4	8	5	10
	Participant 5	5	4	9
Female	Participant 6	8	5	9
	Participant 7	6	6	7
	Participant 8	8	4	8
	Participant 9	7	4	9
	Participant 10	6	6	7

FIGURE 14.3 *Data Layout for a* 2×3 *Mixed ANOVA*

deal with this kind of ANOVA, try to think of (or actually draw) a picture similar to the one displayed in Figure 14.3. This picture is set up for an extremely small-scale study, but it illustrates how each participant is measured repeatedly across levels of the within-subjects factor but not across levels of the between-subjects factor. In this picture, of course, the between-subjects factor is gender and the within-subjects factor is time of day. The scores are hypothetical, and they are meant to reflect the data that might be collected if we asked each of five males and five females to give us a self-rating of his or her typical energy level (on a 0–10 scale) at each of three points during the day: 8 A.M., 2 P.M., and 8 P.M.

Although a two-way mixed ANOVA always involves one between-subjects factor and one within-subjects factor, the number of levels in each factor vary from study to study. Thus, the dimensions and labeling of Figure 14.3 only match our hypothetical two-way mixed ANOVA in which there is a two-level between-subjects factor, a three-level within-subjects factor, and five participants per group. Our picture can easily be adapted to fit *any* two-way mixed ANOVA because we can change the number of main rows and columns, the number of mini-rows (to indicate the number of participants involved), and the terms used to reflect the names of the factors and levels involved in the study.

The purpose of a two-way mixed ANOVA is identical to that of a completely between-subjects two-way ANOVA or of a completely within-subjects two-way ANOVA. In general, that purpose can be described as examining the sample means to see if they are further apart than would be expected by chance. Most researchers take this general purpose and make it more specific by setting up and testing three null hypotheses. These null hypotheses, of course, focus on the populations relevant to the investigation, with the three null statements asserting that (1) the main effect means of the first factor are equal to one another, (2) the main effect means of the second factor are equal to one another, and (3) the two factors do not interact.

To help you understand these three null hypotheses, I have taken the data from Figure 14.3, computed main effect means and cell means, and positioned these means in as shown in following figure:

Time of Day

		8 A.M.	2 P.M.	8 P.M.	
	Male	6	4	9	*6.3*
Gender	Female	7	5	8	*6.7*
		6.5	*4.5*	*8.5*	

One of our three research questions concerns the main effect of gender. To answer this question, the mean of 6.3 (based on the 15 scores provided by the five males) is compared against the mean of 6.7 (based on the 15 scores provided by the 5 females). The second research question, concerning the main effect of time of day, is addressed through a statistical comparison of the column means of 6.5, 4.5, and 8.5 (each based on scores provided by all 10 participants). The third research question, dealing with the interaction between gender and time of day, is dealt with by focusing on the six cell means (each based on five scores). This interaction question asks whether the change in the difference between the male and female means—which remains the same at 8 A.M. and 2 P.M. but then reverses itself at 8 P.M.—is greater than would be expected by chance sampling.

Presentation of Results

If the results of a two-way mixed ANOVA are presented in an ANOVA summary table, three *F*-values are presented—two for the main effects and one for the interaction—just as is the case in the ANOVA summary tables for completely between-subjects and completely within-subjects ANOVAs. However, the summary table for **mixed ANOVAs** is set up differently from those associated with the ANOVAs considered earlier. To illustrate these differences, I analyzed the energy level data originally shown in Figure 14.3, and the results of this two-way mixed ANOVA are found in Table 14.3.

As Table 14.3 shows, the summary table for a mixed ANOVA has an upper section and a lower section. These two sections are often labeled **between subjects** and **within subjects,** respectively. In the upper section, there are two rows of

TABLE 14.3   *ANOVA Summary Table of the Energy Level Data Shown in Figure 14.3*

Source	SS	df	MS	F
Between Subjects		9		
Gender	0.83	1	0.83	0.50
Error (between)	13.34	8	1.67	
Within Subjects		20		
Time of day	80.00	2	40.00	28.17[*]
Gender $\times$ Time of day	6.67	2	3.33	2.35
Error (within)	22.66	16	1.42	
Total	123.50	29		

*$p < 0.05$.

information, one concerning the main effect of the between-subjects factor and the other for the error that goes with the between-subjects main effect. In the lower section of the summary table, there are three rows of information. The first of these rows is for the main effect of the within-subjects factor, the second is for the interaction between the two factors, and the third provides information for the within-subjects error term. As you can see from Table 14.3, the *MS* for the first error was used as a denominator in computing the *F*-value in the upper section of the summary table, whereas the *MS* for the second error was used as the denominator in computing the two *F*-values in the lower section.

Table 14.3 contains information that allows you to understand the structure of the study that provided the data for the two-way mixed ANOVA. To illustrate how you can extract this information from the table, pretend that you have not read anything about the study connected with Table 14.3. In other words, imagine that your first encounter with this study is this ANOVA summary table.

First, the *df* value for the between-subjects row of the table shows that data were gathered from $9 + 1 = 10$ individuals. Second, the name of and the *df* value for the main effect in the upper portion of the table show that there were two groups in the study, with gender being the independent variable associated with this main effect. Third, the name of and *df* for the main effect in the lower portion of the table show that each of the 10 individuals was measured on $2 + 1 = 3$ occasions, each being a particular time of the day. The table's title gives you a hint as to the kinds of scores used in this study, because it states that this ANOVA was conducted on energy level data.

Table 14.3, of course, also contains the results of the ANOVA. To interpret these results, you must look back and forth between the ANOVA summary table and the 2×3 table of cell and main effect means that we considered earlier. The first *F*-value (0.50) indicates that the two main effect means for males (6.3) and females (6.7) were not further apart from each other than we could expect by chance. Accordingly, the null hypothesis for gender was not rejected. The second *F*-value (28.17), however, shows that the null hypothesis for time of day was rejected. This finding came about because the main effect means of the within-subjects factor (6.5, 4.5, and 8.5) were further apart than could be expected by chance. The third *F*-value (2.35), for the interaction, was not statistically significant. Even though the differences between the male and female cell means were not constant across the data collection points during the day, the three male–female differences (-1, -1, and $+1$) did not vary enough to call into question the interaction null hypothesis.

Now let's consider Excerpt 14.17, which comes from a published study that used a 2×2 mixed ANOVA. The title of the table in this excerpt makes it clear that the presentation is for a mixed ANOVA. Even if the word *mixed* had not been included in the table's title, however, you should be able to tell that this kind of ANOVA was used by noting the way the first column of information was set up.

EXCERPT 14.17 • *Summary Table from Two-Way Mixed ANOVA*

TABLE 5 *Mixed ANOVA Table for Accuracy Scores*

Source	SS	df	MS	F	p	η_p^2
Between Subjects		46				
Group	371.05	1	371.05	0.95	0.33	0.02
Error	17,536.12	45	389.69			
Within Subject		47				
Time	329.01	1	329.01	4.44	0.04	0.09
Time $\times$ Group	908.19	1	908.19	12.26	0.001	0.21
Error	3,333.22	45	74.07			
Total	22,477.59	93				

Source: Hartshorn, K. J., Evans, N. W., Merrill, P. F., Sudweeks, R. R., Strong-Krause, D., & Answerson, N. J. (2010). Effects of dynamic corrective feedback on ESL writing accuracy. *TESOL Quarterly, 44*(1), 84–109.

There are four differences between the table in Excerpt 14.17 and the ANOVA summary table we saw in Table 14.3. First, each of the two error rows in Excerpt 14.17 is simply labeled *error*. Even though these two rows carry the same label, it is important to note that they are not interchangeable. The *MS* associated with the first of these errors was used as the denominator in computing the *F*-value that appears in the upper section of the table; in contrast, the *MS* associated with the second of these error rows was used as the denominator in computing each of the two *F*-values that appear in the lower section of the table.[3]

[3]As illustrated in Table 14.3 and Excerpt 14.17, different terms are sometimes used to label the two rows that contain the *MS* values used as the denominators for the *F*s. You are likely to encounter ANOVA summary tables in which these two rows are labeled error 1 and error 2, error (a) and error (b), or error (b) and error (w). A few researchers label these error rows as *subjects within groups* and _____ $\times$ *subjects within groups*, with the blank filled by the name of the within-subjects factor.

The second thing to note about Excerpt 14.17 concerns the column of df values. The second and third of these numbers are not indented under 46; likewise, the numbers 1, 1, and 45 are not indented under 47. If you add together the first seven df numbers, your total does not match the df located in the bottom row. This discrepancy melts away entirely once you realize that the df values of 46 and 47 are simply headings for the upper and lower portions of the table.

The third and fourth differences between Excerpt 14.17 and Table 14.3 are minor and major, respectively. The minor difference is that a specific p-value is presented in Excerpt 14.17 for each F-value, whereas a note beneath Table 14.3 simply says $p < .05$. The major difference between these two ANOVA summary tables is the extra column of information on the far right in Excerpt 14.17. These estimates of effect size for each of the F-values provides insight into the practical significance of the results.

Although it is helpful to be able to look at ANOVA summary tables when trying to decipher and critique research reports, such tables usually do not appear in journal articles. Instead, the results are typically presented strictly within one or more paragraphs of textual material. To illustrate, consider Excerpt 14.18, wherein the results of a two-way mixed ANOVA are presented without a summary table.

EXCERPT 14.18 • _Results of a Two-Way Mixed ANOVA Presented without a Summary Table_

For [the two-way mixed ANOVA] analysis, the within-subject factor was amplification with three levels (unaided, RITA, and RITE) and the between-subject factor was group with two levels (new users and experienced users). . . . Results for the HINT revealed a significant amplification main effect [$p = .016$, partial $\eta^2 = .165$]. Paired samples t-tests were conducted to further investigate the amplification main effect controlling for family-wise error rate across the tests at the 0.05 level, using the Holm's sequential Bonferroni procedure. Results indicated that unaided scores were significantly better than both the RITA and RITE scores; however, scores were not significantly different between the RITA and RITE devices [$ps = 0.018, 0.009$, and 0.898, respectively].

Source: Alworth, L. N., Plyler, P. N., Reber, M. B., & Johnstone, P. J. (2010). The effects of receiver placement on probe microphone, performance, and subjective measures with open canal hearing instruments. _Journal of the American Academy of Audiology, 21_(4), 249–266.

Excerpt 14.18 comes from a study in which two kinds of hearing aids were compared against the unaided ear among people with a hearing loss. In this excerpt, RITE and RITA are acronyms for _received in the ear_ and _received in the aid,_ respectively. The HINT—one of the study's dependent variables—is a test designed to measure each participant's ability to hear in noisy conditions. Results of the mixed ANOVA on the HINT data yield a significant F for the within-subjects main effect, but nonsignificant Fs for the between-subjects main effect and the interaction.

Excerpt 14.18 deserves your attention because of the way the post hoc tests were conducted. Using a popular analytic strategy, the researchers connected with this hearing aid study compared the main effect means for amplification in a post hoc investigation because (1) there are more than two means in that main effect and (2) the interaction is not significant. The post hoc comparisons were conducted in a pairwise fashion using paired *t*-tests. However, the alpha level for these three tests was made more rigorous to head off the inflated Type I error problem. This was done by means of the **Holm's sequential Bonferroni procedure.**

If the regular **Bonferroni** procedure had been used in Excerpt 14.18, each of the three pairwise comparisons would have been tested with alpha set at .05/3, or .0167. With Holm's-sequential Bonferroni procedure, the comparison yielding the smallest *p* (.009) is first compared against the regular Bonferroni alpha level (.0167). Next, the comparison yielding the second smallest *p* (.018) is compared against a slightly more lenient alpha level equal to .05/2, or .025. Finally, the third and last comparison has its *p* (.898) compared against an even more lenient α, computed as .05/1, or .05. As you can see, there is a sequential change in the level of significance across the various tests, with the beginning point in the sequence being the regular Bonferroni-adjusted alpha level and the ending point being the unadjusted alpha level.

Related Issues

Earlier in this chapter, I indicated how the levels of within-subjects factors sometimes can and should be presented to subjects in varying orders. That discussion applies to mixed ANOVAs as well as to fully repeated measures ANOVAs. Excerpt 14.19 shows how the technique of counterbalancing can be used to avoid the bias that might exist if the levels of the within-subjects factor are presented in the same order to all participants. By counterbalancing the order of the spoon-weight factor, the

EXCERPT 14.19 • *Counterbalancing the Levels of the Within-Subjects Factor*

We used a counterbalanced repeated-measures design. Each participant performed the experimental task using spoons of three different weights: (A) lightweight: 35 g, (B) control: 85 g, and (C) weighted: 135 g. Each participant was randomly assigned to one of three experimental sequences (ABC, BCA and CAB) by means of sealed envelopes. . . . Three (weight condition: lightweight vs. control vs. weighted) $\times$ 2 (group: Parkinson's disease vs. control) mixed analyses of variance (ANOVAs) were computed on the kinematic scores.

Source: Ma, H. I., Hwang, W. J., Tsai, P. L., & Hsu, Y. W. (2009). The effect of eating utensil weight on functional arm movement in people with Parkinson's disease: A controlled clinical trial. *Clinical Rehabilitation, 23*(12), 1086–1092.

researchers arranged their study's experiment so that one-third of the members of each group were assigned to each of the different *spoon orders*. By doing this, the researchers made sure that level of the weight condition factor was not confounded with the three trials of the experiment.

A second issue you should keep in mind when examining the results of two-way mixed ANOVAs is the important assumption of **sphericity.** I discussed this assumption earlier when we considered fully repeated measures ANOVAs. It is relevant to mixed ANOVAs as well—but only the *F*-tests located in the within-subjects portion of the ANOVA summary are based on the sphericity assumption. Thus, the *F*-value for the main effect of the between-subjects factor is unaffected by a lack of sphericity in the populations connected to the study. In contrast, the *F*-values for the main effect of the within-subjects factor and for the interaction are positively biased (i.e., turn out larger than they ought to) to the extent that the sphericity assumption is violated.

Well-trained researchers do not neglect the sphericity assumption when they use two-way mixed ANOVAs. Instead, they carefully attend to this assumption. One option used by many researchers is to adjust the degrees of freedom associated with the critical values (using the Geisser–Greenhouse or the Huynh–Feldt procedures), thereby compensating for possible or observed violation of the sphericity assumption. Another option is to apply Mauchly's test to the sample data to see if the assumption appears to be violated; depending on how this test turns out, the regular *F*-tests are or are not examined. In Excerpts 14.20 and 14.21, we see examples where researchers used these two options. Both sets of researchers deserve high marks for demonstrating a concern for the sphericity assumption.

EXCERPTS 14.20–14.21 • *Options for Dealing with the Sphericity Assumption*

A 2 (condition) $\times$ 3 (time) mixed-model analysis of variance (ANOVA) with the IRMA-SF as a dependent variable revealed a significant time effect, $F(1.84, 150.58) = 8.17 p < .01$, partial $\eta^2 = 0.09$, and a Condition $\times$ Time interaction effect, $F(1.84, 150.58) = 6.10, p < .01$, partial $\eta^2 = 0.07$, with Huynh-Feldt corrections.

Source: Hillenbrand-Gunn, T. L., Heppner, M. J., Mauch, P. A., & Park, H.-Y. (2010). Men as allies: The efficacy of a high school rape prevention intervention. *Journal of Counseling & Development, 88*(1), 43–51.

Prior to conducting [the two-way mixed] ANOVAs, Mauchley's test for sphericity was conducted. In those cases where the Mauchley's *W* was significant, the Geisser–Greenhouse conservative *F*-test was interpreted as a safeguard against type I error.

Source: Konnert, C., Dobson, K., & Stelmach, L. (2009). The prevention of depression in nursing home residents: A randomized clinical trial of cognitive–behavioral therapy. *Aging & Mental Health, 13*(2), 288–299.

A third issue to keep in mind when you encounter the results of two-way mixed ANOVAs concerns the distinction between statistical significance and practical significance. I first discussed this distinction in Chapter 7, and I purposely have brought up this issue as often as possible. I have done this, because far too many researchers conduct studies that yield one or more findings that have very little practical significance even though a very low probability level is associated with the calculated value produced by their statistical test(s).

As I indicated earlier in this chapter, the criteria for judging the magnitude of post hoc estimates of effect size (such as d or η^2) are the same in a two-way mixed ANOVA as they are in a two-way ANOVA having no repeated measures. If you must refresh your memory regarding these criteria, take a look at Table 13.2. Recall, too, that researchers have the option of discussing practical significance without using standardized estimates of effect size; instead, they can use their knowledge and expertise to look at, and assess the worth of, raw-score differences between means.

Many researchers who conduct two-way mixed ANOVAs fail to address the question of practical significance. However, a growing number of researchers do this—and they deserve credit for performing a more complete analysis of the study's data than is usually the case. We see illustrations of this good practice in Excerpts 14.17, 14.18, and 14.20.

Three Final Comments

As we near the end of this chapter, I have three final comments. In each case, I argue that you must be alert as you read or listen to formal summaries of research studies so you can (1) know for sure what kind of analysis the researcher *really* used, and (2) filter out unjustified claims from those that warrant your close attention. If you do not put yourself in a position to do these two things, you are likely to be misled by what is contained in the research reports that come your way.

What Kind of ANOVA Was It?

In this chapter, we have considered three different kinds of ANOVAs: a one-way ANOVA with repeated measures, a two-way ANOVA with repeated measures on both factors, and a two-way ANOVA having repeated measures on just one factor. These three ANOVAs are different from each other, and they are also different from the one-way and two-way ANOVAs focused on in Chapters 10, 11, and 13. Thus, for you to understand the structure and results of any one-way or two-way ANOVA, you must know whether it involves repeated measures and, in the case of a two-way ANOVA having repeated measures, you must know whether one or both of the factors involved repeated measures.

As indicated earlier, most researchers clarify what kind of one-way or two-way ANOVA they have used. For example, if repeated measures are involved, they typically use special terms—such as *within-subjects* or *repeated measures*—to describe the factor(s) in the ANOVA, or you may see the term *mixed* used to describe the kind of ANOVA considered near the end of this chapter. If no such term is used, this usually means that no repeated measures were involved.

Unfortunately, not all descriptions of one-way or two-way ANOVAs are clear as to the nature of its factor(s). At times, you may be told that a one-way ANOVA was used when in reality it was a one-way ANOVA with repeated measures. Occasionally, this same thing happens with two-way ANOVAs. Or, you may be told that a two-way repeated measures ANOVA was used, thus causing you to think that there were two within-subjects factors, which is wrong, because only one of the factors actually had repeated measures.

Because the presence or absence of repeated measures does not affect the null hypothesis of a one-way ANOVA or the three null hypotheses of a two-way ANOVA, someone might argue that it really does not matter whether you can tell for sure if the factor(s) of the ANOVA had repeated measures. To that person I ask just one simple question: "Do you know about the sphericity assumption and under what circumstances this assumption comes into play?"

Practical versus Statistical Significance

At various points in this chapter, I tried to help you understand that statistical significance does not always signify practical significance. I did this by means of the words I have written and the excerpts I have chosen to include. (Estimates of effect size appear in Excerpts 14.6, 14.7, 14.13, 14.17, 14.18, and 14.20.)

There is a growing trend for researchers to do something in their studies, either as they choose their sample sizes or as they go about interpreting their results, so they and the recipients of their research reports do not make the mistake of thinking that statistical significance means big and important. However, you are bound to come across research claims that are based on the joint use of (1) one or more of the ANOVAs considered in this chapter and (2) the six-step version of hypothesis testing. When that happens, I hope you remember two things. First, a very small *p may* indicate that nothing big was discovered, only that a big sample can make molehills look like mountains, and, conversely, a large *p may* indicate that something big was left undetected because the sample size was too small. The second thing to remember is that studies can be planned, using the nine-step version of hypothesis testing, such that neither of those two possible problems is likely to occur. When this expanded (but better) version of hypothesis testing is *not* used, results can be murky, as indicated in Excerpt 14.22.

EXCERPT 14.22 • *The Inferential Dilemma When Statistical Power Is Inadequate*

One of the main limitations of the present study [using a mixed ANOVA] is the small sample size [that] reduces the power of the statistical tests. Even at medium (.50)

(continued)

EXCERPT 14.22 • (*continued*)

and large (.80) effect sizes, the power to detect differences between the treatments in this sample is .32 and .62. Therefore, our null findings (no differences between the treatments) do not necessarily indicate that the treatments are equally effective.

Blocher, W. G., & Wade, N. G. (2010). Sustained effectiveness of two brief group interventions: Comparing an explicit forgiveness-promoting treatment with a process-oriented treatment. *Journal of Mental Health Counseling, 32*(1), 58–74.

On the one hand, the researchers who penned the three sentences in Excerpt 14.22 deserve credit for realizing that their negative finding may have been a Type II error. On the other hand, we might legitimately ask the researchers, "Why didn't you do a power analysis in the planning stage of your study so as have samples large enough to detect non-trivial effects, if they exist?"

The Possibility of Inferential Error

Many researchers discuss the results of their studies in such a way that it appears that they have discovered ultimate truth. Stated differently, the language used in many research reports suggests strongly that sample statistics and the results of inferential tests are being reified into population parameters and indisputable claims. At times, such claims are based on the kinds of ANOVA considered in this chapter.

You must remember that the result of any *F*-test might be a Type I error (if the null hypothesis is rejected) or a Type II error (if the null hypothesis is retained). This is true even if the nine-step version of hypothesis testing is used, and even if attention is paid to all relevant underlying assumptions, and even if the data are collected in an unbiased fashion with valid and reliable measuring instruments from probability samples characterized by zero attrition, and even if all other good things are done so the study is sound. Simply stated, inferential error is *always* possible whenever a null hypothesis is tested.

Review Terms

Between subjects	Huynh-Feldt correction
Bonferroni procedure	Latin square
Carryover	Mauchly test
Confounding	Mixed ANOVA
Conservative *F*-test	One between, one within ANOVA
Counterbalancing	Practice effect
Fatigue effect	Sphericity assumption
Geisser–Greenhouse approach	Split-plot ANOVA
	Two-way mixed ANOVA
Holm's-sequential Bonferroni procedure	Within-subjects factor

The Best Items in the Companion Website

1. An interactive online quiz (with immediate feedback provided) covering Chapter 14.
2. Ten misconceptions about the content of Chapter 14.
3. One of the best passages from Chapter 14: "Practical versus Statistical Significance in Mixed ANOVAs."
4. An online resource entitled "Within-Subjects ANOVA."
5. Two jokes about statistics.

To access chapter outlines, practice tests, weblinks, and flashcards, visit the companion website at http://www.ReadingStats.com.

Review Questions and Answers begin on page 531.

The Analysis
of Covariance

In Chapters 10 through 14, we looked at several different kinds of analysis of variance. We focused our attention on one-way and two-way ANOVAs, with consideration given to the situations where (1) each factor is between subjects in nature, (2) each factor is within subjects in nature, and (3) both between-subjects and within-subjects factors are combined in the same study. We closely examined five different kinds of ANOVAs that are distinguished from one another by the number and nature of the factors. In this book, these five ANOVAs have been referred to as a *one-way ANOVA, a two-way ANOVA, a one-way repeated measures ANOVA, a two-way repeated measures ANOVA,* and *a two-way mixed ANOVA.*

We now turn our attention to an ANOVA-like inferential strategy that can be used instead of any of the ANOVAs examined or referred to in earlier chapters. This statistical technique, called the **analysis of covariance** and abbreviated by the six letters **ANCOVA,** can be used in any study regardless of the number of factors involved or the between-versus-within nature of the factor(s). Accordingly, the analysis of covariance is best thought of as an option to the analysis of variance. For example, if a researcher's study involves one between-subjects factor, data can be collected and analyzed using a one-way ANOVA *or* a one-way ANCOVA. The same option exists for any of the other four situations examined in earlier chapters. Simply stated, there is an ANCOVA counterpart to any ANOVA.

In Excerpts 15.1 through 15.3, we see how researchers typically indicate that their data were subjected to an analysis of covariance. Note how these excerpts illustrate the way ANCOVA can be used as an option to ANOVA regardless of the number of factors involved in the study or the between-versus-within nature of any factor.

EXCERPT 15.1–15.3 • *The Versatility of the Analysis of Covariance*

A one-way ANCOVA was conducted to assess whether the high and low beliefs groups differed in their degree of rumination about the anagram task.

Source: Moulds, M. L., Yap, C. S. L., Kerr, E., Williams, A. D., & Kandris, E. (2010). Metacognitive beliefs increase vulnerability to rumination. *Applied Cognitive Psychology, 24*(3), 351–364.

To determine any differences in comprehension scores between conditions, a 2 (sex) × 3 (image condition) between subjects ANCOVA was computed.

Source: Good, J. J., Woodzicka, J. A., & Wingfield, L. C. (2010). The effects of gender stereotypic and counter-stereotypic textbook images on science performance. *Journal of Social Psychology, 150*(2), 132–147.

The effectiveness of the treatment seminars in promoting forgiveness was explored with a 3 × 2 (Condition: Empathy Forgiveness Seminar, Self-enhancement Forgiveness Seminar, wait-list control × Time) analysis of covariance (ANCOVA) with repeated measures.

Source: Sandage, S. J., & Worthington, E. L. (2010). Comparison of two group interventions to promote forgiveness: Empathy as a mediator of change. *Journal of Mental Health Counseling, 32*(1), 35–57.

The Three Different Variables Involved in Any ANCOVA Study

In any of the ANOVAs considered in earlier chapters, there are just two kinds of variables: independent variables and dependent variables. The data analyzed in those ANOVAs, of course, represent the dependent variable; the factors correspond with the study's independent variables. We have seen how ANOVAs can involve more than one factor, factors made up of different numbers of levels, and different kinds of factors (i.e., between-subjects versus within-subjects factors); nevertheless, each and every factor in the ANOVAs considered in this book represents an independent variable. Thus, such ANOVAs could be said to contain two structural ingredients: one or more independent variables and data on one dependent variable.

In any analysis of covariance, three rather than two kinds of variables are involved. Like the ANOVAs we have considered, there will be scores that correspond with the dependent variable and one or more factors that coincide with the study's independent variable(s). In addition, ANCOVAs involve a third variable called a **covariate variable.**[1] Because the covariate is a variable on which the study's participants

---

[1] The term *concomitant variable* is synonymous with the term *covariate variable*.

are measured, it is more similar to the study's dependent variable than to the independent variable(s). However, the covariate and dependent variables have entirely different functions in any ANCOVA study, as the next section makes clear. Before discussing the covariate's function, however, let's consider a few real studies for the purpose of verifying that ANCOVA studies always have *three* structural ingredients.

Excerpts 15.4 and 15.5 are typical of ANCOVA-related passages in research reports because each contains a clear indication of the three kinds of variables involved in the ANOVA. In the first of these excerpts, the researchers used a one-way ANCOVA, so there is one independent variable, one dependent variable, and one covariate variable. Excerpt 15.5 describes these same three kinds of variables, except here there are two independent variables because the authors conducted a two-way ANCOVA. (In the second of these two excerpts, the dependent variable was a posttest whereas the covariate was a pretest. These two variables were conceptually different and were assessed with different measuring instruments.)

EXCERPTS 15.4–15.5 • *The Three Kinds of Variables in Any ANCOVA Study*

Then, a one-way analysis of covariance (ANCOVA) was conducted to determine if there was a statistically significant difference in [students'] absence frequency between the two sections, with absence as the dependent variable, sections as the independent variable, and GPA as the covariate.

Source: Traphagan, T., Kucsera, J., & Kishi, K. (2010). Impact of class lecture webcasting on attendance and learning. *Educational Technology Research & Development, 58*(1), 19–37.

Then a two-way ANCOVA, taking the "pre-test scores of the summative assessment" as the covariate, the "post-test scores of the summative assessment" as the dependent variable, and the "different types of Web-based assessment" and the "different levels of prior knowledge" as the fixed factors, was used to test the [hypotheses].

Source: Wang, T.-H. (2010). Web-based dynamic assessment: Taking assessment as teaching and learning strategy for improving students' e-learning effectiveness. *Computers & Education, 54*(4), 1157–1166.

The Covariate's Role

Like the analysis of variance, the analysis of covariance allows researchers to make inferential statements about main and interaction effects. In that sense, these two statistical procedures have the same objective. However, an ANCOVA is superior to its ANOVA counterpart in two distinct respects, so long as a good covariate is used.[2]

---

[2]The qualities of a good covariate are described later in the chapter.

To understand what is going on in an analysis of covariance, you must understand this dual role of the covariate.

One role of the covariate is to reduce the probability of a Type II error when main or interaction effects are tested, or when comparisons are made within planned or post hoc investigations. As pointed out on repeated occasions in earlier chapters, this kind of inferential error is committed whenever a false null hypothesis is not rejected. Because the probability of a Type II error is inversely related to statistical **power,** let me restate this first role of the covariate by saying that an ANCOVA is more powerful than its ANOVA counterpart, presuming that other things are held constant and that a good covariate has been used within the ANCOVA.

As you have seen, the F-tests associated with a standard ANOVA are computed by dividing the MS for error into the MSs for main and interaction effects. If MS_{error} can somehow be made smaller, then the calculated Fs are larger, ps are smaller, and there is a better chance that null hypotheses will be rejected. When a good covariate is used within a covariance analysis, this is exactly what happens. Data on the covariate function to explain away a portion of within-group variability, resulting in a smaller value for MS_{error}. This mean square is often referred to as "error variance."

Consider Excerpts 15.6 and 15.7. In the first of these excerpts, the researchers explain that they decided to use the analysis of covariance because of its ability to increase power. In Excerpt 15.7, we see results from a different study that illustrate the impact of this power increase. Comparing the posttest means from the four groups, the covariance analysis produced an F with an accompanying p of .084, whereas the ANOVA comparison of those same posttest means generated an F that

EXCERPTS 15.6–15.7 • *The First Role of the Covariate: Increased Statistical Power*

ANCOVA was selected to reduce the probability of a Type II error [and thus] increase power by reducing the error variance [of] the SAT-9 MP.

Source: Griffin, C. C., & Jitendra, A. K. (2009). Word problem-solving instruction in inclusive third-grade mathematics classrooms. *Journal of Educational Research, 102*(3), 187–202.

[T]here were no significant differences among conditions before instruction, $F(3, 56) = 0.94$, *ns*, and interestingly, there were no differences across the conditions after instruction either, $F(3, 56) = 0.87$, *ns*. However, when running an analysis of covariance (which used the pretest scores as a covariate) on the posttest scores, there is a marginally significant difference among the four groups, $F(3, 55) = 2.33$, $p = .084$.

Source: Hirata, Y., & Kelly, S. D. (2010). Effects of lips and hands on auditory learning of second-language speech sounds. *Journal of Speech, Language & Hearing Research, 53*(2), 298–310.

had a *p* equal to .46. The ANCOVA *p* was close to the .05 alpha level and was referred to by the researchers as being "marginally significant." In contrast, the ANOVA *p* was equal to .46 and reported simply as being *ns* ("not significant").

In addition to its power function, the covariate in an analysis of covariance has another function. This second function can be summed up by the word *control*. In fact, some researchers will refer to the covariate of their ANCOVA studies as the *control variable*. Excerpt 15.8 illustrates nicely the fact that a covariate is sometimes used because of its control (or corrective) capability. It is worth noting, in this excerpt, that the covariate was used (to control for pretreatment group differences) even though the 170 participants were randomly assigned to the study's three comparison groups. (The full research report indicates that statistically significant differences were found, at the time of the pretest, on three of the study's six dependent variables. Though random assignment is an excellent feature of studies designed to investigate cause-and-effect relationships, it does not—as illustrated by Excerpt 15.8—guarantee that comparison groups are identical.)

EXCERPT 15.8 • *The Second Role of the Covariate: Control*

This study used a randomized-groups pretest–posttest research design [with participants] randomly assigned to one of three groups. . . . Pretest data were initially analyzed for group differences [and] on all of the baseline measures the CG [Control Group] had the highest means. . . . To reduce sampling error and control for initial differences, the baseline data were used as covariates when comparing the groups on the follow-up and retention measures.

Source: Parrott, M. W., Tennant, L. K., Olejnik, S., & Poudevigne, M. S. (2008). Theory of planned behavior: Implications for an email-based physical activity intervention. *Psychology of Sport and Exercise, 9*(4), 511–526.

The logic behind the control feature of ANCOVA is simple. The comparison groups involved in a study are likely to differ from one another with respect to one or more variables that the researcher may wish to hold constant. In an attempt to accomplish this objective, the researcher could use participants who have identical scores on the variable(s) where control is desired. That effort, however, usually brings forth two undesirable outcomes. For one thing, only a portion of the available participants are actually used, thus *reducing* the statistical power of inferential tests. Furthermore, the generalizability of the findings is greatly restricted as compared with the situation where the analysis is based on a more heterogeneous group of people.

To bring about the desired control, ANCOVA adjusts each group mean on the dependent variable. Although the precise formulas used to make these adjustments are somewhat complicated, the rationale behind the adjustment process is easy to understand. If one of the comparison groups has an *above-average* mean on the control variable (as compared with other comparison groups in the study), then that group's mean score on the dependent variable is *lowered*. In contrast, any group that has a *below-average* mean on the covariate has its mean score on the dependent variable *raised*. The degree to which any group's mean score on the dependent variable is adjusted depends on how far above or below average that group stands on the control variable (and on the correlation between the covariate and the dependent variable). By adjusting the mean scores on the dependent variable in this fashion, ANCOVA provides the best estimates of how the comparison groups would have performed if they had all possessed identical means on the control variable(s).

To illustrate the way ANCOVA adjusts group means on the dependent variable, let's consider a recent study undertaken to compare two methods of teaching nursing students. A total of 49 individuals in baccalaureate, accelerated baccalaureate, and diploma nursing programs were randomly assigned to work with a human patient simulator (HPS) or with interactive case studies (ICS). Both prior to and following the short-term educational intervention, each student took an examination designed by the Health Education Systems, Inc. (HESI). The pretest scores functioned as the covariate in an analysis of covariance that compared the interventions. In Excerpt 15.9, we see how the two groups performed on the pretest and the posttest. We also see the ANCOVA-generated adjusted posttest means.

EXCERPT 15.9 • *Adjusted Means*

Students from each of the three nursing programs were randomly assigned to one of the two teaching strategy groups: HPS or ICS. . . . The same pretest and posttest were administered to all students, regardless of which educational intervention they received. . . . A one-way, between-subjects analysis of covariance (ANCOVA) was used to compare HPS and ICS posttest HESI scores. . . . The adjusted mean posttest HESI score for the HPS group was significantly higher ($P \leq .05$) than the adjusted mean posttest HESI score for the ICS group (Table 2).

TABLE 2   *ANCOVA Comparison of Pretest and Posttest HESI Scores by Educational Intervention*

	Pretest HESI Scores		Posttest HESI Scores		Adjusted Mean Scores
	Mean	SD	Mean	SD	
HPS (simulation group)	713.12	153.56	738.00	131.01	750.42[a]
ICS (case study group)	786.17	184.81	670.08	181.83	657.14[a]

The HPS group scored significantly higher on the posttest than the ICS group did. [a]$P \leq .05$

Source: Howard, V. M., Ross, C., Mitchell, A. M., & Nelson, G. M. (2010). A comparative analysis of learning outcomes and student perceptions. *CIN: Computers, Informatics, Nursing, 28*(1), 42–48.

As you can see from Excerpt 15.9, the two groups began the study with different mean scores on the HESI. With the mean pretest score for all 49 students being 748.90, it is clear that the 24 students in the ICS group started out, on average, with a higher level of proficiency than did the 25 students in the HPS group. In a sense, we could say that the ICS group began the study with an advantage over the HPS group.

If the obtained posttest means on the HESI for the two groups had been directly compared (e.g., with an independent-samples *t*-test or *F*-test), a statistically significant result would be hard to interpret. This is because part or all of the posttest difference between the two groups might simply be a reflection of their different starting points. For this reason, a comparison of the two posttest means is not a fair comparison.

To acknowledge the existence of the difference between the two groups on the covariate, ANCOVA adjusted the posttest means. Examine the three sets of means in Excerpt 15.9 and note the basic logic of this adjustment procedure. The HPS group had a disadvantage at the outset, because its pretest mean (713.12) was lower than the grand average of both groups combined (748.90). Therefore, that group's posttest mean was adjusted upward (from 738.00 to 750.22). The ICS group, in contrast, started out with an advantage, because its pretest mean (786.17) was higher than the grand average of all pretest scores. Consequently, the ICS group's posttest mean was adjusted downward (from 670.08 to 657.14).

In *any* study, this is exactly how the control feature of ANCOVA works. Any group with an above-average mean on the covariate has its mean on the dependent variable adjusted downward, whereas any group with a below-average mean on the covariate has its mean on the dependent variable adjusted upward. These **adjusted means** constitute the best estimates of how groups would have performed on the dependent variable if they had possessed identical means on the control (i.e., covariate) variable used in the study.

Although the logic behind ANCOVA's adjustment of group means on the dependent variable is easy to follow, the statistical procedures used to make the adjustment are quite complicated. The formulas used to accomplish this goal are not presented here because it is not necessary to understand the intricacies of the adjustment formula in order to decipher and critique ANCOVA results. All you must know is that the adjustment process involves far more than simply (1) determining how far each group's covariate mean lies above or below the grand covariate mean and (2) adding or subtracting that difference to that group's mean on the dependent variable. As proof of this, take another look at Excerpt 15.9. Each group's pretest mean in that study was about 36 points away from the grand pretest mean. However, the covariance adjustment caused far less than this amount to be added to or subtracted from the posttest means.

Note that the two purposes of ANCOVA—increased power, on the one hand, and control of extraneous variables, on the other hand—occur simultaneously. If a researcher decides to use this statistical procedure solely to gain the increased power that ANCOVA affords, the means on the dependent variable are automatically adjusted to reflect differences among the group means on the covariate variable. If, however, the researcher applies ANCOVA solely because of a desire to exert statistical control on a covariate variable, there is an automatic increase in the statistical power of the inferential tests. In other words, ANCOVA accomplishes two objectives even though the researcher may have selected it with only one objective in mind.

At the beginning of this section, I stated that ANCOVA allows the researcher to make inferential statements about main and interaction effects in the populations of interest. Because you now know how data on the covariate variable(s) make it possible for the researcher to control one or more extraneous variables, I can now point out that ANCOVA's inferential statements are based on the adjusted means. The data on the covariate and the dependent variable are used to compute the adjusted means on the dependent variable, with ANCOVA's focus resting on these adjusted means whenever a null hypothesis is tested.

Null Hypotheses

As typically used, ANCOVA involves the same number of null hypotheses as is the case in a comparable ANOVA. Hence, you usually find that there are one and three null hypotheses associated with ANCOVAs that have one and two factors, respectively. The nature of these ANCOVA null hypotheses is the same as the null hypotheses I talked about in earlier chapters when we considered various forms of ANOVA, except that the μs in any covariance H_0 must be considered to be adjusted means.[3]

Although null hypotheses rarely appear in research reports that contain ANCOVA results, sometimes researchers refer to them. Excerpt 15.10 provides an

EXCERPT 15.10 • *The Null Hypothesis in ANCOVA*

With ANCOVA, the null hypothesis being tested is that the adjusted population mean [scores] are equal. . . . The observed mean score for the video group on the dialogue sections of the post-test was 9.76 (out of 16, equaling 61.0%), and the adjusted mean score was 9.74 (60.8%). The observed score for the audio-only group on the dialogue sections of the post-test was 8.72 (54.5%), while the adjusted mean score was 8.74 (54.7%). . . . Examination of the [one-way] ANCOVA indicated that the F-value for the dialogue post-test variable reached significance $(F = 5.94, p < 0.05)$, and the null hypothesis was rejected.

Source: Wagner, E. (2010). The effect of the use of video texts on ESL listening test-taker performance. *Language Testing, 27*(3), 1–21.

---

[3]In a completely randomized ANCOVA where each factor is active in nature, the adjusted population means on the dependent variable are logically and mathematically equal to the unadjusted population means.

example of this. Had the null hypothesis in this study been written out symbolically, it have taken the form $H_0: \mu'_{video} = \mu'_{audio}$, where the symbol μ' stands for an adjusted population mean.

The Focus, Number, and Quality of the Covariate Variable(s)

Suppose two different researchers each conduct a study in which they use the analysis of covariance to analyze their data. Further suppose that these two studies are conducted at the same point in time, in the same kind of setting, with the same independent variable(s) [as defined by the factor(s) and levels], and with the same measuring instrument used to collect data on the dependent variable from the same number and type of research participants. Despite all these similarities, these two ANCOVA studies might yield entirely different results because of differences in the focus, number, and quality of the covariate variable(s) used in the two investigations.

Regarding its focus, the covariate in many studies is set up to be simply an indication of each participant's status on the dependent variable at the beginning of the investigation. When this is done, the participant's scores on the covariate are referred to as their *pretest* or *baseline measures*, examples of which can be seen in Excerpts 15.5, 15.7, 15.8, and 15.9. In one sense, this kind of ANCOVA study is simpler (but not necessarily of lower quality) because a single measuring instrument is used to collect data on both the covariate and the dependent variables.

There is no rule, however, that forces researchers to use pretest-type data to represent the covariate variable. In many studies, the covariate variable is entirely different from the dependent variable. For example, consider again Excerpt 15.4. In that study, the dependent variable was student absences from class whereas the covariate variable was grade point average (GPA).

Regardless of whether the covariate and dependent variables are the same or different, the adjustment process of ANCOVA is basically the same. First, the mean covariate score of all subjects in all comparison groups is computed. Next, each comparison group's covariate mean is compared against the grand covariate mean to see (1) if the former is above or below the latter, and (2) how much of a difference there is between these two means. Finally, each group's mean on the dependent variable is adjusted up or down (depending on whether the group was below or above average on the covariate), with larger adjustments made when a group's covariate mean is found to deviate further from the grand covariate mean.

The second way in which two ANCOVA studies might differ—even though they are identical in terms of independent and dependent variables, setting, and participants—concerns the number of covariate variables involved in the study. Simply stated, there can be one, two, or more covariate variables included in any ANCOVA study. Most often, only one covariate variable is incorporated into the researcher's study. Of the excerpts we have considered thus far, most came from studies wherein there was a single covariate variable. In Excerpt 15.11,

EXCERPT 15.11 • *ANCOVA with Multiple Covariates*

A total of 140 students recruited from undergraduate communication courses participated in the experiment in exchange for extra course credit. They were randomly assigned to one of the four experimental conditions. . . . To test H1 and H2, a full-factorial analysis of covariance (ANCOVA) was conducted with two independent variables (Website and Web agent), one dependent variable (amount of systematic processing), and three covariates (institutional trust, prior familiarity, and task importance).

Source: Koh, Y. J., & Sundar, S. S. (2010). Heuristic versus systematic processing of specialist versus generalist sources in online media. *Human Communication Research, 36*(2), 103–124.

we see a case in which three covariates were used in a 2 × 2 ANCOVA. Notice that data on these covariate variables were incorporated into the analysis even though the research participants were randomly assigned to the four treatment conditions.

Although it might seem as if ANCOVA would work better when many covariate variables are involved, the researcher must pay a price for each such variable. We consider this point later in the chapter. For now, all you need to know is that most ANCOVAs are conducted with only a small number covariate variables. You are unlikely to see more than five such variables used in an ANCOVA.

The third way two seemingly similar studies might differ concerns the quality of the covariate variable. Earlier, we considered the two roles of the covariate: power and control. Covariate data do not help in either regard if (1) an irrelevant covariate variable is used or (2) the covariate variable is relevant conceptually but measured in such a way that the resulting data are invalid or unreliable.

In order for a covariate variable to be conceptually relevant within a given study, it must be related to the study's dependent variable. In studies where measurements on the covariate variable are gathered via a pretest, with posttest scores (from the same measuring instrument) used to represent the dependent variable, the conceptual relevance of the covariate is clear. When the covariate and dependent variables are different, it may or may not be the case that the covariate is worthy of being included in the ANCOVA. Later in this chapter, we return to this issue of relevant covariate variables.

Even if the covariate *variable* selected by the researcher is sensible, the *measurement* of the variable must be sound. In other words, the data collected on the covariate variable must be both reliable and valid. Earlier in this book, we considered different strategies available for estimating reliability and validity. Competent researchers use these techniques to assess the quality of their covariate data; moreover, they present evidence of such data checks within their research reports.

Presentation of Results

Most researchers present the results of their ANCOVAs within the text of their research reports. Accordingly, we now look at a few passages that have been taken from recently published articles.

In Excerpt 15.12, we see the results of a one-way analysis of covariance. This passage is very similar to what you might see when a researcher uses a one-way ANOVA to compare the means of four groups. We are given information as to the independent and dependent variables, the calculated F-value, the sample means that were compared, the decision about the null hypothesis ($p < .001$), and an effect size estimate. Moreover, just like a one-way ANOVA, a post hoc investigation was conducted (using Bonferroni-adjusted LSD tests) to make pairwise comparisons among the groups.

EXCERPT 15.12 • *Results of a One-Way ANCOVA*

A one-way analysis of covariance on postexpectancy ratings, with baseline expectancy ratings as the covariate, produced a significant main effect for treatment condition, $F(3,167) = 15.21, p < .001, \text{eta}^2 = .22$. A least significant difference test on estimated marginal means with a Bonferroni adjustment for the number of statistical comparisons revealed that participants in the no-treatment control condition (adjusted mean $= 4.80, SD = 2.95$) expected more pain than those in the hypnotic analgesia (adjusted mean $= 3.41, SD = 2.83$), imaginative analgesia (adjusted mean $= 3.42, SD = 2.92$), and placebo (adjusted mean $= 2.70, SD = 2.99$) conditions. All of the other pairwise comparisons were nonsignificant.

Source: Milling, L. S. (2009). Response expectancies: A psychological mechanism of suggested and placebo analgesia. *Contemporary Hypnosis, 26*(2), 93–110.

There are two things about this passage that make it different from textual presentations of results following a one-way ANOVA. First, note that the four sample means are referred to as *adjusted means*. Sample means are never referred to this way in a one-way ANOVA. Second, note the *df* numbers next to the F-value. If this were a one-way ANOVA, you add those numbers together and then add 1 in order to figure out how many people were involved in the study. Because these results come from an ANCOVA, you must add 2 to the sum of df_{between} and df_{within}, not 1. This is because 1 degree of freedom is used up from the within-groups *df* for each covariate variable included in an ANCOVA study. Knowing this, you can determine that the data for this one-way ANCOVA came from 172 individuals.

Now consider the passage of text in Excerpt 15.13. Here, we can see how a team of researchers reported the results of their two-way mixed ANCOVA. In the study associated with this excerpt, 124 male and female high school students experienced

EXCERPT 15.13 • *Results of a Two-Way Mixed ANCOVA*

[F]or male students, a 2 (condition: experimental, control) $\times$ 3 (time: pretest, posttest, and follow-up) mixed-model analysis of covariance (ANCOVA) with the IRMA-SF as the repeated dependent variable and with the MCSD-C as the covariate revealed a significant condition effect, $F(1,125) = 6.60, p < .01$, partial $\eta^2 = 0.06$, and a Condition $\times$ Time interaction effect, $F(2,250) = 7.19, p < .01$, partial $\eta^2 = 0.05$. . . . Subsequent analyses on the interaction effect revealed that the IRMA-SF scores of the experimental group changed significantly across the three phases [$Ms = 39.49, 36.28,$ and 35.46], $F(2,154) = 3.34, p < .05$, partial $\eta^2 = 0.04$, but those of the control group did not [$Ms = 40.40, 41.30,$ and 41.66]. . . . Specifically, the IRMA-SF scores of the experimental group decreased from pretest to posttest, $F(1,76) = 7.78, p < .001$, partial $\eta^2 = 0.09$, and were significantly lower than those of the control group at posttest, $t(126) = -2.73, p < .001$, and follow-up, $t(126) = -3.76, p < .001$.

Source: Hillenbrand-Gunn, T. L., Heppner, M. J., Mauch, P. A., & Park, H.-J. (2010). Men as allies: The efficacy of a high school rape prevention intervention. *Journal of Counseling & Development, 88*(1), 43–51.

a three-session program on date-rape prevention. A control group of 88 similar students was not given this program. Three times—prior to the intervention, immediately after it was over, and then four weeks later—the students in both groups filled out the IRMA-SF, a questionnaire designed measure one's acceptance of date-rape myths. At the beginning of the study, the students also filled out a personality inventory (the MCSD-C) that assesses one's tendency to answer questions in a socially desirable manner.

The results shown in Excerpt 15.13 are based on the data provided by the 128 male students (78 in the experimental group, 50 in the control group). The researchers associated with this study did two things worth noting. One of these good practices—the estimation of effect size—was done (using partial η^2) in conjunction with every F-ratio presented. The other good practice followed by these researchers is the post-hoc investigation they conducted to probe the significant interaction.

The Statistical Basis for ANCOVA's Power Advantage and Adjustment Feature

In an earlier section of this chapter, you learned that a good covariate variable is both conceptually related to the dependent variable and measured in such a way as to provide reliable and valid data. But how can researchers determine whether existing data (or new data that could be freshly gathered) meet this double criterion? Every researcher who uses ANCOVA ought to be able to answer this question whenever one or more covariate variables are incorporated into a study. With bad covariate variables, nothing of value comes in return for what is given up.

In order for ANCOVA to provide increased power (over a comparable ANOVA) and to adjust the group means, the covariate variable(s) must be correlated with the dependent variable. Although the correlation(s) can be either positive or negative in nature, ANCOVA does not achieve its power and adjustment objectives unless at least a moderate relationship exists between each covariate variable and the dependent variable. Stated differently, nuisance variability within the dependent variable scores can be accounted for to the extent that a strong relationship exists between the covariate(s) and the dependent variable.[4]

There are many ways to consider the correlation in ANCOVA, even when data have been collected on just one covariate variable. Two ways of doing this involve (1) looking at the correlation between the covariate and dependent variables for all participants from all comparison groups thrown into one large group, or (2) looking at the correlation between the covariate and dependent variables separately within each comparison group. The second of these two ways of considering the correlation is appropriate because ANCOVA makes its adjustments (of individual scores and of group means) on the basis of the pooled within-groups correlation coefficient.

One final point is worth making about the correlation between the covariate and dependent variables. Regarding the question of how large the pooled within-groups r must be before the covariate can make a difference in terms of increasing power, statistical theory responds by saying that the absolute value of this r should be at least .20. When r is at least this large, the reduction in the error SS compensates for df that are lost from the error row of the ANCOVA summary table. If r is lower than $\pm.20$, however, the effect of having a smaller number of error df without a proportionate decrease in the error SS is to make the error MS larger, not smaller, a situation that brings about a reduction in power.[5]

Assumptions

The statistical assumptions of ANCOVA include all the assumptions that are associated with ANOVA, plus three that are unique to the situation where data on a covariate variable are used in an effort to make adjustments and increase power. All three of these unique-to-ANCOVA assumptions must be met if the analysis is to function in its intended manner, and the researcher (and you) should consider these assumptions whenever ANCOVA is used—even in situations where the comparison groups are equally large. In other words, equal *n*s do *not* cause ANCOVA to be robust to any of the assumptions we now will consider.

[4]When two or more covariate variables are used within the same study, ANCOVA works best when the covariates are *unrelated* to each other. When the correlations among the covariate variables are low, each such variable has a chance to account for a different portion of the nuisance variability in the dependent variable.
[5]Although we use r in this paragraph, it is actually the population parameter ρ that must exceed $\pm.20$ in order for ANCOVA to have a power advantage.

The Independent Variable Should Not
Affect the Covariate Variable

The first of these three new assumptions stipulates that the study's independent variable should not affect the covariate variable. In an experiment (where the independent variable is an active factor), this assumption clearly is met if the data on the covariate variable are collected before the treatments are applied. If the covariate data are collected after the treatments have been doled out, the situation is far murkier—and the researcher should provide a logical argument on behalf of the implicit claim that treatments do not affect the covariate. In non-experimental (i.e., descriptive) studies, the situation is even murkier because the independent variable very likely may have influenced each participant's status on the covariate variable prior to the study. We return to this issue—of covariance being used in nonrandomized studies—in the next major section.

To see an example of how the covariate can be affected by the independent variable in a randomized experiment, consider Excerpt 15.14. In this study, data on the covariate variable (time-on-task) were collected *after* the independent variable had been applied. In a preliminary analysis, the researchers confirmed what you probably are thinking: mean time-on-task was significantly greater for those participants who saw animated rather than static visuals and who experienced the

EXCERPT 15.14 • *The Independent Variable Should Not Affect the Covariate*

The instructional material used in this study consisted of a 2,000-word physiology unit focusing on the human heart, its parts, locations, and functions during the diastolic and systolic phases [with] 20 visuals of the human heart, which were designed and positioned utilizing the principles of instructional consistency and congruency. . . . This instructional unit was selected to explore the effect of two types of visuals, i.e., static versus animated, and three types of instructional strategies, i.e., no strategy, questions, and questions plus feedback.

Time-on-task was employed as a control variable in [the analysis of covariance portion of] this study. It was defined as the total time students spent studying the respective treatment material, and was recorded by the computer from the moment students began the first frame of the instructional material to the moment they clicked on a button to indicate that they had finished studying the material and were ready to take the tests. The amount of time was collected in seconds. Time-on-task was used as a covariate rather than a dependent variable because the study was not intended to measure the amount of time that each treatment group spent studying the material and to compare if there was a significant difference.

Source: Lin, H., & Dwyer, F. M. (2010). The effect of static and animated visualization: a perspective of instructional effectiveness and efficiency. *Educational Technology Research & Development, 58*(2), 155–174.

questions-plus-feedback learning strategy. Clearly, scores on the covariate variable were affected by the treatment.

Homogeneity of Within-Group Correlations (or Regression Slopes)

The second unique assumption associated with ANCOVA stipulates that the correlation between the covariate and dependent variables is the same within each of the populations involved in the study. This assumption usually is talked about in terms of regression slopes rather than correlations, and therefore you are likely to come across ANCOVA research reports that contain references to the **assumption of equal regression slopes** or to the **homogeneity of regression slope assumption.** The data of a study can be employed to test this assumption—and it should *always* be tested when ANCOVA is used. As is the case when testing other assumptions, the researcher will be happy when the statistical test of the equal slopes assumption leads to a fail-to-reject decision, an outcome interpreted as a signal that it is permissible to analyze the study's data using ANCOVA procedures.

Consider Excerpt 15.15, which shows how a team of researchers attended to ANCOVA's assumption of equal regression slopes. Although this passage is only two sentences in length, it is a highly important element of the research report. If the researchers had not provided the results of this preliminary check on their data's suitability to go into a covariance analysis, we (and, it is hoped, they too) would need to be wary of the ANCOVA results.

EXCERPT 15.15 • *The Assumption of Equal Regression Slopes*

To be valid, ANCOVA assumes homogeneity among within-treatment regressions. For this analysis, within-treatment regressions were not significantly different from each other ($F = 1.34, p = .25358$) and can therefore be considered homogeneous, meeting the assumption of ANCOVA.

Source: Marcus, J. M., Hughes, T. M., McElroy, D. M., & Wyatt, R. E. (2010). Engaging first-year undergraduates in hands-on research experiences: The Upper Green River Barcode of Life Project. *Journal of College Science Teaching, 39*(3), 39–45.

If the equal-slopes H_0 is rejected, there are several options open to the researcher. In that situation, the data can be transformed and then the assumption can be tested again using the transformed data. Or, the researcher can turn to one of several more complicated analyses (e.g., the Johnson–Neyman technique) developed specifically for the situation where heterogeneous regressions exist. Or, the researcher can decide to pay no attention to the covariate data and simply use an ANOVA to compare groups on the dependent variable. These various options come into play only rarely, either because the equal-slopes assumption is not rejected when tested or because the researcher wrongfully bypasses testing the assumption.

Linearity

The third assumption connected to ANCOVA (but not ANOVA) stipulates that the within-group relationship between the covariate and dependent variables should be linear.[6] There are several ways this assumption can be evaluated, some involving a formal statistical test and some involving visual inspections of scatter plots. (A special type of scatter diagram, involving plots of residuals, is used more frequently than plots of raw scores.) Regardless of how researchers might choose to assess the **linearity** assumption, I salute their efforts to examine their data to see if it is legitimate to conduct an analysis of covariance.

Consider Excerpt 15.16. First, note what is said in the first two sentences. Those points apply to any situation in which ANCOVA is used. Next, make sure you see what the researchers said about their test of linearity and what they did when the linearity assumption was deemed untenable. These researchers deserve high praise for attending to ANCOVA's linearity assumption.

EXCERPT 15.16 • The Linearity Assumption

The statistical requirements for ANCOVA are stricter than for ANOVA. In addition to the requirements of ANOVA, ANCOVA requires that the relationship between the dependent variable and the covariate be linear (assumption of linearity). . . . In the current study, *GPA* violates the linearity assumption, and transformations of the *GPA* variable did not resolve the issue. Consequently, ANOVA was used despite its limitations [e.g., lower power] compared to ANCOVA.

Source: Jones, S. H., & Wright, M. E. (2010). The effects of a hypertext learning aid and cognitive style on performance in advanced financial accounting. *Issues in Accounting Education, 25*(1), 35–58.

Other Standard Assumptions

As indicated earlier, the standard assumptions of ANOVA (e.g., normality, homogeneity of variance, sphericity) underlie ANCOVA as well. You should upgrade or downgrade your evaluation of a study depending on the attention given to these assumptions in the situations where F-tests are biased because assumptions are violated. Unfortunately, you are likely to come across *many* ANCOVA studies in which there is absolutely no discussion of linearity, equal regression slopes, normality, or homogeneity of variance.

---

[6] I first discussed the notion of linearity in Chapter 3; you may want to review that earlier discussion if you have forgotten what it means to say that two variables have a linear relationship.

ANCOVA When Comparison Groups Are Not Formed Randomly

In a randomized experiment, the various population means on the covariate variable can be considered identical. This is the case because of the random assignment of research participants to the comparison groups of the investigation. Granted, the sample means for the comparison groups on the covariate variable probably vary, but the corresponding population means are identical.

When ANCOVA is used to compare groups that are formed in a nonrandomized fashion, the population means on the covariate variable cannot be assumed to be equal. For example, if a study is set up to compare sixth-grade boys with sixth-grade girls on their ability to solve word problems in mathematics, a researcher might choose to make the comparison using ANCOVA, with reading ability used as the covariate variable. In such a study, the population means on reading ability might well differ between the two gender groups.

Although my concern over the equality or inequality of the covariate population means may initially seem silly (because of the adjustment feature of ANCOVA), this issue is far more important than it might at first appear. I say this because studies in theoretical statistics have shown that ANCOVA's adjusted means turn out to be biased in the situation where the comparison groups differ with respect to population means on the covariate variable. In other words, the sample-based adjusted means on the dependent variable do not turn out to be accurate estimates of the corresponding adjusted means in the population when the population means on the covariate variable are dissimilar.

Because ANCOVA produces adjusted means, many applied researchers evidently think that this statistical procedure was designed to permit nonrandomly formed groups to be compared. Over the years, I have repeatedly come across research reports in which the researchers talk as if ANCOVA has the magical power to equate such groups and thereby allow valid inferences to be drawn from comparisons of adjusted means. Excerpts 15.17 and 15.18 illustrate this view held by many applied researchers that ANCOVA works well with nonrandomly formed groups. In the first of these excerpts, the clear impressive given is that ANCOVA would not have been used if the groups could have been formed randomly. In Excerpt 15.18, the researchers claim more directly that ANCOVA can control for initial group differences.

EXCERPTS 15.17–15.18 • *Use of ANCOVA with Groups Formed Nonrandomly*

The intervention group $(n = 11)$ received approximately 10 hours per week of behavioral intervention; the eclectic comparison group $(n = 14)$ received treatment as usual. . . . To evaluate of effectiveness of behavioral intervention, we used ANCOVA models. Because the children were not randomly assigned to groups or actively

(continued)

EXCERPTS 15.17–15.18 • (*continued*)

matched, the intake score for the specific outcome measure was entered as a covariate in each analysis. ANCOVAs were run for IQ and adaptive behavior, including all sub domains (except for motor skills) and the adaptive composite scores.

Source: Eldevik, S., Jahr, E., Eikeseth, S., Hastings, R. P., & Hughes, C. J. (2010). Cognitive and adaptive behavior outcomes of behavioral intervention for young children with intellectual disability. *Behavior Modification, 34*(1) 16–34.

Because of the need to accept pre-grouped sets of children, the experimental and control [groups] were not necessarily equivalent in terms of ability level. To compensate for this, . . . analyses of covariance were used to control for pre-existing differences between the control and experimental groups.

Source: Boakes, N. J. (2009). Origami instruction in the middle school mathematics classroom: Its impact on spatial visualization and geometry knowledge of students. *Research in Middle Level Education, 32*(7), 1–12.

Besides ANCOVA's statistical inability to generate unbiased adjusted means when nonrandomly formed groups are compared, there is a second, logical reason why you should be on guard whenever you come across a research report in which ANCOVA was used in an effort to equate groups created without random assignment. Simply stated, the covariate variable(s) used by the researcher may not address one or more important differences between the comparison groups. Here, the problem is that a given covariate variable (or set of covariate variables) is limited in scope. For example, the covariate variable(s) used by the researcher might address knowledge but not motivation (or vice versa).

Consider, for example, the many studies conducted in schools or colleges in which one intact group of students receives one form of instruction whereas a different intact group receives an alternative form of instruction. In such studies, it is common practice to compare the two groups' posttest means via an analysis of covariance, with the covariate being IQ, GPA, or score on a pretest. In the summaries of these studies, the researchers may say that they used ANCOVA "to control for initial differences between the groups." However, it is debatable whether academic ability is reflected in any of the three covariates mentioned (or even in all three used jointly). In this and many other studies, people's motivation plays no small part in how well they perform.

In summary, be extremely cautious when confronted with research claims based on the use of ANCOVA with intact groups. If an important covariate variable was overlooked by those who conducted the study, pay no attention whatsoever to the conclusions based on the data analysis. Even in the case where data on all important covariate variables were collected and used, you *still* should be tentative in your inclination to buy into the claims made by the researchers.

Related Issues

Near the beginning of this chapter, I asserted that any ANCOVA is, in several respects, like its ANOVA counterpart. We have already considered many of the ways in which ANOVA and ANCOVA are similar, such as the way post hoc tests are typically used to probe significant main effects and interactions. At this point, we ought to consider three additional ways in which ANCOVA is akin to ANOVA.

As with ANOVA, the Type I error rate is inflated if separate ANCOVAs are used to analyze the data corresponding to two or more dependent variables. To deal with this problem, the conscientious researcher implements one of several available strategies. The most frequently used strategy for keeping the operational Type I error rate in line with the stated alpha level is the **Bonferroni adjustment technique,** and it can be used with ANCOVA as easily as it can with ANOVA.

In Excerpt 15.19, we see a case in which the Bonferroni adjustment technique was used in conjunction with a one-way ANCOVA that was applied seven times. After dividing their desired study-wide alpha level (of .05) by 7, the researchers used .007 to evaluate the *F*-ratio generated by each ANCOVA.

EXCERPT 15.19 • *Use of the Bonferroni Adjustment Technique*

Correlations and one-way analyses of covariance (ANCOVAs) were conducted. . . . Because seven ANCOVAs were [run], a Bonferroni adjustment required an alpha level of .007 (i.e., .05/7) to be used.

Source: Pulido, D. (2009). How involved are American L2 learners of Spanish in lexical input processing tasks during reading? *Studies in Second Language Acquisition, 31*(1), 31–58.

The second issue that has a common connection to both ANOVA and ANCOVA is the important distinction between **statistical significance** and **practical significance.** Because it is possible, in either kind of analysis, for the data to produce a finding that is significant in a statistical sense but not in a practical sense, you should upgrade your evaluation of any ANCOVA study wherein the researcher conducts an a priori power analysis or estimates the effect size from the sample data.

Earlier (in Excerpts 15.12 and 15.13), we saw examples in which eta squared and partial eta squared were used to estimate the practical significance of ANCOVA results. Now, in Excerpt 15.20, we see a case in which the issue of practical significance is addressed by means of a different way of estimating effect size. As you can see, the researchers here used *d*. This excerpt is especially nice because it shows the good practice of connecting estimates of effect size to *adjusted* means.

EXCERPT 15.20 • *Effect Size Estimation in ANCOVA*

Three one-way analyses of covariance (ANCOVAs) were employed to examine between-group differences on the three dependent variables of home literacy practices, the number of minutes parents read to children, and child emergent literacy skills. [No] significant differences were found between children with developmental disabilities and TD children with regard to home literacy practices, $M_{D\,ADJ} = 20.22$; $M_{T\,ADJ} = 20.45$; $F(1, 952) = .32, p = .57, d = .04$, or the number of minutes parents read to their children, $M_{D\,ADJ} = 22.56$; $M_{T\,ADJ} = 23.89$; $F(1, 952) = 1.28$, $p = .26, d = .07$. However, for children at the average age and SES of the sample, parents reported significant differences in child emergent literacy skills, with TD children ($M_{T\,ADJ} = 3.90$) outperforming children with disabilities, $M_{D\,ADJ} = 3.45$; $F(1, 952) = 10.76, p = .001, d = .18$.

Source: Breit-Smith, A., Cabell, S. Q., & Justice, L. M. (2010). Home literacy experiences and early childhood disability: A descriptive study using the National Household Education Surveys (NHES) Program database. *Language, Speech, & Hearing Services in Schools, 41*(1), 96–107.

The criteria for judging effect size indices in any ANCOVA study are the same as the standards for these indices when they are computed for any ANOVA study. To see the criteria for judging the magnitude of d, η^2, and other ways of estimating effect size, refer to Table 13.2. Be sure to read the note beneath that table. Also, remember that certain experienced researchers choose, with justification, to discuss effect size by focusing on the raw-score difference between means (rather than to use a standardized estimate of effect size).

In Excerpt 15.21, we see a case in which a research team conducted an a priori power analysis in conjunction with their ANCOVA study. This was done to determine

EXCERPT 15.21 • *A Power Analysis to Determine Sample Size*

An a priori power analysis was conducted to determine the minimum number of participants that would need to be enrolled. The criterion for significance was set at $p = .05$. It was assumed that the analysis of covariance would be non-directional (i.e., two-tailed), thus an effect in either direction would be interpreted. A medium effect size was specified (i.e., $f = .25$, as per Cohen 1988). The study included two covariates (age and an expressive communication indicator) that accounted for 80% of the variance in the dependent variable (i.e., $r = .90$ with the dependent variables). Lastly, power was set to .80, meaning that the study would have an 80% probability of finding differences among the three groups if such differences exist in the population. The power analysis indicated that the study required a minimum of 13 cases per group for a total of 39 cases. Actual enrollment was 14 cases per group, for a total of 42 cases.

Source: Mineo, B. A., Ziegler, W., Gill, S., & Salkin, D. (2009). Engagement with electronic screen media among students with autism spectrum disorders. *Journal of Autism & Developmental Disorders, 39*(1), 172–187.

how large the study's sample sizes needed to be. As you can see, the power analysis indicated that the researchers could utilize a small number of people ($n = 13$) in each of the three comparison groups. The main reason for this is the large amount of variability (80%) in the dependent variable predicted to be associated with the study's covariates. If that ingredient for the power analysis had been lower, the needed sample size would have been greater. Based on the way the researchers described their power analysis, it seems that they were not guessing about the effectiveness of their covariates, but rather had firm knowledge that those two variables jointly explained much of the variability in their study's dependent variable.

The third point I want to make in this section is simple: Planned comparisons can be used in ANCOVA studies just as easily as they can be used in ANOVA studies. In Excerpt 15.22, we see an example of this being done. This excerpt not only illustrates the use of planned ANCOVA comparisons, it also shows how a study can be conducted with a two-factor design in which certain of the cells are completely empty!

EXCERPT 15.22 • *ANCOVA and Planned Comparisons*

In order to investigate the specific impact of multimodality and interactivity in the context of common media application formats, we developed a partial 2 × 3 (interactive, noninteractive by high, moderate, and low in multimodality) factorial between-participants follow-up design with four conditions: (a) game (interactive, high multimodality), (b) game replay (noninteractive, high multimodality), (c) hypertext (interactive, medium multimodality), and (d) text (noninteractive, low multimodality). . . . Analysis of covariance (ANCOVA) with a priori contrasts was chosen as the main analysis approach. To explore the impact of multimodality and interactivity on educational outcome, two sets of planned contrasts were performed; the first set used game condition, and the second used text condition as the reference category. . . . [R]elative to text condition, replay conditions yielded higher knowledge gain (especially knowledge gain of definitions) and higher interest in learning. Relative to game condition, hypertext condition had lower knowledge gain of definition items in the posttest.

Source: Ritterfeld, U., Shen, C., Wang, H., Nocera, L., & Wong, W. L. (2009). Multimodality and interactivity: Connecting properties of serious games with educational outcomes. *CyberPsychology & Behavior, 12*(6), 691–697.

A Few Warnings

Before concluding our discussion of ANCOVA, I want to offer a few warnings about deciphering research reports that present results from this form of statistical analysis. As you consider these comments, however, do not forget that ANCOVA legitimately can be thought of as a set of statistical procedures made possible by adding covariate data to an ANOVA-type situation. Because of this fact, all the tips and warnings offered at the conclusions of Chapters 10 through 14 should be kept in mind when you consider the results from a study that used ANCOVA. In addition to being aware of the concerns focused on in those earlier chapters, you should also remain sensitive to the following three unique-to-ANCOVA cautions when considering research claims based on covariance analyses.

The Statistical Focus: Adjusted Means

In a covariance analysis, all F-tests (other than those concerned with underlying assumptions) deal with adjusted means on the dependent variable, not the unadjusted means. This holds true for the F-tests contained in the ANCOVA summary table, the F-tests involved in any planned comparisons, and the F-tests involved in any post hoc investigation. For this reason, adjusted means should be presented—either in a table or within the textual discussion—whenever the researcher attempts to explain the meaning of any F-test result. It is helpful, as we have seen, to have access to the means on the covariate variable and the unadjusted means on the dependent variable. However, the adjusted means on the dependent variable constitute the central statistical focus of any ANCOVA.

Unfortunately, many researchers fail to present the adjusted means in their research reports. When this happens, you are boxed into a corner in which you cannot easily decide for yourself whether a statistically significant finding ought to be considered significant in a practical sense. Because making this kind of decision is one of the things consumers of the research literature ought to do on a regular basis, I must encourage you to downgrade your evaluation of any ANCOVA-based study that fails to contain the adjusted means that go with the F-test(s) focused on by the researcher.

The Importance of Underlying Assumptions

ANCOVA's F-tests that compare adjusted means function as they are supposed to function only if various underlying assumptions are valid. Moreover, the condition of equal sample sizes does not bring about a situation where the assumptions are rendered unimportant. In other words, equal ns do not cause ANCOVA to become robust to its underlying assumptions.

Whenever you consider research claims based on ANCOVA, check to see whether the researcher says anything about the statistical assumptions on which the analysis was based. Upgrade your evaluation of the research report when there is expressed concern over the assumption of equal regression slopes, the assumption of a linear relationship between the covariate and dependent variables, and the assumption that scores on the covariate variable are not influenced by the independent variable. If these assumptions are not discussed, you should downgrade your evaluation of the study.

If an assumption is tested, give the researchers some bonus credit if they use a lenient rather than rigorous alpha level in assessing the assumption's H_0. I say this because researchers deserve credit if they perform statistical tests in such a way that the "deck is stacked against them" in terms of what they would like to show. Because the typical researcher who uses ANCOVA wants the linearity and equal-slopes assumptions to be met, a lenient level of significance (e.g., .10, .15, .20, or even .25) gives the data more of a chance to reveal an improper situation than would be the case if alpha is set equal to .05, .01, or .001. When testing assumptions, Type II errors are generally considered to be more serious than errors of the first kind, and the level of significance should be set accordingly.

ANCOVA versus ANOVA

My final warning has to do with your general opinion of ANCOVA-based studies as compared with ANOVA-based studies. Because ANCOVA is more complex (due to the involvement of a larger number of variables and assumptions), many consumers of the research literature hold the opinion that data-based claims are more trustworthy when they are based on ANCOVA rather than ANOVA. I strongly encourage you to *refrain* from adopting this unjustified point of view.

Although ANCOVA (as compared with ANOVA) does, in fact, involve more complexities in terms of what is involved both on and beneath the surface, it is an extremely delicate instrument. To provide meaningful results, ANCOVA must be used very carefully—with attention paid to important assumptions, with focus directed at the appropriate set of sample means, and with concern over the correct way to draw inferences from ANCOVA's F-tests. Because of its complexity, ANCOVA affords its users more opportunities to make mistakes than does ANOVA.

If used skillfully, ANCOVA can be of great assistance to applied researchers. If not used carefully, however, ANCOVA can be dangerous. Unfortunately, many people think of complexity as being an inherent virtue. In statistics, that is often *not* the case. As pointed out earlier in the chapter, the interpretation of ANCOVA F-tests is problematic whenever the groups being compared have been formed in a nonrandom fashion—and this statement holds true even if (1) multiple covariate variables are involved, and (2) full attention is directed to all underlying assumptions. In contrast, it would be much easier to interpret the results generated by the application of ANOVA to the data provided by participants who have been randomly assigned to comparison groups. Care is required, of course, whenever you attempt to interpret the outcome of *any* inferential test. My point is simply that ANCOVA, because of its complexity as compared to ANOVA, demands a higher—not lower—level of care on your part when you encounter its results.

Review Terms

Adjusted means	Covariate variable
Analysis of covariance	Homogeneity of regression slopes
ANCOVA	Linearity
Assumption of equal regression slopes	Power
Bonferroni adjustment technique	Practical significance
Concomitant variable	Statistical significance

The Best Items in the Companion Website

1. An interactive online quiz (with immediate feedback provided) covering Chapter 15.
2. Nine misconceptions about the content of Chapter 15.
3. One of the best passages from Chapter 15: "Are ANCOVA Studies Better Than ANOVA Studies?"
4. The first two e-articles.
5. The first of the two jokes, because it deals with the analysis of covariance.

To access chapter outlines, practice tests, weblinks, and flashcards, visit the companion website at http://www.ReadingStats.com.

Review Questions and Answers begin on page 531.

Bivariate, Multiple, and Logistic Regression

In Chapter 3, we considered how bivariate correlation can be used to describe the relationship between two variables. Then, in Chapter 9, we looked at how bivariate correlations are dealt with in an inferential manner. In this chapter, our focus is on a topic closely related to correlation. This topic is called *regression*.

Three different kinds of regression are considered here: **bivariate regression, multiple regression,** and **logistic regression.** Bivariate regression is similar to bivariate correlation, because both are designed for situations in which there are just two variables. Multiple and logistic regression, however, were created for cases in which there are three or more variables. Although many other kinds of regression procedures have been developed, the three considered here are by far the ones used most frequently by applied researchers.

The three regression procedures considered in this chapter are like correlation in that they are concerned with relationships among variables. Because of this, you may be tempted to think that regression is simply another way of talking about, or measuring, correlation. Resist that temptation, because these two statistical procedures differ in three important respects: their purpose, the way variables are labeled, and the kinds of inferential tests applied to the data.

The first difference between correlation and regression concerns the purpose of each technique. As indicated in Chapter 3, bivariate correlation is designed to illuminate the relationship, or connection, between two variables. The computed correlation coefficient may suggest that the relationship being focused on is direct and strong, or indirect and moderate, or so weak that it would be unfair to think of the relationship as being either direct or indirect. Regardless of how things turn out, each of the two variables is equally responsible for the nature and strength of the link between the two variables.

Whereas correlation concentrates on the relationship, or link, that exists *between* variables, regression focuses on the variable(s) that exist on one or the other *ends* of the link. Depending on which end is focused on, regression tries to accomplish one or the other of two goals: **prediction** or **explanation.**

In some studies, regression is utilized to predict scores on one variable based on information regarding the other variable(s). For example, a college might use regression in an effort to predict how well applicants will handle its academic curriculum. Each applicant's college grade point average (GPA) might be the main focus of the regression, with predictions made on the basis of available data on other variables (e.g., an entrance exam, the applicant's essay, and the recommendations written by high school teachers). If used in this manner, regression's focus would be on the one variable toward which predictions are made: college GPA.

In other investigations, regression is used in an effort to explain why the study's people, animals, or things score differently on a particular variable of interest. For example, a researcher might be interested in why people differ in the degree to which they seem satisfied with life. If such a study were to be conducted, a questionnaire might be administered to a large group of individuals for the purpose of measuring life satisfaction. Those same individuals would also be measured on several other variables that might explain why some people are quite content with what life has thrown at them whereas others seem to grumble incessantly because they think life has been cruel and unfair to them. Such variables might include health status, relationships with others, and job enjoyment. If used in this manner, regression's focus would be on the variables that potentially explain why people differ in their levels of life satisfaction.

Excerpts 16.1 and 16.2 illustrate the two different purposes of regression. In the first of these excerpts, the clear objective was to use regression analyses to help predict people's adjustment to living in a nursing home. In Excerpt 16.2, the goal

EXCERPTS 16.1–16.2 • *The Two Purposes of Regression: Prediction and Explanation*

Relocation to a nursing home is regarded as one of the most stressful events a person can experience. . . . The purpose of this study was to identify predictors of nursing home adjustment for elderly residents using a direct measure of nursing home adjustment. . . . Descriptive analysis was used and multiple linear regression was performed to identify the predictors of adjustment for nursing home residents.

Source: Lee, G. E. (2010). Predictors of adjustment to nursing home life of elderly residents: A cross-sectional survey. *International Journal of Nursing Studies, 47*(6), 1–8.

Need for recovery (NFR) after work is an indicator for work-related fatigue. . . . This study aims to establish the prevalence of high work-related fatigue [and] explain group differences categorized by gender, age, and education. The study

(continued)

EXCERPTS 16.1–16.2 • (*continued*)

particularly aims to clarify prevalence and explanatory factors in highly educated women. [Our regression] analyses give an indication of the factors that may explain the difference in the prevalence of high NFR between the compared groups, and of the degree to which the combination of [several] demographic, health, and work-related factors can explain the difference in the prevalence of high NFR.

Source: Verdonk, P., Hooftman, W. E., van Veldhoven, M. J. P. M., Boelens, L. R. M., & Koppes, L. L. J. (2010). Work-related fatigue: The specific case of highly educated women in the Netherlands. *International Archives of Occupational and Environmental Health, 83*(3), 309–321.

was explanation, not prediction. Here, the researchers wanted to know which factors explain why some people are afflicted by work-related fatigue whereas others are not. The focus was primarily on women workers, a trait called *need for recovery* after a day's work, and demographic variables.

The second difference between regression and correlation concerns the labels attached to the variables. This difference can be seen most easily in the case in which data on just two variables have been collected. Let's call these variables A and B. In a correlation analysis, variables A and B have no special names; they are simply the study's two variables. With no distinction made between them, their location in verbal descriptions or in pictorial representations can be switched without changing what is being focused on. For example, once *r* becomes available, it can be described as the correlation between A and B *or* it can be referred to as the correlation between B and A. Likewise, if a scatter diagram is used to show the relationship between the two variables, it does not matter which variable is positioned on the abscissa.

In a regression analysis involving A and B, an important distinction between the two variables must be made. In regression, one of the two variables needs to be identified as the **dependent variable** and the other variable must be seen as the **independent variable.**[1] This distinction is important because (1) the scatter diagram in bivariate regression always is set up such that the vertical axis corresponds with the dependent variable whereas the horizontal axis represents the independent variable, and (2) the names of the two variables cannot be interchanged in verbal descriptions of the regression. For example, the regression of A on B is not the same as the regression of B on A.[2]

Excerpts 16.3 and 16.4 come from two studies that were quite different. In the first study, only two variables were involved in the single regression that was

---

[1]The terms **criterion variable, outcome variable,** and **response variable** are synonymous with the term **dependent variable,** whereas the terms **predictor variable** or **explanatory variable** mean the same thing as **independent variable.**

[2]When the phrase "regression of _____ on _____" is used, the variable appearing in the first blank is the dependent variable whereas the variable appearing in the second blank is the independent variable.

EXCERPTS 16.3–16.4 • *Dependent and Independent Variables*

Stress reactions to uncertainty were measured [via] the Physicians' Reactions to Uncertainty Scale (PRUS). . . . Epistemology was measured using the Physicians' Belief Scale (PBS). . . . Our primary hypothesis was tested [by means of] a simple bivariate regression with PRUS scores as the dependent variable and PBS scores as the independent variable.

Source: Evans, L., & Trotter, D. R. M. (2009). Epistemology and uncertainty in primary care: An exploratory study. *Family Medicine, 41*(5), 319–326.

To assess the extent to which students' perceptions of parenting style predicted evaluations of their parents' preferred music we performed a multiple regression analysis using the mean rating score of parent music (ratings given to only those pieces indicated as a favorite by a student's parent) as the dependent variable and scores on the caring and the autonomy dimensions of the PBI, as well as students' age and gender, as independent variables.

Source: Serbun, S. J., & DeBono, K. G. (2010). On appreciating the music of our parents: The role of the parent–child bond. *North American Journal of Psychology, 12*(1), 93–102.

conducted. In the second excerpt, there was one dependent variable and four independent variables. Despite these differences, notice how the researchers associated with each excerpt clearly designate the status of each variable as being a dependent variable or an independent variable.

The third difference between correlation and regression concerns the focus of inferential tests and confidence intervals. With correlation, there is just one thing that can be focused on: the sample correlation coefficient. With regression, however, inferences focus on the correlation coefficient, the regression coefficient(s), the intercept, the change in the regression coefficient, and something called the *odds ratio*. We consider these different inferential procedures as we look at bivariate regression, multiple regression, and logistic regression.

Although correlation and regression are not the same, correlational concepts serve as some (but not all) of regression's building blocks. With that being the case, you may wonder why this chapter is positioned here rather than immediately after Chapter 9. If this question has popped into your head, there is a simple answer. This chapter is located here because certain concepts from the analysis of variance and the analysis of covariance also serve as building blocks in some regression analyses. For example, researchers sometimes base their regression predictions (or explanations) on the interactions between independent variables. Also, regressions are sometimes conducted with one or more covariate variables controlled or held constant. Without knowing about interactions and covariates, you would be unable to understand these particular components of regression analyses.

We now turn our attention to the simplest kind of regression used by applied researchers. Take good mental notes as you study this material, for the concepts you now encounter provide a foundation for the other two kinds of regression to be considered later in the chapter.

Bivariate Regression

The simplest kind of regression analysis is called **bivariate regression.** First, we must clarify the purpose of and the data needed for this kind of regression. Then, we consider scatter diagrams, lines of best fit, and prediction equations. Finally, we discuss inferential procedures associated with bivariate regression.

Purpose and Data

As you would suspect based on its name, bivariate regression involves just two variables. One of the variables serves as the dependent variable whereas the other functions as the independent variable. The purpose of this kind of regression can be either prediction or explanation; however, bivariate regression is used most frequently to see how well scores on the dependent variable can be predicted from data on the independent variable.

To illustrate how bivariate regression can be used in a predictive manner, imagine that Sam, a 41-year-old tennis player, has been plagued by a knee injury that for months has failed to respond to nonsurgical treatment. Consequently, arthroscopic surgery is scheduled to repair Sam's bad knee. Even though he knows that arthroscopic procedures usually permit a rapid return to usual activity, Sam would like to know how long he will be out of commission following surgery. His presurgery question to the doctor is short and sweet: "When will I be able to play again?" Clearly, Sam wants his doctor to make a prediction.

Although Sam's doctor might be inclined to answer this question concerning down-time by telling Sam about the *average* length of convalescence for tennis players following arthroscopic knee surgery, that is really not what Sam wants to know. Obviously, some people bounce back from surgery more quickly than do others. Therefore, Sam wants the doctor to consider his (i.e., Sam's) individual case and make a prediction about how long *he* will have to interrupt his on-court activity. If Sam's doctor is aware of what has happened with other tennis players who have had arthroscopic knee surgery, and if the doctor has a computer program that can perform a bivariate regression, he could provide Sam with a better-than-average answer to the question about postsurgical down time.

In the study conducted with people like Sam, imagine that there are 12 tennis players who had one of their knees repaired via arthroscopic surgery. Also imagine that data exist on each person regarding two variables: age and number of postsurgical days of down-time. Table 16.1 shows the data on these two variables.

TABLE 16.1   Data for Bivariate Regression Example

Post-Surgical Down-Time and Age for 12 Adult Tennis Players

Player	Age	Down-Time (days)
Kathy	41	7
Alex	47	6
Nancy	36	5
David	29	5
Pat	41	6
Andrew	22	3
Allison	21	4
Gary	38	6
Emily	19	5
Candace	31	4
Ted	32	6
Barbara	24	4

Scatter Diagrams, Regression Lines, and Regression Equations

The component parts and functioning of regression can best be understood by examining a scatter diagram. In Figure 16.1, such a picture has been generated using the data from Table 16.1. There are 12 dots in this "picture," each positioned so as to reveal the age and postsurgical convalescent time for one of the tennis players.

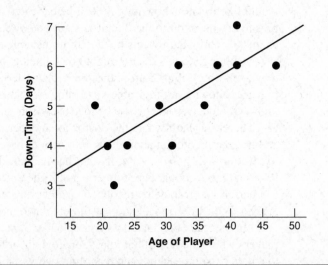

FIGURE 16.1   Scatter Plot with Regression Line

The scatter diagram in Figure 16.1 was set up with days of convalescence on the ordinate and age on the abscissa. These two axes of the scatter diagram were labeled like this because it makes sense to treat convalescence as the dependent variable. It is the variable toward which predictions will eventually be made for Sam and other tennis players who are similar to those who supplied the data we are currently considering. Age, however, is positioned on the abscissa because it is the independent variable. It is the variable that "supplies" data used to make the predictions.[3]

As you can see, a slanted line passes through the data points of the scatter diagram. This line is called the **regression line** or the **line of best fit,** and it functions as the tool our hypothetical doctor will use in order to predict how long Sam will have to refrain from playing tennis. As should be apparent, the regression line is positioned so as to be as close as possible to all of the dots. A special formula determines the precise location of this line; however, you do not need to know anything about that formula except that it is based on a statistical concept called *least squares*.[4]

Let's make a prediction for Sam, pretending now that we are his doctor. All we must do is turn to the scatter diagram and take a little trip with our index finger or our eyes. Our trip begins on the abscissa at a point equal to Sam's age. (Remember, Sam is 41 years old.) We move vertically from that point up into the scatter diagram until we reach the regression line. Finally, we travel horizontally (to the left) from that point on the regression line until reaching the ordinate. Wherever this little trip causes us to end up on the ordinate becomes our prediction for Sam's down time. According to our information, our prediction is that Sam will be out of commission for approximately six days.

Notice that our prediction of Sam's down time would have been shorter if he had been younger and longer if he had been older. For example, we would have predicted about four days if he had been 21 years old, or five days if he had been 31. These alternative predictions for a younger Sam are brought about by the tilt of the regression line. Because there is a positive correlation between the independent and dependent variables, the regression line tilts from lower left to upper right.

Although it is instructive to see how predictions are made by means of a regression line that passes through the data points of a scatter diagram, the exact same objective can be achieved more quickly and more scientifically by means of something called the **regression equation.** In bivariate, linear regression, this equation always has the following form:

$$Y' = a + b \cdot X$$

---

[3]Because we are dealing with regression (and not correlation), it would be improper to switch the two variables in the scatter diagram. The dependent variable always goes on the ordinate; the independent variable always goes on the abscissa.

[4]The *least squares principle* simply means that when the squared vertical distances of the data points from the regression line are added together, they yield a smaller sum than would be the case for any other straight line that could be drawn through the scatter diagram's data points.

where Y' stands for the predicted score on the dependent variable, a is a constant, b is the **regression coefficient,** and X is the known score on the independent variable. This equation is simply the technical way of describing the regression line. For the data shown in Table 16.1 (and Figure 16.1), the regression equation turns out like this:

$$Y' = 1.978 + 0.098 \cdot X$$

To make a prediction for Sam by using the regression equation, we simply substitute Sam's age for X and then work out the simple math. Because the numbers in the regression equation are so close to being whole numbers, let's round things off a bit and rewrite the regression equation as $Y' = 2 + (0.10)X$. We now have a simple prediction model that says to add 2 to one-tenth of person's age, with the result being a guess as to that individual's down-time. When we do this for Sam, $Y' = 6.1$. If we don't round off, $Y' = 5.996$. The fact that these values are very similar to what we predicted from the scatter diagram should not be at all surprising. This is because the regression equation is nothing more than a precise mechanism for telling us where we should end up if, in a scatter diagram, we first move vertically from some point on the abscissa up to the regression line and then move horizontally from the regression line to the ordinate.

Whereas scatter diagrams with regression lines appear only rarely in research reports, regression equations show up more frequently. We will see an example shortly; first, however, let's consider the study that was conducted. In this investigation, the researchers collected data from 67 individuals with multiple sclerosis (MS) who had been on a home-based self-medication program. First, the researcher asked each patient to estimate the percentage of self-injections that had been missed during the previous two months. Then, self-medication adherence was electronically monitored during the next two-month period. These data were used to see if retrospective self-reports could predict prospective (i.e., future) adherence.

The data assessing self-medication adherence during the second half of the study did not come from patients' self-reports. Instead, the researchers provided patients with special containers into which they put their disposable needles after they were used. These containers—called MEMS (Medication Event Monitoring System)—had been designed to record electronically the precise date and time any needle was put into the container. At the end of the full four-month period of the study, the researchers performed a regression analysis in which retrospective self-report of adherence served as the independent (predictor) variable and the electronic MEMS-based measure of adherence was the dependent (criterion) variable.

In Excerpt 16.5, we see the regression equation that appeared in the research report that summarized the self-medication study. As indicated in the excerpt, the data used to create this regression equation came from all research participants except four who reported very poor adherence over the first two-month period of the study.

EXCERPT 16.5 • *The Regression Equation in Bivariate Regression*

[We assessed] the relationship between self-reported retrospective adherence and prospective electronic monitoring among [63] patients who did not report poor adherence at the outset of the study. The data fit the regression line $y = 2.00x + 0.42$ (x = retrospective self-report and y = MEMS % days missed), suggesting that the prediction estimates of poor objective prospective adherence may be achieved by doubling patient reports of retrospective missed doses.

Source: Bruce, J. M., Hancock, L. M., & Lynch, S. G. (2010). Objective adherence monitoring in multiple sclerosis: Initial validation and association with self-report. *Multiple Sclerosis, 16*(1), 112–120.

It should be noted that there are two kinds of regression equations that can be created in any bivariate regression analysis. One of these is called an **unstandardized regression equation.** This is the kind we have considered thus far, and it has the form $Y' = a + b \cdot X$. The other kind of regression equation (that can be generated using the same data) is called a **standardized regression equation.** A standardized regression equation has the form $z'_Y = \beta \cdot z_X$. These two kinds of regression equations differ in three respects. First, a standardized regression equation involves z-scores on both the independent and the dependent variables, not raw scores. Second, the standardized regression equation does not have a constant (i.e., a term for a). Finally, the symbol β (called a **beta weight**) is used in place of the regression coefficient, b.

Interpreting a, b, r, and r^2 in Bivariate Regression

When used for predictive purposes, the regression equation has the form $Y' = a + b \cdot X$. Now that you understand how this equation works, let's take a closer look at its two main ingredients, a and b. In addition, let's now pin down the regression meaning of r and r^2.

Earlier, I referred to a as the *constant*. Alternatively, this component of the regression equation is called the **intercept.** Simply stated, a indicates where the regression line in the scatter diagram would, if extended to the left, intersect the ordinate. It indicates, therefore, the value of Y' for the case where $X = 0$. In many studies, it may be quite unrealistic (or downright impossible) for there to be a case where $X = 0$; nonetheless, $Y' = a$ when $X = 0$.

Earlier, we considered data concerning the post-surgical down-time of 12 adult tennis players. In the regression equation based on those data, the constant was equal to 1.978. That is not a very realistic number, for it indicates the predicted number of down-time days for a tennis player whose age is 0! Clearly, a may be totally devoid of meaning within the context of a study's independent and dependent variables. Nevertheless, it has an unambiguous and not-so-nonsensical meaning within a scatter diagram, because *a* indicates the point where the regression line intercepts the ordinate.

The other main component of the regression is *b*, the regression coefficient. When the regression line has been positioned within the data points of a scatter diagram, *b* simply indicates the **slope** of that line. As you probably recall from your high school math courses, *slope* means "rise over run." In other words, the value of *b* signifies how many predicted units of change (either up or down) in the dependent variable there are for any one unit increase in the independent variable. In Figure 16.1, the regression equation has a slope equal to .098. This means that the predicted down time for our hypothetical patient Sam would be about one-tenth of a day longer if the surgery is put off a year (assuming Sam's knee problem, health status, and fitness level do not change).

When researchers use bivariate regression, they sometimes will focus on either *b* or β more than anything else. Consider, for example, Excerpt 16.6. In the study associated with this excerpt, several college-age men and women were measured on several traits, one of which was aggression. In addition, the research participants had their pain tolerance measured via a device that sent increasing levels of electrical current into their non-dominant hands. After collecting the data, the researchers did a bivariate regression within each gender group to investigate the connection between trait aggression and pain tolerance. Notice how the researchers focused their attention on the beta weights when comparing the men versus the women.

EXCERPT 16.6 • *Focusing on the Regression Coefficient*

A sample of 195 collegiate men and women completed trait measures and a laboratory assessment of pain tolerance. . . . To determine whether the relationship between pain tolerance and aggression differed by sex of participant, we . . . computed simple regression coefficients of pain tolerance and trait aggression for men and women. Analyses indicated that while there was no relationship between pain tolerance and trait aggression for women [$\beta = -.02$], there was a significant positive relationship for men [$\beta = .31$]. . . . Pain tolerance was significantly and positively related to trait aggression in men, but in women the relation between pain tolerance and trait aggression was nil and nonsignificant.

Source: Reidy, D. E., Dimmick, K., MacDonald, K., & Zeichner, A. (2009). The relationship between pain tolerance and trait aggression: Effects of sex and gender role. *Aggressive Behavior, 35*(5), 422–429.

When summarizing the results of a regression analysis, researchers will normally indicate the value of *r* (the correlation between the independent and dependent variables) or r^2. You already know, of course, that such values for *r* and r^2 measure the strength of the relationship between the independent and dependent variables. However, each has a special meaning, within the regression context, that is worth learning.

As you might expect, the value of r is high to the extent that the scatter diagram's data points are located close to the regression line. Although that is undeniably true, there is a more precise way to conceptualize the regression meaning of r. Once the regression equation has been generated, that equation could be used to predict Y for each person who provided the scores used to develop the equation. In one sense, that would be a very silly thing to do, because predicted scores are unnecessary in light of the fact that *actual* scores on the dependent variable are available for these people. However, by comparing the predicted scores for these people against their actual scores (both on the dependent variable), we would be able to see how well the regression equation works. The value of r does exactly this. It quantifies the degree to which the predicted scores correlate, or match up with, the actual scores.

Just as r has an interpretation in regression that focuses on the dependent variable, so it is with r^2. Simply stated, the coefficient of determination indicates the proportion of variability in the dependent variable that is explained by the independent variable. As illustrated in Excerpt 16.7, r^2 is usually turned into a percentage when it is reported in research reports.

EXCERPT 16.7 • *Variability in the Dependent Variable Explained by Variability in the Independent Variable*

Bivariate regression analyses [revealed] MLSS [running speed at maximal lactate steady state] as the strongest individual predictor [$r = 0.93, r^2 = 0.87$] for 2-mile running performance. [This predictor] explained . . . 87% of the variance in running performance.

Source: Tolfrey, K., Hansen, S. A., Dutton, K., McKee, T., & Jones, A. M. (2009). Physiological correlates of 2-mile run performance as determined using a novel on-demand treadmill. *Applied Physiology, Nutrition & Metabolism, 34*(4), 763–772.

Inferential Tests in Bivariate Regression

The data used to generate the regression line or the regression equation are typically considered to have come from a sample, not a population. Thus the component parts of a regression analysis—a, b, and r—are usually viewed as sample statistics, not population parameters. Accordingly, it should not come as a surprise that researchers conduct one or more inferential tests whenever they perform a regression analysis.

In bivariate regression, a test on r is mathematically equivalent to a test on b or β. Therefore, you are unlikely to see a case where both r and b (or r and β) are tested, because such tests would be fully redundant with each other. Researchers have the freedom to have their test focus on r or b or β. If r is tested, try to remember the things we considered in Chapter 9. In particular, keep in mind that the null hypothesis in such a test will probably be set up to say that the correlation in the population is equal to 0.00. Also keep in mind that a test of this null hypothesis operates properly only if certain assumptions (e.g., linearity) are met.

When bivariate regression is involved in a study, you are likely to see a test of significance on either b or β rather than r. The null hypothesis in this alternative (but equivalent) kind of test says that the population value of the regression or beta weight is 0. Stated differently, the null hypothesis in such tests is that the regression line has no tilt, thus meaning that the independent variable provides no assistance in predicting scores on the dependent variable. In Excerpt 16.8, we see a case where such a test was applied. This excerpt is worth considering for two reasons. First, can you tell which of the two variables was the dependent variable? Second, do you agree the researchers deserve a pat on the back for presenting what they did immediately after they cite the test's p-level?

EXCERPT 16.8 • *Testing a Regression Coefficient*

The aim of this study, part of a cross-sectional blood pressure survey, was to study the influence of ambient temperature on blood pressure in a rural West African adult population. . . . Blood pressure, anthropometric, time of blood pressure and room temperature measurements were taken in 574 adult males and females. . . . Linear regression analysis showed that SBP [systolic blood pressure] was significantly and inversely related to ambient temperature ($b = -0.98, p = 0.02$, 95% confidence interval: -1.19 to -0.11).

Source: Kunutsor, S. K., & Powles, J. W. (2010). The effect of ambient temperature on blood pressure in a rural West African adult population: A cross-sectional study. *Cardiovascular Journal of Africa, 21*(1), 17–20.

Multiple Regression

We now turn our attention to the most popular regression procedure of all, **multiple regression.** This form of regression involves, like bivariate regression, a single dependent variable. In multiple regression, however, there are two or more independent variables. Stated differently, multiple regression involves just one Y variable but two, three, or more X variables.[5]

In three important respects, multiple regression is identical to bivariate regression. First, a researcher's reason for using multiple regression is the same as the reason for using bivariate regression, either prediction (with a focus on the dependent variable) or explanation (with a focus on the independent variables). Second, a regression equation is involved in both of these regression procedures. Third, both bivariate and multiple regression almost always involve inferential tests and a

---

[5]Recall that the dependent variable (Y) is sometimes referred to as the *criterion, outcome,* or *response variable,* whereas the independent variable (X) is sometimes referred to as the *predictor* or *explanatory variable.*

measure of the extent to which variability among the scores on the dependent variable has been explained or accounted for.

Although multiple regression and bivariate regression are identical in some respects, they also differ in three extremely important ways. As you will see, multiple regression can be done in *different ways* that lead to different results, it can be set up to accommodate *covariates* that the researcher wishes to control, and it can involve (as predictor variables) one or more *interactions* between independent variables. Bivariate regression has none of these characteristics.

In upcoming sections, these three unique features of multiple regression are discussed. We begin, however, with a consideration of the regression equation that comes from the analysis of data on one dependent variable and multiple independent variables. This equation functions as the most important stepping stone between the raw scores collected in a study and the findings extracted from the investigation.

The Regression Equation

When a regression analysis involves one dependent variable and two independent variables, the regression equation takes the form

$$Y' = a + b_1 X_1 + b_2 X_2$$

where Y' stands for the predicted score on the dependent variable, a stands for the constant, b_1 and b_2 are regression coefficients, and X_1 and X_2 represent the two independent variables. As indicated previously, multiple regression can accommodate more than two independent variables. In such cases, the regression equation is simply extended to the right, with an extra term (made up of a new b multiplied by the new X) added for each additional independent variable. The presence of these extra terms, of course, does not alter the fact that the regression equation contains only one Y' term (located on the left side of the equation) and only one a term (located on the right side of the equation).

In Excerpts 16.9 and 16.10, we see regression equations that were created for the situations where there were two or three independent variables, respectively. In

EXCERPTS 16.9–16.10 • *Regression Equations with Different Numbers of Independent Variables*

Multiple regression analysis was conducted to evaluate how well overall contact conditions predicted post-test scores on the MGUDS-S. . . . The raw coefficients for the predictive equation were as follows:

MGUDS-S post-test score = 24.09 + .67(MGUDS-S pretest) + 1.08(SICS total).

Source: Seaman, J., Beightol, J., Shirilla, P., & Crawford, B. (2010). Contact theory as a framework for experiential activities as diversity education: An exploratory study. *Journal of Experiential Education, 32*(3), 207–225.

(*continued*)

EXCERPTS 16.9–16.10 • (*continued*)

Results showed that social systems did have an impact on nurses' spiritual intelligence. . . . In general, the study yielded a regression equation associated with independent variables as follows: spiritual intelligence = 4.10 + 3.32 (childhood spirituality) + 1.01 (social system) + .03 (age).

Source: Yang, K-P., & Wu, X. J. (2009). Spiritual intelligence of nurses in two Chinese social systems: A cross-sectional comparison study. *Journal of Nursing Research, 17*(3), 189–198.

these regression equations, note that the numerical values of 24.09 (in Excerpt 16.9) and 4.10 (in Excerpt 16.10) represent *a* (i.e., the constants). The other numerical values in each equation are the *b*s (i.e., the regression coefficients), each of which is paired with a particular independent variable.

In each of the regression equations shown in Excerpts 16.9 and 16.10, the algebraic sign between any two adjacent terms on the right side of the equation is positive, meaning that the sign of every regression coefficient was positive. In some multiple regression equations, one of more of the *b*s ends up being negative. The sign of a regression coefficient simply indicates the nature of the relationship between that particular *X* variable and the dependent variable. Thus, if the nurses in the study that gave us Excerpt 16.10 had also been measured on how extensively they feel independently in control of their own lives, this predictor variable's regression coefficient would likely have a negative sign in front of it, thereby implying an inverse relationship between feeling independently powerful and level of spiritual intelligence.

Regardless of whether the multiple regression is being conducted for predictive or explanatory purposes, the researcher is usually interested in comparing the independent variables to see the extent to which each one helps the regression analysis achieve its objective. In other words, there is usually interest in finding out the degree to which each independent variable contributes to successful predictions or valid explanations. Although you (as well as a fair number of researchers) may be tempted to look at the *b*s in order to find out how well each independent variable works, this should not be done because each regression coefficient is presented in the units of measurement used to measure its corresponding *X*. Thus, if one of the independent variables in a multiple regression is height, its *b* will differ in size depending on whether height measurements are made in centimeters, inches, feet, or miles.

To determine the relative importance of the different independent variables, the researcher must look at something other than an unstandardized regression equation like those we have seen thus far. Instead, a standardized regression equation can be examined. This kind of regression equation, for the case of three independent variables, takes the form

$$z'_Y = \beta_1 z_{X_1} + \beta_2 z_{X_2} + \beta_3 z_{X_3}$$

Note that this equation presents the dependent and independent variables in terms of z, it has no constant term, and it uses the symbol β instead of b. These βs are like standardized regression coefficients, and they are called **beta weights.**

Although standardized regression equations are rarely included in research reports, researchers often extract the beta weights from such equations and present the numerical values of these βs. In Excerpt 16.11, we see an instance in which this was done. Notice that the beta weights in this excerpt were compared, with the researchers pointing out that the first beta weight was more than twice as large as the second beta, and more than three times as large as the third. Unstandardized regression coefficients cannot be compared like this.

EXCERPT 16.11 • *Beta Weights*

The betas from regression model were used to determine the relative weights of each factor. [Results indicated that] attitude toward the behavior had the most substantial impact ($\beta = 0.569$) on teachers' intentions to use computers to create and deliver lessons, producing a change of 0.569 units in behavioral intention for each unit change in attitude. This influence on intention is more than twice that of subjective norm ($\beta = 0.229$) and more than three times that of perceived behavioral control ($\beta = 0.144$).

Source: Lee, J., Cerreto, F. A., & Lee, J. (2010). Theory of planned behavior and teachers' decisions regarding use of educational technology. *Journal of Educational Technology & Society, 13*(1), 152–164.

Before concluding our discussion of regression equations, three important points must be made. First, one or more of the independent variables in a regression analysis can be categorical in nature. For example, gender is often used in multiple regression to help accomplish the researcher's predictive or explanatory objectives. As you see the technique of multiple regression used in different studies, you are likely to see a wide variety of categorical independent variables included, such as marital status (single, married, divorced), highest educational degree (high school diploma, bachelor's degree, Master's degree, Ph.D.), and race (Black, White, Hispanic). Such variables are sometimes referred to as **dummy variables.**

Second, researchers often include a term in the regression equation that represents the interaction between two independent variables. Just as two independent variables in a two-way ANOVA can be examined to see if they interact, so too can the interaction of independent variables be assessed in regression contexts. We consider this feature of multiple regression later in the chapter; for now, all I want to do is alert you to the fact that interactions are often used as independent variables in multiple regression analyses.

My third and final comment about regression equations is an important warning. Simply stated, be aware that the regression coefficients (or beta weights) associated with the independent variables can change dramatically if the analysis is repeated with one of the independent variables discarded or another independent variable added. Thus, regression coefficients (or beta weights) do not provide a pure and absolute assessment of any independent variable's worth. Instead, they are *context dependent*.

Three Kinds of Multiple Regression

Different kinds of multiple regression exist because there are different orders in which data on the independent variables can be entered into the analysis. In this section, we consider the three most popular versions of multiple regression: simultaneous multiple regression, stepwise multiple regression, and hierarchical multiple regression.

In **simultaneous multiple regression,** the data associated with all independent variables are considered at the same time. This kind of multiple regression is analogous to the process used in preparing vegetable soup where all ingredients are thrown into the pot at the same time, stirred, and then cooked together. In Excerpt 16.12, we see an example of simultaneous multiple regression.

EXCERPT 16.12 • *Simultaneous Multiple Regression*

A series of [bivariate] linear regression analyses was conducted and analyses indicated that all of the predictor variables independently predicted rebuilding the marriage relationship [i.e., mid-life marital satisfaction]. . . . A simultaneous multiple regression analysis was then conducted with adaptive appraisal, social support, and compensating experiences as predictor variables and rebuilding the marriage relationship as the criterion variable ($n = 476$).

Source: Huber, C. H., Navarro, R. L., Womble, M. W., & Mumme, F. L. (2010). Family resilience and midlife marital satisfaction. *The Family Journal: Counseling and Therapy for Couples and Families, 18*(2), 136–145.

Stepwise multiple regression analysis is analogous to the process of preparing a soup in which the ingredients are tossed into the pot based on the amount of each ingredient. Here the stock goes in first (because there is more of that than anything else), followed by the vegetables, the meat, and finally the seasoning. Each of these different ingredients is meant to represent an independent variable, with "amount of ingredient" equated, somewhat, to the size of the bivariate correlation between a given independent variable and the dependent variable. Here, in **stepwise multiple regression,** the computer determines the order in which the independent variables become a part of the regression equation. In Excerpt 16.13, we see an example of this kind of multiple regression.

EXCERPT 16.13 • *Stepwise Multiple Regression*

Patients completed a symptom-limited exercise treadmill test. . . . Exercise test time (ETT) was recorded in seconds and taken as a measure of exercise capacity. . . . Stepwise multiple regression analysis was performed to examine predictors of exercise test time in our cohort. Variables entered into the model included traditional cardiovascular risk factors (age, sex, presence/absence of hypertension, diabetes, hyperlipidemia, family history of cardiovascular disease (CVD), and BMI), BAD, FMD and NMD.

Source: Heffernan, H. S., Karas, R. H., Patvardhan, E. A., & Kuvin, J. T. (2010). Endothelium-dependent vasodilation is associated with exercise capacity in smokers and non-smokers. *Vascular Medicine, 15*(2), 119–125.

Instead of preparing our vegetable soup by simply tossing everything into the pot at once or by letting the amount of an ingredient dictate its order of entry, we could put things into the pot on the basis of concerns regarding flavor and tenderness. If we wanted garlic to flavor everything else, we would put it in first, even though there is only a small amount of it required by the recipe. Similarly, we would hold back some of the vegetables (and not put them in with the others) if they are tender to begin with and we want to avoid overcooking them. **Hierarchical multiple regression** is like cooking the soup in this manner, for in this form of regression the independent variables are entered into the analysis in stages. Often, as illustrated in Excerpt 16.14, the independent variables that are entered first correspond with things the researcher wishes to control. After these **control variables** are allowed to explain as much variability in the dependent variable as they can, then the other variables are entered to see if they can contribute above and beyond the independent variables that went in first.

EXCERPT 16.14 • *Hierarchical Multiple Regression*

Hierarchical multiple regression was used to analyze the relative importance of personal and peer attitudes supporting sexual aggression in predicting men's willingness to intervene against sexual aggression. We included several demographic variables that were potentially or theoretically related to our variables of interest [year in school, race, fraternity membership, sports team membership, and sexual orientation]. These demographic control variables were entered on the first step of the multiple regression. MCSDS scores were entered on the second step to control for social desirability. Personal and peer attitudes supporting sexual aggression were entered simultaneously on the third step.

Source: Brown, A. L., & Messman-Moore, T. L. (2009). Personal and perceived peer attitudes supporting sexual aggression as predictors of male college students' willingness to intervene against sexual aggression. *Journal of Interpersonal Violence, 25*(3), 503–517.

$R, R^2, \Delta R^2,$ *Adjusted* $R^2,$ *and* sr^2 *in Multiple Regression*

In multiple regression studies, the extent to which the regression analysis achieves its objective is usually quantified by means of $R, R^2,$ or adjusted R^2. Sometimes two of these will be presented, and occasionally you will see all three reported for the same regression analysis. These elements of a multiple regression analysis are not superficial and optional add-ons; instead, they are as central to a regression analysis as the regression equation itself.

In bivariate regression, r provides an indication of how well the regression equation works. It does that by quantifying the degree to which the predicted scores match up with the actual scores (on the dependent variable) for the group of individuals used to develop the regression equation. The R of multiple regression can be interpreted in precisely the same way. **Multiple** R is what we get if we compute Pearson's r between Y and Y' scores for the individuals who provided scores on the independent and dependent variables.

Although the value of R sometimes appears when the results of a multiple regression are reported, researchers are far more likely to report the value of R^2 or to report the percentage equivalent of R^2. By so doing, the success of the regression analysis is quantified by reporting the proportion or percentage of the variability in the dependent variable that has been accounted for or explained by the study's independent variables. Excerpt 16.15 illustrates the way researchers use R^2 in an explained-variance manner.

EXCERPT 16.15 • R^2 *as an Index of Explained Variance*

The correlation coefficient resulting from the [multiple regression] analysis shows that there is a relatively strong correlation $(R = .79)$ between the four relationship dimensions and giving history. The coefficient of determination is relatively strong $(R^2 = .62)$ and shows moderate strength in predicting past giving. Thus, 62% of the variance in the number of years the participants have donated to the organization is explained by the four relationship dimensions.

Source: Waters, R. D. (2010). Increasing fundraising efficiency through evaluation: Applying communication theory to the nonprofit organization–donor relationship. *Nonprofit and Voluntary Sector Quarterly,* in press.

When a multiple regression analysis is conducted with the data from all independent variables considered simultaneously, only one R^2 can be computed. In stepwise and hierarchical regression, however, several R^2 values can be computed, one for each stage of the analysis wherein individual independent variables or sets of independent variables are added. These R^2 values get larger at each stage, and the *increase* from stage to stage is referred to as R^2 change. Another label for the increment in R^2 that's observed as more and more independent variables are used as predictors is ΔR^2 where the symbol Δ stands for the two-word phrase *change in*.

Excerpt 16.16 illustrates nicely the concept of ΔR^2. In the first step of the hierarchical multiple regression, the control variables of SES and gender were entered into the regression model, producing an R^2 of .06. In the second step of the regression analysis, two more independent variables (reading score and reading self-efficacy) entered the model, and they explained an additional 21 percent of variability in the students' English grades. In the last step of the regression analysis, the researcher entered the main variable he was concerned about, a measure of each child's "confidence to manage learning." As indicated in the excerpt, this final independent variable explained an additional eight percent of variability above and beyond what already had been explained by the first four variables.

EXCERPT 16.16 • ΔR^2 in Stepwise or Hierarchical Multiple Regression

[We conducted] hierarchical multiple regression for the LD and NLD groups, with end-of-term English grade as the dependent variable. Control variables of SES (parent education level) and sex were entered at Step 1, followed by reading score and reading self-efficacy at Step 2, and finally SESRL at Step 3. For the LD group, the entry of the control variables at Step 1 did not significantly predict English grade [$R^2 = .06$]. The entry of reading score and reading self-efficacy at Step 2 significantly increased explained variance ($\Delta R^2 = .21$), as did the entry on the final step of SESRL ($\Delta R^2 = .08$), with a final R^2 of .35.

Source: Klassen, R. M. (2010). Confidence to manage learning: The self-efficacy for self-regulated learning of early adolescents with learning disabilities. *Learning Disability Quarterly, 33*(1), 19–30.

Either in place of or in addition to R^2, something called **adjusted R^2** is often reported in conjunction with a multiple regression analysis. If reported, adjusted R^2 takes the form of a proportion or a percentage. It is interpreted just like R^2, because it indicates the degree to which variability in the dependent variable is explained by the set of independent variables included in the analysis. The conceptual difference between R^2 and adjusted R^2 is related to the fact that the former, being based on sample data, always yields an overestimate of the corresponding population value of R^2.

Adjusted R^2 removes the bias associated with R^2 by reducing its value. Thus, this adjustment anticipates the amount of so-called **shrinkage** that would be observed if the study were to be replicated with a much larger sample. As you might expect, the size of this adjustment is inversely related to study's sample size.[6]

---

[6]The size of the adjustment is also influenced by the number of independent variables. With more independent variables, the adjustment is larger.

When reporting the results of their multiple regression analyses, some researchers (who probably do not realize that R^2 provides an exaggerated index of predictive success) report just R^2. Of those who are aware of the positive bias associated with R^2, some include only adjusted R^2 in their reports whereas others include both R^2 and adjusted R^2.

In addition to assessing the effectiveness of the full regression model, many researchers evaluate the worth of each independent variable. Beta weights can help in this regard, but they do not indicate how much variability in the dependent variable is explained uniquely by each independent variable. To accomplish this goal, the square of the semi-partial correlation, symbolized sr^2, is computed for each variable used to help predict (or explain) variability in the dependent variable. In a very real sense, sr^2 is analogous to R^2, with the former index focused on a single predictor, whereas the latter is based on the full set of predictors. Excerpt 16.17 shows how sr^2 can help in the interpretation of results.

EXCERPT 6.17 • Assessing the Worth of Individual Independent Variables with sr^2

The value of the square of the coefficient of semi-partial correlation (sr^2) for each independent variable was also calculated, which allowed us to assess the unique contribution of this variable relative to R^2 in the set of variables included in the model. . . . The regression model found is significant, allowing explanation of 33.9% of the variance in the anxiety/depression symptoms. The only variable with a significant predictive value is the "negative reactivity" temperament dimension, which, once the variance explained by the remainder is controlled, is responsible for 19.5% of the variance.

Source: Lima, L., Guerra, M. P., & de Lemos, M. S. (2010). The psychological adjustments of children with asthma: Study of associated variables. *Spanish Journal of Psychology, 13*(1), 353–363.

Inferential Tests in Multiple Regression

Researchers can apply several different kinds of inferential tests when they perform a multiple regression. The three most frequently seen tests focus on β, R^2, and ΔR^2. Let's consider what each of these tests does and then look at an excerpt in which all three of these tests are used.

When the beta weight for a particular independent variable is tested, the null hypothesis says that the parameter value is equal to 0. If this were true, that particular independent variable would be contributing nothing to the predictive or explanatory objective of the multiple regression. Because of this, researchers frequently test each of the betas in an effort to decide (1) which independent variables should be included in the regression equation being built, or (2) which independent variables included in an already-developed regression equation turned out to be helpful. Beta weights are usually tested with two-tailed t-tests.[7]

When R^2 is tested, the null hypothesis says that none of the variance in the dependent variable is explained by the collection of independent variables. (This H_0, of course, has reference to the study's population, not its sample.) This null hypothesis normally is evaluated via an F-test. In most studies, the researcher hopes that this H_0 will be rejected.[8]

When ΔR^2 is tested, the null hypothesis says that any new independent variables added to an existing regression equation are totally worthless in helping to explain variability in the dependent variable. As with the null hypotheses associated with tests on beta weights and R^2, this particular H_0 has reference to the study's population, not its sample. A special F-test is used to evaluate this null hypothesis. This kind of test, of course, logically fits into the procedures of stepwise and hierarchical multiple regression; however, it is never used within the context of a simultaneous multiple regression.[9]

Consider now Excerpt 16.18 which comes from a study involving a hierarchical multiple regression. Take the time to look at this excerpt closely, because it

EXCERPT 16.18 • *Inferential Tests in Multiple Regression*

Medication adherence was negatively associated with extreme violence ($r = -.21$, $p = .01$) as well as with substance coping ($r = -0.20, p = -.012$). Hierarchical multiple regression was used to assess the predictive ability of extreme violence and substance coping on medication adherence for the total sample, after controlling for both gender and time since diagnosis. Gender and time since diagnosis, which was calculated in months, were entered at Step 1, explaining 3% of the variance in medication adherence. After entry of extreme violence and substance use coping at Step 2, the total variance explained by the model as a whole was 18%, $F(4, 85) = 4.54$, $p = .002$. The two measures accounted for an additional 15% of the variance in adherence, after controlling for gender and time since diagnosis, R^2 change $= .15, F$ change $(2, 85) = 7.56, p = .001$. In the final model, both extreme violence and substance use coping were statistically significant, with extreme violence recording a higher beta value ($beta = -0.29, p = .005$) than substance use coping ($beta = -0.20, p = 0.050$).

Source: Lopez, E. J., Jones, D. L., Villar-Loubet, O. M., Arheart, K. L., & Weis, S. M. (2010). Violence, coping, and consistent medication adherence in HIV-positive couples. *AIDS Education & Prevention, 22*(1), 61–68.

---

[7]The *df* for this kind of t-test is equal to the sample size minus one more than the number of independent variables.
[8]The first *df* for this kind of F-test is equal to the number of independent variables; the second *df* value is equal to the sample size minus one more than the number of independent variables.
[9]The *df* for this kind of F-test is equal to (1) the number of new independent variables and (2) the sample size minus one more than the total number of old and new independent variables.

contains—in the second step of the analysis—tests of R^2, ΔR^2, and the beta weights for the two key independent variables.

Two additional features of Excerpt 16.18 are worth noting. First, the initial sentence contains the bivariate correlations between each of the primary independent variables and the dependent variable. Each of those rs has a negative sign, which is why the two betas, presented in the excerpt's final sentence, turned out to be negative rather than positive. The second thing to notice about Excerpt 16.18 is the fact that the value of R^2 increased dramatically from step 1 to step 2. Clearly, variation in medication adherence is associated with violence and substance abuse, even after gender and time since diagnosis are controlled.

Moderated and Mediated Multiple Regression

Researchers sometimes report that they have conducted a moderated multiple regression or a mediated multiple regression. Despite the similar-sounding names, these two kinds of regression are quite different. Let's briefly consider the goals and the procedure of these two special cases of multiple regression.

When researchers conduct a **moderated multiple regression,** their goal is to see if the findings of the multiple regression are the same (or perhaps different) for different subgroups of people or different settings. For example, suppose data are collected from several young adults on a variety of independent variables that might explain variability in the study's dependent variable: level of satisfaction with a first date. In this situation, the researcher might choose to do a moderated multiple regression due to the thought that men and women could have different reasons for thinking that a first date was a terrific or terrible experience (or anywhere between these extremes).

In Excerpt 16.19, we see an example of a moderated multiple regression conducted in a business setting with data gathered from a large number of employees. The researchers wanted to see if employee commitment to an organization was related to the employee's perception of fit between him or her and the company he

EXCERPT 16.19 • *Moderated Multiple Regression*

A moderated hierarchical multiple regression analysis was computed to predict organizational commitment (*H1*). In Step 1, strategy fit and job alternatives were entered, and in Step 2, the product of strategy fit and job alternatives was entered. The results are presented in Table 2 [not shown here]. Step 1, which includes both main effects (strategy fit and job alternatives), was statistically significant, $R^2 = .11, p < .01$, with strategy fit as the significant predictor, $\beta = .32, p < .01$. The change in R^2 when the product term was entered in Step 2 was also statistically significant ($\Delta R^2 = .02, p < .05$). Therefore, the results suggest that job alternatives moderate the relation between strategy fit and organizational commitment. . . . [I]f respondents

(continued)

EXCERPT 16.19 • (*continued*)

perceive that there are numerous job alternatives, then the correlation between strategy fit and organizational commitment is positive. . . . However, when respondents perceive that there are few job alternatives, then there is no relationship between strategy fit and organizational commitment. When employees do not feel that they have other job alternatives, then their commitment to their organization is the same regardless of whether there is a fit or misfit in strategy. Therefore, as predicted in *H1*, job alternatives moderate the relation between strategy fit and organizational commitment.

Source: Da Silva, N., Hutcheson, J., & Wahl, G. D. (2010). Organizational strategy and employee outcomes: A person–organization fit perspective. *Journal of Psychology, 144*(2), 145–161.

or she worked for. The researchers also wanted to see if the strength of that relationship varies depending on whether other job opportunities exist for the employee. To answer these questions, a moderated multiple regression was conducted. The dependent variable was organizational commitment, with the analytic approach being hierarchical in nature. In step 1, the independent variables were strategy fit and job opportunities. Then in step 2, the researchers added a term to the regression model: the interaction between fit and other opportunities. It was this interaction term that caused the regression analysis to be of the moderated variety. Carefully read the material in Excerpt 16.19 and you get a good feel for the goal and procedures of a moderated multiple regression.

With a mediated multiple regression, a researcher's goal is to see whether the apparent causal influence of one variable on a second variable is attributable—totally, partially, or not at all—to the first variable having an influence through some other third variable. To illustrate, let's consider the plight of graduate students who are on teaching assistantships. These individuals probably feel varying levels of job stress because they are pressured to do three important things in the university setting: (1) earn high grades in the graduate courses they take, (2) perform well as instructors in the undergraduate courses they teach, and (3) get research papers published so they can be competitive in the job market once they complete their graduate degree programs.

We might hypothesize that the graduate students' levels of stress will impact their physical health, with those with more stress being more susceptible to colds, the flu, allergies, and other ailments. At this point in our example, we have an independent variable (job stress) that may have a causal impact on the dependent variable (illness). We could investigate this hypothesized relationship in a simple way by measuring a large group of graduate students on these two variables and then correlating the two sets of scores. By itself, a statistically significant positive correlation from our data would not prove that job stress causes illness; nevertheless, it would constitute a helpful piece of information.

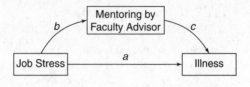

FIGURE 16.2   *Diagram of Three-Variable Mediation Model*

Continuing our stress-and-illness example, suppose we now add a third variable to our causal model: faculty advisor mentoring. The quality of faculty mentoring varies, of course, with some graduate students having faculty advisors who are better than others at handling job stress, more willing to talk with their advisees about the difficulty of being a graduate teaching assistant, and more knowledgeable about the early-warning signals of impending illness. We might hypothesize that the level of job stress felt by graduate students is mediated by the quality of their mentors. In other words, we might conjecture that at least a portion of a graduate student's job stress flows through the relationship he or she has with his or her faculty advisor, with a good advisor functioning to lessen the stress (and ultimately lessen the likelihood of illness), whereas a not-so-good advisor would do little or nothing to mediate the causal impact of stress on illness.

A visual depiction of our example appears in Figure 16.2. The three lines are meant to represent the paths of the causal influence. Line *a* represents the *direct effect* of job stress on illness, whereas lines *b* and *c* represent the path of mediation. As Figure 16.2 illustrates, the effect of job stress may flow through, and may well be reduced by, the quality of mentoring provided by the graduate faculty advisor. Any reduction in the causal impact of the independent variable on the dependent variable is called the *indirect effect*.

To assess the worthiness of a mediated model, researchers do more than just draw diagrams with directional arrows between variable names. To test their hypotheses, they collect and analyze data. The usual data analytic strategy involves a four-step set of regression analyses. The sequence of tests and the needed results to establish mediation are shown in Excerpt 16.20. This excerpt comes from a study

EXCERPT 16.20 • *Mediated Multiple Regression*

To explore whether altruism had an indirect effect on intention we conducted mediation analyses. [M]ediation can be said to occur when four conditions are satisfied: (1) variation in the independent measure (e.g. altruism) accounts for significant variance in the dependent measure (e.g. intention); (2) variation in the independent measure accounts for significant variance in the mediator (e.g. moral norm); (3) variation in the mediator accounts for variance in the dependent measure while controlling for the influence of independent measure; and (4) the significant effect of the independent

(continued)

EXCERPT 16.20 • (*continued*)

measure on the dependent measure is significantly reduced after controlling for the effects of the mediator. . . . Regression analyses showed (1) an effect of altruism on intention, $B = 0.21, t(677) = 3.22, p < .001$, (2) an effect of altruism on moral norm, $B = 0.42, t(677) = 6.26, p < .001$, (3) an effect of moral norm on intention, $B = 0.44, t(677) = 13.13, p < .001$, and (4), the effect of altruism on intention was no longer significant (85% reduction), $B = 0.03, t(677) = 0.43, p = .67$, after including moral norm as additional predictor. This mediation effect was statistically significant, Sobel's $Z = 5.63, p < .001$.

Source: Lemmens, K. P. H., Abraham, C., Ruiter, R. A. C., Veldhuizen, I. J. T., Dehing, C. J. G., Bos, A. E. R., et al. (2009). Modelling antecedents of blood donation motivation among non-donors of varying age and education. *British Journal of Psychology, 100*(1), 71–90.

involving 687 Dutch residents who had never donated blood, even though they were eligible to do so. The researchers hypothesized that their participants' level of intention to donate was influenced by their sense of altruism. However, they also hypothesized that the variable of *norms* (i.e., the degree to which family members and friends donate blood and encourage others to do so) would functioned as a mediator. Examine this excerpt carefully to see the typical approach to a mediated multiple regression.

In the last sentence of Excerpt 16.20, reference is made to the **Sobel test,** one option for seeing if the reduction in the two involved regression coefficients is statistically significant. Another option for making this kind of test is **bootstrapping,** a procedure that can be used in many statistical situations. If bootstrapping is applied in this regression context, a computer is used to develop a sampling distribution (to evaluate the drop in the regression coefficient) by extracting many, many samples from the available data used in the study. This computer-based procedure is considered by some to be superior to the Sobel test.

Logistic Regression

The final kind of regression considered in this chapter is called **logistic regression.** Originally, only researchers from medical disciplines (especially epidemiology) used this form of regression. More recently, however, logistic regression has been discovered by those who conduct empirical investigations in a wide array of disciplines. Its popularity continues to grow at such a rate that it may soon overtake multiple regression and become the most frequently used regression tool of all.

Before considering how logistic regression differs from the forms of regression already considered, let's look at their similarities. First, logistic regression deals with relationships among variables (not mean differences), with one variable being the dependent (i.e., outcome or response) variable whereas the others are the independent (predictor or explanatory) variables. Second, the independent variables can be continuous or categorical in nature. Third, the purpose of logistic regression can be either prediction or explanation. Fourth, tests of significance can be and usually are conducted, with these tests targeted either at each individual independent variable or at the combined effectiveness of the full set of independent variables. Finally, logistic regression can be conducted in a simultaneous, stepwise, or hierarchical manner depending on the timing of and reasons for independent variables entering the equation.

There are, of course, important differences between logistic regression, on the one hand, and either bivariate or multiple regression, on the other hand. These differences are made clear in the next three sections. Logistic regression revolves around a core concept called the **odds ratio** that was not considered earlier in the chapter because it is not a feature of either bivariate or multiple regression. Before looking at this new concept, we must focus our attention on the kinds of data that go into a logistic regression and also the general reasons for using this kind of statistical tool.

Variables

As does any bivariate or multiple regression, logistic regression always involves two main kinds of variables. These are the study's *dependent* and *independent* variables. In the typical logistic regression (as in some applications of multiple regression), a subset of the independent variables is included for control purposes, with the label *control* (or *covariate*) designating any such variable. Data on these three variables constitute the only ingredients that go into the normal logistic regression, and the results of such analyses are inextricably tied, on a conceptual level, to these three kinds of variables. For these reasons, it is important for us to begin with a careful consideration of the logistic regression's constituent parts.

In any logistic regression, as in any bivariate or multiple regression, there is one and only one dependent variable. Here, however, the dependent variable is categorical. Although the dependent variable can have three or more categories, thus making the logistic regression multinomial in nature, we consider here only situations where the dependent variable is dichotomous. Examples of such variables used in recent studies include whether or not a person survives open heart surgery, whether an elderly and ill married person considers his or her spouse to be the primary caregiver, whether a young child chronically suffers from nightly episodes of coughing, and whether an adolescent drinks at least eight ounces of milk a day. As illustrated by these examples, dichotomous dependent variables in logistic regressions can represent either true or artificial dichotomies. Either way, our focus is on what sometimes is referred to as *binary logistic regression*.

In addition to the dependent variable, at least one independent variable is involved in any logistic regression. Almost always, two or more such variables are involved. As in multiple regression, these variables can be either quantitative or qualitative in nature. If of the former variety, scores on the independent variable are construed to represent points along a numerical continuum. With qualitative independent variables, however, scores carry no numerical meaning and only serve the purpose of indicating group membership. In any given logistic regression, the independent variables can be all quantitative, all qualitative, or some of each. Moreover, independent variables can be used individually or jointly as an interaction.

When using logistic regression, applied researchers usually collect data on several independent variables, not just one. In the study alluded to earlier in which the dependent variable dealt with nighttime coughing among preschool children, the independent variables dealt with the child's sex and birth weight, the possible presence of pets and dampness problems in the home, whether the parents smoked or had asthma, and whether the child attended a day care center. It is not unusual to see this many independent variables utilized within logistic regression studies.

As indicated previously, some of the independent variables in a typical logistic regression are control variables. Such variables are included so the researcher can assess the "pure" relationship between the remaining independent variable(s) and the dependent variable. In a very real sense, control variables are included because of suspected confounding that would muddy the water if the connection between the independent and dependent variables were examined directly.

In any given logistic regression wherein control is being exercised by means of the inclusion of covariate variables, it may be that only one such variable is involved, or that two or three are used, or that all but one of the independent variables are covariates. It all depends, of course, on the study's purpose and the researcher's ability to identify and measure potentially confounding variables. In the study concerned with preschoolers and chronic coughing at night, all but one of the independent variables were included for control purposes; by so doing, the researchers considered themselves better able to examine the direct influence of day care versus home care on respiratory symptoms.

In Excerpt 16.21, we see a case in which the three kinds of variables of a typical logistic regression are clearly identified. It is worth the time to read this excerpt closely with an eye toward noting the nature and number of these three kinds of variables.

EXCERPT 16.21 • *Dependent, Independent, and Control Variables*

Logistic regression was used to test the effects of the independent variables while controlling for relevant covariates. . . . The dependent variable [was] whether women completed the 30-day residential treatment program. . . . The independent variables were material and emotional support [from family and friends]. The following categorical demographic variables were included as control variables: marital status, education, drug treatment history, drug use in the past 30 days, ethnicity, and having children.

Source: Lewandowski, C. A., & Hill, T. J. (2009). The impact of emotional and material social support on women's drug treatment completion. *Health & Social Work, 34*(3), 213–221.

Many logistic regression studies are like the one associated with Excerpt 16.21 in that they involve one dichotomous dependent variable, multiple independent variables, and multiple control variables. In some logistic regression studies, there are multiple independent variables and a single control variable. Or, there might be a single independent variable combined with several control variables. It all depends on the goals of the investigation and the researcher's ability to collect data on independent and control variables that are logically related to the dependent variable.

Objectives of a Logistic Regression

Earlier in this chapter, I pointed out that researchers use bivariate and multiple regression in order to achieve one of two main objectives: explanation or prediction. So it is with logistic regression. In many studies, the focus is on the noncontrol independent variables, with the goal being to identify the extent to which each one plays a role in explaining why people have the status they do on the dichotomous dependent variable. In other studies, the focus is primarily on the dependent variable and how to predict whether people end up in one or the other of the two categories of that outcome variable.

In Excerpt 16.22, we see a case in which logistic regression was used for predictive purposes. In the final sentence of this excerpt, the researchers point out which of their independent variables helped predict relapses among patients afflicted with schizophrenia.

EXCERPT 16.22 • *Logistic Regression and Prediction*

Schizophrenia is a severe and chronic mental illness characterized by recurring relapses. . . . To determine predictors of relapse during the 1-year study period, a stepwise logistic regression analyses was conducted. . . . [T]his study identified a small set of variables that help predict subsequent relapse in the usual treatment of schizophrenia, demonstrating the predictive value of prior relapse as a robust marker, along with prior medication nonadherence, younger age at illness onset, having health insurance, and poorer level of functioning.

Source: Ascher-Svanum, H., Baojin, Z., Faries, D. E., Salkever, D., Slade, E. P., Xiaomei, P., et al. (2010). The cost of relapse and the predictors of relapse in the treatment of schizophrenia. *BMC Psychiatry, 10*, 1–7.

Odds, Odds Ratios, and Adjusted Odds Ratios

Because the concept of *odds* is so important to logistic regression, let's consider a simple example that illustrates what this word does (and does not) mean. Suppose you have a pair of dice that are known to be fair and not loaded. If you were to roll these two little cubes and then look to see if you rolled a pair (two of the same number), the answer is either yes or no. Altogether, there are 36 combinations of how the dice might end up, with six of these being pairs. On any roll, therefore, the probability of getting a pair is 6/36, or .167. (Naturally, the probability of not getting a pair is .833.) Clearly, it is more likely that you will fail than succeed in your effort to roll a pair. However, we can be even more precise than that. We could say that the odds are 5 to 1 against you, meaning that you are five times more likely to roll a nonpair than a pair.

Most researchers utilize logistic regression so they can discuss the explanatory or predictive power of each independent variable using the concept of odds. They want to be able to say, for example, that people are twice as likely to end up one way on the dependent variable if they have a particular standing on the independent variable being considered. For example, in a hypothetical study focused on the possible impact of car color on auto accidents, the researchers might summarize their findings by saying that "Red cars are three times as likely to be involved in an accident than white cars." Or, in a different study dealing with exercise and injuries, the research report might include a sentence saying that "Adults who stretched before exercising were found to be one-half as likely to incur a muscle cramp as compared with those who did not stretch."

After performing a logistic regression, researchers often cite the **odds ratio** for each independent variable, or at least for the independent variable(s) not being used for control purposes. The odds ratio is sometimes reported as OR, and it is analogous to r^2 in that it measures the strength of association between the independent variable and the study's dependent variable. However, the odds ratio is considered by many people to be a more user-friendly concept than the Pearson-based coefficient of determination. Because the odds ratio is so central to logistic regression, let's pause for a moment to consider what this index means.

Imagine that two very popular TV programs end up going head-to-head against each other in the same time slot on a particular evening. For the sake of our discussion, let's call these programs A and B. Also imagine that we conduct a survey of folks in the middle of this time slot in which we ask each person two questions: (1) What TV show are you now watching? and (2) Are you a male or a female? After eliminating people who either were not watching TV or were watching something other than program A or B, let's suppose we end up with data like that shown in Figure 16.3.

		TV Program Being Watched	
		Program A	Program B
Gender	Male	200	100
	Female	50	150

FIGURE 16.3 *Hypothetical Data Showing Gender Preferences for Two TV Programs*

As you can see, both TV programs were equally popular among the 500 people involved in our study. Each was being watched by 250 of the people we called. Let's now look at how each gender group spread itself out between the two programs. To do this, we will arbitrarily select Program A and then calculate, first for males and then for females, the odds of watching Program A. For males, the odds of watching Program A are $200 \div 100$ (or 2 to 1); for females, the odds of watching this same program are $50 \div 150$ (or 1 to 3). If we now take these odds and divide the one for males by the one for females, we obtain the ratio of the odds for gender relative to Program A. This OR is equal to $(2 \div 1) \div (1 \div 3)$, or 6. This result tells us that among our sampled individuals, males are six times more likely to be watching Program A than women. Stated differently, gender (our independent variable) appears to be highly related to which program is watched (our dependent variable).

In our example involving gender and the two TV programs, the odds ratio was easy to compute because there were only two variables involved. As we have seen, however, logistic regression is typically used in situations where there are more than two independent variables. When multiple independent variables are involved, the procedures for computing the odds ratio become quite complex; however, the basic idea of the odds ratio stays the same.

Consider Excerpts 16.23 and 16.24. Notice the phrase "about 30% lower" in the first of these excerpts, and the phrases "four times as likely" and "nearly three times as likely" in the second excerpt. Most people can understand conclusions such as these even though they are unfamiliar with the statistical formulas needed to generate an odds ratio type of conclusion. In addition, I suspect you can see, without difficulty, that whether an odds ratio ends up being greater than 1 or less than 1 is

EXCERPTS 16.23–16.24 • *Odds Ratio and Adjusted Odds Ratio*

The odds of graduation for Hispanics are about 30% lower compared to Whites [odds ratio = 0.66].

Source: Jones, M. T., Barlow, A. E. L., & Villarejo, M. (2010). Importance of undergraduate research for minority persistence and achievement in biology. *Journal of Higher Education, 81*(1), 82–115.

There were two important predictors of Emotional Cue Eating in this study: women were four times as likely [AOR = 4.0] to be emotional eaters, and those whose families offered food to comfort were nearly three times as likely (AOR = 2.6) to be emotional eaters.

Source: Brown, S. L., Schiraldi, G. R., & Wrobleski, P. P. (2009). Association of eating behaviors and obesity with psychosocial and familial influences. *American Journal of Health Education, 40*(2), 80–89.

quite arbitrary. It all depends on the way the sentence is structured. For example, the researchers who gave us Excerpt 16.23 would have presented an OR of 1.52 in the final sentence (and they would have said "about 50% higher") if the position of the words *Hispanics* and *Whites* had been reversed.

When the odds ratio is computed for a variable *without* considering the other independent variables involved in the study, it can be conceptualized as having come from a bivariate analysis. Such an OR is said to be a *crude* or *unadjusted odds ratio*. If, as is usually the case, the OR for a particular variable is computed in such a way that it takes into consideration the other independent variable(s), then it is referred to as an **adjusted odds ratio.** By considering all independent variables jointly so as to assess their connections to the dependent variable, researchers often say that they are performing a *multivariate analysis*.

To see an example of an adjusted odds ratio, consider once again Excerpt 16.24. Notice that the letters AOR, the abbreviation for this kind of odds ratio, appears twice in the excerpt. This study's other predictor variables (besides gender and whether a family offered food to comfort) do not appear in this excerpt, but there were many. These included ethnicity, a variety of personality variables, and several family characteristics.

Tests of Significance

When using logistic regression, researchers usually conduct tests of significance. As in multiple regression, such tests can be focused on the odds ratios (which are like regression coefficients) associated with individual independent variables or on the full regression equation. Whereas tests on the full regression equation typically represent the most important test in multiple regression, tests on the odds ratios in logistic regression are considered to be the most critical tests the researcher can perform.

When the odds ratio or adjusted odds ratio associated with an independent variable is tested, the null hypothesis says that the population counterpart to the sample-based OR is equal to 1. If the null hypothesis were true (with OR = 1), it means that membership in the two different categories of the dependent variable is unrelated to the independent variable under consideration. For this null hypothesis to be rejected, the sample value of OR must deviate from 1 further than would be expected by chance.

Researchers typically use one of two approaches when they want to test an odds ratio or an adjusted odds ratio. One approach involves setting up a null hypothesis, selecting a level of significance, and then evaluating the H_0 either by comparing a test statistic against a critical value or by comparing the data-based p against α. The second way a researcher can test an odds ratio is through the use of a confidence interval (CI). The CI rule-of-thumb for deciding whether to reject or retain the null hypothesis is the same as when CIs are used to test means, rs, the difference between means, or anything else. If the confidence interval overlaps the pinpoint number in the null hypothesis, the null hypothesis is retained; otherwise, H_0 is rejected. Excerpts 16.25 and 16.26 illustrate these two ways to test an odds ratio or an adjusted odds ratio.

EXCERPTS 16.25–16.26 • *Testing an OR or an AOR for significance*

Logistic regression was also used to explore what baseline variables could predict psychological distress at six months (predictive model). . . . Stroke severity (Wald's $\chi^2 = 7.95, P < 0.01$) was a significant predictor of psychological distress [OR = 1.24].

Source: Hilari, K., Northcott, S., Roy, P., Marshall, J., Wiggins, R. D., Chataway, J., et al. (2010). Psychological distress after stroke and aphasia: The first six months. *Clinical Rehabilitation, 24*(2), 181–190.

The goal of this study was to determine whether residential exposure to vehicular traffic was associated with SAB [spontaneous abortion]. . . . SAB was examined in relation to the traffic exposure measures using logistic regression adjusting for a number of demographic and lifestyle variables. ... Among women who were non-smokers, significantly increased odds of SAB were observed in the highest traffic exposure group (AOR = 1.47; 95% CI, 1.07–2.04).

Source: Green, R. S., Malig, B., Windham, G. C., Fenster, L., Ostro, B., & Swan, S. (2009). Residential exposure to traffic and spontaneous abortion. *Environmental Health Perspectives, 117*(12), 1939–1944.

Notice in Excerpt 16.25 that a **Wald test** was used to see if the odds ratio was statistically significant. This test is highly analogous to the *t*-test in multiple regression that is used to see if a beta weight is statistically significant. These two tests are only analogous, however, for they differ not only in terms of the null hypothesis but also in the kinds of calculated and critical values used to test the H_0. As illustrated in Excerpt 16.25, the Wald test is tied to a theoretical distribution symbolized by χ^2 rather than *t*. (This is the *chi-square distribution*.)

Excerpt 16.26 illustrates how a CI can be used to test an odds ratio. Take the time to look closely at this excerpt's CI, note its ends, and then recall that the pinpoint number in the null hypothesis being tested is 1.0. I hope that you see why the researchers' AOR of 1.47 was reported to reflect "significantly increased odds of SAB" for one of the study's comparison groups.

As indicated previously, it is possible in logistic regression to assess whether the collection of independent and control variables do a better-than-chance job of accounting for the status of people on the dependent variable. Three popular procedures exist for doing this. One approach involves setting up and testing a single null hypothesis concerning the full equation, with a data-based *p*-level compared against an alpha level to see if the independent variables, as a unified set, are linked to the dependent variable. A second approach involves computing **Nagelkerke's** R^2, an index that is highly analogous to the R^2 used in multiple regression.[10] A third approach involves determining the **hit rate** to see how successful the set of independent variables are at correctly classifying individuals into the categories of the dependent variable. Some researchers use one of these approaches, others use two, and a few, as illustrated in Excerpt 16.27, use all three.

EXCERPT 16.27 • *Evaluating the Full Logistic Regression Model*

The final logistic regression model was significant, $\chi^2(3, N = 31) = 25.48$, $p < .001$, indicating that combined performance on the three tasks distinguished adults with SLI from those with TL. The model explained 75% (Nagelkerke's R^2) of the variance in language status (i.e., affected or unaffected) and correctly classified 87% of cases, with a sensitivity of 85% and a specificity of 89% where cases with a .50 or greater predicted probability were classified as affected.

Source: Poll, G. H., Betz, S. K., & Miller, C. A. (2010). Identification of clinical markers of specific language impairment in adults. *Journal of Speech, Language, and Hearing Research, 53*(2), 414–429.

Excerpt 16.27 contains two new technical terms: sensitivity and specificity. **Sensitivity** is the hit rate for correctly classifying people as being in the first category—usually the category that has a disease or ailment—of the dependent variable, whereas **specificity** is the hit rate for correctly classifying people into the other category. Both sensitivity and specificity are based on the data used to build the logistic regression model, with each index computed as the percentage of people actually in a given category who are correctly classified as being members of that category.

Final Comments

As we conclude this chapter, we must consider four additional regression-related issues: assumptions, control, practical significance, and the inflated Type I error risk. If you keep these issues in mind as you encounter research reports based on bivariate, multiple, and logistic regression, you will be in a far better position to both decipher and critique such reports.

---

[10]Nagelkerke's R^2 sometimes is referred to as an *approximate* or *pseudo-measure of explained variability* because of the way it is computed. (It begins with the Cox and Snell's measure of R^2, which itself is only an approximation, and then rescales that measure so it has minimum and maximum values of 0 and 1, respectively.)

All three forms of regression considered in this chapter carry with them *underlying assumptions*. If these assumptions are violated, regression results can be misleading. Therefore, give credit to researchers who indicate that they attended to their regression's assumptions before analyzing their data to get answers to their research questions. Excerpt 16.28 illustrates this good practice.

EXCERPT 16.28 • *Concern for Assumptions*

The assumptions of multiple regression were examined using a series of diagnostic graphs and tests for outliers, normality of residuals, homoscedasticity, multicollinearity, linearity, model specification, and independence. . . . The regression models provided an acceptable description of the data because no violations of the assumptions were observed.

Source: Cowley, P. M., Ploutz-Snyder, L. L., Baynard, T., Heffernan, K., Jae, S. Y., Hsu, S., et al. (2010). Physical fitness predicts functional tasks in individuals with Down Syndrome. *Medicine and Science in Sports and Exercise, 42*(2), 388–393.

Two important terms in Excerpt 16.28 have not been discussed previously in this book: multicollinearity and model specification. **Multicollinearity** exists if two or more independent variables are too highly correlated with each other. This undesirable situation causes inferences about individual predictor variables to be untrustworthy. Accordingly, regression assumes that multicollinearity *does not* exist. **Model specification** is concerned with the researcher's decision regarding which variables to include in the regression model. If important variables are overlooked or if irrelevant variables are included, the regression model is said to be *misspecified*. Understandably, the assumption here is that the model has been specified properly, thereby avoiding the problem of misspecification.

In the discussions of both hierarchical multiple regression and logistic regression, we saw that researchers often incorporate control or covariate variables into their analyses. Try to remember that such *control* is very likely to be less than optimal for three reasons: First, one or more important confounding variables might be overlooked. Second, potential confounding variables that *are* measured are likely to be measured with instruments possessing less than perfect reliability. Finally, recall that the analysis of covariance undercorrects when used with nonrandom groups that come from populations that differ on the covariate variable(s). Regression suffers from this same undesirable characteristic.

My next concern relates to *the distinction between statistical significance and practical significance*. We considered this issue in connection with tests focused on means and *r*s, and it is just as relevant to the various inferential tests used by researchers within regression analyses. In Excerpts 16.29 and 16.30, we see two cases in which researchers attended to the important distinction between useful and

EXCERPTS 16.29–16.30 • *Practical versus Statistical Significance*

An a priori power analysis was conducted [so as] to determine the needed sample size to answer the research questions. For multiple linear regression analysis, with a significance level of 0.05, 80% power, a total of 15 predictor variables, and an estimated moderate effect size ($R^2 = 0.25$), 70 subjects were needed. A conservative R^2 was estimated from a study using the Quality of Life Index – Dialysis version with persons on hemodialysis, in which the R^2 was reported as 0.28.

Source: Kring, D. L., & Crane, P. B. (2009). Factors affecting quality of life in persons on hemodialysis. *Nephrology Nursing Journal, 36*(1), 15–55.

A doctoral degree in counselor education was the second [significant] contributor to the regression equation. Nevertheless, with an [increase in] R^2 of .04 when this variable is added, and a resulting effect size similar to that of the equation with just one predictor variable, it does not contribute to the prediction ability in a highly meaningful way. Thus, it is important to note that a doctoral degree contributes to cognitive complexity, but the relatively small change in the R^2 means that it does not have a high degree of practical significance.

Source: Granello, D. H. (2010). Cognitive complexity among practicing counselors: How thinking changes with experience. *Journal of Counseling & Development, 88*(1), 92–100.

trivial findings. The researchers associated with the first excerpt set a good example by conducting an a priori power analysis in the planning stage of their investigation. The researchers associated with the second of these excerpts deserve high praise for realizing (and for warning their readers) that inferential tests can yield results that are statistically significant without being important in a practical manner.

In many research reports, researchers make a big deal about a finding that seems small and of little importance. Perhaps such researchers are unaware of the important distinction between practical and statistical significance, or it may be that they know about this distinction but prefer not to mention it due to a realization that their statistically significant results do not matter very much. Either way, it is important that *you* keep this distinction in mind whenever you are on the receiving end of a research report. Remember, you have the right to evaluate a statistical finding as having little or no meaningfulness after you examine the research report's summary statistics, and you can draw such a conclusion even if your opinion is at odds with those of the researcher's.

As we have seen in the excerpts of previous chapters, competent researchers are sensitive to the inflated Type I error risk that occurs if a given level of significance, say .05, is used multiple times within the same study when different null hypotheses are tested. Give credit to researchers when they apply the Bonferroni adjustment procedure (or some other comparable strategy) within their regression studies. Excerpt 16.31 provides an example of this.

EXCERPT 16.31 • *The Bonferroni Adjustment Procedure*

To adjust for alpha inflation with multiple tests, a Bonferroni correction factor was applied to the six multiple regressions conducted, and models were only significant if they reached the $p < .008$ level of significance (i.e., $.05/6 = .008$).

Source: McGinley, M., Carlo, G., Crockett, L. J., Raffaelli, M., Torres Stone, R. A., & Iturbide, M. I. (2010). Stressed and helping: The relations among acculturative stress, gender, and prosocial tendencies in Mexican Americans. *Journal of Social Psychology, 150*(1), 34–56.

Review Terms

Adjusted odds ratio	Multiple regression
Adjusted R^2	Nagelkerke's R^2
Beta weight	Odds ratio
Bivariate regression	Outcome variable
Bootstrapping	Prediction
Control variable	Predictor variable
Criterion variable	Regression coefficient
Dependent variable	Regression equation
Dummy variable	Regression line
Explanation	Response variable
Explanatory variable	Sensitivity
Hierarchical multiple regression	Shrinkage
	Simultaneous multiple regression
Hit Rate	Slope
Independent variable	Sobel test
Intercept	Specificity
Line of best fit	Standardized regression equation
Logistic regression	Stepwise multiple regression
Mediated multiple regression	Unstandardized regression equation
	Wald test
Model specification	R
Moderated multiple regression	R^2
	R^2 Change
Multicollinearity	ΔR^2

The Best Items in the Companion Website

1. An interactive online quiz (with immediate feedback provided) covering Chapter 16.
2. Ten misconceptions about the content of Chapter 16.
3. An interactive online resource entitled "Bivariate Regression."

4. An email message sent by the author to his students entitled "Logistic Regression."
5. Several e-articles illustrating the use of bivariate regression, multiple regression, and logistic regression.

To access chapter outlines, practice tests, weblinks, and flashcards, visit the companion website at http://www.ReadingStats.com.

> **Review Questions and Answers begin on page 531.**

Inferences on Percentages, Proportions, and Frequencies

In your journey through Chapters 9 through 16, you have examined a variety of inferential techniques that are used when at least one of the researcher's variables is quantitative in nature. The bulk of Chapter 9, for example, dealt with inferences concerning Pearson's *r*, a bivariate measure of association designed for use with two quantitative variables. Beginning in Chapter 10, you saw how inferential techniques can be used to investigate one or more groups in terms of means (and variances), with the dependent variable in such situations obviously being quantitative in nature. In Chapter 16, we considered how inferential procedures can be used with regression techniques involving at least one quantitative variable.

In this chapter, your journey takes a slight turn, because we now look at an array of inferential techniques designed for the situation in which *none* of the researcher's variables is quantitative. In other words, the statistical techniques discussed in this chapter are used when all of a researcher's variables involve questions concerning membership in categories. For example, a researcher might wish to use sample data to help gain insights as to the prevalence of AIDS in the general population. Or, a pollster might be interested in using sample data to estimate how each of three political candidates competing for the same office would fare "if the election were to be held today." In these two illustrations as well as in countless real investigations, the study's data do not reflect *the extent* to which each person (or animal) possesses some characteristic of interest but instead reveal how each research participant has been classified into one of the categories established by the researcher.

When a study's data speak to the issue of group membership, the researcher's statistical focus is on frequencies, percentages, or proportions. For example, the hypothetical pollster referred to in the previous paragraph might summarize the study's data by reporting, "Of the 1,000 voters who were sampled, 428 stated that they would vote for candidate A, 381 stated that they would vote for candidate B, and 191 reported that they would vote for candidate C." Instead of providing us with **frequencies** (i.e., the number of people in each response category), the same data could be summarized through **percentages.** With this approach, the researcher would report that "candidates A, B, and C received 42.8 percent, 38.1 percent, and 19.1 percent of the votes, respectively." Or, the data could be converted into **proportions,** with the researcher asserting that "the proportionate popularity of candidates A, B, and C turned out to be .428, .381, and .191, respectively." The same information, of course, is communicated through each of these three ways of summarizing the data.

Regardless of whether the data concerning group membership are summarized through frequencies, percentages, or proportions, it can be said that the level of measurement used within this kind of study is nominal (rather than ordinal, interval, or ratio). As I pointed out in Chapter 3, a researcher's data *can* be nominal in nature. In focusing on inferential techniques appropriate for means, $r, R,$ and R^2, we spent the last several chapters dealing with procedures that are useful when the researcher's data are interval or ratio in nature. In this chapter, however, we restrict our consideration to statistical inferences appropriate for nominal data.

Although a multitude of inferential procedures have been developed for use with nominal-level data, we consider here only six of these procedures that permit researchers to evaluate null hypotheses. These procedures are the *sign test, the binomial test, Fisher's Exact Test, the chi-square test, McNemar's test,* and *Cochran's test.* These are the most frequently used of the test procedures designed for nominal-level data, and a knowledge of these procedures puts you in a fairly good position to understand researchers' results when their data take the form of frequencies, percentages, or proportions.

In this chapter, I illustrate how *z*-tests can be used, in certain situations, to answer the same kinds of research questions as those posed by some of the six basic test procedures we will examine. Moreover, I show how the Bonferroni technique can be used with any of these test procedures to control against an inflated Type I error rate. The distinction between statistical significance and practical significance is also considered. Finally, I point out how important it is to consider the null hypothesis when judging the worth of claims based on this chapter's test procedures.

The Sign Test

Of all inferential tests, the **sign test** is perhaps the simplest and easiest to understand. It requires that the researcher do nothing more than classify the participants of the study into two categories. Each of the participants put into one of these categories receives a plus sign (i.e., a $+$); in contrast, a minus sign (i.e., a $-$) is given to each participant who falls into the other category. The hypothesis testing procedure is then used to evaluate the null hypothesis that says the full sample of participants comes from a population in which there are as many pluses as minuses.

If the sample is quite lopsided with far more pluses than minuses (or far more minuses than pluses), the sign test's H_0 is rejected. However, if the frequencies of pluses and minuses in the sample are equal or nearly equal, the null hypothesis of the sign test is retained.

The sign test can be used in any of three situations. In one situation, there is a single group of people, with each person in the group evaluated as to some characteristic (e.g., handedness) and then given a $+$ or a $-$ depending on his or her status on that characteristic. In the second situation, there are two matched groups; here, the two members of each pair are compared, with a $+$ given to one member of each dyad (and a $-$ given to his or her mate) depending on which one has more of the characteristic being considered. In the third situation, a single group is measured twice, with a $+$ or a $-$ given to each person depending on whether his or her second score is larger or smaller than his or her first score.

In Excerpts 17.1 and 17.2, we see two examples of the sign test. In the first of these excerpts, the sign test is analogous to a one-sample t-test. As you can see, two of these tests were conducted, with the "split" being 13 to 10 across the two categories (right/wrong) with the line discrimination task, and 12 to 8 with the circle discrimination task. In both cases, the split was too close to the null notion of an equal split, so neither sign test was significant. In Excerpt 17.2, the sign test is analogous to a pre–post correlated t-test. In this case, 18 of 26 individuals increased their rating, a result that was beyond the limits of chance sampling. Stated differently,

EXCERPTS 17.1–17.2 • *The Sign Test*

Participants indicated which target line [of the two] was longer or which circle [of the two] was larger, by pointing. . . . For lines, we found that 13 out of 23 people correctly discriminated length (one-tailed sign test, $p = 0.3388$). For circles, we found that 12 out of 20 people correctly discriminated size (one-tailed sign test, $p = 0.2517$). These results suggest that ability of normal adults to discriminate objects with a size difference of 0.0357 deg was not reliably better than chance.

Source: Palomares, M., Ogbonna, C., Landau, B., & Egeth, H. (2009). Normal susceptibility to visual illusions in abnormal development: Evidence from Williams Syndrome. *Perception, 38*(2), 186–199.

The same subjects were measured under the placebo and oxytocin conditions, with each person serving as his own control. . . . Oxytocin increased the rankings of attachment security [as] 18 subjects (69%) out of 26 increased in rating "secure attachment," whereas only 8 subjects (31%) decreased ($p = .038$, one-tailed exact sign test).

Source: Buchheim, A., Heinrichs, M., George, C., Pokorny, D., Koops, E., Henningsen, P., et al. (2009). Oxytocin enhances the experience of attachment security. *Psychoneuroendocrinology, 34*(9), 1417–1422.

the sign test concluded that this kind of split (18 versus 8) is quite unlikely if the sample came from a population in which there was a 50–50 split. Accordingly, the null hypothesis was rejected.

The Binomial Test

The **binomial test** is similar to the sign test in that (1) the data are nominal in nature, (2) only two response categories are set up by the researcher, and (3) the response categories must be mutually exclusive. The binomial test is also like the sign test in that it can be used with a single group of people who are measured just once, with a single group of people who are measured twice (e.g., preintervention and postintervention), or with two groups of people who are matched or are related in some logical fashion (e.g., husbands and wives). The binomial and sign tests are even further alike in that both procedures lead to a data-based *p*-level that comes from tentatively considering the null hypothesis to be true.

The only difference between the sign test and the binomial test concerns the flexibility of the null hypothesis. With the sign test, there is no flexibility. This is because the sign test's H_0 always says that the objects in the population are divided evenly into the two response categories. With the binomial test, however, researchers have the flexibility to set up H_0 with any proportionate split they wish to test.

In Excerpt 17.3, we see a case that shows the versatility of the binomial test. In the study associated with this excerpt, 560 pilots were first put into groups based on the kind of plane they typically flew. Then, within in each group, data were gathered regarding the gender of the pilots' offspring. A binomial test then compared, for each group, the gender split of the offspring against the national average of 51.2 percent males and 48.8 percent females. The E-2 group of pilots had the most lopsided ratio of male-to-female offspring, with 43.7 percent being males. However, this ratio and the others were not significant when compared against the national average.

EXCERPT 17.3 • *The Binomial Test*

Just as for the overall group, there was no significant sex ratio difference found for any of the three subgroups when compared to the general population. Although several of the values for percent male [e.g., 43.7% and 45.3%] appear impressively different from the national average of 51.2% male, none of the groups reach significance via the binomial test.

Source: Baczuk, R., Biascan, A., Grossgold, E., Isaacson, A., Spencer, J., & Wisotzky, E. (2009). Sex ratio shift in offspring of male fixed-wing naval aviation officers. *Military Medicine,* *174*(5), 523–528.

In the study dealing with the pilots and their offspring, H_0 had a null value of 51.2 percent, based on information the researchers acquired from the CDC in Atlanta. If that number from the CDC had been 50 percent, the researchers could have used either a sign test or a binomial test when analyzing the sample data concerning the pilots. Even though the CDC-supplied number was close to 50, the researchers did the right thing by using the binomial test.

Fisher's Exact Test

The sign test and the binomial test, as we have seen, can be used when the researcher has dichotomous data from either a single sample or from two related samples. However, researchers often conduct studies for the purpose of comparing two independent samples with respect to a dichotomous dependent variable. In such situations, **Fisher's Exact Test** often serves as the researcher's inferential tool.[1]

The null hypothesis of Fisher's Exact Test is highly analogous to the typical null hypothesis of an independent-samples t-test. With an independent-samples t-test, most researchers evaluate a null hypothesis that says $H_0: \mu_1 = \mu_2$ (or, alternatively, as $H_0: \mu_1 - \mu_2 = 0$). Using the symbols P_1 and P_2 to stand for the proportion of cases (in the first and second populations, respectively) that fall into one of the two dichotomous categories of the dependent variable, the null hypothesis of Fisher's Exact Test can be expressed as $H_0: P_1 = P_2$ (or, alternatively, as $H_0: P_1 - P_2 = 0$).

In Excerpt 17.4, we see an example of Fisher's Exact Test. As this excerpt makes clear, the raw data of Fisher's Exact Test are the ns (i.e., frequencies) of the two groups, not means. To get a feel for what was happening statistically, it is best to think in terms of percentages (especially when two groups have dissimilar ns, as is the case here). In the experimental group, about 54 percent of the individuals were very satisfied with treatment and outcome; in the control group, only about 22 percent were

EXCERPT 17.4 • *Fisher's Exact Test*

The groups were compared [via] Fisher's exact test. . . . Significantly more patients in the intervention group were very satisfied with the overall treatment and outcome compared with the control group (15 of 28 patients versus 7 of 32, $P = 0.02$).

Source: Nielsen, P. R., Jørgensen, L. D., Dahl, B., Pedersen, T., & Tønnesen, H. (2010). Prehabilitation and early rehabilitation after spinal surgery: Randomized clinical trial. *Clinical Rehabilitation, 24*(2), 137–148.

---

[1]The word *exact* in the title of this test gives the impression that the Fisher's test is superior to other test procedures. This is unfortunate, because many other tests (e.g., the sign test and the binomial test) possess just as much "exactness" as does Fisher's test.

that satisfied. Fisher's Exact Test compared these two percentages and found them to differ more than could be expected by chance. Hence, the null hypothesis was rejected.

It should be noted that the null hypothesis of Fisher's Exact Test does *not* say that each of the study's two populations will be divided evenly into the two dichotomous categories. Rather, it simply says that the split in one population is the same as the split in the other population. Thus, if 21 of the 28 patients in the experimental group had been satisfied along with 24 of the 32 individuals in the control group, each sample would have 75 of its members being satisfied. These two values, being identical, would cause H_0 to be retained even though the percentage split in each group is not even close to being 50–50.

In research reports, it is not unusual to see the term *related* or the term *association* used to describe the goal or the results of a Fisher's Exact Test. This way of talking about Fisher's Exact Test is legitimate and should not throw you when you encounter it. If the two sample proportions turn out to be significantly different, then there is a nonzero relationship (in the sample data) between the dichotomous variable that creates the two comparison groups and the dichotomous dependent variable. Hence, the use of Fisher's Exact Test accomplishes the same basic goal as does a test of significance applied to a phi or tetrachoric correlation coefficient.[2]

Chi-Square Tests: An Introduction

Although inferential tests of frequencies, percentages, or proportions are sometimes made using the sign test, the binomial test, or Fisher's Exact Test, the most frequently used statistical tool for making such tests is called **chi-square.** The chi-square procedure can be used, in certain circumstances, instead of the sign, binomial, or Fisher's tests. In addition, the chi-square procedure can be used to answer research questions that cannot be answered by any of the inferential techniques covered thus far in this chapter. Because the chi-square test is so versatile and popular, it is important for any reader of the research literature to become thoroughly familiar with this inferential technique. For this reason, I feel obliged to consider the chi-square technique in a careful and unhurried fashion.

Different Chi-Square Tests

The term *chi-square test* technically describes any inferential test that involves a critical value being pulled from or a data-based *p*-value being tied to one of the many chi-square distributions.[3] Each such distribution is like the normal and *t* distributions

---

[2]Again we have a parallel between Fisher's Exact Test and the independent-samples *t*-test, because the *t*-test's comparison of the two sample means is mathematically equivalent to a test applied to the point-biserial correlation coefficient that assesses the relationship between the dichotomous grouping variable and the dependent variable.

[3]From a technical standpoint, the term *chi squared* is more accurate than *chi square*. However, most applied researchers use the latter label when referring to this inferential test.

in that it (1) has one mode, (2) is asymptotic to the baseline, (3) comes from a mathematical formula, and (4) helps applied researchers decide whether to reject or fail to reject null hypotheses. Unlike the normal and t distributions (but like any F distribution), each chi-square distribution is positively skewed. There are many chi-square distributions simply because the degree of skewness tapers off as the number of degrees of freedom increases. In fact, the various chi-square distributions are distinguished from one another solely by the concept of degrees of freedom.

Certain of the inferential tests that are called *chi square* (because they utilize a chi-square distribution) have nothing to do with frequencies, proportions, or percentages. For example, a comparison of a single sample's variance against a hypothesized null value is conducted by means of a chi-square test, and, as we saw in Chapter 16, certain tests in logistic regression utilize chi-square (see Excerpts 16.25 and 16.27). However, these kinds of chi-square tests are clearly in the minority. Without a doubt, most chi-square tests *do* involve the types of data being focused on throughout this chapter. In other words, it is likely that any chi-square test you encounter deals with nominal data.

Even when we restrict our consideration of chi square to those cases that involve nominal data, there still are different types of chi-square tests. One type is called a *one-sample chi-square test* (or a *chi-square goodness-of-fit test*), a second type is called an *independent-samples chi-square test* (or a *chi-square test of homogeneity of proportions*), and the third type is called a *chi-square test of independence*. We consider each of these chi-square tests shortly, and then later in the chapter we see how a chi-square test can also be used with related samples. Before we look at any of these chi-square tests, however, it is appropriate first to consider how to tell that a researcher is presenting results of a chi-square test.

Chi-Square Notation and Language

Excerpts 17.5 through 17.7 illustrate the variation in how applied researchers refer to the chi-square tests used in their studies. Although the studies from which these excerpts were taken differ in the number of samples being compared and the number of nominal categories in the data, it should be noted that each of these studies had the same statistical focus as all of the other tests considered in this chapter: frequencies, proportions, or percentages.

EXCERPTS 17.5–17.7 • Different Ways of Referring to Chi-Square

The frequency distributions of PSQI scores among the three groups differed significantly ($\chi^2 = 8.69, df = 2, p = .01$).

Source: Ko, S.-H., Chang, S.-C., & Chen, C.-H. (2010). A comparative study of sleep quality between pregnant and nonpregnant Taiwanese women. *Journal of Nursing Scholarship, 42*(1), 23–30.

(continued)

EXCERPTS 17.5–17.7 • (*continued*)

> Despite very careful examination of the animals involved in the second trial, this proportion [of ticks] was not significantly reduced (26% vs 36%; Chi^2 = 2.3, P > 0.10).
>
> *Source:* Stachurski, F., & Adakal, H. (2010). Exploiting the heterogeneous drop-off rhythm of *Amblyomma variegatum* nymphs to reduce pasture infestation by adult ticks. *Parasitology, 137*(7), 1129–1137.
>
> --
>
> Pearson's χ^2 analysis was applied to assess differences in travel mode by sex.
>
> *Source:* Voss, C., & Sandercock, G. (2010). Aerobic fitness and mode of travel to school in English schoolchildren. *Medicine and Science in Sports and Exercise, 42*(2), 281–287.

In Excerpt 17.5, we see the Greek symbol for chi square, χ^2. Excerpt 17.6 contains the written-out name: chi^2. Finally, in Excerpt 17.7, the phrase *Pearson chi square* is used.

The adjective *Pearson* is the technically correct way to indicate that the chi-square test has been applied to frequencies (rather than, for example, to variances). However, very few applied researchers use the phrase **Pearson chi square** (or the more formal label, *Pearson's approximation to chi square*). Accordingly, it is fairly safe to presume that any chi-square test you encounter is like those considered in this chapter, even though the word *Pearson* does not appear in the test's label. (Of course, this would not be a safe bet if the term *chi-square test* is used within a context in which it is clear that the test's statistical focus deals with something other than frequencies, proportions, or percentages.)

Three Main Types of Chi-Square Tests

We now turn our attention to the three main types of chi-square tests used by applied researchers—the one-sample chi-square test, the independent-samples chi-square test, and the chi-square test of independence. Although applied researchers typically refer to all three using the same label (*chi-square test*), the null hypotheses of these tests differ. Accordingly, you must know which kind of chi-square test has been used in order to understand what is meant by a statistically significant (or nonsignificant) finding.

The One-Sample Chi-Square Test

With this kind of chi-square test, the various categories of the nominal variable of interest are first set up and considered. Second, a null hypothesis is formulated. The H_0 for the **one-sample chi-square test** is simply a specification of what percentage of the population being considered falls into each category. Next, the researcher determines what percentage of the sample falls into each of the established categories. Finally, the hypothesis testing procedure is used to determine whether the discrepancy between the set of sample percentages and those specified in H_0 is large enough to permit H_0 to be rejected.

Excerpt 17.8 illustrates the use of a one-sample chi-square test. The single sample was made up of 62 college students who were near the end of a semester-long course. They were surveyed to find out which of three different student *identifiers* used on the course's written projects—names, ID numbers, and bar codes like the ones used in stores—was most fair to students. The null hypothesis said that the three response options had equal "drawing power," and this H_0 would have been retained if the percentage of students choosing each option had been about the same. However, the chi-square test showed that there was more variability among the three percentages than would be expected by chance. Hence, the null hypothesis was rejected.

EXCERPT 17.8 • *One-Sample Chi-Square Test*

To implement bar code usage in grading written papers, a class of undergraduate students $(n = 62)$ participated in the use of the bar code grading method over one semester. . . . After the final written projects were graded, the instructor surveyed their perceptions of fairness in the grading process. The students were asked, "Which identifier method provides greater anonymity during the grading process?" with the options of "names, social security number, and bar code." . . . The chi-square test was used to examine the null hypothesis: there is no preference among the three methods. The result shows that the null hypothesis is rejected. There is a strong preference among the three methods in which 45 students perceived that the use of bar codes would provide greater anonymity, $\chi^2(2, N = 62) = 44.9, p < .01$.

Source: Jae, H., & Cowling, J. (2009). Objectivity in grading. *College Teaching, 57*(1), 51–55.

If you look again at Excerpt 17.8, you see χ^2's *df* presented as the first number in parentheses following the chi-square symbol. This *df* is not equal to one less than the number of people in the sample. Instead, it is equal to one less than the number of categories. This is the way the *df* for all one-sample chi-square tests are computed, because the sample percentages, across the various categories, must add up to 100. Because of this, you could figure out the final category's percentage once you have been given the percentages for all other categories. The final category's percentage, therefore, is not free to vary but rather has a value that is known as soon as the percentages for the other categories are recorded.

There is one additional thing you should know about the one-sample chi-square test. The null hypothesis concerning the various percentages is usually set up in a "no difference" fashion, as exemplified by Excerpt 17.8. However, the null hypothesis can be set up such that these percentages are dissimilar. For example, in comparing the handedness of pub patrons who play darts, we might set up the H_0 with the right- and left-handed percentages equal to 90 and 10, respectively. These numbers come from census figures, and our little study would be asking the simple question, "Do the census figures seem to hold true for the population of dart throwers represented by the sample used in our study?"

Because the one-sample chi-square test compares the set of observed sample percentages with the corresponding set of population percentages specified in H_0, this kind of chi-square analysis is sometimes referred to as a **goodness-of-fit test.** If these two sets of percentages differ by an amount that can be attributable to sampling error, then there is said to be a *good fit* between the observed data and what would be expected if H_0 were true. In this situation, H_0 is retained. However, if sampling error cannot adequately explain the discrepancies between the observed and null percentages, then a bad fit is said to exist, and H_0 is rejected. The researcher's level of significance, in conjunction with the data-based *p*-value, makes it easy to determine what action should be taken whenever this chi-square goodness-of-fit test is applied.

On occasion, researchers use the chi-square goodness-of-fit test to see if it is reasonable to presume that the sample data have come from a normally distributed population. Of course, for researchers to have this concern, their response variable must be quantitative, not qualitative. If researchers have data that are interval or ratio in nature and if they want to apply this kind of a **test of normality,** the baseline beneath the theoretical normal distribution can be subdivided into segments, with each segment assigned a percentage to reflect the percentage of cases in a true normal distribution that would lie within that segment. These percentages are then put into H_0. Next, the sample is examined to determine what percentage of the observed cases fall within each of the predetermined segments, or categories. Finally, the chi-square goodness-of-fit test compares the observed and null percentages across the various categories to see whether sampling error can account for any discrepancies.[4]

The Independent-Samples Chi-Square Test

Researchers frequently wish to compare two or more samples on a response variable that is categorical in nature. Because the response variable can be made up of two or more categories, we can set up four different kinds of situations to which the **independent-samples chi-square test** can be applied: (1) two samples compared on a dichotomous response variable, (2) more than two samples compared on a dichotomous response variable, (3) two samples compared on a response variable that has three or more categories, and (4) more than two samples compared on a response

---

[4]The Kolmogorov-Smirnov one-sample test is another goodness-of-fit procedure that can be used as a check on normality. It has several properties that make it superior to chi-square in situations where concern rests with the distributional shape of a continuous variable.

variable that has three or more categories. As you will see, considering the first and the fourth of these four situations generates some valuable insights about chi-square and its relationship with other inferential tests we have covered.

When two independent samples are compared with respect to a dichotomous dependent variable, the chi-square test can be thought of as analogous to an independent-samples t-test. With the t-test, the null hypothesis usually tested is $H_0: \mu_1 = \mu_2$. With the chi-square test, the null hypothesis is $H_0: P_1 = P_2$, with P_1 and P_2 representing the percentage of cases (in the first and second populations) that fall into one of the two response categories. Thus, the null hypothesis for this form of the chi-square test simply says that the two populations are identical in the percentage split between the two categories of the response variable.

In Excerpt 17.9, we see an example of this first kind of independent-samples chi-square test. The two groups were Latina and Caucasian women with breast cancer. The two categories of the response variable were set up to correspond with yes and no answers to the question, "Has there been a report of psychiatric illness?"

EXCERPT 17.9 • *Two-Group Independent-Samples Chi-Square Test with a Dichotomous Response Variable*

The findings of this study describe differences between Latina and Caucasian breast cancer survivors in perceived social support, uncertainty, QOL, and selected demographic variables. . . . [D]ata from 181 Caucasian participants and 97 Latina participants were included in the analysis. . . . A significant [difference] was noted [for] the presence of psychiatric illness ($\chi^2 [1, n = 278] = 18.71, p < 0.001$). Latinas reported more psychiatric illness ($n = 13$) than did Caucasians ($n = 2$).

Source: Sammarco, A., & Konecny, L. M. (2010). Quality of life, social support, and uncertainty among Latina and Caucasian breast cancer survivors: A comparative study. *Oncology Nursing Forum, 37*(1), 93–99.

To help you understand the chi-square test that was applied to the data of Excerpt 17.9, I have constructed a **contingency table** and present it in Table 17.1. In such a table, the data of a study are arranged in a 2×2 matrix for the purpose

TABLE 17.1 *Contingency Table Containing Raw Data from Excerpt 17.9*

| | | Diagnosed as Being Psychotic | | |
		Yes	No	
Group	Caucasian	2	179	181
	Latina	13	84	97

of showing how each group split itself up on the dichotomous response variable. Contingency tables are worth looking at (if they are provided in research reports) or creating (if they are not provided), because they make it easier to understand the chi-square null hypothesis and why the data led to the rejection or retention of H_0.

The null hypothesis associated with Excerpt 17.9 did *not* specify that each of the two populations—Caucasian and Latina women with breast cancer—had a 50–50 split between the two categories of the response variable (with half of each population having some form of psychiatric illness). Instead, H_0 said that the two populations were identical to each other in the percentage (or proportion) of women falling into each of the response categories. Thus, the null hypothesis of the study would not have been rejected if about the same percentage of the Caucasian and Latina women had been diagnosed as being psychotic, regardless of whether that percentage was close to 30, 10, 80, or any other value.

Because the null hypothesis deals with percentages (or proportions), it is often helpful to convert each of the cell frequencies of a contingency table into a percentage (or proportion). I have created such a table for Excerpt 17.9 and present it in Table 17.2. As before, the rows and columns correspond to the groups and the response categories, respectively. Now, however, the cells on either row indicate the percentage split of that row's women across the yes and no responses to the question. This contingency table shows why the chi-square null hypothesis was rejected, because the two samples clearly differed in their percentages in either column.

Earlier, I stated that a chi-square test that compares two groups on a dichotomous response variable is analogous to an independent *t*-test. This kind of chi-square is even more similar to Fisher's Exact Test, because these two tests have the same null hypothesis and also utilize the same kind of data. Because of these similarities, you may have been wondering why some researchers choose to use a Fisher's Exact Test whereas others subject their data to an independent-samples chi-square test. Although I address this question more fully later in the chapter, suffice it to say that Fisher's test works better when the researcher has a small number of subjects.

TABLE 17.2 *Contingency Table Containing Percentages from Excerpt 17.9*

		Diagnosed as Being Psychotic		
		Yes	No	
Group	Caucasian	1.1%	98.9%	100%
	Latina	13.4%	86.6%	100%

An independent-samples chi-square test can involve more than two groups or a response variable that has more than two categories. To illustrate, let's consider a study dealing with the game of darts. In this investigation, the research participants were 100 fifth-graders in Greece who had never played darts before. These boys and girls were randomly assigned to four different groups prior to the dart-throwing task. After watching a demonstration of how properly to throw a dart, those in the first three groups were given a goal to focus on while throwing. Members of one group were given instructions to think about the process (i.e., their form) when throwing; those in the second group were told to focus on their performance outcome (i.e., how well they did); participants in the third group were asked to concentrate on both process *and* performance. Members of the fourth group—the control group—were simply asked to do their best. After receiving their instructions, each member of each group performed the dart-throwing task.

After each child missed the bulls-eye on two consecutive throws, he or she was asked why the previous throw had missed its mark. The children's answers to this question were coded into four categories: "Technique," "Focus," "Ability," or "Don't Know." The percentage of children from each group who provided each of these four reasons (i.e., attributions for missing the bulls-eye) is shown in Table 17.3.

A chi-square test compared the four groups in terms of the way the group members explained why they had missed the bulls-eye. The null hypothesis for this test involved four populations: fifth-graders (like those in the study) who are given a process goal before throwing darts for the first time, fifth-graders (like those in the study) who are given a performance goal before throwing darts for the first time, and so on. This H_0 stipulated that each of these populations has the same distribution

TABLE 17.3   *Contingency Table for Dart-Throwing Investigation*

		— *Reason Given (i.e., Attribution) for the Dart Missing the Target's Bulls-eye*				
		Technique	Focus	Ability	Don't Know	
	Process Goal ($n = 29$)	75.9%	13.8%	3.4%	6.9%	100%
	Performance Goal ($n = 29$)	51.7%	27.6%	10.3%	10.3%	100%
Group	Process & Performance Goal ($n = 29$)	65.6%	17.2%	3.5%	13.8%	100%
	Control ($n = 13$)	23.1%	15.4%	7.7%	53.8%	100%

Notes: Percentages in rows 2 and 3 do not add to 100 due to rounding errors. Data from Kolovelonis, Goudas, and Dermitzaki's 2010 article (The effect of different goals and self-recording on self-regulation of learning a motor skill in a physical education setting) that appeared in the journal *Learning and Instruction*.

of attributions across the categories of "Technique," "Focus," "Ability," and "Don't Know." For this null hypothesis to be true, that distribution of percentages could be even (i.e., 25–25–25–25) or uneven (e.g., 50–30–15–5); however, whatever is the case for any one population must be the same for the other three populations. Excerpt 17.10 contains the result of this χ^2 test. As you can see, the chi-square null hypothesis was rejected.[5]

EXCERPT 17.10 • *Four-Group Independent-Samples Chi-Square Test with a Four-Category Response Variable*

After the fifth minute of practice, when a student had missed the centre of the target for two consecutive throws, she or he was asked the single question "Why do you think you missed the centre of the target in your last throw?". . . . The 4 (group) × 4 (attribution) cross-tabulation analysis showed a significant difference, $\chi^2(9, N = 100) = 21.80, p < .01$, [among] the four groups . . . showing that all goal [groups] attributed more frequently the missing throws to their incorrect technique compared to the control group.

Source: Kolovelonis, A., Goudas, M., Dermitzaki, I. (2011). The effect of different goals and self-recording on self-regulation of learning a motor skill in a physical education setting. *Learning and Instruction*, in press.

In Excerpt 17.10, notice that the number 9 is positioned inside a set of parentheses, just to the left of the study's sample size. This number was the chi-square's *df*. With this or any other contingency table, the *df* for χ^2 is determined by multiplying 1 less than the number of rows times 1 less than the number of columns. In this case, $df = (4 \text{ rows} - 1)(4 \text{ columns} - 1) = 9$.

Chi-Square as a Correlational Probe

In many studies, a researcher is interested in whether a nonchance relationship exists between two nominal variables. In such studies, a single sample of subjects is measured, with each research participant classified into one of the available categories of the first variable and then classified once more into one of the categories of the second variable. After the data are arranged into a contingency table, a chi-square test can be used to determine whether a statistically significant relationship exists between the two variables.

In Excerpts 17.11 through 17.13, we see three terms used in conjunction with chi-square tests that let you know the researchers were using chi-square in a

---

[5]The term *cross-tabulation analysis* in Excerpt 17.10 refers to a table that shows the percentage or number of cases that fall into each cell of a two-dimensional chart.

EXCERPTS 17.11–17.13 • *Terms that Indicate Chi-Square Is Used as a Correlational Probe*

Associations between categorical variables were explored using the chi-square test of association.

Source: Decloedt, E., Leisegang, R., Blockman, M., & Cohen, K. (2010). Dosage adjustment in medical patients with renal impairment at Groote Schuur Hospital. *South Afrrican Medical Journal, 100*(5), 304–306.

Chi-square statistics were used to examine relationships between categorical variables or between categorical and ordinal variables.

Source: Harlow, K. C., & Roberts, R. (2010). An exploration of the relationship between social and psychological factors and being bullied. *Children & Schools, 32*(1), 15–26.

A chi-square test of independence demonstrated that presences of salamanders and crayfish were not independent ($\chi^2 = 7.46, df = 1, P = 0.006$); significantly more salamanders and crayfish co-occurred under the same rocks than expected by chance.

Source: Pierce, B. A., Christiansen, J. L., Ritzer, A. L., & Jones, T. A. (2010). Ecology of Georgetown Salamanders (*Eurycea naufragia*) within the flow of a spring. *Southwestern Naturalist, 55*(2), 291–297.

correlational manner. The first two of these terms—*association* and *relationship*—are not new; we saw them used in Chapter 3 while considering bivariate correlation. The third term, however, is new. A chi-square **test of independence** is simply a test to see whether a relationship (or association) exists between the study's two variables.

When a chi-square test is used as a correlational probe, it does not produce an index that estimates the strength of the relationship between the two variables that label the contingency table's rows and columns. Instead, the chi-square test simply addresses the question, "In the population of interest, are the two variables related?" Focusing on the sample data, this question takes the form, "In the contingency table, is there a nonchance relationship between the two variables?"

To illustrate what I mean by *nonchance relationship,* imagine that we go out and ask each of 100 college students to name a relative. (If anyone responds with a gender-free name like Pat, we then ask the respondent to indicate whether the relative is a male or a female.) We also keep track of each respondent's gender. After collecting these two pieces of information from our 100th student, we might end up with sample data that look like this:

		Gender of the Relative	
		Male	*Female*
Gender of the Student	*Male*	30	20
	Female	23	27

In the 2×2 contingency table for our hypothetical study, there is a relationship between the two variables—student's gender and relative's gender. More of the male students responded with the name of a male relative whereas more of the female students thought of a female relative. (Or, we could say that there was a tendency for male relatives to be thought of by male students whereas female relatives were thought of by female students.) But is this relationship something other than what would be expected by chance?

If there were *no* relationship in the population between the two variables in our gender study, the population frequencies in all four cells of the contingency table would be identical. But a sample extracted from that population would not likely mirror the population perfectly. Instead, sampling error would likely be in the sample data, thus causing the observed contingency table to have dissimilar cell frequencies. In other words, we would expect a relationship to pop up in the sample data even if there were no relationship in the population. Such a relationship, in the sample data, would be due entirely to chance. Although we should expect a *null population* (i.e., one in which there is no relationship between the two variables) to yield sample data in which a relationship *does* exist between the two variables, such a relationship ought to be small, or weak. It *is* possible for a null population to yield sample data suggesting a strong relationship between the two variables, but this is *not* very likely to happen. Stated differently, if researchers end up with a contingency table in which there is a meager relationship, they have only weak evidence for arguing that the two variables of interest are related in the population. If, in contrast, a pronounced relationship shows up in the contingency table built with the sample data, the researchers possess strong evidence for suggesting that a relationship does, in fact, exist in the population.

Returning to our little gender study, the chi-square test can be used to label the relationship that shows up in the sample data as being either meager or pronounced. Using the hypothesis testing procedure in which the level of significance is set equal to .05, the null hypothesis of no relationship in the population cannot be rejected. This means that the observed relationship in the contingency table could easily have come from a sample pulled from a population characterized by H_0.

In addition to using a chi-square test to see if a nonchance relationship exists in the sample data, researchers can convert their chi-square calculated value into an index that estimates the strength of the relationship that exists in the population. By making this conversion, the researcher obtains a numerical value that is analogous to the correlation coefficient generated by Pearson's or Spearman's technique. Several different conversion procedures have been developed.

The phi coefficient can be used to measure the strength of association in 2×2 contingency tables. I discussed this correlational procedure in Chapters 3 and 9 and pointed out in those discussions how phi is appropriate for the case of two dichotomous variables. Now, I can extend this discussion of phi by pointing out its connection to chi-square. If a chi-square test has been applied to a 2×2 contingency table, the phi index of association can be obtained directly by putting the chi-square calculated value into this simple formula:

$$\text{phi} = \sqrt{\frac{\chi^2}{N}}$$

where N stands for the total sample size. Researchers, of course, are the ones who use this formula in order to convert their chi-square calculated values into phi coefficients. As illustrated by Excerpt 17.14, you will not have to do this.

EXCERPT 17.14 • *Chi-Square and Phi*

A Chi-square test indicated an association between receipt of ancillary medication and ethnicity ($\chi^2 = 9.94, P < .01, \text{phi} = .28$). Within each ethnicity category, substantially more participants received ancillary medication than did not receive medication (White: 84% vs. 16%; Black: 58% vs. 42%; and Hispanic: 61% vs. 39%, respectively). However, a substantially higher proportion of White participants received ancillary medication compared to other ethnic groups.

Source: Hillhouse, M., Domier, C. P., Chim, D., & Ling, W. (2010). Provision of ancillary medications during buprenorphine detoxification does not improve treatment outcomes. *Journal of Addictive Diseases, 29*(1), 23–29.

For contingency tables that have more than two rows or columns, researchers can convert their chi-square calculated value into a measure of association called the **contingency coefficient.** This index of relationship is symbolized by C, and the connection between C and chi square is made evident by the following formula for C:

$$C = \sqrt{\frac{\chi^2}{N + \chi^2}}$$

In Excerpt 17.15, we see an illustration of how the contingency coefficient can be computed following a chi-square test of independence.[6]

---

[6]A variation of C is called the *mean square contingency coefficient*. This index of relationship uses the same formula as that presented for phi.

EXCERPT 17.15 • *Chi-Square and the Contingency Coefficient*

We categorized participants who had travelled abroad as engaging in heavy episodic drinking (i.e., HED) or not engaging in heavy episodic drinking (i.e., NHD) for analyses. . . . Chi square analysis demonstrated that NHD ($n = 27$) and HED ($n = 16$) participants differed significantly in the frequency with which they described specific motivations or reasons for their alcohol use [cultural experience vs. social enhancement vs. accessibility due to being of age] ($\chi^2(2, N = 43) = 7.74, p < .05$; Contingency Coefficient = .39).

Source: Smith, G., & Klein, S. (2010) Predicting women's alcohol risk-taking while abroad. *Women & Health, 50*(3), 262–278.

The formula for *C* shows that this index of association turns out equal to zero when there is no relationship in the contingency table (because in that case, the calculated value of χ^2 itself turns out equal to zero) and that it assumes larger values for larger values of χ^2. What this formula does not show is that this index usually cannot achieve a maximum value of 1.00 (as is the case with Pearson's *r*, Spearman's rho, and other correlation coefficients). This problem can be circumvented easily if the researcher computes **Cramer's measure of association,** because Cramer's index is simply equal to the computed index of relationship, *C*, divided by the maximum value that the index could assume, given the contingency table's dimensions and marginal totals.

In Excerpt 17.16, we see a case in which Cramer's measure of association, symbolized as *V*, was computed in conjunction with two different 2 × 3 contingency tables. Notice that the larger *V* is paired with the larger χ^2.

EXCERPT 17.16 • *Chi-Square and Cramer's V*

With respect to the academic level, we observed no significant differences in relation to the distribution in the level of motivation [high or low] $\chi^2(2, N = 258) = 3.270; p = .195$ (with Cramer's $V = .113; p = .195$). . . . In contrast to this finding, the academic level resulted in significant differences in value [intrinsic or extrinsic], showing that final-level students are more intrinsic (56%) than intermediate (19%) or initial level students (36%) $\chi^2(2, N = 258) = 17.486; p < .001$ (with Cramer's $V = .260; p < .001$).

Source: Rabanaque, S., & Martínez-Fernández, J. R. (2009). Conception of learning and motivation of Spanish psychology undergraduates in different academic levels. *European Journal of Psychology of Education, 24*(4), 513–528.

Issues Related to Chi-Square Tests

Before we conclude our discussion of chi-square tests, a few related issues must be addressed. Unless you are aware of the connection between these issues and the various chi-square tests we have covered, you will be unable to fully understand and critique research reports that contain the results of chi-square tests. Accordingly, it is important for you to be sensitive to the following issues.

Post Hoc Tests

If an independent-samples chi-square test is used to compare two groups, interpretation of the results is straightforward regardless of what decision is made regarding the null hypothesis. If there are three or more comparison groups involved in the study, the results can be interpreted without difficulty so long as H_0 is not rejected. If, however, the independent-samples chi-square test leads to a rejection of H_0 when more than two groups are contrasted, the situation remains unclear.

When three or more samples are compared, a statistically significant outcome simply indicates that it is unlikely that all corresponding populations are distributed in the same way across the categories of the response variable. In other words, a rejection of H_0 suggests that at least two of the populations differ, but this outcome by itself does not provide any insight as to which specific populations differ from one another. To gain such insights, the researcher must conduct a post hoc investigation.

In Excerpt 17.17, we see a case where a post hoc investigation after an omnibus chi-square test yielded a statistically significant result. The original chi-square test involved a 3 × 2 arrangement of the data, with three groups (men who had

EXCERPT 17.17 • Post Hoc Investigation Following a Significant Chi-Square Test

Furthermore, based on a chi-square test, the three groups differed significantly in reported rates of lifetime STI [sexually transmitted infection], $\chi^2 (2, 1180) = 6.21, p < .05$ . . . Based on follow-up tests, men who perpetrated sexual aggression more than once were significantly more likely than men who had never perpetrated sexual aggression to have contracted an STI in their lifetime, $p < .05$, with 25% of men who perpetrated multiple acts of sexual aggression reporting a history of STI and 16% of nonaggressive men reporting a history of STI. Of the men who perpetrated sexual aggression only once, 21% reported a history of STI. The group of men who perpetrated sexual aggression only once did not differ significantly from either of the other groups in rates of STI.

Source: Peterson, Z. D., Janssen, E. & Heiman, J. R. (2010). The association between sexual aggression and HIV risk behavior in heterosexual men. *Journal of Interpersonal Violence,* 25(3), 538–556.

perpetrated 0, 1, or 2+ sexually aggressive acts) and two categories of STI (either they had or had not contracted a sexually transmitted infection). After the omnibus χ^2 turned out to be significant, the researchers probed their data with three different 2×2 chi-square analyses. In a very real sense, this post hoc investigation had the same goal as would a set of Tukey pairwise comparisons computed after a one-way ANOVA yields a significant F.

Whenever two or more separate chi-square tests are performed within a post hoc investigation, with each incorporating the same level of significance as that used in the initial (omnibus) chi-square test, the chances of a Type I error being made somewhere in the post hoc analysis exceeds the nominal level of significance. This is not a problem in those situations where the researcher judges Type II errors to be more costly than Type I errors. Be that as it may, the scientific community seems to encourage researchers to guard against Type I errors.

In Excerpt 17.17, you saw a case in which a post hoc investigation, involving three pairwise comparisons (each using a reduced 2×2 contingency table) was conducted after an omnibus chi-square test yielded a significant result. In the research report that provided this excerpt, I could not find any indication that the researchers used the Bonferroni adjustment procedure (or some similar device) to protect against an inflated Type I error rate in this post hoc investigation. Perhaps they adjusted their alpha level but just did not report having done so. Or, perhaps they failed to lower α when they conducted their post hoc chi-square tests.

Small Amounts of Sample Data

To work properly, the chi-square tests discussed in this chapter necessitate sample sizes that are not too small. Actually, it is the **expected frequencies** that must be sufficiently large for the chi-square test to function as intended. An expected frequency exists for each category into which sample objects are classified, and each one is nothing more than the proportion of the sample data you would expect in the category if H_0 were true and if there were absolutely no sampling error. For example, if we were to perform a taste test in which each of 20 individuals is asked to sip four different beverages and then indicate which one is the best, the expected frequency for each of the four options would be equal to 5 (presuming that H_0 specifies equality among the four beverages). If this same study were to be conducted with 40 participants, each of the four expected values would be equal to 10.

If researchers have a small amount of sample data, too many groups, or too many categories of the response variable, the expected values associated with their chi-square test will also be small. If the expected values are too small, the chi-square test should not be used. Various rules of thumb have been offered over the years to help applied researchers know when they should refrain from using the chi-square test because of small expected values. The most conservative of these rules says that none of the expected frequencies should be smaller than 5; the most liberal rule stipulates that chi-square can be used so long as the average expected frequency is at least 2.[7]

The option of turning to Fisher's Exact Test when the expected frequencies are too small is available in situations in which the sample data create a 2 × 2 contingency table. This option does not exist, however, if their researcher is using (1) a one-sample chi-square test with three or more categories or (2) chi-square with a contingency table that has more than two rows or more than two columns. In these situations, the problem of small expected frequencies can be solved by redefining the response categories such that two or more of the original categories are collapsed together. For example, if men and women are being compared regarding their responses to a five-option Likert-type question, the researcher might convert the five original categories into three new categories by (1) merging together the "Strongly Agree" and "Agree" categories into a new single category called "Favorable Response," (2) leaving the "Undecided" category unchanged, and (3) merging together the "Disagree" and "Strongly Disagree" categories into a new single category called "Unfavorable Response." By doing so, the revised contingency table might not have any expected frequencies that are too small.

In Excerpt 17.18, we see a case where a pair of researchers decided against using chi-square to analyze their data because of small expected frequencies. They used the strategy of collapsing categories of the response variable, but they still had small expected frequencies. They then employed Fisher's Exact Test. Note the last six words of this excerpt.

EXCERPT 17.18 • *Use of Fisher's Exact Test Rather Than Chi-Square Because of Small Expected Frequencies*

Response categories for self-rated physical and mental health were dichotomized [and then] Fisher's exact tests were used to test hypotheses on associations with these variables due to small expected cell counts.

Source: Berkman, C. S., & Ko, E. (2009). Preferences for disclosure of information about serious illness among older Korean American immigrants in New York City. *Journal of Palliative Medicine, 12*(4), 351–357.

Yates' Correction for Discontinuity

When applying a chi-square test to situations where $df = 1$, some researchers use a special formula that yields a slightly smaller calculated value than would be the case if the regular formula were employed. When this is done, it can be said that the data are being analyzed using a chi-square test that has been *corrected for discontinuity*

---

[7]The contingency table shown earlier in this chapter for the dart throwing study had half of its expected cell frequencies smaller than 5; however, the mean of all 16 values was 6.25.

(also known as **Yates' correction for discontinuity**). This special formula was developed by a famous statistician named Yates, and occasionally the chi-square test has Yates' name attached to it when the special formula is used. Excerpt 17.19 shows that Yates' correction is used in conjunction with chi-square analyses. It is not used with any of the other statistical procedures considered in this book.

EXCERPT 17.19 • *Yates' Correction for Discontinuity*

Prevalences of symptoms between groups (categorical data) were compared with the χ^2-test with Yates' correction. . . . Migraine occurred significantly more often in CPAs [37/200 = 18.5%] than in controls [17/210 = 8.1%] after 3 months.

Source: Stovner, L. J., Schrader, H., Mickevičiene, D., Surkiene, D., & Sand, T. (2009). Headache after concussion. *European Journal of Neurology, 16*(1), 112–120.

Statistical authorities are not in agreement as to the need for using Yates' special formula. Some argue that it should *always* be used in situations where $df = 1$ because the regular formula leads to calculated values that are too large (and thus to an inflated probability of a Type I error). Other authorities take the position that the Yates adjustment causes the pendulum to swing too far in the opposite direction because Yates' correction makes the chi-square test overly conservative (thus increasing the chances of a Type II error). Ideally, researchers should clarify why the Yates formula either was or was not used on the basis of a judicious consideration of the different risks associated with a Type I or a Type II error. Realistically, however, you are most likely to see the Yates formula used only occasionally and, in those cases, used without any explanation as to why it was employed.

McNemar's Chi-Square

Earlier in this chapter, we saw how a chi-square test can be used to compare two independent samples with respect to a dichotomous dependent variable. If the two samples involved in such a comparison are related rather than independent, chi-square can still be used to test the **homogeneity of proportions** null hypothesis. However, both the formula used by the researchers to analyze their data and the label attached to the test procedure are slightly different in this situation where two related samples are compared. Although there is no reason to concern ourselves here with the unique formula used when correlated data have been collected, it *is* important that you become familiar with the way researchers refer to this kind of test.

A chi-square analysis of related samples is usually referred to simply as *McNemar's test*. Sometimes, however, it is called *McNemar's change test, McNemar's chi-square test, McNemar's test of correlated proportions,* or *McNemar's test for paired data*. Occasionally, it is referred to symbolically as $Mc\chi^2$. Excerpt 17.20 illustrate the use of McNemar's test.

EXCERPT 17.20 • *McNemar's Chi-Square Test*

This pilot-study aimed to examine the feasibility, acceptability, and effectiveness of family-based treatment [for anorexia nervosa in adolescent girls] in Brazil. . . . McNemar's test was performed in order to compare menstrual status between assessment points. . . . At first evaluation, eight (89%) of the patients had amenorrhea. At the end of treatment, four (44%) had regular menses, whereas all of the patients evaluated had regular menses at the end of follow-up. When McNemar's test was applied, a significant improvement in menstrual status was found when comparing baseline to the end of follow-up ($p = 0.016$).

Source: Turkiewicz, G., Pinzon, V., Lock, J., & Fleitlich-Bilyk, B. (2010). Feasibility, acceptability, and effectiveness of family-based treatment for adolescent anorexia nervosa: An observational study conducted in Brazil. *Revista Brasileira de Psiquiatria, 32*(2), 169–172.

McNemar's chi-square test is very much like a correlated-samples *t*-test in that two sets of data being compared can come either from a single group that is measured twice (e.g., in a pre/post sense) or from matched samples that are measured just once. Excerpt 17.20 obviously falls into the first of these categories because data from a single group are compared at two points in time, at baseline and then at the end of follow-up. The intervention provided between these two points in time was designed to help adolescent girls who suffered from anorexia nervosa.

Although the McNemar's chi-square is similar to a correlated *t*-test with respect to the kind of sample(s) involved in the comparison, the two tests differ dramatically in terms of the null hypothesis. With the *t*-test, the null hypothesis involves population means; in contrast, the null hypothesis of McNemar's chi-square test is concerned with population percentages. In other words, the null hypothesis of McNemar's test always takes the form $H_0: P_1 = P_2$, whereas the *t*-test's null hypothesis always involves the symbol μ (and it usually is set up to say $H_0: \mu_1 = \mu_2$).

The Cochran Q Test

A test developed by Cochran is appropriate for the situation where the researcher wishes to compare three or more related samples with respect to a dichotomous dependent variable. This test is called the **Cochran Q test,** with the letter Q simply being the arbitrary symbol used by Cochran to label the calculated value produced by putting the sample data into Cochran's formula. This test just as easily could have been called *Cochran's chi-square test* inasmuch as the calculated value is compared against a chi-square critical value to determine whether the null hypothesis should be rejected.

The Cochran Q test can be thought of as an extension of McNemar's chi-square test, because McNemar's test is restricted to the situation where just two correlated samples of data are compared, whereas Cochran's test can be used when there are any number of such samples. Or, the Cochran Q test can be likened to the one-factor repeated-measures analysis of variance covered in Chapter 14; in each case, multiple related samples of data are compared. (That ANOVA is quite different from the Cochran test; however, because the null hypothesis in the former focuses on μs whereas Cochran's H_0 involves Ps.)

In Excerpt 17.21, we see a case where Cochran's Q test was used. This excerpt comes from a study in which nurse trainees used three different kinds of pumps to administer IVs to mannequins. The researchers wondered whether the different pumps might differentially affect the nurses' ability to avoid giving a particular doctor-ordered medicine to the wrong "patient." Because all nurse trainees used all three pumps, the Cochran Q test (rather than a regular chi-square test) was used to compare the pumps.

EXCERPT 17.21 • *The Cochran Q Test*

We, therefore, conducted an experimental study to directly compare pump type [differences] on nurses' ability to safely deliver IV medications. . . . Cochran Q tests were followed by pairwise comparisons (using Bonferroni correction) between different combinations of pump types using the McNemar χ^2 test. . . . There was a significant difference in the resolution of patient ID errors across pumps [Cochran Q = 14.36; $df = 2$; $p < 0.05$]. The number of nurses (out of 24) who remedied patient identification errors was significantly higher with the barcode pump (21 (88%)) than with the traditional pump (11 (46%)) or the smart pump (14 (58%)). The difference between the traditional pump and the smart pump, however, was not significant.

Source: Trbovich, P. L., Pinkney, S., Cafazzo, J. A., & Easty, A. C. (2010). The impact of traditional and smart pump administration performance on nurse medication in a simulated infusion technology. *Quality and Safety in Health Care, 19*, 430–434.

In the study associated with Excerpt 17.21, the null hypothesis for Cochran's Q test could be stated as $H_0: P_1 = P_2 = P_3$, where each P stands for the population percentage of nurse trainees, using one of the three pumps, who remedied a patient ID problem. As you can see, the Cochran Q test led to a rejection of this null hypothesis. From a statistical point of view, the three sample percentages—88 percent, 46 percent, and 58 percent—were further apart than would be expected by chance.

When Cochran's test leads to a rejection of the omnibus null hypothesis, the researcher will probably conduct a post hoc investigation. Within this follow-up investigation, researchers most likely will set up and test pairwise comparisons using McNemar's test. This is what happened in Excerpt 17.21. Note that the Bonferroni adjustment was used in conjunction with the three McNemar tests that were conducted.

The Use of z-Tests When Dealing with Proportions

As you may recall from Chapter 10, researchers sometimes use a z-test (rather than a t-test) when their studies are focused on either the mean of one group or the means of two comparison groups. It may come as a surprise that researchers sometimes apply a z-test when dealing with dependent variables that are qualitative rather than quantitative in nature. Be that as it may, you are likely to come across cases where a z-test has been used by researchers when their data take the form of proportions, percentages, or frequencies.

If a researcher has a single group that is measured on a dichotomous dependent variable, the data can be analyzed by a one-sample chi-square test *or* by a z-test. The choice here is immaterial, because these two tests are mathematically equivalent and always lead to the same data-based p-value. The same thing holds true for the case where a comparison is made between two unrelated samples. Such a comparison can be made with an independent-samples chi-square test or a z-test; the p-value of both tests will be the same.

Whereas the z-tests we have just discussed and the chi-square tests covered earlier (for the cases of a dichotomous dependent variable used with a single sample or two independent samples) are mathematically equivalent, there is another z-test that represents a **large sample approximation** to some of the tests examined in earlier sections of this chapter. To be more specific, researchers sometimes use a z-test, if they have large samples, in places where you might expect them to use a sign test, a binomial test, or a McNemar test. In Excerpts 17.22 and 17.23, we see two examples of a z-test being used in connection with test procedures considered

EXCERPTS 17.22–17.23 • Use of z-Tests with Percentages

A sign test confirmed that participants cooperated significantly more often when their partner was displaying an enjoyment smile than when she was displaying a non-enjoyment smile ($Z = 2.07, p < 0.05$).

Source: Johnston, L., Miles, L., & Macrae, C. N. (2010). Why are you smiling at me? Social functions of enjoyment and non-enjoyment smiles. *British Journal of Social Psychology, 49*(1), 107–127.

(*continued*)

EXCERPTS 17.22–17.23 • (*continued*)

For the group comparison of men versus women we used a binomial test procedure. . . . Because of the large sample size, we used a normal approximation to the binomial distribution.

Source: Maarsingh, O. R., Dros, J., Schellevis, F. G., van Weert, H. C., Bindels, P. J., & van der Horst, H. E. (2010). Dizziness reported by elderly patients in family practice: Prevalence, incidence, and clinical characteristics. *BMC Family Practice, 11*(2), 1–9.

in this chapter. In the first of these excerpts, a large-sample approximation to the sign test was used. In Excerpt 17.23, a large-sample approximation to the binomial test was used.

A Few Final Thoughts

As you have seen, a wide variety of test procedures have been designed for situations where data take the form of frequencies, percentages, or proportions. Despite the differences among these tests (in terms of their names, the number of groups involved, and whether repeated measures are involved), there are many commonalities that cut across the tests we have considered. These commonalities exist because each of these tests involves the computation of a data-based p-value that is then used to evaluate a null hypothesis.

In using the procedures considered in this chapter within an applied research study, a researcher follows the various steps of hypothesis testing. Accordingly, many of the side issues dealt with in Chapters 7 and 8 are relevant to the proper use of any and all of the tests we have just considered. In an effort to help you keep these important concerns in the forefront of your consciousness as you read and evaluate research reports, I feel obliged to conclude this chapter by considering a few of these more generic concerns.

My first point is simply a reiteration that the data-based p-value is always computed on the basis of a tentative belief that the null hypothesis is true. Accordingly, the statistical results of a study are always tied to the null hypothesis. If the researcher's null hypothesis is silly or articulates something that no one would defend or expect to be true, then the rejection of H_0, regardless of how impressive the p-value, does not signify an important finding.

If you think that this first point is simply a "straw man" that has no connection to the real world of actual research, consider this *real* study that was conducted not too long ago. In this investigation, chi-square compared three groups of teachers in terms of the types of instructional units they used. Two kinds of data were collected from the teachers: (1) their theoretical orientation regarding optimal teaching-learning practices and (2) what they actually did when teaching. The results indicated that skill-based instructional units tended to be used more by teachers who had a skill-based theoretical orientation, that rule-based instructional units were used more so by teachers who had a rule-based theoretical orientation, and that function-based instructional units were utilized to a greater extent by teachers who possessed a function-based theoretical orientation. Are you surprised that this study's data brought forth a rejection of the chi-square null hypothesis of no relationship between teachers' theoretical orientation and type of instructional unit used? Was a study needed to reach this finding?

My second point is that the chances of a Type I error increase above the researcher's nominal level of significance in the situation where multiple null hypotheses are evaluated. Although there are alternative ways of dealing with this potential problem, you are likely to see the Bonferroni technique employed most often to keep control over Type I errors. In Excerpt 17.21, we saw a case in which the level of significance was adjusted (and made more rigorous) because the McNemar's test was used three times in a post hoc investigation. Keep this good example in mind as you encounter research reports in which several null hypotheses are evaluated by means of the test procedures considered in this chapter. If the researchers associated with such reports give no indication that they attended to the inflated Type I error-rate problem, accept their claims of statistical significance with a grain of salt.

My third point concerns the distinction between statistical significance and practical significance. As I hope you recall from our earlier discussions, it is possible for H_0 to be rejected, with an impressive data-based p-value (e.g., $p < .0001$), even though the computed sample statistic does not deviate much from the value of the parameter expressed in H_0. I also hope you remember my earlier contention that conscientious researchers either design their studies or conduct a more complete analysis of their data with an eye toward avoiding the potential error of figuratively making a mountain out of a molehill.

There are several ways researchers can demonstrate sensitivity to the distinction between practical significance and statistical significance. In our examination of t-tests, F-tests, and tests on correlation coefficients, we have seen that these options include either (1) conducting, in the design phase of the investigation, a power analysis so as to determine the proper sample size; or (2) calculating, after the data have been collected, an effect size estimate. These two options are as readily available to researchers who use the various test procedures covered in this chapter as they are to those who conduct t-tests, F-tests, or tests involving one or more correlation coefficients.

The main statistical technique discussed in this chapter was chi-square. To judge whether a computed chi-square-based effect is small, medium, or large, researchers usually convert their computed value of χ^2 into phi or Cramer's V. In Excerpt 17.24, we see a case where V was used to estimate the effect size in the populations associated with a 2×3 contingency table. In this study, the research participants were the characters in coloring books, each being a male or female. The

EXCERPT 17.24 • *Chi-Square with Cramer's V Used as an Estimate of Effect Size*

Of the 436 males, 44% engaged in stereotypic behaviors and only 3% in cross-gender behaviors. Interestingly, 53% of the males engaged in gender-neutral behaviors. Thus, males were more likely to engage in gender neutral behaviors than male stereotypic ones, partially refuting our second hypothesis in which we expected both genders to be depicted predominantly in gender-stereotypic behaviors. Of the 306 females, 58% engaged in female-stereotypic behaviors and 6% in crossgender behaviors. In comparison to the 53% of males engaging in gender-neutral behaviors, only 32% of females did. Thus, females were more likely to engage in female stereotypic behavior than either of the other types of behavior. This overall pattern of differences in depictions of gender stereotypes was significant; $\chi^2(2, N = 742) = 310.04, p < .001$. This effect was large (Cramer's $V = .65, p < .001$).

Source: Fitzpatrick, M. J., & McPherson, B. J. (2010). Coloring within the lines: Gender stereotypes in contemporary coloring books. *Sex Roles, 62*(1–2), 127–137.

other variable concerned the activity of each character depicted in the coloring book (and whether the activity was traditionally feminine, traditionally masculine, or gender-neutral).

In the final sentence of Excerpt 17.24, the researchers report that the effect they found was "large." This comment was not based on *p* turning out to be smaller than .001. Instead, the value of *V* was compared against some widely used criteria for evaluating the Cramer's *V*. For ease of reference, Table 17.4 contains the standard criteria for judging *V* as well as two other estimates of effect size: phi and *w*.

In Excerpt 17.25, we see an example of an a priori power analysis that was conducted in conjunction with a study using chi-square. There was a clear advantage of doing this kind of power analysis in the design phase of the ulcer investigation. By determining the sample sizes after considering several statistical components of their planned analysis (e.g., level of significance, desired power, and the dividing point between a trivial versus an important finding), the researchers set up their

TABLE 17.4 *Effect Size Criteria for Use with Tests on Frequencies*

Effect Size Measure	Small	Medium	Large
Cramer's *V*	.10	.30	.50
Phi	.10	.30	.50
w	.10	.30	.50

Note: The standards for judging relationship strength are quite general and should be changed to fit the unique goals of any given research investigation.

EXCERPT 17.25 • *A Power Analysis to Determine the Needed Sample Size*

A power analysis was performed on the basis of a recent pilot study of our group, where the effect size w ($\sqrt{\chi^2/N}$) was 0.39 regarding healing. This revealed the necessity to include 52 ulcers to ascertain a difference between the two groups with a power of 0.8 at a two-sided $P < .05$. . . . Fifty-five patients completed the study protocol; 28 (50.9%) in the stocking and 27 (49.1%) in the bandage group.

Source: Brizzio, E., Amsler, F., Lun, B., & Blättler, W. (2010). Comparison of low-strength compression stockings with bandages for the treatment of recalcitrant venous ulcers. *Journal of Vascular Surgery, 51*(2), 410–416.

study so that the chances were minimal that they would either (1) end up with statistical significance but not practical significance or (2) end up without statistical significance when a meaningful effect existed in the study's populations. I salute these researchers for having conducted their a priori power analysis!

Review Terms

Binomial test

Chi-square

Cochran Q test

Contingency coefficient

Contingency table

Cramer's measure of
   association

Expected frequency

Fisher's Exact Test

Frequencies

Goodness-of-fit test

Homogeneity of
   proportions

Independent-samples chi-square test

Large-sample approximation

McNemar's chi-square test

One-sample chi-square test

Pearson chi-square

Percentages

Proportions

Sign test

Test of independence

Test of normality

Yates' correction for discontinuity

The Best Items in the Companion Website

1. An interactive online quiz (with immediate feedback provided) covering Chapter 17.
2. Ten misconceptions about the content of Chapter 17.
3. One of the best passages from Chapter 17: "Consider the Null Hypothesis before Looking at the *p*-Level."

4. Four interactive online resources.
5. The first of the two jokes, because it deals with one of the statistical tests covered in Chapter 17.

To access the chapter outline, practice tests, weblinks, and flashcards, visit the companion website at http://www.ReadingStats.com.

Review Questions and Answers begin on page 531.

Statistical Tests on Ranks (Nonparametric Tests)

In Chapter 17, we examined a variety of test procedures designed for data that are qualitative, or nominal, in nature. Whether dealing with frequencies, percentages, or proportions, those tests involved response categories devoid of any quantitative meaning. For example, when a chi square test was used in Excerpt 17.10 with data from fifth-graders who threw darts for the first time, neither the grouping variable (defined by the goals the students were told to focus on) nor the response variable (reasons why they missed the bulls-eye) involved categories that had any numerical meaning.

We now turn our attention to a group of test procedures that utilize the simplest kind of quantitative data: ranks. In a sense, we are returning to this topic (rather than starting from scratch), because in Chapter 9, I pointed out how researchers can set up and evaluate null hypotheses concerning Spearman's rho and Kendall's tau. As I hope you recall from Chapter 3, each of these correlational procedures involves an analysis of ranked data.

Within the context of this chapter, we consider five of the many test procedures that have been developed for use with ordinal data: the median test, the Mann–Whitney *U* test, the Kruskal–Wallis one-way analysis of variance of ranks, the Wilcoxon matched-pairs signed-ranks test, and the Friedman two-way analysis of variance of ranks. Most people refer to these statistical tools as **nonparametric test** procedures.[1]

The five tests considered in this chapter are not the only ones that involve ranked data, but they are the ones used most frequently by applied researchers. Because

---

[1] The term *nonparametric* is simply a label for various test procedures that involve ranked data. In contrast, the term **parametric** is used to denote those tests (e.g., *t*, *F*) that are built on a different statistical view of the data—and usually a more stringent set of assumptions regarding the population(s) associated with the study's sample(s).

these five tests are used so often, we examine each one separately in an effort to clarify the research setting for which each test is appropriate, the typical format used to report the test's results, and the proper meaning of a rejected null hypothesis. First, however, we must consider the three ways in which a researcher can obtain the ranked data needed for any of the five tests.

Obtaining Ranked Data

One obvious way for a researcher to obtain ranked data is to ask each research participant to rank a set of objects, statements, ideas, or other things. When this is done, numbers get attached to the things by each person doing the ranking, with the numbers 1, 2, 3, and so on used to indicate an ordering from best to worst, most important to least important, strongest to weakest, and the like. The resulting numbers are **ranks.**[2]

In Excerpt 18.1, we see a case where ranks were used in a research study. In the study associated with this excerpt, the researchers collected survey data from a random sample of 1,076 Finnish adolescents. The researchers' goal was to better understand the lifestyle, values, and behavior of the target group. Several of the survey's questions asked the adolescents to rank a set of items listed by the researchers.

EXCERPT 18.1 • *Obtaining Ordinal Data by Having People Rank a Set of Things*

One question in our survey was "Which qualities/characteristics do you prefer most?" There were 16 characteristics [e.g., having friends, honesty, brand name clothes] that students had to rank in order from 1 (most important) to 16 (least important) that they valued most.

Source: Soininen, M., & Merisuo-Storm, T. (2010). The life style of the youth, their every day life and relationships in Finland. *Procedia Social and Behavioral Sciences, 2*(2), 1665–1669.

A second way for a researcher to obtain ranks is to observe or arrange the study's participants such that each one has an ordered position within the group. For example, we could go to the Boston Marathon, stand near the finish line while holding a list of all contestants' names, and then record each person's standing (first, second, third, or whatever) as he or she completes the race. Or, we might go into a

---

[2]Ranks are often confused with ratings. *Ranks* indicate an ordering of things, with each number generated by having a research participant make a *relative* comparison of the things being ranked. **Ratings,** however, indicate amount, and they are generated by having a research participant make an *independent* evaluation of each thing being rated (perhaps on a 0 to 100 scale).

classroom, ask the students to line up by height, and then request that the students count off beginning at the tall end of the line.[3] In Excerpt 18.2, we see an example of this method for obtaining ranks.

EXCERPT 18.2 • *Obtaining Ordinal Data by Noting the "Order of Finish"*

Three researchers were present in the room. One researcher was blind to the purpose of the experiment and instructed to catch all the birds as quickly as possible using a large, padded net. This researcher caught the birds in all trials. The second researcher took the birds out of the net and returned them to their cages, and the third researcher recorded the identification numbers and order captured. After all birds were captured they were returned to their colony rooms. One week later, this procedure was repeated with the same [birds]. . . . We used Spearman correlations to assess 'repeatability' [because] our dependant measure is a rank order.

Source: Guillette, L. M., Bailey, A. A., Reddon, A. R., Hurd, P. L., & Sturdy, C. B. (2010). A brief report: Capture order is repeatable in chickadees. *International Journal of Comparative Psychology, 23*(2), 216–224.

The third way for a researcher to obtain ranks involves a two-step process. First, each participant is independently measured on some variable of interest with a measuring instrument that yields a score indicative of that person's absolute standing with respect to the numerical continuum associated with the variable. Then, the scores from the group of participants are compared and converted into ranks to indicate each person's relative standing within the group.

In Excerpt 18.3, we see a case in which this two-step process was used. Ranks were used in this study because the researchers wanted to see if the "top" applicants to

EXCERPT 18.3 • *Converting More Refined Measurements into Ranks*

The United Kingdom Clinical Aptitude Test (UK-CAT) was introduced for the purpose of student selection by a consortium of 23 UK University Medical and Dental Schools, including the University of Aberdeen. . . . The applicants [to UoA] were ranked on the basis of UK-CAT score.

Source: Fernando, N., Prescott, G., Cleland, J., Greaves, K., & McKenzie, H. (2009). A comparison of the United Kingdom Clinical Aptitude Test (UK-CAT) with a traditional admission selection process. *Medical Teacher, 31*(11), 1018–1023.

---

[3]Although none of the tests discussed in this chapter could be applied to just the ranks obtained in our running or line-up-by-height examples, two of the tests could be used if we simply classified each subject into one of two or more subgroups (e.g., gender) in addition to noting his or her order on the running speed or height variable.

the University of Aberdeen (UoA) medical school fare well on both the standardized entrance examination—used throughout the United Kingdom—and the University's own applicant evaluation criteria. In the year the study was conducted, there were 1,538 applicants. Surprisingly, only 101 of the 314 applicants admitted to UoA's medical school were among the 318 most-qualified applicants based on the entrance exam.

Reasons for Converting Scores on a Continuous Variable into Ranks

Sometimes, as in the study associated with Excerpt 18.3, raw scores are converted into ranks because the core essence of the research question involves ranks. In most cases, however, that is not the reason why researchers engage in the two-step, data-conversion process whereby scores on a variable of interest are converted into ranks. Because the original scores typically are interval or ratio in nature, whereas the ranks are ordinal, such a conversion might appear to be ill-advised in that it brings about a loss of information. There are, however, three reasons why researchers might consider the benefits associated with the scores-to-ranks conversion to outweigh the loss-of-information liability.

One reason why researchers often change raw scores into ranks is that the test procedures developed for use with ranks involve fewer assumptions than do the test procedures developed for use with interval- or ratio-level data. For example, the assumptions of normality and homogeneity of variance that underlie t- and F-tests do not serve as the basis for some of the tests considered in this chapter. As Excerpts 18.4 and 18.5 make clear, researchers sometimes convert their raw scores into ranks because the original data involved nonnormality or heterogeneity of variance.

EXCERPTS 18.4–18.5 • *Nonnormality and Heterogeneous Variances as Reasons for Converting Scores to Ranks*

Differences in reported time spent online for fertility issues versus cancer issues were tested with nonparametric methods (Wilcoxon signed-rank tests) because of the observed skewness of the distribution of reported times.

Source: Meneses, K., McNees, P., Azuero, A., & Jukkala, A. (2010). Development of the Fertility and Cancer Project: An Internet approach to help young cancer survivors. *Oncology Nursing Forum, 37*(2), 191–197.

[T]he assumption of homogeneity of variance failed for all variables, as the Levene's tests turned out to be significant. Because parametric testing was not justified, nonparametric tests (Kruskal-Wallis) were conducted.

Source: Terband, H., Maassen, B., Guenther, F. H., & Brumberg, J. (2009). Computational neural modeling of speech motor control in childhood apraxia of speech (CAS). *Journal of Speech, Language & Hearing Research, 52*(6), 1595–1609.

A second reason why researchers convert raw scores to ranks is related to the issue of sample size. As you may recall, t- and F-tests tend to be robust to violations of underlying assumptions when the samples being compared are the same size and large. When the ns differ or are small, however, nonnormality or heterogeneity of variance in the population(s) can cause the t- or F-test to function differently than intended. For this reason, some researchers turn to one of the five test procedures discussed in this chapter if they have small samples or if their ns differ. In Excerpt 18.6, we see a case where concerns about sample size prompted the researchers to use nonparametric tests.

EXCERPT 18.6 • *Sample Size as a Reason for Converting Scores to Ranks*

Six elite sprint cross-country skiers from the Austrian national and student national teams (mean age $= 27 \pm 4$ yr, body weight $= 77 \pm 6$ kg, body height $= 181 \pm 8$ cm) volunteered to participate in the study. . . . Owing to the subject size [i.e., sample size], nonparametric statistical techniques were adopted in the present study.

Source: Stoggl, T., Kampel, W., Muller, E., & Lindinger, S. (2010). Double-push skating versus V2 and V1 skating on uphill terrain in cross-country skiing. *Medicine and Science in Sports and Exercise, 42*(1), 187–196.

Regarding sample size, it is legitimate to ask the simple question, "When are samples so small that parametric tests should be avoided even if the ns are equal?" Unfortunately, there is no clear-cut answer to this question because different mathematical statisticians have responded to this query with conflicting responses. According to one point of view, nonparametric tests should be used if the sample size is 6 or less, even if all samples are the same size. A different point of view holds that parametric tests can be used with very small samples as long as the ns do not differ. I mention this controversy simply to alert you to the fact that some researchers use nonparametric tests because they have small sample sizes, even though the ns are equal.

The third reason for converting raw scores to ranks is related to the fact that raw scores sometimes appear to be more precise than they really are. In other words, a study's raw scores may provide only ordinal information about the study's participants even though the scores are connected to a theoretical numerical continuum associated with the dependent variable. In such a case, it is improper to treat the raw scores as if they indicate the absolute distance that separates any two participants that have different scores, when in fact the raw scores only indicate, in a relative sense, which person has more of the measured characteristic than the other.

Consider, for example, the popular technique of having participants respond to a **Likert-type attitude inventory.** With this kind of measuring device, the respondent indicates a level of agreement or disagreement with each of several statements by selecting one of several options that typically include "Strongly Agree" and "Strongly Disagree" on the ends. In scoring a respondent's answer sheet, consecutive integers are typically assigned to the response options (e.g., 1, 2, 3, 4, 5) and then the respondent's total score is obtained by adding together the individual scores earned on each of the inventory's statements. In this fashion, two people in a study might end up with total scores of 32 and 29.

With Likert-type attitude inventories, the total scores derived from the participant responses are probably only ordinal in nature. For one thing, the arbitrary assignment of consecutive integers to the response options does not likely correspond to any participant's view of how the response options relate to another. Moreover, it is probably the case that certain of the inventory's statements are more highly connected than others to one's reason for holding a positive or negative attitude toward the topic being focused on—yet all statements are equal in their impact on a respondent's total score. For these reasons, it is not very plausible to presume that the resulting total scores possess the characteristic of equal intervals that is embodied in interval (and ratio) levels of measurement.

Excerpts 18.7 and 18.8 illustrate how a concern for level of measurement sometimes prompts researchers to use nonparametric tests. The word *ordinal* that we see in these excerpts was used because data in the studies came from Likert-type scales. However, other kinds of data can be ordinal if the measurement scale lacks the quality of equal intervals. (Rulers and thermometers yield numbers on a scale that has equal intervals because a difference of 2 inches or 10° means the same thing anywhere along the scale; in contrast, the numbers associated with most psychological inventories are not characterized by equal intervals.) For example, the research

EXCERPTS 18.7–18.8 • *"Scale" Reasons for Treating Data as Ordinal in Nature*

Nonparametric analyses (Mann–Whitney *U* test and Wilcoxon signed-ranks test) were used for analyses comparing the TBI and control group data [because] the item response format for individual items yields ordinal, not ratio, data.

Source: Douglas, J. M. (2010). Relation of executive functioning to pragmatic outcome following severe traumatic brain injury. *Journal of Speech, Language & Hearing Research, 53*(2), 365–382.

Because the data were ordinal, a Kruskal-Wallis test was done to assess for any differences in ease of observation among the 3 types of dressings.

Source: McIe, S., Petitte, T., Pride, L., Leeper, D., & Ostrow, C. L. (2009). Transparent film dressing vs. pressure dressing after percutaneous transluminal coronary angiography. *American Journal of Critical Care, 18*(1), 14–20.

participants in the study associated with Excerpt 18.7 were classified into nine disability levels with these labels: none, mild, partial, moderate, moderately severe, severe, extremely severe, vegetative state, and extreme vegetative state. These categories had increasing levels of disability, but they were not characterized by the notion of *equal intervals*.

Now that we have considered how and why a researcher might end up with ranked data, let's take a look at each of the five test procedures that deserve the label *popular nonparametric test*. As noted earlier, these test procedures are the median test, the Mann–Whitney U test, the Kruskal–Wallis one-way ANOVA, the Wilcoxon matched-pairs signed-ranks test, and the Friedman two-way ANOVA. In looking at each of these test procedures, I want to focus our attention on the nature of the research setting for which the test is appropriate, the way in which the ranked data are used, the typical format for reporting results, and the meaning of a rejected null hypothesis.

The Median Test

The **median test** is designed for use when a researcher wishes to compare two or more independent samples. If two such groups are compared, the median test is a nonparametric analog to the independent-samples t-test. With three or more groups, it is the nonparametric analog to a one-way ANOVA.

A researcher might select the median test in order to contrast two groups defined by a dichotomous characteristic (e.g., male versus female, experimental versus control) on a dependent variable of interest (e.g., throwing ability, level of conformity, or anything else the researcher wishes to measure). Or, the median test might be selected if the researcher wishes to compare three or more groups (that differ in some qualitative fashion) on a measured dependent variable. An example of this latter situation might involve comparing football players, basketball players, and baseball players in terms of their endurance while riding a stationary bicycle.

The null hypothesis of the two-group version of the median test can be stated as H_0: $Mdn_1 = Mdn_2$, where the abbreviation Mdn stands for the median in the population and the numerical subscripts serve to identify the first and second populations. If three or more groups are compared using the median test, the null hypothesis takes the same form except that there would be additional Mdns involved in H_0. The alternative hypothesis says that the two Mdns differ (if just two groups are being compared) or that at least two of the Mdns differ (in the situation in which three or more groups are being contrasted).

To conduct a median test, the researcher follows a simple three-step procedure. First, the comparison groups are temporarily combined and a single median is determined for the entire set of scores. (This step necessitates that ranks be assigned either to all participants or at least to those who are positioned near the middle of the pack.) In the second step, the comparison groups are reconstituted so that a contingency table can be set up to indicate how many people in each comparison group lie above and below the grand median identified in the first step. This contingency table has as many columns as there are comparison groups, but it always has two rows (one labeled *above the median,* the other labeled *below the median*). Finally, an independent-samples chi-square test is applied to the data in the contingency table to see if the samples differ (in the proportion of cases falling above the combined median) by more than what would be expected by chance alone, presuming that H_0 is true.

In Excerpt 18.9, we see a case where a median test was used in a study of 604 men who were concerned with hair loss. Of the full group, 321 of the research participants had consulted a doctor about their hair loss problem; the other 283 men had not yet sought out medical help but were planning to do so. These two groups were compared in terms of how many self-treatments—such as vitamins, special shampoo, and over-the-counter medications—had been tried. The symbol χ^2 is in the excerpt because the median test involved analyzing the study's data via a 2×2 contingency table.

EXCERPT 18.9 • *The Median Test Used to Compare Two Groups*

Regarding self-treatments prior to medical consultation, respondents were provided a list of 27 possible treatments, including specific prescription and nonprescription medicines, alternative treatments (e.g., vitamins, shampoos, etc.), and at-home devices (e.g., laser comb, wig, etc.). . . . Men reported a median of two to three such treatments, with more prior treatments reported by men who had not yet consulted a doctor but were likely to so in the near future (χ^2 median test $= 26.30, p < 0.001$).

Source: Cash, T. F. (2009). Attitudes, behaviors, and expectations of men seeking medical treatment for male pattern hair loss: Results of a multinational survey. *Current Medical Research and Opinion, 25*(7), 1811–1820.

In the study associated with Excerpt 18.9, perhaps the researchers set up the contingency table so that the columns corresponded with the groups (seen a doctor: yes or no) and the rows corresponded to being above or below the grand median, with each of the 604 men positioned in one of the four cells. If the two frequencies in each column had been about the same, the null hypothesis would have been retained. However, the actual frequencies in the contingency table produced a statistically significant value for chi square, with a greater-than-chance number of people from the yet-to-see-a-doctor group positioned above the grand median (and a greater-than-chance number of people from the other group below that median).

Excerpt 18.9 is instructive because the fourth word in the second sentence is *median.* Note that this is singular, not plural. Many people mistakenly think that a median test involves a statistical comparison of two sample medians to see if they are far enough apart from each other to permit a rejection of H_0. However, there is only one sample median involved in a median test (the grand median based on the data from all groups), and the statistical question being asked is whether the comparison groups differ significantly in terms of the percentage of each group that lies above this single median. Given any set of scores, it would be possible to change a few scores and thereby change the group medians (making them closer together or further apart) *without* changing the median test's calculated value or *p*. To me, this constitutes proof that the median test is *not* focusing on the individual medians of the two samples.

As mentioned earlier, the median test can compare two groups or more than two groups. In Excerpt 18.10, we see a case where the median test was used to compare three groups. These three groups were patients with leg ulcers who received different treatments to help them heal. The median test did not produce a statistically significant result when the groups were compared in terms of the dependent variable, healing time.

EXCERPT 18.10 • *The Median Test Used to Compare Three Groups*

An open, randomized, prospective, single-center study was performed in order to determine the healing rates of VLU [venous leg ulcers] when treated with different compression systems and different sub-bandage pressure values. . . . To compare the median healing times from the three groups, a median test was performed. . . . Median healing time in group A was 12 weeks (range, 5–24 weeks), 11 weeks (range, 3–25 weeks) in group B, and 14 weeks (range, 5–24 weeks) in group C (median test: $P > .05$).

Source: Milic, D. J., Zivic, S. S., Bogdanovic, D. C., Jovanovic, M. M., Jankovic, R. J., Milosevic, Z. D., et al. (2010). The influence of different sub-bandage pressure values on venous leg ulcers healing when treated with compression therapy. *Journal of Vascular Surgery, 51*(3), 655–661.

There is one final point to be made about the median test. You may see or hear the terms *Mood's median test* and *Levene's median test*. The first of these is the test we have been considering, and the name of its inventor is sometimes used when referring to this statistical procedure. Levene's median test is altogether different; it is a parametric test used to compare samples in terms of their variances.

The Mann–Whitney U Test

The **Mann–Whitney *U* test**[4] is like the two-sample version of the median test in that both tests allow a researcher to compare two independent samples. Although these two procedures are similar in that they are both considered to be nonpara-

---

[4]This test is also referred to as the *Wilcoxon test,* the *Wilcoxon rank-sum test*, and the *Wilcoxon–Mann–Whitney test*.

metric tests, the Mann–Whitney U test is the more powerful of the two. In other words, if the two comparison groups truly do differ from each other, the Mann–Whitney U test (as compared to the median test) is less likely to produce a Type II error. This superiority of the Mann–Whitney test comes about because it utilizes more information from the data than does the median test.

When using the Mann–Whitney U test, the researcher examines the scores of the research participants on the variable of interest. Initially, the two comparison groups are lumped together. This is done so that each person can be ranked to reflect his or her standing within the combined group. After the ranks have been assigned, the researcher reconstitutes the two comparison groups. The previously assigned ranks are then examined to see if the two groups are significantly different.

If the two samples being compared come from identical populations, then the **sum of ranks** in one group ought to be approximately equal to the sum of ranks in the other group. For example, if there were four people in each sample and if H_0 were true, we would not be surprised if the ranks in one group were 2, 4, 5, and 8 whereas the ranks in the other group were 1, 3, 6, and 7. Here, the sum of the ranks are 19 and 17, respectively. It *would* be surprising, however, to find (again assuming that H_0 is true) that the sum of the ranks are 10 and 26. Such an extreme outcome would occur if the ranks of 1, 2, 3, and 4 were located in one of the samples whereas the ranks of 5, 6, 7, and 8 were located in the other sample.

To perform a Mann–Whitney U test, the researcher computes a sum-of-ranks value for each sample and then inserts these two numerical values into a formula. It is not important for you to know what that formula looks like, but it *is* essential that you understand the simple logic of what is going on. The formula used to analyze the data produces a calculated value called U. Based on the value of U, the researcher (or a computer) can then derive a p-value that indicates how likely it is, under H_0, to have two samples that differ as much or more than do the ones actually used in the study. Small values of p, of course, are interpreted to mean that H_0 is unlikely to be true.

In Excerpt 18.11, we see a case in which the Mann–Whitney U test was used to compare the amount of alcohol consumed by two groups of college students.

EXCERPT 18.11 • *The Mann-Whitney U Test*

A Mann-Whitney U test was used to compare the difference between the mean alcohol consumption scores at the 2 institutions. The Mann-Whitney U was used as a substitute for a Student t test because the assumption of the normality of the distribution alcohol consumption scores is questionable. . . . The mean alcohol consumption for the students at the religious institution was 11.9 ($SD = 27.6$) drinks in the 30 days prior to the survey, which was significantly lower than the 26.9 ($SD = 53.1$) drinks per 30 days for students attending the secular university ($U = -7.55, p < .05$).

Source: Wells, G. M. (2010). The effect of religiosity and campus alcohol culture on collegiate alcohol consumption. *Journal of American College Health, 58*(4), 295–304.

Although the means of the two samples appear in this excerpt, the test's calculated value, U, was based on the sum of ranks.

Although it is quite easy for a researcher to obtain a calculated value for U from the sample data and to compare that data-based number against a tabled critical value, the task of interpreting a statistically significant result is a bit more difficult, for two reasons. First, the null hypothesis being tested deals not with the ranks used to compute the calculated value but rather with the continuous variable that lies behind or beneath the ranks. For example, if we used a Mann–Whitney U test to compare a sample of men against a sample of women with respect to their order of finish after running a 10-kilometer race, the data collected might very well simply be ranks, with each person's rank indicating his or her place (among all contestants) on crossing the finish line. The null hypothesis, however, would deal with the continuous variable that lies beneath the ranks, which in our hypothetical study is running speed.

The second reason why statistically significant results from Mann–Whitney U tests are difficult to interpret is related to the fact that the null hypothesis says that the two populations have identical distributions. Consequently, rejection of H_0 could come about because the populations differ in terms of their central tendencies, their variabilities, or their distributional shapes. In practice, however, the Mann–Whitney test is far more sensitive to differences in central tendency, so a statistically significant result is almost certain to mean that the populations have different average scores. But even here, an element of ambiguity remains because the Mann–Whitney U test could cause H_0 to be rejected because the two populations differ in terms of their means, or in terms of their medians, or in terms of their modes.

In the situation where the two populations have identical shapes and variances, the Mann–Whitney U test focuses on means, and thus H_0: $\mu_1 = \mu_2$. However, applied researchers rarely know anything about the populations involved in their studies. Therefore, most researchers who find that their Mann–Whitney U test yields a statistically significant result legitimately can conclude only that the two populations probably differ with respect to their averages. Another way of drawing a proper conclusion from a Mann–Whitney U test that causes H_0 to be rejected is to say that the scores in one of the populations tend to be larger than scores in the other population. This statement could only be made in a tentative fashion, however, because the statistically significant finding might well represent nothing more than a Type I error.

With most of the tests we have considered so far (such as t-tests, F-tests, and chi-square tests), large calculated values cause the p-level to be small whereas small calculated values cause p to be large. With the Mann–Whitney U test, however, there is a direct relationship between p and U. With this nonparametric test, it is small rather than large values of U that make the sample data improbable when compared to what we would expect if the null hypothesis were true.

The Kruskal–Wallis H Test

In those situations in which a researcher wishes to use a nonparametric statistical test to compare two independent samples, the Mann–Whitney U test is typically used to analyze the data. When researchers wish to compare three or more such groups, they more often than not utilize the Kruskal–Wallis H test. Hence, the Kruskal–Wallis procedure can be thought of as an extension of the Mann–Whitney procedure in the same way that a one-way ANOVA is typically considered to be an extension of an independent-samples' t-test.[5]

The fact that the Kruskal–Wallis test is like a one-way ANOVA shows through when one considers the mathematical derivation of the formula for computing the test's calculated value. On a far simpler level, the similarity between these two test procedures shows through when we consider their names. The parametric test we considered in Chapter 11 is called a one-way ANOVA whereas the nonparametric analog to which we now turn our attention is called the **Kruskal–Wallis one-way ANOVA of ranks.**

The Kruskal–Wallis test works very much as the Mann–Whitney test does. First, the researcher temporarily combines the comparison groups into a single group. Next, the people in this one group are ranked on the basis of their performance on the dependent variable. Then, the single group is subdivided so as to reestablish the original comparison groups. Finally, each group's sum of ranks is entered into a formula that yields the calculated value. This calculated value, in the Kruskal–Wallis test, is labeled H. When the data-based H beats the critical value or when the p-value associated with H turns out to be smaller than the level of significance, the null hypothesis is rejected.

In Excerpts 18.12 and 18.13, we see two examples of the Kruskal–Wallis test being used in applied studies. In the first of these excerpts, three groups of parents

EXCERPTS 18.12–18.13 • *The Kruskal–Wallis H Test*

Quantitative analyses substantiated racial/ethnic differences; black parents placed significantly higher demands on children for the amounts ($H = 5.89, 2\,df, P < .05$; Kruskal–Wallis) and types ($H = 8.39, 2\,df, P < .01$; Kruskal–Wallis) of food eaten compared to parents of other races/ethnicities.

Source: Ventura, A. K., Gromis, J. C., & Lohse, B. (2010). Feeding practices and styles used by a diverse sample of low-income parents of preschool-age children. *Journal of Nutrition Education and Behavior, 42*(4), 242–249.

(continued)

---

[5]When just two groups are compared, the ANOVA F-test and the independent-samples t-test yield identical results. In a similar fashion, the Kruskal–Wallis and Mann–Whitney tests are mathematically equivalent when used to compare two groups.

EXCERPTS 18.12–18.13 • (*continued*)

A comparison of these [pre–post] difference scores between groups with a Kruskal–Wallis one-way analysis of variance by ranks revealed no significant differences, $H(3) = 0.89, p > .05$.

Source: Marshall, P., Cheng, P. C.-H., & Luckin, R. (2010). Tangibles in the balance: A discovery learning task with physical or graphical material. *Proceedings of the Fourth International Conference on Tangible, Embedded, and Embodied Interaction*, 153–160.

were compared separately on two different dependent variables. We know that there were three groups because each *H* test had 2 *df*. In Excerpt 18.13, we can figure out that there were four comparison groups by paying attention to the *df* number that appears next to letter *H*. Notice that the full name of this test procedure is used in Excerpt 18.13, whereas the shortened version of this name is used in Excerpt 18.12.

The Kruskal–Wallis *H* test and the Mann–Whitney *U* test are similar not only in how the subjects are ranked and in how the groups' sum-of-ranks values are used to obtain the test's calculated value, but also in the null hypothesis being tested and what it means when H_0 is rejected. Technically speaking, the null hypothesis of the Kruskal–Wallis *H* test is that the populations associated with the study's comparison groups are identical with respect to the distributions on the continuous variable that lies beneath the ranks used within the data analysis. Accordingly, a rejection of H_0 could come about because the population distributions are not the same in central tendency, in variability, or in shape. In practice, however, the Kruskal–Wallis test focuses primarily on central tendency.

Although the Mann–Whitney and Kruskal–Wallis tests are similar in many respects, they differ in the nature of the decision rule used to decide whether H_0 should be rejected. With the Mann–Whitney test, H_0 is rejected if the data-based *U* turns out to be smaller than the critical value. In contrast, the Kruskal–Wallis H_0 is rejected when the researcher's calculated *H* is larger than the critical value. In Excerpt 18.12, note that the larger of the two calculated values of *H* is paired with the smaller of the two *p*-values.

Whenever the Kruskal–Wallis *H* test leads to a rejection of H_0, there remains uncertainty as to which specific populations are likely to differ from one another. In other words, the Kruskal–Wallis procedure functions very much as an omnibus test. Consequently, when such a test leads to a rejection of H_0, the researcher usually turns to a post hoc analysis so as to derive more specific conclusions from the data. Within such post hoc investigations, comparison groups are typically compared in a pairwise fashion.

The post hoc procedure used most frequently following a statistically significant *H* test is the Mann–Whitney *U* test. Excerpt 18.14 illustrates the use of the *U*

EXCERPT 18.14 • *Use of the Mann–Whitney U Test within a Post Hoc Investigation*

A Kruskal–Wallis test revealed a significant difference between the [four] student groups. . . . Mann–Whitney post hoc analyses revealed, as expected, that physics students reported significantly less familiarity with self-harm behaviour, than either medical ($U = 177.000, p < 0.001$), clinical psychology ($U = 94.500, p < 0.001$) or nursing students ($U = 105.000, p < 0.001$). However, both nursing and clinical psychology students reported significantly more familiarity than medical students did ($U = 324.500, p = 0.001$ and $U = 310.000, p = 0.004$, respectively).

Source: Law, G. U., Rostill-Brookes, H., & Goodman, D. (2009). Public stigma in health and non-healthcare students: Attributions, emotions and willingness to help with adolescent self-harm. *International Journal of Nursing Studies, 46*(1), 108–119.

test in post hoc investigation following rejection of the Kruskal–Wallis null hypothesis. When used in this capacity, most researchers use the Bonferroni procedure for adjusting the level of significance of each post hoc comparison.

The Wilcoxon Matched-Pairs Signed-Ranks Test

Researchers frequently wish to compare two related samples of data generated by measuring the same people twice (e.g., in a pre/post sense) or by measuring two groups of matched individuals just once. If the data are interval or ratio in nature and if the relevant underlying assumptions are met, the researcher will probably utilize a correlated *t*-test to compare the two samples. On occasion, however, that kind of parametric test cannot be used because the data are ordinal or because the *t*-test assumptions are untenable (or considered by the researcher to be a nuisance). In such situations, the two related samples are likely to be compared using the **Wilcoxon matched-pairs signed-ranks test.**

In conducting the **Wilcoxon test,** the researcher (or a computer) must do five things. First, each pair of scores is examined so as to obtain a *change* score (for the case where a single group of people has been measured twice) or a *difference* score (for the case where the members of two matched samples have been measured just once). These scores are then ranked, either from high to low or from low to high. The third step involves attaching a $+$ or a $-$ sign to each rank. (In the one-group-measured-twice situation, these signs indicate whether a person's second score turned out to be higher or lower than the first score. In the two-samples-measured-once situation, these signs indicate whether the people in one group earned higher or lower scores than their counterparts in the other group.) In the fourth step, the researcher simply looks to see which sign appears less frequently and then adds up the ranks that have that sign. Finally, the researcher labels the sum of the ranks that have the least frequent sign as T, considers T to be the calculated value, and compares T against a tabled critical value.

With computers readily available to do the computations, the researcher has a much easier task when conducting a Wilcoxon test. The raw data are simply entered into the computer and then, in a flash, the calculated value appears on the computer screen. The way many statistics programs are set up, this calculated value for the Wilcoxon test is a z-score rather than a numerical value for T.

In Excerpt 18.15, we see a case in which the Wilcoxon matched-pairs signed-ranks test was used in a study dealing with middle school teachers. The researchers wanted to know if teachers' perception of digital mini-games (designed to enhance academic learning) would change if the teachers used and then discussed such games. This excerpt is instructive because it contains two sum-of-ranks values, one for those teachers who improved their scores and the other for the teachers whose scores decreased. As shown in the excerpt, the sum-of-ranks value for the smaller T-value became the calculated value.

EXCERPT 18.15 • *Use of the Wilcoxon Matched-Pairs Signed-Ranks Test*

The Wilcoxon Signed Ranks Test for matched pairs was used to assess whether $N = 18$ preservice teachers perceptions' were modified by participation in a series of digital mini-games based on the ranked order magnitude of the change between their before and after responses. The results revealed a statistically significant difference in perception after the preservice teachers played the games ($T = 5.5, p < .01$ [two tailed]). The sum of the ranked increases totaled 72.50, and the sum of the ranked decreases totaled 5.50. Because higher scores indicated more positive perceptions, the results revealed that participation in the digital mini-games modified preservice teachers' views by positively improving their perceptions regarding their efficacy.

Source: Ray, B., & Coulter, G. A. (2010). Perceptions of the value of digital mini-games: Implications for middle school classrooms. *Journal of Digital Learning in Teacher Education, 26*(3), 92–100.

When the Wilcoxon test leads to a numerical value for T, the researcher's conclusion either to reject or to retain H_0 is based on a decision-rule like that used within the Mann–Whitney U test. Simply stated, that decision-rule gives the researcher permission to reject H_0 when the data-based value of T is equal to or smaller than the tabled critical value (because a direct relationship exists between T and p). However, if the Wilcoxon test's calculated value is z, the decision-rule is just the opposite. Here, it is large values of z that permit the null hypothesis to be rejected.

Although it is easy to conduct a Wilcoxon test, the task of interpreting the final result is more challenging. The null hypothesis says that the populations associated with the two sets of sample data are each symmetrical around the same common point. This translates into a statement that the population of change (or difference) scores is symmetrical around a median value of zero. Interpreting the outcome of a Wilcoxon matched-pairs signed-ranks test is problematic, because the null hypothesis could be false because the population of change/difference scores is not symmetric, because the population median is not equal to zero, or because the population is not symmetrical around a median other than zero. Accordingly, if the Wilcoxon test leads to a statistically significant finding, neither you nor the researcher will know the precise reason why H_0 has been rejected.

There are two different ways to clarify the situation when one wants to interpret a significant finding from the Wilcoxon test. First, such a test can be interpreted to mean that the two populations, each associated with one of the samples of data used to compute the difference/change scores, are probably not identical to each other. That kind of interpretation is not too satisfying, because the two populations could differ in any number of ways. The second interpretation one can draw if the Wilcoxon test produces a small p-value is that the two populations probably have different medians. (This is synonymous to saying that the population of difference/change scores is probably not equal to zero.) This interpretation is legitimate, however, only in the situation where it is plausible to assume that both populations have the same shape.

Friedman's Two-Way Analysis of Variance of Ranks

The Friedman test is like the Wilcoxon test in that both procedures were developed for use with related samples. The primary difference between the Wilcoxon and Friedman tests is that the former test can accommodate just two related samples whereas the Friedman test can be used with two or more such samples. Because of this, **Friedman's two-way analysis of variance of ranks** can be thought of as the nonparametric equivalent of the one-factor repeated-measures ANOVA that we considered in Chapter 14.[6]

To illustrate the kind of situation to which the Friedman test could be applied, suppose you and several other individuals are asked to independently evaluate the quality of the five movies that previously have won the Best Picture award from the Academy of Motion Pictures. I might ask you and the other people in this study to rank the five movies on the basis of whatever criteria you typically use when evaluating movie quality. Or, I might ask you to rate each of the movies (possibly on a 0 to 100 scale), thus providing me with data that I could convert into ranks. One way or the other, I could end up with a set of five ranks from each person indicating his or her opinion of the five movies.

If the five movies being evaluated are equally good, we would expect the movies to be about the same in terms of the sum of the ranks assigned to them. In

---

[6]Although the Friedman and Wilcoxon tests are similar in that they both were designed for use with correlated samples of data, the Friedman test actually is an extension of the sign test.

other words, movie A ought to receive some high ranks, some medium ranks, and some low ranks if it is truly no better or worse than movies B, C, D, and E. That would also be the case for each of the other four movies. The Friedman test treats the data in just this manner, because the main ingredient is the sum of ranks assigned to each movie.

Once the sum of ranks are computed for the various things being compared, they are inserted into a formula that yields the test's calculated value. I do not discuss here the details of that formula, or even present it. Instead, I want to focus on three aspects of what pops out of that formula. First, the calculated value is typically symbolized as χ_r^2 (or sometimes simply as χ^2). Second, large values of χ_r^2 suggest that H_0 is not true. Third, the value of χ_r^2 is referred to a null distribution of such values so as to determine the data-based p-value and decide whether the null hypothesis should be rejected.

Excerpt 18.16 illustrates the use of the Friedman test. In the study associated with this excerpt, the researchers wanted to see if women seemed more attractive to men when the women were wearing an outfit that made them, the women, feel attractive. The 49 men who did the evaluation saw three pictures of each of 25 women dressed in outfits she had selected that made her feel attractive, unattractive, or just *comfortable*. Each man ranked the three pictures of each woman, and then his ranks given to all 25 photographs where the women felt attractive were summed. Likewise, each man's sums of ranks for the unattractive and comfortable pictures were computed. These three sum-of-rank scores from each man were then ranked 1, 2, and 3 to indicate preferences for the three photographic conditions of the study: attractive, unattractive, and comfortable. These three simple ranks for the 49 men were analyzed via Friedman's test, with the result being a statistically significant result. The interesting twists to this study are twofold: (1) the photographs showed only the women's faces and not the outfits they were wearing, and (2) an effort was made to have each woman's facial expression be the same in each of her three pictures.

EXCERPT 18.16 • *Friedman's Two-Way Analysis of Variance of Ranks*

The faces of 25 females volunteers between the ages of 22 and 28 . . . were photographed [while] the women were wearing clothes in which they felt (1) attractive, (2) unattractive and (3) comfortable. . . . The women were asked to have a neutral facial expression and always look in the same direction when the photographs were taken to avoid effects caused by differences in smiles or eye contact. . . . The men were asked to rank the three [face-only] images of each woman in order of attractiveness. . . . The face considered to be most attractive by men was the one in which the women were dressed in clothes that made them feel attractive (Friedman two-way ANOVA: $\chi^2 = 57.8, df = 2, p < 0.001$).

Source: Lõhmus, L., Sundström, L. F., and Björklund, M. (2009). Dress for success: Human facial expressions are important signals of emotion. *Annales of Zoologici Fennici, 46*(1), 75–80.

If the Friedman test leads to a rejection of the null hypothesis when three or more things (such as movies in our earlier hypothetical example) are compared, you are likely to see a post hoc follow-up test utilized to compare the things that have been ranked. Although many test procedures can be used within such a post hoc investigation, you will likely see the Wilcoxon matched-pairs signed-ranks test employed to make all possible pairwise comparisons. In using the Wilcoxon test in this fashion, the researcher should use the Bonferroni adjustment procedure to protect against an inflated Type I error rate.

Large-Sample Versions of the Tests on Ranks

Near the end of Chapter 17, I pointed out how researchers sometimes conduct a z-test when dealing with frequencies, percentages, or proportions. Whenever this occurs, researchers put their data into a special formula that yields a calculated value called z, and then the data-based p-value is determined by referring the calculated value to the normal distribution. Any z-test, therefore, can be conceptualized as a *normal curve test*.

In certain situations, the z-test represents nothing more than an option available to the researcher, with the other option(s) being mathematically equivalent to the z-test. In other situations, however, the z-test represents a **large-sample approximation** to some other test. In Chapter 17, I pointed out how the sign, binomial, and McNemar procedures can be performed using a z-test if the sample sizes are large enough. The formula used to produce the z calculated value in these large-sample approximations varies across these test procedures, but that issue is of little concern to consumers of the research literature.

Inasmuch as tests on nominal data can be conducted using z-tests when the sample(s) are large, it should not be surprising that large-sample approximations exist for several of the test procedures considered in this chapter. To be more specific, you are likely to encounter studies in which the calculated value produced by the Mann–Whitney U test is not U, studies in which the calculated value produced by the Kruskal–Wallis one-way analysis of variance of ranks is not H, and studies in which the calculated value produced by the Wilcoxon matched-pairs signed-ranks test is not T. Excerpts 18.17 through 18.19 illustrate such cases.

EXCERPTS 18.17–18.19 • *Large-Sample Versions of the Mann–Whitney, Kruskal–Wallis, and Wilcoxon Tests*

In this sample, the mean number of indoor tanning sessions during the past year was 12 (SD = 21), and women reported more sessions during the past year relative to men ($Ms = 16$ vs 5, respectively; Mann–Whitney U test; $z = -7.15, p < .001$).

Source: Mosher, C. E., & Danoff-Burg, S. (2010). Indoor tanning, mental health, and substance use among college students: The significance of gender. *Journal of Health Psychology, 15*(6), 819–827.

(continued)

EXCERPTS 18.17–18.19 • (*continued*)

There were significant differences between species (Kruskal–Wallis test: $\chi^2 = 17.52$; $df = 3; P = 0.001$).

Source: Vlamings, P. H. J. M., Hare, B. & Call, J. (2010). Reaching around barriers: The performance of the great apes and 3–5-year-old children. *Animal Cognition, 13*(2), 273–285.

The mean pleasantness score was larger for the blue lighting (mean rating 3.5 ± 0.94, where 5 was the most unpleasant and 1 is the most pleasant) than for the red lighting (mean rating 2.1 ± 0.85). The Wilcoxon test showed that this difference between the two colours is highly statistically significant ($z = -2.72, p < 0.007$).

Source: Laufer, L., Láng, E., Izsó, L., & Németh, E. (2009). Psychophysiological effects of coloured lighting on older adults. *Lighting Research & Technology, 41*(4), 371–378.

In Excerpts 18.17 and 18.19, we see that the calculated value in the large-sample versions of the Mann–Whitney and Wilcoxon tests is a z-value. In contrast, the calculated value for the large-sample version of the Kruskal–Wallis test is a chi-square value. These excerpts thus illustrate nicely the fact that many of the so-called large-sample versions of nonparametric tests yield a p-value that is based on the normal distribution. Certain of these tests, however, are connected to the chi-square distribution.

The Friedman test procedure—like the Mann–Whitney, Kruskal–Wallis, and Wilcoxon procedures—can be conducted using a large sample approximation. Most researchers do this by comparing their calculated value for χ^2_r against a chi-square distribution in order to obtain a p-value. If you look again at Excerpt 18.16, you see a case in which the Friedman test was conducted in this fashion.

It should be noted that the median test is inherently a large-sample test to begin with. That is the case because this test requires that the data be cast into a 2×2 contingency table from which a chi-square calculated value is then derived. Because this chi-square test requires sufficiently large expected cell frequencies, the only option to the regular, large-sample median test is Fisher's Exact Test. Fisher's test, used within this context, could be construed as the small-sample version of the median test.

Before concluding this discussion of the large-sample versions of the tests considered in this chapter, it seems appropriate to ask the simple question, "How large must the sample(s) be in order for these tests to function as well as their more exact, small-sample counterparts?" The answer to this question varies depending on the test being considered. The Mann–Whitney z-test, for example, works well if both ns are larger than 10 (or if one of the ns is larger than 20), whereas the Wilcoxon z-test performs adequately when its n is greater than 25. The

Kruskal–Wallis chi-square test works well when there are more than three comparison groups or when the *n*s are greater than 5. The Friedman chi-square test functions nicely when there are more than four things being ranked or more than 10 research participants doing the ranking.

Although not used very often, other large-sample procedures have been devised for use with the Mann–Whitney, Kruskal–Wallis, Wilcoxon, and Friedman tests. Some involve using the ranked data within complex formulas. Others involve using the ranked data within *t*- or *F*-tests. Still others involve the analysis of the study's data through two different formulas, the computation of an average calculated value, and then reference to a specially formed critical value. Although not now widely used, some of these alternative procedures may gain popularity among applied researchers in the coming years.

Ties

Whenever researchers rank a set of scores, they may encounter the case of **tied observations.** For example, there are two sets of ties in this hypothetical set of 10 scores: 8, 0, 4, 3, 5, 4, 7, 1, 4, 5. Or, ties can occur when the original data take the form of ranks. Examples here would include the tenth and eleventh runners in a race crossing the finish line simultaneously, or a judge in a taste test indicating that two of several wines equally deserve the blue ribbon.

With the median test, tied scores do not create a problem. If the tied observations occur within the top half or the bottom half of the pooled group of scores, the ties can be disregarded because all of the scores are easily classified as being above or below the grand median. If the scores in the middle of the pooled data set are tied, the *above* and *below* categories can be defined by a numerical value that lies adjacent to the tied scores. For example, if the 10 scores in the preceding paragraph had come from two groups being compared using a median test, high scores could be defined as anything above 4, whereas low scores could be defined as less than or equal to 4. (Another way of handling ties at the grand median is simply to drop those scores from the analysis.)

If tied observations occur when the Mann–Whitney, Kruskal–Wallis, Wilcoxon, or Friedman tests are being used, researchers typically do one of three things. First, they can apply mean ranks to the tied scores. (The procedure for computing mean ranks was described in Chapter 3 in the section dealing with Kendall's tau.) Second, they can drop the tied observations from the data set and subject the remaining, untied scores to the statistical test. Third, they can use a special version of the test procedure developed to handle tied observations.

In Excerpts 18.20 and 18.21, we see two cases in which the third of these three options was selected. In both of these cases, the phrase *corrected for ties* is an unambiguous signal that the tied scores were left in the data set and that a special formula was used to compute the calculated value.

EXCERPTS 18.20–18.21 • *Using Special Formulas to Accommodate Tied Observations in the Data*

Mann–Whitney U test was used to compare means between groups, and the p-values are the exact two-tailed significance values corrected for ties.

Source: Frich, P. S., Kvestad, C. A., & Angelsen, A. (2009). Outcome and quality of life in patients operated on with radical cystectomy and three different urinary diversion techniques. *Scandinavian Journal of Urology & Nephrology, 43*(1), 37–41.

When the presence or absence of gonads for all collected anemones is analyzed using the Kruskal–Wallis test there is no significant difference between seasons ($P = 0.42$ when corrected for ties).

Source: Lombardi, M. R., & Lesser, M. P. (2010). The annual gametogenic cycle of the sea anemone *Metridium senile* from the Gulf of Maine. *Journal of Experimental Marine Biology and Ecology, 390*(1), 58–64.

Ties can also occur within the Friedman test. This could happen, for example, if a judge were to report that two of the things being judged were equally good. Such tied observations are not discarded from the data set, because that would necessitate tossing out all the data provided by that particular judge. Instead, the technique of assigning average ranks is used, with the regular formula then employed to obtain the calculated value for the Friedman test.

A Few Final Comments

As we approach the end of this chapter, five final points must be made. These points constitute my typical end-of-chapter warnings to those who come into contact with technical research reports. By remaining sensitive to these cautions, you will be more judicious in your review of research conclusions that are based on nonparametric statistical tests.

My first warning concerns the quality of the research question(s) associated with the study you find yourself examining. If the study focuses on a trivial topic, no statistical procedure has the ability to "turn a sow's ear into a silk purse." This is as true of nonparametric procedures as it is of the parametric techniques discussed earlier in the book. Accordingly, I once again urge you to refrain from using data-based p-levels as the criterion for assessing the worth of empirical investigations.

My second warning concerns the important assumptions of random samples and independence of observations. Each of the nonparametric tests considered in this chapter involves a null hypothesis concerned with one or more populations. The null hypothesis is evaluated with data that come from one or more samples that are assumed to be representative of the population(s). Thus, the notion of randomness is just as essential to any nonparametric test as it is to any parametric procedure. Moreover, nonparametric tests, like their parametric counterparts, are based on an assumption of **independence,** simply meaning that the data provided by any individual are not influenced by what happens to any other person in the study.[7]

My third warning concerns the term **distribution-free,** a label that is sometimes used instead of the term *nonparametric.* As a consequence of these terms being used as if they were synonyms, many applied researchers are under the impression that nonparametric tests work equally well no matter what the shape of the population distribution(s). This is not true. As indicated earlier in the chapter, the proper meaning of a rejected null hypothesis is frequently influenced by what is known about the distributional shape of the populations.

My fourth warning is really a reiteration of an important point made earlier in this book regarding **overlapping distributions.** If two groups of scores are compared and found to differ significantly from each other (even at impressive *p*-levels), it is exceedingly likely that the highest scores in the low group are higher than the lowest scores in the high group. When this is the case, as it almost always is, a researcher should not claim—or even suggest—that each of the individuals in the high group had a higher score than any of the individuals in the other group. What legitimately *can* be said is that people in the one group, on the average, did better. Those three little words *on the average* are essential to keep in mind when reading research reports.

To see clearly what I mean about *overlapping distributions,* consider Excerpt 18.22. In the study associated with this excerpt, the researchers compared two groups of individuals with acute cardiovascular symptoms. One group had a prior history of cerebrovascular disease; the other group did not. In the full research report, the

EXCERPT 18.22 • *Overlapping Distributions*

We stratified the study patients according to a history of CVD [cerebrovascular disease] and compared their baseline demographic characteristics, treatments, and outcomes. Continuous data are reported as the median and interquartile range, and categorical data are reported as percentages. The Mann-Whitney *U* test was used for comparison of continuous variables. . . . Patients with CVD were older [median = 75, IQR = 67 − 81] than their counterparts without CVD [median = 67, IQR = 57 − 76].

Source: Lee, T. C., Goodman, S. G., Yan, R. T., Grondin, F. R., Welsh, R. C., Rose, B., et al. (2010). Disparities in management patterns and outcomes of patients with non–ST-elevation acute coronary syndrome with and without a history of cerebrovascular disease. *American Journal of Cardiology, 105*(8), 1083–1089.

[7]With the median, Mann–Whitney, and Kruskal–Wallis tests, independence is assumed to exist both within and between the comparison groups. With the Wilcoxon and Friedman tests, the correlated nature of the data causes the independence assumption to apply only in a between-subjects sense.

researchers indicate that they found a statistically significant difference (with $p < .001$ using the Mann–Whitney U test) between the ages of the two groups of individuals. Because of this, they stated, in the research report's abstract and in the report's full text, that "patients with a history of CVD were older." But is this really true? Were all of the patients with a history of CVD older than those in the comparison group?

By looking at the information contained in Excerpt 18.22, you can see the presence of overlapping distributions. Because the variability of ages in each group are reported via the interquartile range (which focuses on the middle 50 percent of the scores), we do not know precisely how young the youngest person was in the "older" CVD group. However, that person had to be no older than 67. Similarly, we do not know the exact age of the oldest person in the "younger" non-CVD comparison group; however, that person had to be at least 76 years old.

I think Excerpt 18.22 provides a powerful example of why you must be vigilant when reading or listening to research reports. Researchers frequently say that the members of one group outperformed the members of one or more comparison groups. When the researchers fail to include the two important words *on average* when making such statements, you should mentally insert this phrase into the statement that summarizes the study's results. You can feel safe doing this, because nonoverlapping distributions are very, very rare.

My final warning concerns the fact that many nonparametric procedures have been developed besides the five focused on within the context of this chapter. Such tests fall into one of two categories. Some are simply alternatives to the ones I have discussed, and they utilize the same kind of data to assess the same null hypothesis. For example, the Quade test can be used instead of the Friedman test. The other kind of nonparametric test not considered here has a different purpose. The Jonckheere–Terpstra test, for instance, allows a researcher to evaluate a null hypothesis that says a set of populations is ordered in a particular way in terms of their average scores. I have not discussed such tests simply because they are used infrequently by applied researchers.

Review Terms

Distribution-free	Nonparametric test
Friedman two-way analysis of variance of ranks	Overlapping distributions
	Parametric test
Independence	Ranks
Kruskal–Wallis one-way ANOVA of ranks	Ratings
	Sum of ranks
Large-sample approximation	Tied observations
	Wilcoxon–Mann–Whitney test
Likert-type attitude inventory	Wilcoxon matched-pairs signed-ranks test
	Wilcoxon rank-sum test
Mann–Whitney U test	Wilcoxon test
Median test	

The Best Items in the Companion Website

1. An interactive online quiz (with immediate feedback provided) covering Chapter 18.
2. Ten misconceptions about the content of Chapter 18.
3. One of the best passages from Chapter 18: "The Importance of the Research Question(s)."
4. The interactive online resource entitled "Wilcoxon's Matched-Pairs Signed-Ranks Test."
5. The website's final joke: "The Top 10 Reasons Why Statisticians Are Misunderstood."

To access the chapter outline, practice tests, weblinks, and flashcards, visit the companion website at http://www.ReadingStats.com.

Review Questions and Answers begin on page 531.

Multivariate Tests on Means

In previous chapters, we looked at several different kinds of analyses that involve tests on means. We focused our attention in those earlier chapters on three kinds of t-tests: one-way ANOVAs with and without repeated measures; post hoc and planned comparisons; two-way ANOVAs with zero, one, or two between-subjects factors; and several different kinds of analysis of covariance. All these test procedures are similar in that the statistical focus is on one or more population means.

A second common denominator of the test procures mentioned in the preceding paragraph is the fact that they are univariate in nature. Despite variations in the number of independent variables involved, in the between-versus-within kind of independent variables involved, or in the presence or absence of covariates, any t-test, any ANOVA, any ANCOVA, any post hoc investigation, and any planned comparison is applied to the data corresponding to a single dependent variable. We have seen many cases, of course, where a study had two or more dependent variables; however, the statistical tests in those situations that we examined earlier were always applied separately to each of the study's dependent variables.

We now turn our attention to statistical procedures that deal simultaneously with the means of two or more dependent variables. Such tests are considered to be **multivariate** in nature. In several respects, these procedures that we now consider are similar to the ones we have examined previously. Like their univariate "cousins," multivariate tests have null and alternative hypotheses, they involve a level of significance and produce a p-value, a decision about the tested H_0 can be wrong and constitute a Type I or a Type II error, assumptions are important, post hoc tests are often conducted, and statistical significance does not necessarily imply practical (i.e., clinical) significance.

Despite the similarities between univariate and multivariate tests, there are some important differences between these two kinds of tests. Three of these differences are worth noting here. First, a multivariate null hypothesis is more complicated than a univariate H_0 because there are two or more comparison groups along with data on two or more dependent variables. Second, certain of the assumptions that underlie multivariate test are altogether different from the assumptions of univariate tests. Finally, the post hoc procedures that work well with univariate tests are not recommended for use with multivariate tests. As this chapter unfolds, we consider each of these points in more detail.

The Versatility of Multivariate Tests

For any of the tests on means considered in earlier chapters, a multivariate procedure exists to handle the case of multiple dependent variables. Stated differently, for any ANOVA that has been designed for data on a single dependent variable, there is a **MANOVA** (with the letter *M* standing for *multivariate*) available for use. The same is true for the analysis of covariance; for any ANCOVA, there is a parallel **MANCOVA** that is appropriate for the situation where the researcher has data on two or more dependent variables.

In Excerpts 19.1 and 19.2, we see two examples of a one-way MANOVA being used. In the first of these excerpts, there were two comparison groups (ecstasy users and nonusers), just like the ANOVA we considered in Chapter 10. Here, however, the members of two groups were measured on four dependent variables: rash-impulsivity, reward-drive, postive effect, and negative effect. In Excerpt 19.2, there are three comparison groups (practicum students, pre-doctoral interns, and licensed professional staff), and a one-way ANOVA (like that considered in Chapter 11) would have been used if there had been just one dependent variable. Here, however, a one-way MANOVA was used because there were 10 dependent variables: the OQ-45 and the nine scales of the CAS.

EXCERPTS 19.1–19.2 • *Two- and Three-Group One-Way MANOVAs*

Participants were divided into two groups; those who have never taken ecstasy (termed non-users) and those who had taken ecstasy at any point in their life (users). . . . To explore group differences in impulsivity [both rash-impulsivity and reward-drive] and affect [both positive and negative], a between groups one-way multivariate analysis of variance (MANOVA) was performed.

Source: Egan, S. T., Kambouropoulos, N., & Staiger, P. K. (2010). Rash-impulsivity, reward-drive and motivations to use ecstasy. *Personality and Individual Differences, 48*(5), 670–675.

As a preliminary step, it was important to verify whether clients assigned to practicum students, predoctoral interns, and licensed professional staff differed in

(*continued*)

EXCERPTS 19.1–19.2 • (*continued*)

> their initial symptom severity. We therefore conducted a multivariate analysis of variance (MANOVA) using intake scores on all dependent measures (the OQ-45 and all CAS subscales) across counselor training levels.
>
> *Source:* Nyman, S. J., Nafziger, M. A., & Smith, T. B. (2010). Client outcomes across counselor training level within a multitiered supervision model. *Journal of Counseling and Development,* 88(2), 204–209.

Excerpts 19.3 and 19.4 illustrate the application of two-way MANOVAs. The first of these excerpts comes from a study in which both factors (type of friendship and gender) were between in nature, just like the two-way ANOVAs considered in Chapter 13. Here, however, there were two dependent variables: relational and physical victimization. The study associated with Excerpt 19.4 involved a 2 × 3 mixed design (in which the between factor was Condition and the within factor was Time), making the data set somewhat like the two-way mixed ANOVAs we looked at in Chapter 14. In this study, however, there were three outcome variables: SIF, Empathy, and TRIM.

EXCERPTS 19.3–19.4 • *Two-Way MANOVAs with and without a Within Factor*

> The 384 children in the present study [included] 293 (76.3%) who had a reciprocated best friendship (162 girls, 131 boys) and 91 (40 girls, 51 boys) who identified a unilateral best friend. A two-way MANOVA was conducted to determine if the levels of relational and physical victimization differed as a function of whether the friendship was mutual or unilateral and the child's sex.
>
> *Source:* Daniels, T., Quigley, D., Menard, L., & Spence, L. (2010). "My best friend always did and still does betray me constantly": Examining relational and physical victimization within a dyadic friendship context. *Canadian Journal of School Psychology,* 25(1), 70–83.
>
> ----
>
> First, we conducted a 2 (Immediate-Treatment Condition and Waiting-List Condition) × 3 (Time 1, Time 2, Time 3) repeated measures MANOVA on the three outcome variables related to forgiveness of a target offense (i.e., SIF, Empathy, and TRIM).
>
> *Source:* Kiefer, R. P., Worthington, E. L., Myers, B. J., Kliewer, W. L., Berry, J. W., Davis, D. E., et al. (2010). Training parents in forgiving and reconciling. *American Journal of Family Therapy,* 38(1), 32–49.

In Excerpts 19.5 and 19.6, we see examples of one-way and two way multivariate analyses of covariance. If there had been just one dependent variable in each of these studies, they would have been very much like the ANCOVAs we considered in Chapter 15. However, these excerpts are included in this chapter because each is multivariate in nature due to three dependent variables involved in Excerpt 19.5 (dissatisfaction with body fat, waist-to-hip ratio, and breast size) and 17 such variables involved in Excerpt 19.6.

EXCERPTS 19.5–19.6 • *Multivariate Analyses of Covariance*

In this study, heterosexual ($n = 95$) and nonheterosexual ($n = 84$) women were asked to rate figure drawings and computer-generated images of women that varied in body fat, waist-to-hip ratio, and breast size in terms of self, ideal, and cultural ideal; discrepancy indices, indicating body dissatisfaction, were created for each body aspect. . . . Because BMI was correlated with self-ideal discrepancies in both groups, a one-way MANCOVA, controlling for BMI, with sexual orientation as the independent variable, was performed.

Source: Koff, E., Lucas, M., Migliorini, R., & Grossmith, S. (2010). Women and body dissatisfaction: Does sexual orientation make a difference? *Body Image, 7*(3), 255–258.

A two-way MANCOVA was employed in which playing status (professional, amateur) and position (goalkeeper, defender, midfielder and forwards) were the between-participant factors and maturation (the difference between skeletal and chronological age) the covariate. All 17 dependent measures [e.g., height, mass, percentage body fat] were included in the analysis.

Source: le Gall, F., Carling, C., Williams, M., & Reilly, T. (2010). Anthropometric and fitness characteristics of international, professional and amateur male graduate soccer players from an elite youth academy. *Journal of Science and Medicine in Sport, 13*(1), 90–95.

The Multivariate Null Hypothesis

In the typical univariate situation where two groups are compared with a *t*-test or a one-way ANOVA, the null hypothesis is fairly easy to conceptualize. To do this, we imagine two populations of scores, each with the same mean. As indicated in Chapter 10, the null hypothesis in this situation states that the two population means are equal: $H_0: \mu_1 = \mu_2$. Staying with the univariate case, situations with three or more groups require only that we add a new μ for each additional population. We considered such null hypotheses in Chapter 11.

If data exist on multiple dependent variables, and if a multivariate approach is taken to compare group means, there are two legitimate ways to conceptualize the null hypothesis. The first of these ways of thinking about, or actually defining, the null hypothesis involves just three familiar concepts: groups, dependent variables, and population means. The alternative conceptualization requires us to understand terminology (e.g., *linear combination of dependent variables*, *group separation*) not yet considered in this book. Let us now consider these two ways of stating the multivariate H_0.

It is possible to conceptualize the MANOVA null hypothesis as an extension of the univariate H_0. From this perspective, the MANOVA's null hypothesis states that the study's populations have the same mean on the first dependent variable, the same mean on the second dependent variable, and so on. For example, if a MANOVA were to compare undergraduate students from two different universities in terms of height and IQ, the null hypothesis would state, simultaneously, two things: the two populations of students have the same mean height, and these two populations have the same mean IQ. (Notice that the null hypothesis does *not* say that the means of the dependent variables—height and IQ—are equal; the multivariate H_0 stipulates equality of means across populations on each dependent variable, not equality of means across dependent variables.)

In Excerpt 19.7, we see a case where a team of researchers articulated their study's MANOVA null hypothesis in this way. In the symbolic representation of this null hypothesis, each column of μs corresponds to a different population, while each row of μs corresponds to a different dependent variable. If the null hypothesis were true, the means in any row (e.g., $\mu_{11}, \mu_{12}, \ldots, \mu_{1g}$) would be identical, as those means correspond to different populations; however, the means in any column (e.g., $\mu_{11}, \mu_{21}, \mu_{31}$) could differ because those μs corresponds to the study's dependent variables of yield, plant height and harvested head weight.

EXCERPT 19.7 • *The Null Hypothesis in a One-Way MANOVA*

The data to be analysed [involve] dependent variables represented by the response namely: yield, harvested head weight, and plant height. We wish to observe the effects of the treatments on three dependent variables simultaneously. . . . MANOVA is an extension of analysis of variance [that can] test whether there are differences between the means of the identified groups of subjects on a combination of dependent variables, that is, it is used to test the null hypothesis.

$$H_o = \begin{pmatrix} \mu_{11} \\ \mu_{21} \\ \mu_{31} \end{pmatrix} = \begin{pmatrix} \mu_{12} \\ \mu_{22} \\ \mu_{32} \end{pmatrix} = \cdots = \begin{pmatrix} \mu_{1g} \\ \mu_{2g} \\ \mu_{3g} \end{pmatrix}$$

Where $\mu_{ig}(i = 1, 2, 3)$ is the population mean for the variable yield, plant height and harvested head weight respectively for all g [i.e., for all treatment groups].

Source: Maposa, D., Mudimu, E., & Ngweny, O. (2010). A multivariate analysis of variance (MANOVA) of the performance of sorghum lines in different agroecological regions of Zimbabwe. *African Journal of Agricultural Research, 5*(3), 196–203.

When one or more covariates are involved in a multivariate test of means, one way to conceptualize the null hypothesis is to think of it as an extension of the H_0 we considered in Chapter 15. Whereas the null hypothesis of a univariate ANCOVA has a single set of adjusted population means, the MANCOVA null hypothesis can be thought of as having multiple sets of adjusted means, one set for each dependent variable. In Excerpt 19.8, we see a case where a researcher used this approach to describe the null hypothesis evaluated by his multivariate analysis of covariance. In this particular MANCOVA, there was a single independent variable (corresponding to the two comparison groups), three dependent variables (corresponding to the posttest scores called *overall, dialogue text,* and *lecturette*), and three covariates (corresponding to the pretest measure of each dependent variable).

EXCERPT 19.8 • *The Null Hypothesis in a One-Way MANCOVA*

For this study, the multivariate null hypothesis that was tested in covariance was that the adjusted population mean vectors [on the dependent variables] for the two groups were equal. . . . MANCOVA accounts for the differences in ability (as measured by the pre-test) between the two groups by adjusting the means on the post-test to account for the differences on the covariates. . . . The dependent variables related to the two groups' scores on the post-test: overall post-test scores, dialogue text post-test scores, and lecturette post-test scores.

Source: Wagner, E. (2010). The effect of the use of video texts on ESL listening test-taker performance. *Language Testing, 27*(3) 1–21.

Excerpt 19.8 contains the word **vectors.** In this study there were two vectors, or sets, of adjusted population means, one for each comparison group. Each of these vectors contained the hypothesized adjusted population means on the three dependent variables. To see a display of mean vectors, take another look at Excerpt 19.7. Each vector in that excerpt is a vertical set of μs contained inside a set of brackets.

The alternative way of conceptualizing the null hypothesis of a multivariate test of means requires us to do some upside-down (but useful) thinking and create, in our minds, a new variable. Three steps are involved. First, we view the study's multiple dependent variables as if they were the independent variables in a multiple-regression-like equation, with weights attached to these variables to indicate their relative usefulness to the explanatory goal of this regression. Second, we insert each person's data into this equation to get a predicted score on our equation's dependent variable. In a very real sense, these predicted scores correspond to a new variable that we have created by statistically combining the original dependent variables into a single variable. Finally, we look to see if the comparison groups have different mean scores on the newly created dependent variable.

In the first step of the procedure just described, the weights for the original dependent variables are determined statistically so as to maximize differences between the groups in their scores on the newly created dependent variable. In other words, those weights are chosen to achieve the goal of **group separation.** Scores on the newly created variable designed to show this group separation come into existence via a **linear combination of dependent variables.** The weights used within this equation are called **discriminant coefficients,** because their function, collectively, is to discriminate the comparison groups from each other on the newly created dependent variable.[1]

In Excerpt 19.9, we see a passage from a research report that contains the phrase, "a linear combination of the measures." In the MANOVA conducted in this study, there was a single independent variable (gender) and 10 dependent variables (associated with different subscales of four instruments designed to assess personality and attitudinal traits). Notice how the final sentence in this excerpt begins. The researchers state that they used MANOVA to see if there was any group difference—that is, group separation—on the newly created dependent variable generated by combining the study's 10 measured variables.

EXCERPT 19.9 • *The Multivariate Notion of a Linear Combination of Dependent Variables*

The aim of the present study was to add to existing research focusing on the reasons for gender differences in help-seeking by comparing the frequency among men and women of a number of individual and socioculturally influenced attitudinal factors that may influence help-seeking for mental health problems. . . . A multivariate analysis of variance was performed to find any group differences based on a linear combination of the measures of interest, that is, LSS, TAS-26, the six facets of the NEO-O, and the two subscales of the DSS.

Source: Judd, F., Komiti, A., & Jackson, H. (2008). How does being female assist help-seeking for mental health problems? *Australian and New Zealand Journal of Psychiatry,* 42(1), 24–29.

Now that we see how MANOVA actually works, it is possible to consider the second way of conceptualizing the multivariate null hypothesis. This H_0 says that each population's mean on the newly created dependent variable is the same regardless of the discriminant coefficients used in an effort to create group separation. In other words, the null hypothesis is that there is no group separation, among the

---

[1]If there are more than two dependent variables, MANOVA actually generates additional new variables, each orthogonal to (i.e., independent from) the others, representing alternative ways of combining the dependent variables. Usually, however, these additional linear combinations contribute little to the goal of group separation as compared with the first one that's created.

populations, in terms of the new variable created to show group differences. Stated differently, this null hypothesis says that each of the study's dependent variables contributes nothing to group separation because the various populations have identical means on each dependent variable.

It may appear that the two conceptualizations of the multivariate null hypothesis are the same, as both stipulate that the various populations have identical means on each of study's dependent variables. Although similar in that respect, we have seen that the second of the two conceptualizations also involves concepts such as group separation, a linear combination of dependent variables, and discriminant coefficients. These three concepts are not superfluous; instead, they lie at the core of MANOVAs logic and computations.

Testing the Multivariate Null Hypothesis

Several test procedures have been developed to test MANOVA's omnibus null hypothesis. Although researchers sometimes use tests referred to as **Hotelling's trace** or **Roy's largest root,** the test procedures used most often are called **Wilks' lambda** and **Pillai's trace.** In Excerpts 19.10 and 19.11, we see examples of these latter two tests being used in applied studies. As you can see, the two multivariate tests in the first of these excerpts, and the single multivariate test in the second excerpt, all led to a rejection of the omnibus null hypothesis.

EXCERPTS 19.10–19.11 • *Wilks' Lambda and Pillai's Trace*

Data analyses were performed using multivariate analysis of variance (MANOVA) because the response we measured (survival, length of larval period and body mass at metamorphosis) is inherently multivariate. Inclusion of all three response variables in the analyses provided the maximum amount of information regarding the effects of our experimental treatments. . . . The overall MANOVAs showed that our treatments significantly affected traits of both cane toads (Wilks' lambda = 0.18, $P < 0.04$) and ornate burrowing frogs (Wilks' lambda = $0.025, P < 0.001$).

Source: Crossland, M. R., Alford, R. A., & Shine, R. (2009). Impact of the invasive cane toad (*Bufo marinus*) on an Australian frog (*Opisthodon ornatus*) depends on minor variation in reproductive timing. *Oecologia, 158*(4), 625–632.

A multivariate analysis of variance examining all adherence variables combined revealed a significant omnibus effect of anxiety on adherence (Pillai's trace = $2.44, p = 0.01$).

Source: Kuhl, E. A., Fauerbach, J. A., Bush, D. E., & Ziegelstein, R. C. (2009). Relation of anxiety and adherence to risk-reducing recommendations following myocardial infarction. *American Journal of Cardiology, 103*(12), 1629–1634.

In Excerpts 19.10 and 19.11, the calculated values for Wilks' lambda (0.18 and 0.025) and Pillai's trace (2.44) are reported in their raw forms.[2] Although the calculated values of multivariate tests are occasionally reported like that, it is more common to see the calculated values converted into an *F*-value. Excerpts 19.12 and 19.13 illustrate this kind of conversion. These excerpts contain two *df* values located next to the *F*-value, as would have been the case if a univariate analysis had been conducted. With most multivariate tests, however, the first of each *F*'s two *df* values is determined by multiplying the number of dependent variables by the 1 less than the number of groups. Knowing this, we can figure out that there were two comparison groups in the study from which Excerpt 19.13 was taken.[3]

EXCERPTS 19.12–19.13 • *Converting Multivariate Calculated Values into F-Ratios*

A MANOVA was used to test for gender differences in mean levels of support from the various sources. The five CASSS support subscale scores were entered as dependent variables, and [results indicated] Wilks' lambda = $.865, F(5, 632) = 19.65, p < .001$.

Source: Rueger, S. Y., Malecki, C. K., & Demaray, M. K. (2010). Relationship between multiple sources of perceived social support and psychological and academic adjustment in early adolescence: Comparisons across gender. *Journal of Youth and Adolescence, 39*(1), 47–61.

A multivariate analysis of variance (MANOVA) was conducted to determine whether there were significant between-group differences in DP-gram peak occurrence, peak height, and peak width. The Hotelling's trace multivariate test of overall between-group differences was not significant, $F(3, 40) = 0.321, p = .81$.

Source: Bhagat, S. (2009). Analysis of distortion product otoacoustic emission spectra in normal-hearing adults. *American Journal of Audiology, 18*(1), 60–68.

Researchers frequently report only one of the four popular multivariate test statistics, as was the case in Excerpts 19.10 through 19.13. Sometimes, however, more than one is reported for the same data that have been analyzed. This occurs because computer programs typically conduct all four tests on any set of data that's analyzed, because the results of all four tests usually lead to the same decision regarding the multivariate null hypothesis, and because Pillai's trace is more robust than the other test procedures.

---

[2]Wilks' lambda is often reported via the upper- or lower-case letter for lambda, Λ or λ.

[3]The formula for determining the second *df* associated with the MANOVA *F*-test is complicated and varies depending on which test procedure is used to test the omnibus null hypothesis.

Assumptions

Like the analysis of variance, MANOVA and MANCOVA have assumptions. Competent researchers attend to these assumptions when conducting multivariate tests on means, and they make adjustments in the planned analytic strategy if any important underling assumption appears to be untenable.

When conducting a MANOVA, researchers should be aware of seven main assumptions. These are: (1) random samples from the relevant populations, (2) independence of observation, (3) multivariate normality, (4) homogeneity of variance–covariance matrices, (5) linear relationships between dependent variables, (6) no outliers, and (7) no multicollinearity. An additional assumption, equality of regression slopes, comes into play with any MANCOVA.

In research reports, multivariate assumptions are dealt with by applied researchers in different ways. In many cases, researchers say absolutely nothing about any of the assumptions, thereby giving the impression that they were unaware that MANOVA and MANCOVA do not operate as intended if important assumptions are violated. In other cases, researchers report that they checked the assumptions and discovered one or more of them to be untenable; then, they move right ahead with their multivariate analysis and warn readers of the research report "to be careful when interpreting the findings." Both of these ways of dealing with assumptions leaves much to be desired.

In Excerpt 19.14, we see a passage from a research report that indicates three things. Via two relatively short sentences, the researcher makes it clear that he is aware of MANOVA's underlying assumptions, that these assumptions were tested in a preliminary phase of the investigation's data analysis, and that none of the assumptions seemed to be violated. Give credit to researchers when they incorporate these three bits of information into their research reports.

EXCERPT 19.14 • *Attending to Assumptions*

In order to check whether the assumptions of MANOVA were met, preliminary assumption testing for normality, linearity, univariate and multivariate outliers, homogeneity of variance–covariance matrices and multi-collinearity were conducted. No significant violation was found.

Source: Sahin, M. (2010). The impact of problem-based learning on engineering students' beliefs about physics and conceptual understanding of energy and momentum. *European Journal of Engineering Education, 35*(5), 519–537.

When researchers assess the veracity of assumption underlying their planned multivariate analyses, one or more of the assumptions may seem untenable. When that occurs, several legitimate and helpful options exist. In situations like this, the researcher can eliminate problematic scores (or even entire variables) from the analysis, use data transformations in an effort to reduce nonnormality or variance–covariance heterogeneity, choose a more robust test procedure, make the level of significance more rigorous, or decide not to use MANOVA or MANCOVA.

Excerpt 19.15 comes from a study in which the researchers attended to the assumptions associated with the one-way MANOVA they planned to use. In checking to see if their study's data conformed to the assumptions, the researchers discovered various problems. This excerpt deserves your close attention, because it illustrates several different options that exist when assumptions appear to be violated. The researchers associated with this excerpt earn high marks for clearly indicating what they did as a consequence of their "preliminary assumption testing."

EXCERPT 19.15 • *MANOVA Options when Assumptions Seem Untenable*

A one-way between groups MANOVA was performed to investigate the significance of the suggestibility score differences between the HGSHS [groups]: A, the GSHA, and the CURSS. . . . Preliminary assumption testing was conducted to check for normality, linearity, multicollinearity, univariate and multivariate outliers, and homogeneity of variance–covariance matrices. All dependent variables except voluntary responding met the assumption of normality. A square-root transformation was used to normalize the voluntary responding distribution. The assumptions of linearity and multicollinearity were met. Four participants were found to be multivariate outliers and were therefore deleted from the analyses (two participants from the GSHA condition and two participants from the CURSS condition). Box's test of equality of covariances indicated a violation of the assumption of homogeneity of variance–covariance matrices. In order to ensure the robustness of Pillai's statistic despite this violation, cases were randomly deleted so that all three sample sizes were equal, $n = 103$ [in size].

Source: Barnes, S. M., Lynn, S. J., & Pekala, R. J. (2009). Not all group hypnotic suggestibility scales are created equal: Individual differences in behavioral and subjective responses. *Consciousness and Cognition, 18*(1), 255–265.

Some researchers subject their data to a nonparametric analysis when MANOVA assumptions seem untenable. Several such procedures have been developed to do this, and they occasionally are referred to by the acronyms **NPMANOVA** (in which the letters NP stand for *nonparametric*) or **PERMANOVA** (in which the letters PER stand for *permutation*). In Excerpt 19.16, we see a case in which one of these nonparametric procedures was used. Notice that a regular MANOVA was used with the data from one set of minnows, as the assumptions there did not seem to be violated. However, the data from the second set of minnows failed to support MANOVA's assumptions; accordingly, those data were converted into ranks and an extension of the nonparametric Kruskal–Wallis test was applied to the ranked data.

EXCERPT 19.16 • *Nonparametric MANOVA*

The [two-factor] experiment consisted of a conditioning phase, during which minnows were taught to recognize a brown trout as a predator in clear water, followed by a testing phase, where minnows were exposed to brown trout, rainbow trout or perch [factor 1] in clear or turbid water [factor 2]. We calculated the change in shelter use and time moving from the prestimulus baseline. Because the two variables were not independent from each other, we analysed them simultaneously using a MANOVA procedure. Behavioural data from the control minnows followed parametric assumptions and were analysed with a two-way MANOVA. Behavioural data from the alarm cue minnows did not meet homoscedasticity assumptions (nonhomogeneity of variances). Hence, the data were rank-transformed prior to performing a nonparametric MANOVA using the Sheirer–Ray–Hare extension of the Kruskal–Wallis test [for ranks].

Source: Ferrari, M. C. O., Lysak, K. R., & Chivers, D. P. (2010). Turbidity as an ecological constraint on learned predator recognition and generalization in a prey fish. *Animal Behaviour, 79*(2), 515–519.

Statistical Significance and Practical Significance

One of the purposefully recurring themes in this book has been the distinction between statistical significance and practical significance. Repeatedly, I have made the point that a finding can end up being statistically significant even though there is little or no practical importance associated with the claimed "discovery." This is just as true for multivariate tests on means as it is for the univariate tests we considered in earlier chapters. Accordingly, it should come as no surprise that techniques exist that can help researchers—as well as the readers of their research reports—avoid the mistake of thinking that a statistically based finding is "big" when in fact it is "small" (or perhaps even smaller-than-small).

One way to assess the practical significance of a multivariate result is through a data-based estimate of effect size. When the omnibus test from a MANOVA or a MANCOVA is computed, an estimate of effect size can also be computed. Although there are several different indices available for use, the two that are used most often are eta squared and partial eta squared. These two kinds of effect size estimates are illustrated in Excerpts 19.17 and 19.18.

EXCERPTS 19.17–19.18 • *Estimating Effect Size in One-Way and Two-Way MANOVAs*

Results of the [one-way] MANOVA revealed significant, but weak, differences between the control and experimental groups, Wilks' Lambda = 0.96, $F = 2.59$, $p < 0.05$, $\eta^2 = 0.04$. Thus, initial analysis revealed that offering students the choice

(continued)

EXCERPTS 19.17–19.18 • (*continued*)

among differing types of examination had a small effect on their perception of both fairness and learning.

Source: Mauldin, R. K. (2009). Gendered perceptions of learning and fairness when choice between exam types is offered. *Active Learning in Higher Education, 10*(3), 253–264.

To examine differences between types of fans, a 2 (Gender of Participant) $\times$ 4 (Type of Interest) MANOVA was conducted using fanship, entitativity, identification with the group, and collective happiness as dependent variables. The omnibus results show a main effect of type of interest (Wilks' $\lambda = .730, F(12,897) = 8.96$, $p < .001, \eta_p^2 = .100$), a main effect of participant gender (Wilks' $\lambda = .960$, $F(4,339) = 3.50, p = .008, \eta_p^2 = .040$), and an interaction between gender and type of interest (Wilks' $\lambda = .929, F(12,897) = 1.85, p = .015, \eta_p^2 = .024$).

Source: Reysen, S., & Branscombe, N. R. (2010). Fanship and fandom: Comparisons between sport and non-sport fans. *Journal of Sport Behavior, 33*(2), 176–193.

In Excerpt 19.17, eta squared was used to estimate effect size. As you may recall from Chapter 11, eta squared and partial eta squared are equal to the same value in a one-way ANOVA. This is the case in a one-way MANOVA only when there are just two comparison groups. If three or more groups are involved in the one-way MANOVA, partial eta squared turns out smaller than eta squared.[4] Excerpt 19.18 contains the results of a two-way MANOVA, with a partial eta squared provided for each *F*-value. In two-way ANOVAs, values of eta squared and partial eta squared are different due to the way they are computed. However, that is not the case in a two-way MANOVA. If there are just two levels of a factor, the computational formulas for partial eta squared and eta squared cause these two estimates of effect size for that factor's main effect to be identical. That is also the case for the interaction effect in any 2 $\times$ 2 MANOVA.

The popular criteria for assessing standardized estimates of effect size, such as partial eta squared, are the same in a multivariate analysis as they are in a univariate analysis. For partial eta squared (and eta squared), the lower limits for the labels *small, medium,* and *large* are .01, .06, and .14, respectively. If these criteria are used to evaluate the effect size estimates in Excerpt 19.18, two of the statistically significant findings would be called *small* whereas one would be classified as being *medium* in size.

The computation of effect size estimates represents a post hoc way of dealing with the possibility that statistical significance might exist even though the true

---

[4]Eta squared = $1 - \lambda$; partial eta squared = $1 - \lambda^{1/s}$, where $s =$ the smaller of two things: (1) the *df* for the effect being tested, if that effect were being tested univariately, or (2) the number of dependent variables.

effect has little or no practical significance. An alternative way of dealing with this same issue is a priori in nature. By performing a power analysis in the planning stage of an investigation, a researcher determines the proper sample size for his or her study, thus reducing the chances that too much data will cause a small effect to look big, or that not enough data will cause a big effect to be overlooked.

In Excerpt 19.19, we see an example of an a priori power analysis being used in conjunction with a four-group one-way MANCOVA. In this excerpt, notice that the effect size is said to be .50. This is not a data-based, estimated effect size, like those we saw in Excerpts 19.17 and 19.18. Rather, the effect size in Excerpt 19.19 was selected by the researchers to indicate the dividing line between real effects that are of trivial magnitude and those that are large enough to be considered large, noteworthy, and important.

EXCERPT 19.19 • *A Priori Power Analysis in a MANOVA Study*

The purpose of this study was to examine the impact of co-morbid disorders in the area of behavior problems within a specific population of adults residing at two state-run facilities. This was achieved by using multivariate analyses to compare groups of participants with ID [intellectual disability] alone, ID and epilepsy alone, ID and ASD [autism spectrum disorders] alone, and finally, a combined group with ID, ASD, and epilepsy, which are disorders commonly found among individuals residing at state-run residential facilities. . . . An a priori power analysis was conducted to determine the total sample size required for the present study. [W]hen alpha (α) is set at .05 [along] with a medium effect size set at .50 and power set at .80, . . . it was determined that a total sample of 80 participants was required for a MANOVA with four groups (i.e., $n = 20$).

Source: Smith, K. R. M., & Matson, J. L. (2010). Behavior problems: Differences among intellectually disabled adults with co-morbid autism spectrum disorders and epilepsy. *Research in Developmental Disabilities, 31*(5), 1062–1069.

Post Hoc Investigations

If a MANOVA or MANCOVA omnibus null hypothesis is rejected, the researcher will likely conduct some form of post hoc investigation in order to understand the forces at play in the data that have been analyzed. Such investigations are appropriate even in studies where just two groups have been compared, due to their multivariate nature.

My review of a large number of journals shows that the most popular strategy for probing a significant multivariate finding is a set of univariate tests, one conducted for each dependent variable. For example, if a researcher achieves significance with a one-way MANOVA, he or she is likely to conduct a one-way ANOVA (or a *t*-test, perhaps, if there are just two groups of scores) on the data corresponding to the first dependent variable, a similar analysis on the data corresponding to the second dependent variable, and so on until each and every dependent variable has been dealt with. Then, any of these univariate ANOVAs, if significant, is itself probed, perhaps with Tukey's HSD, if there are three or more comparison groups.

In a similar fashion, the most common strategy for probing one or more significant findings from a two-way MANOVA, or from any kind of MANCOVA, involves multiple univariate analyses, each focused on a different dependent variable. Then, any significant result from any of these univariate ANOVAs or ANCOVAs is probed, if necessary, as was illustrated earlier in Chapters 13 through 15. Such post hoc probing is intended to discover what caused the initial multivariate result to be significant.

This extremely popular strategy for probing significant multivariate findings is illustrated in Excerpt 19.20. The MANOVA referred to in this excerpt was a 2×2 multivariate analysis of variance in which the factors were Gender (Males versus Female) and Culture (Chinese versus Canadian). The research participants were adolescent males and females, ages 16 through 18, from each culture who completed a questionnaire regarding their current romantic (i.e., dating) relationships. Their responses produced data on the study's three dependent variables: trust, intimacy, and companionship.

EXCERPT 19.20 • *Probing the Results of a Multivariate Analysis with Univariate Tests*

A MANOVA was conducted with partner's trust, intimacy, and companionship as dependent variables, and culture and gender as between-subjects factors. The results indicated a significant multivariate effect of culture, $F(3, 211) = 15.35, p < .001$, and a significant culture by sex interaction, $F(3, 211) = 6.03, p < .001$. Follow-up univariate ANOVAs [showed that] Chinese adolescents reported less trust and less companionship in their romantic relationships than did Canadian daters. A gender interaction with culture was significant for intimacy. In China, the boys reported greater intimacy with their romantic partner than did girls ($t = 3.06, p < .01$). In Canada, a reverse trend was found, with girls reporting greater romantic intimacy than boys ($t = 2.64, p < .01$).

Source: Li, Z. H., Connolly, J., Jiang, D., Pepler, D., & Craig, W. (2010). Adolescent romantic relationships in China and Canada: A cross-national comparison. *International Journal of Behavioral Development, 34*(2), 113–120.

As indicated in Excerpt 19.20, the two-way MANOVA produced two significant results. To probe these results, the researchers conducted three separate two-way ANOVAs, one for each dependent variable. The univariate ANOVA on the trust scores produced a significant effect only for the main effect of culture. That was also the case for the univariate ANOVA of the companionship scores. The univariate analysis of the intimacy data, however, produced a significant effect for the gender-by-culture interaction. To probe this interaction, tests of simple main effects were conducted (via *t*-tests) in which the two genders within each culture were compared.

In the study associated with Excerpt 19.20, the interpretation of the three *F*s in each of these two-way univariate ANOVAs, and the simple effects investigation associated with the analysis of the intimacy data, parallel exactly what the interpretations and post hoc testing would have been if three separate two-way ANOVAs had been conducted without any preliminary MANOVA. The essence of this popular strategy for performing the analysis of multivariate data, illustrated in Excerpt 19.20 and in thousands of other studies, is captured nicely by the four-word phrase, *multivariate first, then univariate.*

Several statistical authorities frown on the "multivariate first, then univariate" strategy, despite its popularity. Alternative procedures exist for probing multivariate data sets after the omnibus null hypothesis is rejected, and we will examine a few of these momentarily. Before doing that, however, let's consider why the post hoc strategy used widely across many disciplines has its detractors. There are two reasons, both connected to any study's dependent variables.

First, the dependent variables in any MANOVA or MANCOVA are likely to be correlated.[5] Because of this, the multivariate world of any study—referred to technically as the study's **multivariate space**—cannot be described well by thinking about the dependent variables one at a time, in a univariate manner. Proof of this fact can be found in the assumption of homogeneous variance–covariance matrices. To define or evaluate this assumption, we must consider the interdependence among the full set of dependent variables. Moreover, the most appropriate techniques available for screening data for potential outliers evaluate each data point within the multivariate space.

To understand what a multivariate space is, first imagine that we measure each of 100 adult workers on three dependent variables: weight, salary, and intelligence. Next, imagine we take our data into a square room that has its four walls facing north, south, east, and west. Further imagine that the numerical values for the continuum associated with the weight variable have been marked along the bottom edge of the south wall, that the numerical values for the continuum associated with the intelligence variable have been marked along the bottom edge of the west wall, and that the numerical values for the continuum associated with the salary variable have been marked along the vertical joint where the south and west walls meet.

Now we display the data collected from our sample of 100 workers by means of 100 ping pong balls. For each worker, we tie one end of a piece of string to that worker's ball, and then we tie the other end of the string to a tack that we push into the ceiling. By carefully measuring the length of each piece of string, and by carefully finding the right spot to push each tack into the ceiling, we could suspend the 100 ping pong balls such that any given ball has a position in the room that

---

[5]The dependent variables not only are *likely* to be correlated; they *ought* to be correlated. If the dependent variables are all independent from one another, a multivariate analysis has disadvantages as compared with a strategy that involves only separate univariate tests.

corresponds exactly to a particular worker's weight, intelligence, and salary. The room containing these hanging ping pong balls is our multivariate space.

To make our imaginary room with the ping pong balls fit a MANOVA situation, imagine that 50 of the ping pong balls are blue because they represent male workers, whereas the other half of the ping pong balls are pink because they represent female workers. If our room contains a single cloud of 100 balls, with blue and pink interspersed randomly, the multivariate null hypothesis is retained. However, if the blue balls tend to be located away from the pink balls, then a test such as Wilks' lambda might cause the null hypothesis test to be rejected. The important thing to see here is that our comparison of blue and pink balls is being made inside a three-dimensional room. If you looked at the balls from only one angle (e.g., the south wall), you might well miss a difference between the two clouds of data points.

Discussions and pictures of multivariate space typically are found only in textbooks dealing with intermediate or advanced quantitative techniques. In Excerpt 19.21, we see a rare reference to the notion of multivariate space that appeared in an applied research report.

EXCERPT 19.21 • *Multivariate Space*

MANOVA tests for differences among groups in the multivariate space defined by the original set of outcome measures (TV, BV, BV/TV, TMD, σTMD, and BMC). In this space, each of the k experimental groups is described in terms of a vector of means, rather than a single mean value. MANOVA [creates] linear combinations of the original outcome measures. These linear combinations are constructed such that the separation among groups is maximized.

Source: Morgan, E. F., Mason, Z. D., Chien, K. B., Pfeiffer, A. J., Barnes, G. L., Einhorn, T. A., et al. (2009). Micro-computed tomography assessment of fracture healing: Relationships among callus structure, composition, and mechanical function. *Bone, 44*(2), 335–344.

To understand the second reason why certain statistical authorities argue against using univariate tests to probe a significant multivariate finding, recall that the dependent variables are used jointly to create a new variable designed to maximize group separation. The new variable comes into existence by means of the linear combination of dependent variables, and the MANOVA or MANCOVA null hypothesis stipulates that the study's population means are located at the same position on the continuum corresponding to this newly created variable. If the sample data produce group means, derived from the linear combination of dependent variables, that are further apart than would be expected by chance, the multivariate H_0 is rejected. Statistical authorities argue, therefore, that a post hoc investigation ought to answer the question, "Which dependent variable(s), *within the linear combination of such variables*, played a major role in causing the multivariate null hypothesis to be rejected?"

We now look at two procedures that can be used in a post hoc investigation to probe a significant MANOVA or MANCOVA. These procedures are admittedly used infrequently as compared with the more popular univariate strategy we first considered. Nevertheless, it is appropriate that we consider these two procedures, as their popularity is likely to increase in the coming years.

The first of the two post hoc procedures to be discussed here was used in a study involving young elite female basketball players. Three groups were compared: centers, forwards, and guards. There were eight dependent variables involved, each of which related to some sort of athletic ability (e.g., sprinting, jumping, throwing). The multivariate analysis produced a significant value for Wilks' lambda, so the researchers conducted a post hoc investigation. As shown in Excerpt 19.22, they did this by computing and comparing **discriminant ratio coefficients** from the main linear combination of variables that contributed to the group separation in the study's multivariate space.

EXCERPT 19.22 • *Post Hoc Investigation Using Discriminant Ratio Coefficients*

The differences among the groups were examined using multivariate analysis of variance (MANOVA) and descriptive discriminant analysis (DDA) as a follow-up procedure (Huberty, 2006). To find out which variables distinguished the groups the most, discriminant ratio coefficients (DRC) were considered. . . . The initial MANOVA was significant (Wilks $\lambda = .53, F(16,100) = 2.35, p = .005$) which shows there were differences between the positions. To further study the resulting differences, linear discriminant functions were obtained. The test of dimensionality revealed one significant discriminant function (canonical $R = .65; p = .005$) which accounted for 87.7% of the variance. . . . The discriminant ratio coefficients (Table 3) suggest that the best variable for distinguishing between the positions is the *20m sprint*, followed by the *basketball throw, 6 × 5m sprint* and *medicine ball throw*.

TABLE 3   *Discriminant Ratio Coefficients (DRC)*

Variable	DRC
S20	.398
D20	.077
BBT	.200
MBT	.128
S6X5	.185
D6X5	.041
CMJ	.008
DJ25	− .037

Legend: S20: 20 m sprint; D20: 20 m sprint dribble; BBT: basketball throw; MBT: medicine ball throw; S6X5: 6 × 5 m sprint; D6X5: 6 × 5 m sprint dribble; CMJ: countermovementjump; DJ25: drop jump 25 cm height.

Source: Erčulj, F., Blas, M., Čoh, M., & Bračič, M. (2009). Differences in motor abilities of various types of European young elite female basketball players. *Kinesiology, 41*(2), 203–211.

The second recommended post hoc procedure used by some researchers to probe a significant MANOVA or MANCOVA is called *the Roy–Bargman stepdown analysis,* or simply the **stepdown *F*-test procedure.** This procedure requires the researcher to first prioritize the dependent variables based on practical or theoretical considerations. Then, univariate comparisons among the groups take place, primarily via the analysis of covariance, to see if each new variable considered explains a significant amount of group separation above and beyond the amount already explained by the variables initially considered. In Excerpt 19.23, we see an example of this kind of post hoc strategy being used in a study in which gifted male and female college students were compared on the various components of a personality trait called *overexcitability.*

EXCERPT 19.23 • *Post Hoc Investigation Using Stepdown F-Tests*

The OEQ-II is a 50-item, self-rating questionnaire to measure OE [overexcitability]. Ten items that assess each of the five OEs (emotional, intellectual, imaginational, sensual, and psychomotor) are randomly distributed throughout the instrument. . . . A MANOVA to determine whether OE profiles differed by sex found an overall difference between males and females ($\Lambda = .649, F = 12.1, df = 5, 112, p < .01$). Stepdown F tests indicated which OE means were significantly different for the group variable sex. Males scored higher on intellectual OE (3.85 vs. 3.52), whereas females scored higher on emotional (4.21 vs. 3.42) and sensual OE (3.58 vs. 3.18).

Source: Miller, N. B., Falk, R. F., & Huang, Y. (2009). Gender identity and the overexcitability profiles of gifted college students. *Roeper Review, 31*(3), 161–169.

The two post hoc strategies illustrated in Excerpts 19.22 and 19.23 are similar in that they help researchers achieve the same goal: identification of the dependent variable(s) responsible for an initial rejection of the multivariate null hypothesis. That is the same goal researchers have when they use the more popular post hoc strategy of applying a univariate test to each dependent variable. Although discriminant ratio coefficients and stepdown *F*-tests, on the one hand, and univariate *t*- and *F*-tests, on the other hand, look the same in terms of the kind of information provided, these two classes of post hoc strategies are dissimilar in their philosophical approach to the way data should be analyzed. That difference is not irrelevant; what is revealed in a post hoc investigation using univariate tests can be different from what is discovered via multivariate analyses.

Three Final Comments

Before we finish our consideration of multivariate tests on means, I want to pose (and then answer) three questions: (1) How much sample data is needed to perform a multivariate analysis? (2) Should the alpha level be modified in a post hoc investigation if univariate tests are used to probe a significant multivariate result? (3) What, if anything, can cause a multivariate analysis to produce murky or misleading results?

Regarding the issue of sample size, there is one mathematical-based ground-floor requirement, several rules-of-thumb for what is needed above that mathematical minimum, and one optimal way to answer the question, "How large should the samples be?" As for the mathematical requirement, the computational formulas of a multivariate analysis mandate that the number of scores in each comparison group must exceed the number of dependent variables involved in the study. More data than that, however, is needed because of concerns for avoiding Type II errors. I have seen rules of thumb that deal with the minimum size of comparison groups (some say n should be at least 20; others say 30), the minimum size of the total data set (some say at least 100 cases are needed), and the minimum cases per dependent variable (some say the ratio should be at least 10 to 1). The best way for a researcher to determine the minimum sample size, of course, is to have an a priori power analysis answer the question, "How large should n be?"

Should alpha be adjusted in a post hoc investigation? Most applied researchers do not do this. This is due, I think, to a widespread belief that an initial multivariate test has some form of built-in feature that holds down the Type I error risk when post hoc tests are conducted. In reality, the risk of rejecting true null hypotheses is inflated unless the Bonferroni or some other procedure is used to make the separate post hoc tests more rigorous. This is true if the initial multivariate test is followed by a set of univariate tests; it is also true if the post hoc investigation involves a series of Roy–Bargman stepdown F-tests.

The last of our three concluding questions asks, "What, if anything, can cause the results of a multivariate to produce murky or misleading results?" The truthful answer requires just two words: "Many things." Problems arise, for instance, if samples are not random subsets of populations, if dependent variables are measured with unreliable or invalid instruments, if important assumptions are violated, if irrelevant dependent variables are included, if important and relevant dependent variables are overlooked, if the sample size is inadequate, and if no effort is made to assess the practical significance of results.

In addition to the list of items included in the preceding paragraph, we must remember that a multivariate analysis is inferential in nature, and thus the decision to reject or not reject any null hypothesis does not *prove* anything. Unfortunately, many researchers talk about their multivariate findings as if indisputable facts have been unveiled. Although we have no control over what those researchers write or say, we most certainly *do* have control over how we interpret the claims they make. Remember, therefore, that any statistical test that causes its null hypothesis to be rejected might indicate nothing more than a Type I error. Similarly, any statistical test that causes its null hypothesis to be retained might represent nothing more than a Type II error.

Review Terms

Discriminant ratio coefficients	Multivariate space
Group separation	NPMANOVA
Linear combination of dependent variables	PERMANOVA
Hotelling's trace	Pillai's trace
MANCOVA	Roy's largest root
MANOVA	Stepdown F-test procedure
Multivariate	Vectors
	Wilks' lambda

The Best Items in the Companion Website

1. An interactive online quiz (with immediate feedback provided) covering Chapter 19.
2. Nine misconceptions about the content of Chapter 19.
3. One of the best passages from Chapter 19: "What, if anything, can cause the results of a multivariate analysis to produce murky or misleading results?"

To access chapter outlines, practice tests, weblinks, and flashcards, visit the companion website at http://www.ReadingStats.com.

Review Questions and Answers begin on page 531.

Factor Analysis

We have considered the notion of correlation in several previous chapters. The central focus of Chapters 3, 9, and 16 was on descriptive and inferential procedures that assess the strength of relationships. In Chapter 4, we considered cases in which correlation is used to estimate reliability and validity. In Chapter 17, we noticed how chi square and associated techniques are sometimes used in a correlational fashion. In Chapters 15 and 19, we saw how the concept of correlation is embedded in the techniques of ANCOVA, MANOVA, and MANCOVA. Because correlation is involved, either directly or indirectly, in so many kinds of data analysis, it is no exaggeration to say that correlation is the single most important statistical instrument in the applied researcher's toolkit.

We now turn our attention to another statistical procedure that has the concept of correlation as its core: the technique of factor analysis. As with most of the statistical procedures considered in this book, we do not examine the detailed formulas that come into play when data are factor analyzed. Instead, we concentrate on three things: the goals of a factor analysis, the way researchers report the results of their factor analytic studies, and the reasons why some factor analyses deserve the label *well done,* whereas others are deficient in small or large ways.

The Goal (and Basic Logic) of Factor Analysis

Factor analysis is a procedure that attempts to reduce the complexity of a multi-variable data set so it becomes easier for people to use the data in applied settings or in the development/refinement of theory. In a word, the goal of factor analysis is *parsimony.* The main question is simple: Can the people, animals, or things that have been measured on several variables be described, accurately, by means of a small number of numerical descriptors rather than by scores on each of the initial variables?

Suppose, for example, that each of 100 job applicants is measured in terms of 15 different traits: creativity, vocabulary, honesty, age, math ability, perseverance, social skills, attractiveness, general intelligence, physical stamina, height, education, tact, writing ability, and kindness. Having 15 scores per applicant makes it extremely difficult to decide who has the most assets and the least liabilities. A factor analysis might help in this situation, because it attempts to reduce the initial set of 15 variables into a more manageable set of descriptors. Results might indicate that each applicant could be described fairly well by scores on just three mega-traits: academic characteristics, physical characteristics, and interpersonal characteristics. In this example, there has been a parsimonious reduction of 15 variables into just three.

The basic logic of factor analysis is simple. If two of the initial variables are highly related with each other but largely unrelated to any of the other variables, then those two variables should be merged together, with a new variable created so a single score can represent a person's standing on the two combined variables. Doing this achieves the goal of parsimony by reducing the redundancy between those two initial variables being combined. Similarly, if three of the other initial variables are highly related with one another (but not related to any of the other initial variables), then those three variables also can be combined, with a second, newly created variable standing in for the three that have been merged. Again, reduction in redundancy produces parsimony.

Each of the newly created variables in a factor analysis is called a **factor.**[1] In any given study, there may end up being one, two, or more factors. The number of factors that emerge from a factor analysis depends on the network of relationships among the original variables. The results of a factor analysis also depend on certain decisions the researcher makes when conducting the analysis. Before considering some of those needed decisions facing a researcher who performs a factor analysis, let's look at two tables that help to illuminate the goal (and logic) of factor analysis.

The two tables we next consider come from a study concerned with children's ability to write. The study's participants were 120 school children ages 8 through 11. First, the children were told that they would hear an adult read a paragraph and then their task would be to write down everything they could remember.[2] Next, the paragraph was read twice, after which the children were given as much time as they wanted to perform the writing task. Finally, the researchers evaluated each student's paper in terms of nine criteria: total number of words (TNW), number of ideas from the paragraph (IDEAS), number of clauses beginning with the coordinating conjunctions *and, but,* or *or* (T-UNIT), percentages of

---

[1] This kind of factor should not be confused with kind of factor involved in an ANOVA, ANCOVA, MANOVA, or MANCOVA. In those analyses, factors designate the independent (i.e., grouping) variables.

[2] The paragraph contained 227 words and 20 sentences, and it answered the question, "Where do people live?" This paragraph was carefully selected to be age-appropriate for the study's children.

words misspelled (SPELL), conventional punctuation (CONVEN), and four other indicators of quality writing. These nine criteria were the original variables in the factor analysis.

After the nine scores became available for each child's written response, the researchers computed the 36 possible bivariate correlations among the nine variables. Those correlations appear in Excerpt 20.1. In a very real sense, a correlation matrix such as this contains the ingredients for the factor analysis. In other words, any factor analysis begins with an examination of bivariate correlations. Take a moment to examine this excerpt's correlations, and see if you can identify any subsets of variables that are characterized by (1) high correlations among the variables within the subset and (2) low correlations between the subset's variables and variables outside the subset.

EXCERPT 20.1 • *The Starting Point of a Factor Analysis: Correlation Coefficients*

TABLE 4 *Intercorrelations among writing measures*

Variable	1	2	3	4	5	6	7	8	9
1. TNW	—	.83**	.88**	.43**	.93**	.25**	−.19*	−.39**	−.007
2. IDEAS		—	.79**	.24**	.81**	.15	−.19*	−.35**	−.06
3. T-UNIT			—	−.024	.83**	−.10	.13	−.27**	−.06
4. MLT-UNIT				—	.38**	.78**	.17	−.33**	.07
5. CLAUSES					—	.43**	−.16	−.36**	−.05
6. C-DENSITY						—	−.05	−.21*	.02
7. GRAM T-UNIT							—	.21*	−.20*
8. SPELL								—	−.30*
9. CONVEN									—

*p < .05; **p < .01.

Source: Puranik, C. S., Lombardino, L. J., & Altmann, L. J. P. (2008). Assessing the microstructure of written language using a retelling paradigm. *American Journal of Speech-Language Pathology, 17*(2), 107–120.

After the factor analysis in the writing study had been performed, the researchers reported that the nine variables could be represented well by three factors, which they labeled *productivity, complexity,* and *accuracy.* To defend their claim that these three new variables, or factors, represented a sensible and statistically defensible reduction of the nine original variables, the researchers computed the bivariate correlation between each of the nine original variables and each of the three factors. These correlations appear in Excerpt 20.2.

EXCERPT 20.2 • *The End Point of a Factor Analysis: Derived Factors*

TABLE 6 *Intercorrelations among factors and writing measures*

Writing variable	Factor		
	Productivity	*Complexity*	*Accuracy*
1. TNW	.78	.30	−.22
2. IDEAS	.74	.16	−.18
3. T-UNIT	.77	−.13	−.11
4. MLT-UNIT	.19	.91	−.24
5. CLAUSES	.77	.39	−.17
6. C-DENSITY	.16	.95	−.14
7. GRAM T-UNIT	.23	.13	−.55
8. SPELL	−.19	−.24	.69
9. CONVEN	−.27	−.008	−.79

Source: Puranik, C. S., Lombardino, L. J., & Altmann, L. J. P. (2008). Assessing the microstructure of written language using a retelling paradigm. *American Journal of Speech-Language Pathology, 17*(2), 107–120.

If you look closely at the information in Excerpt 20.2, you will be able to understand why the three factors in this study were called Productivity, Complexity, and Accuracy. Within each row of correlations, one of the three *r*s was quite large compared with the other two correlations on the same row. This means that each of the nine original variables was found to be mainly associated with just one of the three derived factors. The writing variables that had their highest correlation with the first factor were variables that dealt with *how much* the students wrote (e.g., total number of words and total number of ideas), those that correlated most with the second factor assessed *how sophisticated* the students' written responses were (e.g., clause density), and those that correlated most with the third factor measured *how careful* the students were in following writing rules (e.g., correct spelling).

The Three Main Uses of Factor Analysis

Although factor analysis is used in applied research investigations for many reasons, it seems that most researchers utilize this statistical procedure in an effort to achieve one of three goals. These goals can be described as *data reduction, instrument development,* and *trait identification.* Before we look at the specific steps that researchers take when doing a factor analysis, let's briefly consider what they hope to achieve by using this statistical procedure.

In some studies, factor analysis is used to see if a small number of factors can adequate represent a larger number of original variables. This is precisely why the researchers associated with the writing study subjected their data to a factor analysis. Other researchers have the same goal, as illustrated in Excerpt 20.3.

Besides being used because of its data-reduction capability, factor analysis is frequently used in studies designed to develop, refine, or assess questionnaires, surveys, and tests. In some cases, factor analysis helps a researcher assign individual items to different subscales of the instrument (and to identify poorly performing items that should be discarded). In other cases, the goal is validation. Excerpt 20.4 represents the popular usage of factor analysis to assess construct validity.

Finally, factor analysis is often used to help identify underling personality constructs that do not manifest themselves totally in any test or questionnaire. These *latent traits,* as they are sometimes called, lie below the surface of typical measuring instruments.[3] In Excerpt 20.5, we see an example where the stated purpose in using factor analysis was to identify "underlying dimensions."

EXCERPTS 20.3–20.5 • *Different Reasons for Using Factor Analysis*

The goal of our factor analysis is to find the smallest number of interpretable factors that explain the correlations among the set of variables. . . . By the help of factor analysis, reducing a large amount of data to identify the common characteristics of a group of variables will facilitate to interpret the results of the research.

Source: Aydin, B., & Ceylan, A. (2009). The effect of spiritual leadership on organizational learning capacity. *African Journal of Business Management, 3*(5), 184–190.

Factor analysis was used to test construct validity and discriminant validity.

Source: Bechor, T., Neumann, S., Zviran, M., & Glezer, C. (2010). A contingency model for estimating success of strategic information systems planning. *Information & Management, 47*(1), 17–29.

Factor analysis was employed to determine the underlying dimensions of lifestyle variables.

Source: Hur, W. M., Kim, H. K., Park, J. K. (2010). Food- and situation-specific lifestyle segmentation of kitchen appliance market. *British Food Journal, 112*(3), 294–305.

Exploratory and Confirmatory Factor Analysis

Several different kinds of factor analysis exist because there are different ways to perform the computations. Some of these are discussed in the next major section when we focus our attention on the six main steps of a factor analysis. At this point,

---

[3]The measured variables of this kind of factor analysis are sometimes referred to as *indicators, observed variables,* or *manifest variables.*

I simply want to distinguish between the two overarching categories of factor analysis. One of these categories involves factor analyses that are exploratory in nature; in the other category, factor analyses are designed to be confirmatory.

In an **exploratory factor analysis (EFA),** the researcher has little or no idea as to number or nature of factors that will emerge from the analysis. With this kind of factory analysis, it is as if the researcher is about to visit a new art museum that has just been built and filled with artistic treasures. Once inside, the researcher discovers, for the first time, how to navigate through the different rooms, where the different installations are located, and what specific items of art have the strongest personal appeal. This metaphor is a bit exaggerated, of course, because a researcher performing an EFA knows several things about the study, such as the instruments used to measure each variable, the nature of the research participants, related research findings, and, perhaps, theory-based hypotheses. However, the hallmark of an EFA is the lack of any a priori constraints on the number or nature of factors that are identified.

Excerpt 20.6 comes from a study in which an EFA was conducted. In this excerpt, notice how the researchers provide an explanation as to why they chose to use this kind of analysis.

EXCERPT 20.6 • *Exploratory Factor Analysis (EFA)*

The objectives of this study were to assess the validity and reliability of a Setswana translation of the Perceived Wellness Survey (PWS) in the South African Police Service and to investigate differences in the perceived wellness of police members, based on gender, qualification, age and rank. . . . The PWS was translated into Setswana for purposes of this study. . . . [T]he current authors used exploratory factor analysis because the PWS is a recently developed measuring instrument, and no studies regarding its validity in South Africa were found. Exploratory factor analysis was therefore used to examine construct equivalence.

Source: Rothman, S., & Ekkerd, J. (2007). The validation of the Perceived Wellness Survey in the South African Police Service. *Journal of Industrial Psychology, 33*(3), 35–42.

In a **confirmatory factor analysis (CFA),** the researcher plays a more active role than in an exploratory factor analysis. Guided by theory or the findings from previous research, the researcher in this kind of analysis specifies, on the front end, the desired number of factors and how measured variables are related to those factors. Returning to our museum metaphor, the researcher here is like a person entering a museum that has been previously visited by one of the researcher's friends. That friend has described the museum's floor plan, where the best pieces of art are located, when the crowds will be gone, and how to join a docent-guided tour. Armed with this information, our museum visitor has various expectations as to what things are on display, where they are located, and how to navigate around the museum.

Those expectations may or may not be met, however, due to changes that may have been made since that friend visited the museum.

In addition to having control over the number of factors that emerge from the analysis, researchers who conduct CFAs can utilize statistical tests to assess null hypotheses. We see examples of this in our next section and in Chapter 21. For now, suffice it to say that many people consider this hypothesis-testing opportunity to be an asset of confirmatory factor analyses.

Excerpt 20.7 comes from a study in which a new, shorter version of an existing test was created, used on a trial basis, and evaluated. A CFA was used because the researchers had information about the longer instrument's factors. Notice that one of the researcher's goals in creating the BQ-13 was to "maintain the original's theoretical constructs."

EXCERPT 20.7 • *Confirmatory Factor Analysis (CFA)*

The 27-item Barriers Questionnaire [called the BQ-27] is a valid and reliable measure of patients' beliefs about pain and analgesics. . . . The specific aims of this [study] were to use statistical and analytical approaches to create a shortened BQ tool, maintain the original's theoretical constructs, and determine the new tool's validity, internal consistency, stability, and sensitivity. . . . The BQ-27 was reduced to 13 items using data from 259 patients [who recommended retention of certain items]. . . .Confirmatory factor analysis was used to evaluate the construct validity of the BQ-13 [showing that] the BQ-13 is valid, [with] seven items that measure barriers related to pain management and six items specifically related to analgesic side effects.

Source: Boyd-Seale, D., Wilkie, D. J., Kim, Y. O., Suarez, M. L., Lee, H., Molokie, R., et al. (2010). Pain barriers: Psychometrics of a 13-item questionnaire. *Nursing Research, 59*(2), 93–101.

Many studies are characterized by the joint use of both EFA and CFA. Such studies typically are two-stage investigations in which EFA is used in the initial portion of the investigation. Then, in the second part of the study, CFA is performed, with guidance provided by the findings in the first stage. Excerpt 20.8 illustrates this popular was of incorporating both kinds of factor analysis into the same study.

EXCERPT 20.8 • *Exploratory and Confirmatory Factor Analyses Used Together*

Research has identified a large number of strategies that people use to self-enhance or self-protect. We aimed for an empirical integration of these strategies. Two studies used [newly created] self-report items to assess all commonly recognized self-

(continued)

EXCERPT 20.8 • (*continued*)

enhancement or self-protection strategies. In Study 1 (N = 345), exploratory factor analysis identified 4 reliable factors. In Study 2 (N = 416), this model was validated using confirmatory factor analysis.

Source: Hepper, E. G., Gramzow, R. H., & Sedikides, C. (2010). Individual differences in self-enhancement and self-protection strategies: An integrative analysis. *Journal of Personality, 78*(2), 781–814.

In the next section, we consider the various steps researchers take when they perform an EFA. After looking at this first main kind of factor analysis, we then focus our attention on how a CFA differs from the kind of factor analysis that is exploratory.

Exploratory Factor Analysis

In this section, we consider the various steps researchers take when doing an EFA. Because this statistical technique is complex, with a variety of options available as the individual steps are taken, the description presented here is only an overview of the path taken by the typical researcher. Moreover, our focus here, as we look at the sequential process of doing a factor analysis, is on what applied researchers do and how they report what they discovered.[4]

With the preceding paragraph indicating what is (and is not) included in the coming presentation, let's now consider the various steps researchers take—or *should* take—when they perform an EFA.

Step 1: Checking the Suitability of Data for a Factor Analysis

The initial step in any factor analysis involves checking to see if certain important features of the data set meet basic requirements for this kind of statistical analysis. What is done here is a bit like conducting a preliminary check on assumptions, a step taken by researchers when using many of the test procedures considered in earlier chapters. However, applied researchers say that they have checked the *suitability* of their data to indicate that they have examined a specific subset of assumptions. Later, I point out that there are other assumptions connected to a factor analysis beside the ones focused on in this initial step of a factor analysis.

---

[4]Following the tradition of this book's earlier chapters, the presentation here does not focus on formulas or statistical theory. These two things—theory and data manipulation—are not unimportant, and the interested reader is encouraged to use other available resources to become knowledgeable of (1) the statistical rationale for each options when doing a factor analysis and (2) the way data are treated when a computer performs the analysis.

One basic feature of a study that can make a set of data unsuitable for factor analysis is sample size. Simply put, factor analysis does not work well with small samples. The issue here is not the absolute size of *n* but rather the size of the sample relative to the number of original variables. Many statistical authorities argue that the *n*-to-variables ratio should be at least 10 to 1, with even larger sample sizes than that encouraged by other authorities. Factor analysis can be (and frequently is) used with smaller-than-recommended sample sizes; however, empirical studies have shown that the identified factor structure is likely to be *in*accurate if the sample size is too small.

In addition to considering the sample size, researchers typically do three other things when checking to see if their data are suitable for a factor analysis. They inspect the determinant of the correlation matrix, they compute the Kaiser–Meyer–Olkin (KMO) measure of sampling adequacy, and they apply Bartlett's chi-square test of sphericity. The data are judged to be factorable if the determinant is greater than .00001, if the **KMO measure of sampling adequacy** is greater than .60, and if **Bartlett's test of sphericity** is significant.[5]

Excerpt 20.9 illustrates how researchers typically report their efforts to assess the suitability of their data for a factor analysis. This excerpt deserves your close attention, because it indicates the desirable outcomes when checks are made on the suitability of data for EFA.

EXCERPT 20.9 • *Assessing the Suitability of Data*

The determinant of the correlation matrix as an indictor of multicollinearity was .007, which is substantially greater than the minimum recommended value of .00001. . . . The Kaiser–Meyer–Olkin (KMO) coefficient of sampling adequacy fell within the excellent range at .84 [and further analysis] showed that all KMO values for individual variables were greater than .700. . . . The Bartlett's Test of Sphericity, which examines whether the matrix is different from the identity matrix, was significant ($\chi^2(171) = 1921.83, p < .0001$), indicating that the matrix does not resemble an identity matrix, further supporting the existence of factors within the data.

Source: Randolph, K. A., & Radey, M. (2009). Measuring parenting practices among parents of elementary school-age youth. *Research on Social Work Practice,* in press.

The first sentence in Excerpt 20.9 contains the word **multicollinearity.** The unfavorable condition of multicollinearity exists if two or more of the original variables are too highly correlated with each other. However, the correlations between variables should not all be very small, otherwise no factors will be identified. Thus, some researchers hope that all bivariate correlations will be moderate in size (and they inspect the correlation matrix to see if all correlations fall between .30 and .80).

[5]The null hypothesis of Bartlett's test states that all population correlations among the original variables are equal to zero.

If preliminary checks indicate that data are unsuitable for a factor analysis, researchers typically do one of two things. One option is to delete problematic variables, recheck for suitability, and then proceed with a factor analysis if no new red flags appear. The other option is to decide not to do a factor analysis. Give researchers credit when they do either of these things after detecting a problem when assessing the factorability of their data.

Step 2: Selecting a Method of Factor Extraction

Presuming that no initial problems with the data have been identified, the second step in a factor analysis involves selecting a statistical method that extracts the factors from the correlation matrix. In choosing a method of **factor extraction,** the researcher functions like a spelunker who must choose what kind of flashlight to take into a dark cave. Just as there are many different kind of flashlights—head-mounted versus handheld, spotlight versus lantern-like, and so on—that could help illuminate the cave's interior, there are several different kinds of techniques that can be used to help researchers "see" the factors that exist among, or perhaps beneath, the full set of bivariate correlations.

The factor extraction methods used most often by applied researchers are called **maximum likelihood, principal components analysis,** and **principal axis factoring.**[6] The first of these works best if the assumptions underlying factor analysis are met; otherwise, either of the other two extraction procedures is recommended. In Excerpts 20.10, 20.11, and 20.12, we see examples of these three extraction methods being used in applied studies.

EXCERPTS 20.10–20.12 • *Factor Extraction*

An exploratory Factor Analysis (FA) was employed in order to determine which of the thirty items formed related subsets. . . . The maximum likelihood extraction was used to find the factor solution which would best fit the observed correlations.

Source: Baytiyeh, H., & Pfaffman, J. (2010). Volunteers in Wikipedia: Why the community caters. *Educational Technology & Society, 13*(2), 128–140.

Principal Components Analysis (PCA) method was [used] to extract the underlying factors affecting the attitudes of librarians toward information technology.

Source: Ramzan, M., & Singh, D. (2010). Factors affecting librarians' attitudes toward IT application in libraries. *The Electronic Library, 28*(2), 334–344.

(continued)

---

[6]Some of the other extraction procedures are called ordinary least squares, alpha factoring, generalized least squares, and image factoring.

EXCERPTS 20.10–20.12 • (*continued*)

Applying factor analysis to symptom cluster research is an effective statistical approach to identifying common factors that explain the correlation between symptoms and finding the communality that "binds" 2 or more symptoms together into a common concept. Accordingly, exploratory factor analysis with principal axis factoring was used to identify symptom clusters.

Source: Ryu, E., Kim, K., Cho, M. S., Kwon, I. G., Kim, H. S., & Fu, M. R. (2010). Symptom clusters and quality of life in Korean patients with hepatocellular carcinoma. *Cancer Nursing, 33*(1), 3–10.

Step 3: Deciding How to Rotate Factors

The factors extracted from the correlation matrix are better if they are rotated. This is because rotated factors help the researcher achieve the goals of simplification and clarity when trying to understand and describe the factor structure of the data. **Factor rotation** helps achieve these goals by reducing the number of factors required to explain any given amount of variance contained in the original variables. For example, with unrotated factors, it might take four factors to account for 90 percent of the variability in the data set; after rotation, just two factors might be able to achieve this same level of explanatory power.

Returning to our cave exploration analogy, deciding on factor rotation is like choosing a place to stand in the cave when holding the light in an effort to see the cave's interior. Certain vantage points are better than others in illuminating geological formations, fossils, and drawings on the walls. Similarly, factor rotation makes a difference in what the researcher "sees" when trying to identify connections among the original variables. Neither the position of the light in the cave nor the selected method of rotation in factor analysis changes the reality of what is being looked at, however. It is only the interpretation of that reality that is affected by where the light is or which rotational method is used.

There are two main categories of factor rotation and several specific methods within each category. One category involves an **orthogonal rotation** of factors, thereby keeping the factors statistically independent (i.e., uncorrelated). The most popular rotation used in this category is called *varimax* rotation; alternative frequently used rotational procedures go by the names *quartimax* and *equamax*. The other category involves an **oblique rotation** of factors, thereby allowing the factors to be correlated. Specific methods of rotation in this category are called *direct oblimin, quartimin,* and *promax.*

In Excerpts 20.13 and 20.14, we see examples of how researchers typically indicate whether they used an orthogonal rotation or an oblique rotation of extracted factors. Excerpt 20.15 illustrates the joint use of both rotational procedures within the same investigation. (As illustrated by Excerpt 20.15, it is not uncommon for

EXCERPTS 20.13–20.15 • *Factor Rotation*

The construct validity of the measures was tested using exploratory factor analysis (principal component analysis and varimax orthogonal rotation method).

Source: Filiz, Z. (2010). Service quality of travel agents in Turkey. *Quality & Quantity, 44*(4), 793–805.

We performed exploratory factor analysis using the principal component analysis extraction method with an oblique (promax) rotation on the 17-item instrument.

Source: Krauss, S. E., Hamid, J. A., & Ismail, I. A. (2010). Exploring Trait and Task Self-awareness in the Context of Leadership Development among Undergraduate Students from Malaysia. *Leadership, 6*(1), 3–19.

Because we had reason to suspect that the four factors would be correlated, we ran the exploratory factor analysis using oblique rotation as well as varimax (orthogonal) rotation. The results were virtually identical with regard to which items loaded on the four factors.

Source: Darling, R. B., & Heckert, D. A. (2010). Orientations toward disability: Differences over the lifecourse. *International Journal of Disability, Development and Education, 57*(2), 131–143.

similar results to be obtained when both orthogonal and oblique rotations are computed for the same data.)

Step 4: Determining the Number of Useful Factors

Both before and after rotation, there are as many factors as there are variables. However, these factors vary in how useful they are in accounting for variability among the original variables. Simply put, certain factors are better than others. Therefore, the researcher's next task when conducting an EFA is to determine which of the factors should be retained and which should be discarded.

Four different strategies are used by applied researchers in their effort to identify useful factors and thereby "separate the wheat from the chaff." These strategies involve using Kaiser's criterion, examining a scree plot, conducting a parallel analysis, or applying the 5 percent rule. These strategies have the same goal, but they try to achieve that common goal in different ways.

After factor extraction and rotation has taken place, a single numerical value called an **eigenvalue** is associated with each factor.[7] In any given analysis, the sum of the eigenvalues is equal to the number of variables. Therefore, a factor analysis of four variables might produce four factors with eigenvalues equal to 2.0, 1.5, 0.4, and 0.1. Any factor's eigenvalue is large to the extent that it accounts for variance in the

---

[7]Eigenvalues are sometimes called *characteristic roots*.

full set of original variables. Thus, big eigenvalues imply useful factors, whereas small eigenvalues imply superfluous factors. When researchers apply **Kaiser's criterion,** factors are retained only if they have eigenvalues larger than 1.0. Excerpt 20.16 shows an example of this eigenvalue-greater-than-1 strategy being used.

EXCERPT 20.16 • *Kaiser's Criterion*

We used principal components factor analysis in order to collapse the motivational and attitudinal data into indices representing their underlying constructs. The indices were created separately for students and employee samples. Following the Kaiser criterion, we retained factors with an eigenvalue greater than 1.

Source: Komarek, T., Lupi, F., Kaplowitz, M., & Thorp, L. (2010). Institutional management of greenhouse gas emissions: How much does "green" reputation matter? Paper presented at the Agricultural & Applied Economics Association AAEA, CAES, & WAEA Joint Annual Meeting, Denver, Colorado.

The second and third strategies for identifying useful factors are similar in that they both use a graph. An example of this kind of visual aid is contained in Excerpt 20.17. As you can see, the vertical axis on the left represents the numerical

EXCERPT 20.17 • *Scree plot and Parallel Analysis*

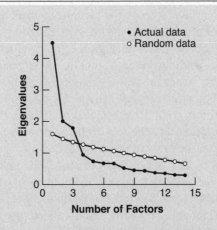

FIGURE 1 *Plot of eigenvalues from principal components analysis of the actual questionnaire version of the Picture Story Exercise data versus the 95th percentile of eigenvalues derived from a parallel principal components analysis of random data.*

Source: Schultheiss, O. C., Yankova, D., Dirlikov, B., & Schad, D. J. (2009). Are implicit and explicit motive measures statistically independent? A fair and balanced test using the Picture Story Exercise and a cue- and response-matched questionnaire measure. *Journal of Personality Assessment, 91*(1), 72–81.

values of eigenvalues, whereas the horizontal axis represents the factors that have been identified. In this particular study, there are 14 points along the baseline because there were 14 personality variables involved in the factor analysis.

The second strategy for identifying useful factor involves plotting a single set of dots in the graph, with adjacent dots connected by line segments. The resulting jagged line is called a **scree plot,** and it appears in Excerpt 20.17 as the set of darkened dots labeled *actual data*. The left end of this jagged line is positioned high off the baseline because the first factor's eigenvalue was equal to 4.48. Moving from that starting point to the right, the line connecting the solid dots drops sharply, because the eigenvalue for the second identified factor turned out equal to 2.00. The line then drops slightly and then sharply again (because the eigenvalues for the third and fourth factors were 1.79 and 0.95, respectively). Thereafter, the line connecting the darkened dots seems to decrease in a more constant and gradual fashion.[8]

When only a scree plot is in view (as would be the case if just the darkened dots had been plotted in Excerpt 20.17), the researcher decides how many useful factors there are by looking to see where the jagged line begins to have a constant and slow rate of decline. This is called the graph's *elbow.* In Excerpt 20.17, the elbow seems to occur at the dot associated with the fourth factor's eigenvalue. Factors associated with dots to the left of the elbow are deemed to be useful.

The third strategy for identifying useful factors involves something called a **parallel analysis.** With this approach, two jagged lines are put into the graph. One of these jagged lines is simply the scree-plot line. The second line comes from a factor analysis of a new set of random numbers set up to have the same sample size and number of variables as the data actually being analyzed in the study. Because this second, "parallel" factor analysis uses random data, any correlations between variables exist because of chance, and thus the eigenvalues from the second analysis should all be low.

Once the two jagged lines have been created, the place where the lines cross allows the researcher to quickly determine how many useful factors there are. Such factors are those associated with the dots in the jagged line from the first analysis positioned to the left of the point where the two lines intersect. In Excerpt 20.17, this rule leads to three useful factors.

The last of the four strategies for identifying useful factors is the *5 percent rule*. This strategy says to maintain any factor so long as its eigenvalue represents no less that 5 percent of the total eigenvalue "pie." Earlier, I used an example of a factor analysis of four variables that produced factors with eigenvalues equal to 2.0, 1.5, .4, and .1. The last of these eigenvalues is smaller than 5 percent of all four added together; accordingly, its factor would be deemed too weak to be retained.

Some researchers use just one of the four different strategies—Kaiser's criterion, a scree plot, parallel analysis, or the 5 percent rule—to decide which factors are important enough to be retained. Many researchers use a combination of these strategies. If applied to the data in Excerpt 20.17, three of the strategies—Kaiser's

---

[8]The line connecting the dots will never veer upwards because the factors along the baseline are ordered based on the sizes of their eigenvalues.

criterion, a scree plot, and parallel analysis—would lead to a retention of just the first three factors; however, the strategy based on the 5 percent rule would also retain the fourth factor. This example illustrates the value in using more than one approach to deciding how many useful factors underlie the data.

Step 5: Determining the Variable Make-Up of Each Factor

After identifying a subset of factors based on an examination of eigenvalues, the researcher's next task is to figure out which of the original variables go with each of these strong factors. Researchers do this by computing and then carefully examining a set of **factor loadings.** As illustrated in Excerpt 20.18, factor loadings are almost always displayed in a chart that has columns corresponding to factors and rows corresponding to variables. (Excerpt 20.18 has an extra column on the right and two extra rows at the bottom; we'll consider this extra information shortly.)

EXCERPT 20.18 • *Factor Loadings*

TABLE 2 *The Factor Loading Matrix of the Consideration of Future Consequences Scale: Principal Components Analysis With Varimax Rotation*

Item	Factor			Communality
	A1	*A2*	*A3*	
4[a]	**.713**	−.019	.174	.539
3[a]	**.662**	.189	.299	.563
11[a]	**.647**	.418	−.005	.594
5[a]	**.575**	−.175	.153	.385
9[a]	**.523**	.461	−.136	.505
10[a]	**.488**	.402	−.004	.400
12[a]	**.478**	.124	−.100	.254
8	−.027	**.692**	.186	.515
7	.118	**.636**	.059	.422
6	.102	**.601**	.149	.393
2	.023	.059	**.839**	.707
1	.163	.295	**.701**	.605
Eigenvalue	2.490	1.966	1.427	5.877
% of variance	20.8	16.4	11.9	49.0

Note: Loadings in bold are values greater than .40 and are retained for that factor. Underlined values indicate a multiple loading on two factors. Eigenvalues and percentage of variance are after rotation.

[a]Reverse-scored items.

Source: Rappange, D. R., Brouwer, W. B. F., & van Exel, N. J. A. (2009). Back to the Consideration of Future Consequences Scale: Time to reconsider? *Journal of Social Psychology, 149*(5), 562–584.

In the study from which Excerpt 20.18 was taken, the researchers factor analyzed a personality inventory designed to measure people's inclination to consider the future consequences of their current behavior. This inventory contained 12 items—e.g., "I am willing to do something I find not much fun if it pays off later on"—with a five-option format available so each respondent could indicate how accurately each item described him or her. Each of these items was treated as a variable in the factor analysis, with the original data coming from 2,006 young adolescents (ages 11–15) in the Netherlands.

Each row in Excerpt 20.18 shows the factor loadings of a given item on each of the three factors that had been identified. Each factor loading is simply the product–moment correlation between the adolescents' scores on the item and their scores on the factor. For example, the information for item 2 (the 11th item in the list) shows that this item had a correlation of .023 with factor A1, a correlation of .059 with factor A2, and a correlation of .839 with factor A3. Using the language of factor analysis, we can interpret these correlations by saying that this item loaded heavily on factor A3, but hardly at all on factors A1 or A2. This simply means that item 2 belongs in, or is a component of, the third factor.

To help us see which items belong to each of the factors, the researchers who prepared the table in Excerpt 20.18 bolded any factor loading that was greater than .40. To help us even more, the researchers juggled the order of the items such that those items loading most on factor A1 are listed first, followed by items that loaded most on Factors A2 and A3. Finally, the researchers underlined three factor loadings that were higher than .40 but not the highest for these items; this calls our attention to the fact that those three items have nontrivial loadings on two of the factors. (Cases like these of *double loading* are not uncommon.)

In Excerpt 20.18, the final column contains the **communality** for each variable (i.e., each item in the personality inventory). These communalities, which can range in size from 0 to 1, indicate the relative value of a variable to the factor structure being created. A large communality indicates that a variable is useful; conversely, small communalities may prompt the researcher to drop the variable from future analyses (or, if the variables are items in a test being developed, eliminate the item from the test).[9]

The final thing to note about Excerpt 20.18 is beneath each column of factor loadings. In that spot in the table, the researchers presented each factor's eigenvalue. We considered eigenvalues previously, but now we can see how they are computed. If we first square the individual factor loadings for factor A1 and then add up those resulting squares, we get 2.490. Likewise, the sum of the squared factor loadings for factor A2 (or factor A3) is equal to 1.966 (1.427). The percentage beneath each of these values was computed by dividing each eigenvalue by 12, the number of variables (i.e., items). Collectively, the three factors account for about 49 percent of the variance in the original data.

---

[9]A communality is computed as the sum of the squared factor loadings. In Excerpt 20.18, for example, the first communality equals $(.713)^2 + (-.019)^2 + (.174)^2$.

Step 6: Naming Factors

The factors identified in a factor analysis initially have no names. They are just factors, as exemplified by the table we saw in Excerpt 20.18. In an EFA, the names of factors are established in a post hoc manner by the researcher looking to see which variables load most heavily on each factor. An illustration of this being done appears in Excerpt 20.19.

EXCERPT 20.19 • *Naming Factors in an Exploratory Factor Analysis*

In Study 1, exploratory factor analyses revealed four factors. . . . Items with salient loadings on Factor I corresponded to parental behavior which shamed the child, made the child feel unsafe and which placed developmentally inappropriate demands on the child; items with salient loadings on Factor II described insensitive parental behavior; items with loadings on Factor III described physically violent parental behavior which would terrorize the child and items with loadings on Factor IV described rejecting/isolating parental behavior. Thus, the four factors were named as "Inappropriate Expectations," "Insensitivity," "Terrorizing," and "Rejecting/Isolating," respectively.

Source: Uslu, R. I., Kapci, E. G., Yildirim, R., & Oney, E. (2010). Sociodemographic characteristics of Turkish parents in relation to their recognition of emotional maltreatment. *Child Abuse & Neglect, 34*(5), 345–353.

In this kind of factor analysis, the naming of factors is a subjective process. The names given to the factors by one person who looks at a table of factor loadings might differ from the names created by someone else looking at the same table. Accordingly, it is prudent in this kind of factor analysis to have different people independently establish factor names. If the resulting sets of factor names are similar, we can be more confident that the factors have been properly labeled. Unfortunately, few researchers take the time to do this.

Confirmatory Factor Analysis

Confirmatory factor analysis (CFA) is similar to exploratory factor analysis (EFA) in several respects. CFA involves measured variables, an initial matrix of intercorrelations among those variables, factors, and factors loadings. In addition, this form of factor analysis is often used for the purpose of assessing the construct validity of measuring instruments designed to assess personality traits.

Although EFA and CFA are alike in certain respects, they differ in two main ways. First, there is the issue of hypotheses. Whereas a researcher can conduct an EFA without having any hunches as to how the analysis will turn out, CFA demands that hypotheses guide the way the data are analyzed. Second, there is the issue of *model fit*. With EFA, the results (in terms of the number of factors, factor names, and factor composition) present the researcher with what we might call a *model,* but there is no feature of the analytic procedure that allows that model to be tested. In contrast, CFA allows the researcher to statistically test the fit of any proposed model(s). These two differences make it legitimate to think that EFA is a theory-*generating* activity whereas CFA is a theory-*testing* endeavor.

With the preceding two paragraphs providing a brief introduction, let's now consider the various steps researchers usually take when they perform a bare-bones CFA. As noted in Chapter 21, a CFA can be, and often is, performed within the context of a complex statistical procedure called *structural equation modeling (SEM)*. Our current consideration of confirmatory factor analysis provides an overview of this kind of factor analysis when it is not conducted within the context of a full-blown SEM analysis.

Step 1: Articulation of Hypotheses and the Model

Based on previous research or existing theory, a researcher begins a CFA by specifying the factors that presumably exist beneath the variables that as yet have not been measured. The researcher's hypothesis may be that a single factor exists, or multiple factors may be hypothesized. Regardless of how many factors are predicted to exist, each one is often referred to as a **latent variable.**

After the researcher has specified the latent variable(s) thought to exist, he or she next makes plans to collect data on each of the observed variables. In many studies, these **observed variables** are individual items in a questionnaire or survey; however, such variables can be anything the researcher thinks is a good proxy for, or representative of, the underlying hypothesized latent variable(s). The measured variables are often referred to as the study's **indicators** or its **manifest variables.**

Once the latent and observed variables have been specified, the researcher's next task involves creating a model that predicts which of the observed variables will load on each of the hypothesized factors. In an EFA, this pairing of observed variables to factors comes after data have been collected, and it is influenced largely by the computed correlations among the observed variables. In a CFA, this pairing is done within the model that is articulated prior to any data collection.

Excerpt 20.20 illustrates how latent variables are pre-specified in a CFA study. In this investigation, a personality inventory dealing with eating disorders was administered to 203 females attending college in Canada. Because this instrument initially had been validated with a clinical sample of individuals known to be suffering from eating disorders, the researchers wanted to see if the two factors of the questionnaire held up in a nonclinical sample. (The observed variables referred to in this excerpt were the 23 items in the EDRSQ, the Eating Disorder Recovery Self-Efficacy Questionnaire).

EXCERPT 20.20 • *The A Priori Nature of the Factors in a CFA*

The original version of the EDRSQ was created to account for two important aspects of eating disorder recovery self-efficacy: NESE and BISE. The two constructs were thought to be different, yet associated. . . . The NESE scale measures the confidence in the ability to adopt healthy eating habits without becoming anxious and without restricting, bingeing, purging and exercising excessively. The BISE scale measures the confidence in the ability to maintain a realistic body image that is not overshadowed by an unhealthy drive for thinness. . . . In order to test this model, a CFA was conducted. The model was composed of two latent variables (NESE and BISE) with 14 and nine observed variables, respectively.

Source: Couture, S., Lecours, S., Beaulieu-Pelletier, G., Philippe, F. L., & Strychar, I. (2010). French adaptation of the eating disorder recovery self-efficacy questionnaire (EDRSQ): Psychometric properties and conceptual overview. *European Eating Disorders Review, 18*(3), 234–243.

Step 2: Collection of Data

After the researcher's model has been specified, the next phase of a CFA involves the collection of data on the observed variables. This form of statistical analysis does not work well with data from small samples, so the researcher must gather a sufficient amount of data. Various rules of thumb for establishing the minimum sample size exist. Some rules stipulate that *n* must be at or above some absolute level, other rules say that *n* should be a multiple of the number of observed variables, and a few rules dictate that the minimum *n* should be determined by a power analysis or a consideration of desired accuracy in parameter estimation.

When conducting CFAs, researchers should indicate what rule of thumb they used to determine that the sample size was large enough to proceed with the analysis. However, simply citing a rule of thumb is not good enough. Researchers should indicate the reason(s) why they decided to use one particular rule of thumb rather than other available ones.

Step 3: Concern for Missing Data

After the data become available, the researcher's next task is to screen the data for missing observations. When questionnaires or personality inventories are administered to large groups of individuals, certain people may purposefully or inadvertently fail to answers one or more questions. These omissions create a problem for CFA. Consequently, either a hypothetical score must be inserted for each piece of missing data, or the individuals who supplied only partial data must be expunged from the sample.

Excerpt 20.21 illustrates a concern for missing data. This excerpt is instructive because it shows both options for dealing with missing data: the imputing of hypothetical scores and the deletion of individuals from the sample.

EXCERPT 20.21 • *Concern for Missing Data*

Confirmatory factor analysis (CFA; using AMOS 5.0) was conducted on the 14 items ($N = 331$), to test the fit of the data to the two-factor model. . . . If a participant was missing data for just one of the 14 items, the mean of the scale was inserted. For two or more, the participant was deleted from the analyses. This is because CFA is problematic when there are missing data.

Source: Fox, C. L., Elder, T., Gater, J., & Johnson, E. (2010). The association between adolescents' beliefs in a just world and their attitudes to victims of bullying. *British Journal of Educational Psychology, 80*(2), 183–198.

Step 4: Assessment of Model Fit

The results of a CFA permit the researcher to evaluate how well the model fits the data. This is not done by evaluating the factors individually, as is done in an EFA. Instead, the entire set of relationships among the observed (i.e., manifest) variables and the hypothesized latent factors is examined in a holistic fashion. This goal is accomplished via the simultaneous inspection of several **goodness-of-fit indices.**

In Excerpt 20.22, we see how several goodness-of-fit indices were used to see if a four-factor model fit the data. Included in this excerpt are the widely used criteria

EXCERPT 20.22 • *Assessing Model Fit*

The adequacy of the four-factor structure of the MASC-T was examined using confirmatory factor analysis (CFA) for the community sample of 12,536 children and adolescents. We also examined the adequacy of the four-factor structure of the MASC-T in six subgroups of participants grouped according to gender (boys and girls) and age (8–11, 12–15, and 16–19 years). Four indices including the root mean square error of approximation (RMSEA), standardized root mean square residual (SRMR), non-normed fit index (NNFI), and comparative fit index (CFI) were [computed]. RMSEA values larger than .10 are typically considered poor and values smaller than .10 are acceptable. An SRMR $<$ .08, NNFI $>$ .90, and *CFI* $>$ .90 indicate a good fit. . . . The adequacy of the four-factor structure of the MASC-T (Physical Symptoms, Harm Avoidance, Social Anxiety, and Separation/Panic) was examined using CFA. . . . The values of all indices [RMSEA = .050; SRMR = .053; NNFI = .944; CFI = .948] met our goodness-of-fit standards and were invariant across gender and age. The results indicated that the four-factor model was well fitted for Taiwanese children and adolescents.

Source: Yen, C.-F., Yang, P., Wu, Y.-Y., Hsu, F.-C., & Cheng, C.-P. (2010). Factor structure, reliability and validity of the Taiwanese version of the Multidimensional Anxiety Scale for Children. *Child Psychiatry and Human Development, 41*(3), 342–352.

for determining whether the fit is good or poor. In this case, all four goodness-of-fit indices suggested that the four-factor model fit the data. Note that this support for the model came from two fit indices (RMSEA and SRMR) being small and the other two indices (NNFI and CFI) being large. This is because certain goodness-of-fit indices measure the degree to which the model and the data depart from each other, whereas certain other goodness-of-fit indices assess the degree to which the model and the data coincide. Taken together, a variety of these fit indices provides a better assessment of model fit than any one looked at by itself.

Step 5: Inspecting Factor Loadings and Correlations Among Factors

In addition to assessing the fit of a model, researchers also examine the factor loadings and correlations among factors to see if the a priori hypotheses are supported by the results. This is done to establish *convergent* and *discriminant* *validity*. Convergent validity is shown when the factor loadings for a given latent variable's indicator variables are high. The other kind of validity, discriminant validity, is shown via small factor loadings for other indicator variables on that latent variable.

   Excerpt 20.23 illustrates this step of inspecting factor loadings to assess convergent and discriminant validity. This excerpt also shows how researchers examine the correlations among the factors.

EXCERPT 20.23 • *Examining Factor Loadings and Correlations*

The Multidimensional Perfectionism Cognitions Inventory (MPCI) is a promising new instrument developed at the University of Tokyo, Japan, for assessing the frequency of cognitions associated with dispositional perfectionism along three dimensions [i.e., factors]: personal standards, pursuit of perfection, and concern over mistakes. . . . The aim of the present research was to provide a first investigation of the reliability and validity of the English version of the MPCI, the MPCI-E, using a large English-speaking sample. First, a confirmatory factor analysis was conducted to investigate the factorial validity with the aim to replicate the original measure's three-factor oblique structure. . . . All items displayed substantial loadings on their target factor. Moreover, as was expected, all three factors showed substantial intercorrelations [because] the factor representing pursuit of perfection showed high correlations with the factor representing personal standards and the factor representing concern over mistakes, whereas the latter two factors showed a more modest correlation.

Source: Stoeber, J., Kobori, O., & Tanno, Y. (2010). The Multidimensional Perfectionism Cognitions Inventory–English (MPCI-E): Reliability, validity, and relationships with positive and negative affect. *Journal of Personality Assessment, 92*(1), 16–25.

Step 6a: Model Modification

It often is the case that the fit of the initial model is inadequate. This situation might be caused by an observed factor loading equally on more than one factor, or it could be caused by the model itself having too many (or not enough) factors. When the model fit turns out to be less than ideal, the researcher usually modifies the model in some fashion and then repeats steps 1, 3, 4, and 5. One frequently seen type of modification in the model is the elimination of one or more problematic observed variables. Excerpt 20.24 illustrates this kind of model modification.

EXCERPT 20.24 • *Model Modification*

In evaluating construct validity, we ran a confirmatory factor analysis (CFA) on our three constructs (instruction, curriculum, and ecology). CFA [substantiated] the three constructs with model trimming used to eliminate any indicators that did not contribute significantly to each construct. In an attempt to achieve the most parsimonious model, [we] trimmed the 26 total indicators to 14 (five for instruction, four for curriculum, and five for ecology).

Source: Marshall, J. C., Smart, J., & Horton, R. M. (2010). The design and validation of EQUIP: An instrument to assess inquiry-based instruction. *International Journal of Science and Mathematics Education, 8*(2), 299–321.

On occasion, the hypothesized model is deemed to be fully inadequate. If this happens, most researchers return to their data and perform an exploratory factor analysis. Then, based on the findings of that investigation, the researcher will likely conduct a new confirmatory factor analysis, with the model this time based on the discoveries of the exploratory factor analysis.

Step 6b: Comparison of Different Models

One option available in CFA is the comparison of different models. Usually, this kind of comparison contrasts two or more models having different numbers of latent variables. Excerpt 20.25 comes from a study in which this kind of comparison was the driving force behind the investigation.

EXCERPT 20.25 • *An Option in CFA: Model Comparison*

This study assesses the Shodan survey as an instrument for measuring an individual's or a team's adherence to the extreme programming (XP) methodology.

(continued)

EXCERPT 20.25 • (*continued*)

> Specifically, we hypothesize that the adherence to the XP methodology is not a unidimensional construct as presented by the Shodan survey but a multidimensional one reflecting dimensions that are theoretically grounded in the XP literature. Using data from software engineers in the University of Sheffield's Software Engineering Observatory, two different models were thus tested and compared using confirmatory factor analysis: a uni-dimensional model and a four-dimensional model.
>
> *Source:* Michaelides, G., Thomson, C., Wood, S. (2010). Measuring fidelity to extreme programming: A psychometric approach. *Empirical Software Engineering, 15*(6), 599–617.

When different models are compared, the fit statistics for each model are usually examined. In addition, a chi-square test can be applied to see if there is a statistically significant difference between the two models. This test first involves the computation of separate model fit chi-square values for each model, and then it determines whether the difference between the two χ^2 values is beyond chance expectation. These two kinds of comparisons were made in the study from which Excerpt 20.25 was taken, and the results appear in Excerpt 20.26. In this excerpt, note that the better of the two models was indicated by the model chi-square value that was *smaller,* because each separate χ^2 value measured the degree to which the model *failed* to fit the mode.

EXCERPT 20.26 • *Statistical Comparison of Different Models*

> Finally, a comparison between the uni-dimensional and the four-dimensional models was performed using the chi-square differences between the two models. . . . If the two models are significantly different, the model with the smaller chi-square is significantly better than the first. The comparison indicated that the four-dimensional model was significantly better than the uni-dimensional model ($\Delta \chi^2 = 73.2, \Delta df = 6, p < .001$). Comparisons of all other fit indices corroborate this result. These results provide strong support for the first hypothesis (*H1*) indicating that the four-dimensional model can better explain the variability in the Shodan questionnaire items, whilst at the same time providing evidence about its construct validity.
>
> *Source:* Michaelides, G., Thomson, C., & Wood, S. (2010). Measuring fidelity to extreme programming: A psychometric approach. *Empirical Software Engineering, 15*(6), 599–617.

Three Final Comments

Before concluding our consideration of factor analysis, three final comments are necessary. These deal with the concepts of sampling, factor names, and replication. In a sense, each of the following end-of-chapter comments is a warning.

As is the case with most other statistical procedures, the nature of the results that emanate from a factor analysis are tied to the nature of the sample(s) from which data are gathered. Some researchers take pride in discussing the size of their samples without acknowledging that their samples are quite homogeneous, not random, or limited by low response rates. Do not be lulled into thinking that the results of a factor analysis can be trusted simply because the data have come from hundreds (or thousands) of individuals. Look carefully at the source of a study's input data before thinking that the findings generalize to every person on the face of the Earth!

My second warning about factor analysis concerns the names of factors. Regardless of whether a research report is based on an EFA or a CFA, the names of the factors were decided on in a subjective manner. Give EFA researchers credit when they point out that two or more people independently arrived at the same names for the factors that popped out of their analyses, and give CFA researchers bonus points when they discuss the convergent and discriminant validity of the factors in the models they propose as having the best fit. Keep in mind that a factor name can be poorly named even though, in an EFA, its eigenvalue is large, or, in a CFA, it is part of a model that has optimal fit indices.

Finally, remember that replication of statistical findings is persuasive. This is just as true for factor analysis as it is for any other statistical procedure. Given credit to those researchers who report having conducted a formal cross-validation of their study or who present evidence of the invariance of findings across different kinds of samples.

Review Terms

Bartlett's test of sphericity	KMO measure of sampling adequacy
Communality	Latent variable
Confirmatory factor analysis (CFA)	Manifest variable
Eigenvalue	Maximum likelihood
Exploratory factor analysis (EFA)	Multicollinearity
Factor	Oblique rotation
Factor extraction	Observed variable
Factor loading	Orthogonal rotation
Factor rotation	Parallel analysis
Goodness-of-fit indices	Principal axis factoring
Indicator	Principal components analysis
Kaiser's criterion	Scree plot

The Best Items in the Companion Website

1. An interactive online quiz (with immediate feedback provided) covering Chapter 20.
2. Five misconceptions about factor analysis.
3. One of the best passages from Chapter 20: "The Basic Logic of Factor Analysis."

To access chapter outlines, practice tests, weblinks, and flashcards, visit the companion website at hrrp://www.ReadingStats.com.

Review Questions and Answers begin on page 531.

C H A P T E R
21

Structural Equation Modeling[1]

Reading a research report that deals with structural equation modeling (SEM) can be intimidating, even for experienced consumers of research reports. This statistical procedure's name appears ominous, as are its synonyms: *covariance structure analysis, covariance structure modeling,* and *analysis of covariance structures.* Results of SEM studies are presented not just with tables, but with seemingly complex diagrams containing geometric shapes and arrowed lines. Researchers refer to certain variables as being *endogenous* and others as being *exogenous.* Measurement error comes into play, but in a different way than we saw in Chapter 4. Despite these and other features of SEM, the goals of and logic behind this statistical procedure, as well as the way researchers present their SEM findings, can be understood by anyone who is willing to move slowly through this chapter.

Structural equation modeling is like factor analysis in that the focus is on variables that lie beneath the surface of characteristics that can be observed and directly measured. However, factor analysis and SEM are different in terms of their goals. Factor analysis is used in an effort to identify unseen variables (i.e., factors) or to confirm the existence of such variables. In contrast, SEM is used in an effort to illuminate any causal connections that may exist among a study's unseen variables.[2] Once an SEM data analysis is completed, these causal links between variables frequently are displayed pictorially by means of a path diagram.

Factor analysis and SEM also differ in terms of the need for theory and hypotheses. One type of factor analysis—the exploratory kind—can be used without

---

[1]The initial draft of this chapter was written by Shelley Esquivel and Amy Beavers. Additional assistance with this chapter was provided by Hongwei Yang.

[2]The words *cause* and *causal* appear frequently throughout this chapter so as to make the language of this discussion of SEM coincide with that contained in typical research reports. It should be noted, however, that SEM applied to cross-sectional data is unable, by itself, to address cause-and-effect questions.

theory being a guiding force as the analysis is conducted. SEM, however, is not exploratory at all; instead, it is used to assess the researcher's conceptualization, or model, of causal relationships dictated by theoretically-based hypotheses. Moreover, SEM can accommodate the simultaneous testing and comparison of multiple models. The analytic capacity of SEM extends beyond factor analysis or other statistical analyses, thus necessitating the additional terminology referred to earlier.

Although the goals, terminology, and reporting procedures of SEM may be different from other statistical procedures, many of SEM's underlying components and concepts are quite similar to those considered earlier in this book. At its most basic level, structural equation modeling is the simultaneous analysis of relationships among variables using regression and correlation techniques, and it provides sets of weights which can be thought of as relationship strength indications. SEM then goes on to compare the actual relationships among variables to the theorized relationships hypothesized by the researcher. In other words, it evaluates how well the theoretical model explains the collected data. The difference between the actual data and the theoretical predictions provides an index of model worthiness via the notion of "fit," introduced in Chapter 20.

SEM Diagrams, Terms, and Concepts

To understand SEM, one must become familiar with a set of terms and concepts, as well as with the way models are depicted graphically. In this section, we consider these building blocks of SEM. We begin by looking at the diagram of a hypothetical SEM model, and then we use this model to help pin down the meaning of various terms and concepts.

A Hypothetical SEM Model

The diagram in Figure 21.1 is for a hypothetical structural equation model concerned with members of high school swim teams. In this fictitious study, our imaginary researcher is trying to see if some theory-based predictions can explain why certain swimmers seem to have more competitive drive than their peers. The researcher's two main thoughts are simple and straightforward: (1) both nature (i.e., genetics) and nurture (i.e., experiences) have a causal impact on a young athlete's swimming ability, and (2) swimming ability is positively related to, and has causal influence on, the degree to which the athlete has competitive drive.

A quick glance at Figure 21.1 shows that it is composed of geometric shapes that take the form of ovals, rectangles, and circles. The ovals are positioned near the center because they represent the most important elements of the model. Also note that every oval, rectangle, and circle has at least one arrow leading to it or away from it. In the coming paragraphs, I refer to this diagram frequently as we deconstruct this SEM model by considering what each shape and line represents.

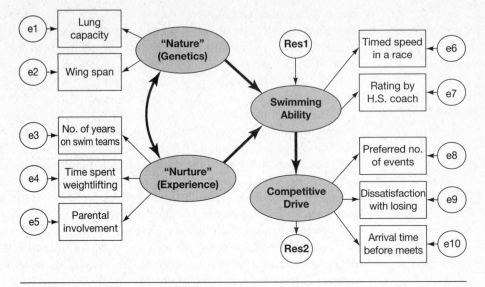

FIGURE 21.1 *Diagram for Hypothetical Swimming SEM Study*

Latent and Observed Variables

The ovals and rectangles in an SEM diagram represent variables. Two kinds of shapes are needed to represent a study's variables because there are two main types of variables in an SEM study. These are the study's latent and observed variables.

The **latent variables** of an SEM study are traits or constructs that cannot be observed or measured directly. Examples of such variables would be your level of test anxiety, the amount of intelligence you possess, your trustworthiness, your fear of spiders, how much you enjoy hiking in the woods, and so on. As indicated earlier, researchers who conduct SEM studies are mainly interested in latent variables—identifying them, determining the relationships among them, and (especially) illuminating any case(s) where one latent variable has a causal impact on some other latent variable.

In Figure 21.1, there are four latent variables: Nature, Nurture, Swimming Ability, and Competitive Drive. The thick arrows that connect these latent variables indicate the researcher's theoretical predictions that (1) Nature and Nurture are related, (2) Nature and Nurture each have a causal impact on Swimming Ability, and (3) Swimming Ability has a causal impact on Competitive Drive.

SEM models involve observed variables that can be measured. Examples of such variables include your pulse rate, the score you earn on a test, your age, the number of siblings you have, how frequently you blink during a 60-second inter-val, and how many calories you typically consume in a day. In SEM studies, each of these observed variables is technically referred to as an **observed variable,** as a **manifest variable,** or as an **indicator variable.** Such variables can be a single item in a test or questionnaire, a subscale score based on a collection of items, a full-scale score based on all of the instrument's items, or a measurement of something (e.g., school attendance) not based on a test or questionnaire.

In Figure 21.1, the 10 observed variables of our hypothetical swimming study are located in rectangles near the left and right sides of the diagram. The five measures on the left were selected because they are the visible manifestations of the study's Nature and Nurture latent variables. The physiological measures of lung capacity and wing span (i.e., arm length) get at a person's genetic predisposition to be a good swimmer. Life experiences are indexed by the number of years a person has been on swim teams, the number hours per month a swimmer spends lifting weights, and the level of parental support as indicated by a parent's response to the single question, "On a scale from 0 to 10, how much do you *show* your child that you support him/her being on the high school swim team?"

Five additional observed variables are located near the right side of Figure 21.1. The first two of these—timed speed and coach's rating—were chosen because they are reasonable ways to measure a person's swimming ability. The final triad of observed variables represents the researcher's three-pronged way of getting at a person's competitive drive. They involve measuring the number of swimming events an individual would like enter, his or her rating given to the item "not winning makes me mad" in an attitude inventory, and the amount of time he or she shows up at the swim meet venue *prior to* the coach's stated time-to-arrive.

As shown in Figure 21.1, arrows extend outward from each latent variable to a subset of the observed variables. These directional arrows are meant to convey the notion that any given latent variable is likely to have an impact on the observed variables that serve as measurable proxies for the latent variable. For example, the degree to which you enjoy hiking in the woods (a latent variable) ought to influence the way you answer questions about your hobbies and how you spend your leisure time (two observed variables). Or, the degree to which you are scared of spiders (a latent variable) likely influences how close you get to an open jar containing a tarantula and how you respond to a question that asks you to rate how much spiders make you uncomfortable.[3]

In the diagram we have been considering, there are 10 manifest variables but only five latent variables. In the typical SEM study, researchers use multiple manifest variables in an effort to get at each latent variable. This is a prudent practice for two reasons. First, multiple measures usually provide a more reliable assessment of a latent variable. Second, multiple measures tend to tap into different features of a construct, thereby increasing validity.

It is worth noting that in the typical SEM study, researchers concentrate first on the hypothesized latent variables, not the observed variables. This is the opposite

---

[3]Manifest variables that are influenced by latent variables are called *reflective* measures. In Figure 21.1 (and in the typical SEM study), all the observed variables are reflective. However, the direction of influence can go the other way, with the observed variable having an impact on a latent variable. In this latter situation, the manifest variable is called a *formative* measure.

of what happens in an exploratory factor analysis where factors emerge and are named *after* the researcher has first decided on the study's measured variables. In most SEM studies, researchers *start* by using their knowledge of theory or previous research to carefully define the latent variables; then, they select measurable variables that are thought to reflect well the unseen, latent factors. If successful in choosing the right observed variables, the researcher can peer through the rectangle variables in order to see what's going on with the oval variables.

Excerpt 21.1 illustrates the way researchers always clarify the observed and latent variables that comprise their structural equation models. This excerpt is especially useful, because it shows how the indicator variables for a latent trait can be created via a process called **parceling.** This simple process involves clustering the items of a lengthy measuring instrument into small subsets of items, with a summary score on each subset of items serving as the observed variable. Parceling usually involves an odd–even split of the instrument's items or a random subdivision of the full set of items.[4]

EXCERPT 21.1 • *Indicator and Latent Variables*

We used structural equation modeling for our main analyses. For the TPV [targeted peer victimization], we used the three first items in the TPV scale as the indicator variables for the "Relational TPV" latent variable, and the last three last TPV items were used as the indicator variables for the "Physical TPV" latent variable. For indicators of the Depressive Symptoms latent variable, we divided the CDI into two parts, based on odd and even numbered items, with each part containing 13 items.

Source: Van Tran, C. (2010). Longitudinal relations between targeted peer victimization and depression. Unpublished doctoral dissertation, Vanderbilt University, Nashville, Tennessee, p. 14.

The Measurement Model and the Structural Model

A complete structural equation model is made up of two sub-models: the measurement model and the structural model. Both models are concerned mainly with a study's latent variables. These models differ, however, in what is specified in each model.

The **measurement model** does two things. First, it posits the existence of the study's latent variables. Second, it asserts that these latent variables manifest themselves in the study's observed variables. As applied to the diagram in Figure 21.1, the measurement model says, in essence, that the four hypothesized constructs— Nature, Nurture, Swimming Ability, and Competitive Drive—really exist, and that the study's 10 measured variables capture those four latent variables.

---

[4]Manifest variables based on parceling often have three advantages compared to the use of individual items as indicators: higher reliability, less skewness or kurtosis, and fewer parameters in the model.

The **structural model** extends beyond the measurement model and posits the way in which the latent variables are related to each other. This model stipulates which pairs of latent variables have a causal connection, which pairs of variables are related but not in a causal manner, and which pairs of variables are independent of each other. Such relationships are depicted in Figure 21.1 via the thick lines with arrows. For example, one of those lines displays graphically the researcher's hypothesis that Swimming Ability has a causal impact on Competitive Drive.

In Excerpt 21.2, we see reference made to the measurement and structural models of a recent study. This investigation was focused on the diets of rural adults, with a main concern for the amount of dietary fat consumed by these individuals.

EXCERPT 21.2 • *Measurement and Structural Models*

This study tested a multi-group structural equation model to explore differences in the relative influence of individual, social, and physical environment factors on dietary fat intake amongst adults aged 40–70 years. . . . First, a measurement model using confirmatory factor analysis (CFA) was used to confirm the relationship between the latent variables (i.e., theoretical constructs) and their indicator (observed) variables. . . . Second, the structural model was tested to estimate the strength of the relationships between latent variables.

Source: Hermstad, A. K., Swan, D. W., Kegler, M. C., Barnette, J. K., & Glanz, K. (2010). Individual and environmental correlates of dietary fat intake in rural communities: A structural equation model analysis. *Social Science & Medicine, 71*(1), 93–101.

Exogenous and Endogenous Variables

Latent variables are often described as being either exogenous or endogenous. These two SEM terms are analogous to the notions of independent and dependent variables in a regression setting. **Exogenous variables** are considered to be the independent variables in an SEM model, because they are thought to be on the front end of a causal relationship. **Endogenous variables,** however, are considered to be the model's dependent variables, because they are believed to be affected by one or more of the other latent variables.

In Figure 21.1, the Nature and Nurture variables are exogenous, for the diagram depicts each of these latent variables as having an impact on a different latent variable, Swimming Ability. The variable Competitive Drive, however, is endogenous, because it is on the receiving end of another latent variable's influence. Swimming Ability has a dual role in the model, because it functions both as a dependent variable and as an independent variable.[5]

---

[5]A variable that serves both as a dependent and as an independent variable is classified as an *endogenous variable.* This convention is probably because variables with dual roles function first as dependent variables.

Excerpt 21.3 comes from a recent study concerned with the determinants of health-related quality of life among patients with liver disease. In the SEM model investigated in this study, there was one exogenous variable (self-efficacy), one endogenous variable (health-related quality of life, HRQoL), and several other variables—such as depression—that had a dual role in the model because they were first influenced by self-efficacy and then had an impact on HRQoL.

EXCERPT 21.3 • *Exogenous and Endogenous Variables*

The exogenous variable was Self-Efficacy, and the endogenous variable was HRQoL [health-related quality of life]. The other variables in the model were both endogenous and exogenous.

Source: Gutteling, J. J., Duivenvoorden, H. J., Busschbach, J. J. V., de Man, R. A., & Darlington, A. E. (2010). Psychological determinants of health-related quality of life in patients with chronic liver disease. *Psychosomatics, 51*(2), 157–165.

Correlations, Causal Paths, and Independence

In a structural model, the researcher hypothesizes relationships among some or all of the variables. Two variables, of course, can be related without one having a causal impact on the other. Such relationships are referred to simply as *correlations* and are represented via curved lines, with arrows on each end, connecting the two variables. In Figure 21.1, the relationship between Nature and Nurture is hypothesized to be simply correlational, not causal.

As we have seen, causal relationships in a structural model are represented by directional lines. Each of these lines represents a **causal path.** In Figure 21.1, three causal paths are hypothesized: one leading from Nature to Swimming Ability, a second leading from Nurture to Swimming Ability, and the final one leading from Swimming Ability to Competitive Drive.[6]

It is possible, of course, for a researcher to hypothesize that two latent variables are not connected in either a correlational or a causal manner. When this is the case, no line in the diagram connects the two variables. In Figure 21.1, therefore, the hypothesized model specifies no correlation between either Nature or Nurture and Competitive Drive.

Excerpt 21.4 comes from a study concerned with hotel workers and their satisfaction with flexible schedules. Three hypothesized causal paths are described in this excerpt, two of which were supported by the study's data.

---

[6]When a causal path exists between an exogenous and an endogenous variable, this type of causal path is called a *Gamma (Γ) path.* When a causal path exists between two endogenous variables, this type of causal path is called a *Beta (β) path.* Thus, the path between Nature and Swimming Ability in Figure 21.1 is a gamma path, whereas the path between Swimming Ability and Competitive Drive is a beta path.

EXCERPT 21.4 • *Causal Paths*

Hypothesis 1 predicted that a hotel worker's level of emotional intelligence has a positive influence on his/her satisfaction with schedule flexibility. Our findings [revealed that] H1 is supported. Similarly, H2 posited that overall job satisfaction of hotel workers has a positive influence on their satisfaction with schedule flexibility [and] H2 is supported. Our model then hypothesized a positive causal path between satisfaction with schedule flexibility and organizational citizenship behavior (H3) [but] H3 is not supported.

Source: Lee, G., Magnini, V. P. & Kim, B. (2010). Employee satisfaction with schedule flexibility: Psychological antecedents and consequences within the workplace. *International Journal of Hospitality Management, 30*(1), 22–30.

Mediator Variables

As indicated earlier, some latent variables function simultaneously as exogenous (independent) variables and endogenous (dependent) variables. This type of variable is often referred to as a **mediator variable.**[7] Mediation is the result of an exogenous variable having an influence that passes through an endogenous variable (the mediator) before affecting another endogenous variable. In the diagram for the swimming example we have been considering, Swimming Ability is a mediator variable, because the effects of Nature and Nurture pass through it as they influence Competitive Drive.

   Excerpt 21.5 comes from a study in which structural equation modeling was used to investigate a mediation hypothesis. As indicated in this excerpt, the study's

EXCERPT 21.5 • *Mediator Variable*

Early Head Start children may be more likely to exhibit difficulties with social–emotional functioning due to the high-risk environments in which they live. However, positive parenting may serve as a protective factor against the influence of risk on children's outcomes. The current study examines the effects of contextual and proximal risks on children's social–emotional outcomes and whether these effects are mediated by maternal sensitivity. . . . A theoretically derived structural equation model was tested to examine the direct paths from family risk variables to children's social–emotional functioning and the indirect paths by way of the mediator variable, maternal sensitivity. Support was found for a model that identified maternal sensitivity as a mediator of the relationship between parenting stress and children's social–emotional functioning.

Source: Whittaker, J. E. W., Harden, B. J., See, H. M., Meisch, A. D., & Westbrook, T. R. (2010). Family risks and protective factors: Pathways to Early Head Start toddlers' social–emotional functioning. *Early Childhood Research Quarterly, 26*(1), 74–86.

---

[7]This kind of mediation is the same as what we considered in Chapter 16 when we looked at multiple regression.

results showed that maternal sensitivity (a latent variable having three indicators: warmth, acceptance, and responsiveness) was a mediator between parenting stress (a latent variable with three indicators gleaned from a 38-item questionnaire) and the children's social–emotional functioning (a latent variable measured via rating scales that asked parents to evaluate their child's social competence, problem behaviors, and aggression).

A mediator variable can take one of two forms: it can be a *complete* (or *full*) *mediator* or a *partial mediator*. This distinction is most easily understood by examining the diagrams for two different SEM studies.

In Figure 21.1 (from our swimming study), both Nature and Nurture are hypothesized to have an influence on Competitive Drive that flows totally through the Swimming Ability variable. This makes the mediator variable, Swimming Ability, a complete mediator. It is possible, however, that only part of an independent variable's impact passes through the mediator variable, with the remaining portion of the exogenous variable's influence going directly to the dependent variable. When this is the case, mediation is partial.

Excerpt 21.6 comes from a study in which the researchers evaluated a structural equation model containing partial mediation. In this excerpt's diagram, notice

EXCERPT 21.6 • *Partial Mediation*

[T]his study chose structural equation modeling to test hypotheses [wherein] job rotation and role stress among nurses are independent variables, and organizational commitment is a dependent variable, while job satisfaction is the mediating variable. . . . The overall research framework is shown as Figure 1.

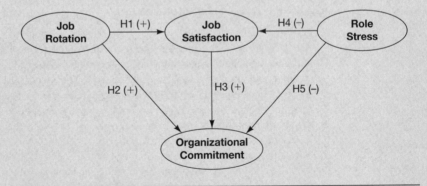

FIGURE 1   Conceptual framework for the relationship among job rotation, job satisfaction, organizational commitment, and role stress.

Source: Ho, W.-H., Chang C. S., Shih, Y. L., & Liang, R. D. (2009). Effects of job rotation and role stress among nurses on job satisfaction and organizational commitment. *BMC Health Services Report, 9*(8), 1–10.

that each of the study's independent variables (Job Rotation and Role Stress) is hypothesized to have an influence that flows through the mediator variable (Job Satisfaction) as well as an influence that moves directly into the dependent variable (Organizational Commitment). Note also that each of the five arrows is labeled to show whether the hypothesized causal influence is positive or negative.

Measurement Error and Residual Error

A strength of structural equation modeling over some other statistical techniques is that it is able to account for and remove the effects of two types of error: measurement error and residual error. By including these errors in the model, SEM has a better chance of revealing the true pattern of relationships among the observed and latent variables.

Measurement error is created whenever data are gathered by means of a measuring instrument or process that has less than perfect reliability. Because perfectly reliable measuring instruments are few and far between, measurement error is almost always embedded in the scores created when researchers try to tap into observed variables. Accounting for and removing measurement error in SEM is analogous to the technique we saw in Chapter 9 of correcting a correlation coefficient for attenuation.

Whereas measurement error is connected to observed variables, **residual error** is associated with latent variables, but only latent variables that function as dependent variables. This kind of error can be thought of as what is left after the relevant independent (exogenous) variable(s) explain, or account for, as much variability in the dependent (endogenous) variable as it or they can. Accounting for and removing residual error in SEM is analogous to the technique we saw in Chapter 15 of using a covariate to decreases error variance in an ANCOVA study.

The pictorial representation of measurement and residual errors is handled in a variety of ways. Often, as in Figure 21.1, these errors show up in SEM diagrams as small circles, each with an arrow pointing to its relevant indicator variable or endogenous latent variable. When this is done, the measurement errors usually are abbreviated as e1, e2, and so on, whereas the residual errors are labeled Res1, Res2, and so on. Sometimes, researchers put the error abbreviations into the diagram without enclosing them in small circles. Occasionally, researchers choose not to include any reference at all to these two kinds of errors in their SEM diagrams; this is done, most likely, to make the diagrams less cluttered and easier to understand.

Assessing SEM Hypotheses: A Brief Overview

The creation of a logical, theoretically-based model involving observed and latent variables is only the first phase of an SEM study. A second and equally important task involves assessing the quality of the model and, if necessary, revising the model in small or large ways. The assessment-of-the-model phase of the investigation typically involves doing four things: checking on important assumptions, evaluating the measurement model, determining if the structural model fits the data, and modifying the model to achieve a better fit.

As we now turn our attention to the statistical techniques used in a typical SEM investigation, we presume that the model being evaluated truly is theory based. We also presume that the study's population(s) have been defined appropriately, that large enough samples have been extracted properly from the relevant populations, that no problems of low response rate or refusal to participate exist, and (most important of all) that the topic under investigation is relevant to the work of practitioners or other researchers. As is the case with many other statistical procedures, SEM modeling cannot magically undo the fatal limitations caused by biased samples or irrelevant research questions.

Steps in Assessing Model Worth

In the following paragraphs, we follow the statistical route researchers usually take when they conduct an SEM study. Not every researcher does these things, for alternative procedures are used in certain applied studies. Be that as it may, the following steps represent the analytic strategy most likely to be included in a journal article, convention paper, doctoral dissertation, or other report.

Checking on Assumptions

The statistical procedures used to assess the measurement and structural models of an SEM study are based on important assumptions. Accordingly, the conscientious researcher begins his or her data analysis by checking to see if these prerequisite conditions seem tenable. The researcher's hope, of course, is that no violations of the assumptions crop up.

One important assumption is concerned with normality. This assumption says that the scores on the study's set of continuous variables in the relevant population(s) form a multivariate normal distribution. Such a distribution is analogous to a three-dimensional bell-shaped object that maintains that shape no matter which side-view angle is used to look at the bell. This assumption is important, because multivariate nonnormality can disrupt tests of **model fit** and bias the parameter estimates used to assess path strength in the structural model.

In Excerpt 21.7, we see a case where a team of researchers attended to the multivariate normality assumption in their SEM study. This assumption was evaluated by means of **Mardia's test.** This test procedure yields two numbers that are often reported: the critical ratio (CR) and the normalized estimate. The normalized estimate is like a z-score, and it is the index of the two reported numbers examined to see if the normality assumption seems tenable.

EXCERPT 21.7 • *Concern for Normality and Outliers*

The assumption of multivariate normality which underlies the use of statistical modeling was assessed using Mardia's coefficient of multivariate kurtosis. The analysis revealed that the data did not violate the multivariate normality assumption (multivariate kurtosis normalized estimate = 1.63; CR = 1.29); univariate kurtosis values ranged from −0.159 to 0.771 (mean = 0.46, SD = 0.31), and univariate skewness values ranged from −0.078 to 0.723 (mean = 0.39; SD = 0.34). The presence of potential outliers was tested according to Mahalanobis distance [measure]. According to that criterion, no case was considered an outlier.

Source: Nuevo, N., Wetherell, J. L., Montorio, I., Ruiz, M. A., & Cabrera, I. (2009). Knowledge about aging and worry in older adults: testing the mediating role of intolerance of uncertainty. *Aging & Mental Health, 13*(1), 135–141.

There are two nice features of Excerpt 21.7. First, a test of multivariate normality was conducted even though the univariate indices of skewness and kurtosis suggested that each individual variable was approximately normal in shape. The multivariate test was needed because univariate normality is a necessary but not sufficient condition for multivariate normality. Second, notice that a check for multivariate outliers was conducted. This was done via the **Mahalanobis distance measure,** one of the most popular techniques for identifying abnormal scores.

When researchers discover that the assumption of multivariate normality seems untenable, they typically do one of several things. One option is to use a robust statistical procedure for testing model fit. The Satorra–Bentler scaled chi-square test, often reported as S-Bχ^2, is an example of such a procedure. Another robust procedure, illustrated in Excerpt 21.8, is the mean- and variance-adjusted weighted least square parameter estimator.

EXCERPT 21.8 • *Use of a Robust Approach to SEM*

As in other multivariate techniques, maximum likelihood (ML) method is a generally used estimating procedure in SEM. A basic assumption of this ML-estimator is the multivariate normal distribution of all continuous endogenous variables in the model [but] in reality this assumption is not always fulfilled. Our models include several not-normally distributed variables and, moreover, our final outcome variable car use is categorical. An alternative estimator in such circumstances is a mean- and variance-adjusted weighted least square parameter estimator (WLSMV) which we used instead. WLSMV is a robust estimator yielding robust standard errors that does not require extensive computations and does not require enormously large sample sizes.

Source: Van Acker, V., & Witlox, F. (2010). Car ownership as a mediating variable in car travel behaviour research using a structural equation modelling approach to identify its dual relationship. *Journal of Transport Geography, 18*(1), 65–74.

Two other options exist when the multivariate normality assumption is untenable. One of these is to use bootstrapping to create a sampling distribution tailor-made for the study's data. The other option is to delete or replace one or more of the original variables, with the normality assumption retested to see if the corrective action achieved its goal.

Although multivariate normality is important, it is not the only important assumption. Because SEM relies heavily on the technique of multiple regression, it should not be surprising that two additional assumptions ought to be checked out in the preliminary phase of the data analysis. One of these assumptions says that there is a linear relationship between variables; the other says that multicollinearity does not exist. In Excerpt 21.9, we see a case where both of these assumptions were evaluated.

EXCERPT 21.9 • *Concern for Multicollinearity and Linearity*

Multicollinearity was examined with a variation inflation factor (VIF) value, and linearity of such relationships was inspected indirectly, with the expected scatter plot of the residual indicating homoscedasticity, the normal P-P plot of the regression standardized residual showing the normality of the residual, and R^2. The results of the regression analyses showed that the latent variables were not multicollinear ($1.253 \leq$ VIF ≤ 2.316). The results also indicated that each relationship satisfied homoscedasticity and normality of the residual, and had a considerably high R^2 value ($0.202 \leq R^2 \leq 0.625$). Based on the results, it can be argued that the assumption of linearity is met considerably in such relationships.

Source: Yoo, S. H. (2010). Exploring an integrated model of governmental agency evaluation utilization in Korea: Focusing on executive agency evaluation. *International Review of Public Administration, 15*(1), 35–49.

Evaluating the Measurement Model

After important assumptions have been considered, the next order of business is an evaluation of the measurement model. The concern here is whether the study's hypothesized constructs (latent variables) are adequately caught by the study's observable measures. For obvious reasons, it is important to have confidence that the hypothesized latent variables exist and can be measured before trying to determine causal paths that may exist between such variables.

In an SEM study, the measurement model is usually evaluated by means of a confirmatory factor analysis (CFA). The statistical procedures for conducting a CFA were outlined in Chapter 20, and that discussion is not repeated here. However, it may be helpful if we do two things now that were not done previously. First, we look at the diagram of a study's measurement model that contains results produced by the CFA. Second, we examine the more typical text-only description of a study's measurement model.

The first study we now consider was concerned with dental anxiety. To measure this kind of anxiety, the researchers had 783 individuals respond to the five items of a questionnaire called the Modified Dental Anxiety Scale (MDAS). The items deal with a person's emotional reaction to these five aspects of an upcoming appointment with a dentist: the visit itself, being in the waiting room, drilling, scaling, and receiving a local anesthetic injection. Each of these items used a 1 to 5 rating scale, extending from "no anxiety" to "extreme anxiety." These items were considered to be the manifest variables in the SEM study, and they were labeled mdas1, mdas2, and so on.

Excerpt 21.10 contains the diagram the researchers created after conducting a confirmatory factor analysis on their data. Based on a consideration of theories of anxiety and earlier studies focused on differentiating different kinds of anxiety, the researchers hypothesized two latent variables: anticipatory dental anxiety, on the one hand, and treatment dental anxiety, on the other. The researchers also hypothesized that two of the MDAS items—mdas1 and mdas2—could measure the anticipatory kind of dental anxiety whereas the other three items—mdas3, mdas4, and mdas5—could measure the treatment portion of dental anxiety.

EXCERPT 21.10 • *The Measurement Model in an SEM Study*

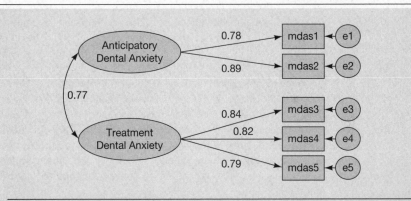

FIGURE 2 Measurement model for the two-factor version of the MDAS with standardized parameter estimates.

Source: Yuan, S., Freeman, R., Lahti, S., Lloyd-Williams, F., & Humphris, G. (2008). Some psychometric properties of the Chinese version of the Modified Dental Anxiety Scale with cross validation. *Health and Quality of Life Outcomes, 22*(6), 1–11.

Excerpt 21.10 presents some of the statistical findings from the CFA. The number next to each straight line in the diagram shows the factor loading of the manifest variable on the latent variable. Being standardized (and thus having an upper limit of 1.0), all five of these standardized parameter estimates are quite high. Collectively, these five numerical findings from the CFA provide support for the researchers' a priori belief that there are two kinds of dental anxiety and that the items of the MDAS instrument are adequate indicators of these two latent variables.[8]

Excerpt 21.11 contains a text-only discussion of an SEM's measurement model. This excerpt's content is instructive, because it shows how researchers focus on three psychometric components of the measurement model: reliability, convergent validity, and discriminant validity. This excerpt also reveals how these three aspects of the measurement model were assessed.

EXCERPT 21.11 • *Using CFA to Assess the Measurement Model*

A confirmatory factor analysis was conducted to test the measurement model. Six common model-fit measurements were used to assess the model's fit [indicating] most of the model-fit indices exceed the respective common acceptance levels suggested by previous research, demonstrating that the measurement model exhibited a good fit with the data collected. Therefore, we proceeded to evaluate the psychometric properties of the measurement model in terms of reliability, convergent validity, and discriminant validity.

Moreover, we evaluate reliability and convergent validity of the factors estimated by composite reliability and average variance extracted. . . . Composite reliability (CR) for all factors in our measurement model was above 0.70, which meant that more than one-half of the variances observed in the items were accounted for by their hypothesized factors. Thus, all factors in the measurement model had adequate reliability and convergent validity. To examine discriminate validity, we compared the shared variances between factors with the average variance extracted of the individual factors. The average variances extracted (AVE) were all above the recommended 0.50 level to be considered reliable. This showed that the shared variance between factors were lower than the average variance extracted of the individual factors, confirming discriminate validity. . . . In summary, the measurement model demonstrated adequate reliability, convergent validity, and discriminate validity.

Source: Ouyang, Y. (2010). A relationship between the financial consultants' service quality and customer trust after financial tsunami. *International Research Journal of Finance and Economics, 36,* 75–86.

If the measurement model does not seem reasonable, the researcher refrains from moving forward. Instead, he or she changes the way the original manifest variables are measured, changes the study's manifest variables, or changes the hypothesized latent variables.

---

[8]The researchers also gained support for their two-factor theory of dental anxiety by examining several fit statistics.

Evaluating the Structural Model

The second thing researchers focus on in the assessment phase of an SEM investigation is the structural model. Here, the network of causal paths and correlations hypothesized to exist among the latent variables is compared against the study's empirical evidence. The overarching research question addressed at this point in the SEM study is simple and straightforward: Does the model fit the data? If the answer to this question is negative, researchers usually pose a new question: Can the model be revised such that the modified structural model and empirical evidence are aligned?

The assessment of model fit is done in a holistic manner, with the full model considered to be a single entity. This assessment typically leads to one of three conclusions about how well the model coincides with the data: a good fit, a moderate fit, or a poor fit. In Excerpts 21.12 through 21.14, we see statements from three different studies that produced these three kinds of fit.

EXCERPTS 21.12–21.14 • *Good, Moderate, and Poor Model Fit*

The results of the estimation of the hypothesized causal structure model indicated that the model fit well the sample.

Source: Akbar, S., Som, A. P. M., Wadood, F., & Alzaidiyeen, N. J. (2010). Revitalization of service quality to gain customer satisfaction and loyalty. *International Journal of Business and Management, 5*(6), 113–122.

The initial [structural] model had a moderate fit to the data.

Source: Aspden, T., Ingledew, D. K., & Parkinson, J. A. (2010). Motives and health-related behaviours: An investigation of equipotentiality and equifinality. *Journal of Health Psychology, 15*(3), 7–79.

The [structural] model provided a poor fit to the data.

Source: Longbottom, J. L., Grove, J. R., & Dimmock, J. A. (2010). An examination of perfectionism traits and physical activity motivation. *Psychology of Sport and Exercise, 11*(6), 574–581.

The degree to which a structural model fits a study's data is not measured by a single number, as is the case when a single bivariate correlation coefficient is used to measure the degree of relationship between two sets of measured variables. Instead, researchers involved in SEM studies compute several fit indices, examine each of them individually (with guidelines as to what indicates good, moderate, or poor fit), and then merge together the individual findings to reach an overall assessment of model fit. This approach is necessary because the different fit indices assess different aspects of the model-versus-data match-up. The value in having multiple fit indices is analogous to having an opportunity to view a sculpture from different vantage points.

Most of the more popular fit indices have labels that are reduced to acronyms: TLI, RMSEA, GFI, CFI, NNFI, AGFI, and SRMR.[9] Another index is referred to as the *relative chi-square index;* this is simply the computed value of chi square divided by its degrees of freedom (i.e., χ^2/df). Rules of thumb have been proposed for evaluating the numerical values for each fit index, and it is helpful when a researcher cites in a research report the criteria that were used to evaluate the fit indices computed and examined in an SEM study. Excerpt 21.15 illustrates this good practice of citing the fit criteria along with the computed fit indices, thus permitting readers of the research report to see why a model is described as having a good (or poor) fit with the data.

EXCERPT 21.15 • *Model Fit Criteria*

The adequacy of the model fit was ascertained using Chi-square test (χ^2), Goodness-of-Fit Index (GFI), Comparative Fit Index (CFI), Root-Mean-Square of Approximation (RMSEA), and Chi-square to Degrees of freedom ratio (χ^2/df). A properly fit model must have the following fit characteristics: RMSEA < 0.08; chi-square to degrees of freedom ratio < 3.5; GFI > 0.9, and CFI > 0.9.... The fit indices for the estimated model in Fig. 2 [not shown here] are $\chi^2/df = 2.523$, GFI = 0.97, CFI = 0.95, and RMSEA = 0.053. They all satisfy the criteria for a well fit model.

Source: Amah, O. E. (2010). Multi-dimensional leader member exchange and work attitude relationship: The role of reciprocity. *Asian Journal of Scientific Research, 3*(1), 39–50.

It is worth noting that many of the fit indices assess the degree to which the model and the data coincide, whereas other fit indices measure the degree to which the model and the data differ. In Excerpt 21.15, the GFI and CFI indices do the former, whereas RMSEA and χ^2/df do the latter. This is why the CFI and GFI signify a good fit when they are close to the maximum value of 1.0, whereas the other two fit indices turn out to be small when a good fit exists.

In addition to assessing model fit, researchers usually examine the standardized regression coefficient associated with each of the causal paths in the model to see if it is statistically significant. Moreover, if directional hypotheses have been articulated—indicating whether an exogenous variable's influence on an endogenous variable

---

[9]TLI, Tucker–Lewis Index; RMSEA, Root Mean Square Error of Approximation; GFI, Goodness-of-Fit Index; CFI, Comparative Fit Index; NNFI, Non-Normed Fit Index; AGFI, Adjusted Goodness-of-Fit Index; and SRMR, Standardized Root Mean Square Residual.

is positive or negative—those hypotheses can also be evaluated. To see an example of a model containing such hypotheses, take another look at Excerpt 21.6. That model contains five directional hypotheses concerning the causal paths. Each of those hypotheses was tested individually, and the results are contained in Excerpt 21.16.

EXCERPT 21.16 • *Testing Individual Path Coefficients*

The Linear Structural Relationship Model was employed to examine the relationships among nurses' job rotation, role stress, job satisfaction, and organizational commitment. Hypotheses 1 to 5 in this study were demonstrated to be significant [each with $p < .01$]. Nurses' job rotation had a positive influence on job satisfaction ($\gamma_{11} = 0.51$) and organizational commitment ($\gamma_{21} = 0.46$). Nurses' job satisfaction ($\beta_{21} = 0.63$) had a positive influence on organizational commitment. Nurses' role stress had a negative influence on job satisfaction ($\gamma_{12} = -0.52$) and organizational commitment ($\gamma_{22} = -0.79$).

Source: Ho, W. H., Chang C. S., Shih, Y. L., & Liang, R. D. (2009). Effects of job rotation and role stress among nurses on job satisfaction and organizational commitment. *BMC Health Services Report, 9*(8), 1–10.

Model Modification

In SEM studies, it is usually the case that a researcher's initial model constitutes a mediocre (or poor) fit to the data, even when the model has been developed carefully from theory or previous research. This is especially true when the model is complex, with a variety of model parameters. However, most researchers use SEM as a model-generating tool. Consequently, it is not at all uncommon to see a research report wherein there is a discussion of model 1, followed by a refinement of that initial model so as to create model 2, with a third (and even better) model created to replace model 2.

The process of model modification involves changing the model in some fashion. This occurs if initial latent or indicator variables are eliminated or new ones added, if the network of causal paths is changed, if exogenous variables are converted into endogenous variables (or vice versa), or if elements of the model—such as the measurement errors connected to two indicator variables—are considered to be correlated rather than independent (or vice versa). When any of these things occurs, the model changes. Using technical SEM lingo, such changes cause the initial model to be *respecified*.

When engaged in **model respecification,** the researcher has two goals. One goal is to have a revised model that fits the data better. The other goal is to have a modified model that is parsimonious. This latter goal can be achieved by reducing the number of model parameters—such as eliminating one or more causal paths—in the model, a procedure referred to as **model trimming.** In Excerpt 21.17, we see

EXCERPT 21.17 • *Trimming to Achieve Parsimony*

Structural modeling presents a set of relationships between exogenous and endogenous variables with causal effects. The initial model (MI) was revised three times. . . . Our initial model (MI) included all unidirectional paths relating latent constructs. In the next model (MT+X), we excluded direct paths between accountability measures and employee performance, tantamount to including an extra path (professional accountability → workload) not part of our theoretical model (MT). The final model (MT-X) excluded nonsignificant paths in MT. The final model (MT-X) shows acceptable fit: the chi-square value ($\chi^2 = 135.05, df = 125, p < .254$) is not significant and the relative chi-square ($\chi^2/df = 1.08 < 2$) is below the conservative rule-of-thumb criteria. All other practical indices are within the acceptable fit ranges ($NFI > .9, TLI > .9, CFI > .9, PMSEA < .05$). In addition, the final model has the highest parsimony ratio (.595) and parsimony comparative fit index (.592). Although the chi-square value has increased in the process of model trimming, the chi-square difference between the initial and final model is trivial (4.25), and values for the other fit indices remained virtually unchanged within acceptable ranges. This suggests that the final model represents the most parsimonious and best fit model overall.

Source: Kim, S. E., & Lee, J. W. (2010). Impact of competing accountability requirements on perceived work performance. *American Review of Public Administration, 40*(1), 100–118.

a case where model trimming brought about a more parsimonious model without causing a precipitous decrease in model fit.

Researchers who engage in model modification typically use one or both of two kinds of statistical information to help them decide how exactly to change the model. One kind of guidance comes from the standardized parameter estimate associated with each causal path in the model. If one of these estimates turns out to be nonsignificant (when tested against a null value of zero), the path associated with that parameter estimate can be dropped from the model. The revised model, containing fewer model parameters, is said to be a **nested model** of the previous model.[10] Reference to such a model appears in Excerpt 21.18.

EXCERPT 21.18 • *A Nested Model*

Several strategies were employed to evaluate the IMB model of ART adherence in relation to rates of self-reported adherence. The first involved evaluation of the full IMB model, depicted in Fig. 1 [not shown here], which was assessed in terms of inspection of standardized path estimates and with standard model fit indices (e.g., χ^2, CFI,

(continued)

[10]In such a case, the first model is usually called the *full model,* whereas the second model is usually called the *reduced* or *restricted model relative to the full model.*

EXCERPT 21.18 • (*continued*)

RMSEA). A second, nested model was analyzed to evaluate the mediation hypothesis where the full IMB model was compared to a restricted IMB model where the non-mediated paths from information and all motivation constructs were set to zero. The restricted model fit was evaluated with standard fit indices and a χ^2 difference test. . . . The fit indices generated [by the nested model] compared favorably to those generated by the full model [and] Chi-Square Difference ($df = 3$) = 2.59, $p < .46$. Thus, the mediated model was supported as providing a comparable fit to the sample data that is more parsimonious.

Source: Amico, K. R., Barta, W., Konkle-Parker, D. J., Fisher, J. D., Cornman, D. H., Shuper, P. A., et al. (2009). The Information–Motivation–Behavioral Skills Model of ART adherence in a deep south HIV+ clinic sample. *AIDS and Behavior, 13*(1), 66–75.

When engaged in model respecification, researchers can get help from the **modification indices** produced in the analysis of the initial model. One modification index is generated for any possible line that *was not* included when the initial model was diagrammed. The absence of such connecting lines—such as between a pair of variables or between errors associated with two indicator variables—caused the two elements at the opposite ends of the line to be uncorrelated in the SEM analysis. Each modification index shows how much the overall model χ^2 would decrease if this constraint on the model were to be removed.

Excerpt 21.19 illustrates how modification indices can help researchers when they modify their models. Note the phraseology: "a structural path was found." It was not found by searching aimlessly in the dark. Rather, the researchers had a light focused on it by means of a large modification index.

EXCERPT 21.19 • *Modification Indices*

Based on information provided by structural estimates and modification indices greater than five, a modified model was built. . . . The Safety dimension of product perceived personality was removed from the analysis, as it turned out not to play any role in the model; while a structural path was found from subjective norms to moral norms. Estimation of this modified model showed much better fit statistics, which reached minimum thresholds for acceptable model's fit ($\chi^2(11) = 26.308$, $p < 0.01$; $\chi^2/df = 2.392$; GFI = 0.965; AGFI = 0.912; CFI = 0.942; NFI = 0.908; RMSEA = 0.082). The χ^2 difference test also confirmed that the modified model performed better than the basic one ($\Delta\chi^2(5) = 39.355$, $p < 0.001$).

Source: Guido, G., Prete, M. I., Peluso, A. M., Maloumby-Baka, R. C., & Buffa, C. (2010). The role of ethics and product personality in the intention to purchase organic food products: A structural equation modeling approach. *International Review of Economics, 57*(1), 79–102.

Other Uses of SEM

Structural equation modeling can be used to address many types of questions. The examples used in previous sections illustrated the use of SEM to determine the fit of a single model (or a respecified model) for a single sample. In this section, we consider how SEM can be used to (1) determine whether an a prior model applies to multiple groups and (2) compare groups in terms of their means on latent variables.

Assessing Model Invariance

So far, we have discussed the use of structural equation modeling to answer questions about measurement models and structural models for a single sample. Often, however, researchers are interested in comparing two (or more) groups of people. For instance, a researcher might want to know if a model accurately depicts the relationships among variables for multiple groups of people who differ according to age, gender, ethnicity, or political affiliation. SEM can be used to answer these questions about whether measurement or structural models are equivalent (or *invariant*) between or among groups.

Researchers who investigate the **model invariance** of their results can perform one or more of three post hoc tests. These tests focus on different kinds of invariance: measurement invariance, configural invariance, and structural invariance. All three are important.

A model has measurement invariance if the indicator variables tap into the same latent variables for different groups of individuals. Suppose a survey that measures teacher satisfaction is administered to a sample of public school teachers. Would it be appropriate to compare the results from this survey to the results of the same survey administered to private school teachers? Before making such a comparison, a researcher should first determine whether the measuring instrument is operating in a similar manner for both groups of teachers.

Configural invariance exists if the network of causal paths and correlations is similar across different groups of individuals. Here, the focus is on the set of straight and curved arrows in the SEM diagram. If the set of arrows are identical across the different comparison groups, researchers can say that they have established configural invariance.

Structural invariance exists if the size of each parameter associated with an arrow remains stable, even if the model is recreated for a different group of individuals. Because configural invariance is concerned with the model's arrows whereas structural invariance focuses on the *strength* of the paths, it is possible to have configural invariance without structural invariance. There reverse, however, is not possible.

In Excerpt 20.20, we see a case in which all three kinds of invariance tests were applied. Notice that the results supported two kinds of invariance (measurement and configural) but not the third kind (structural).

EXCERPT 21.20 • *Invariance Tests*

We performed two-group invariance tests across the two independent survey samples to establish whether relationship perceptions vary between buyers and suppliers. . . . First, we performed configural invariance test [and found] the structure of the model is optimally represented with the pattern of paths and factor loadings specified. Second, we performed the measurement invariance test. [Results] showed that only five items were noninvariant across the two groups. In effect, buyers and suppliers generally interpret measurement items equivalently in most cases. Finally, we performed structural invariance testing. [Three] tests (LM Test, $\Delta\chi^2$ and ΔCFI) indicate that noninvariance is supported for 5 paths, and confirm that there are differences in relationship perceptions between the buyer and supplier samples.

Source: Nyaga, G. N., Whipple, J. M., & Lynch, D. F. (2010). Examining supply chain relationships: Do buyer and supplier perspectives on collaborative relationships differ? *Journal of Operations Management, 28*(2), 101–114.

Comparing Group Means on Latent Variables

Although SEM is primarily concerned with the analysis of relationships (i.e., variances and covariances), it also can be used to address questions about mean differences in latent variables. Once measurement invariance has been confirmed, it is reasonable to ask questions about group differences on the latent variable(s) of interest. A question about group mean differences of latent variables takes the form, "Is the mean of construct *x* for one group the same as the mean for that construct in another group?"

Excerpt 21.21 illustrates how groups can be compared on a latent variable. The two groups in this study came from Thailand and the United States. The construct

EXCERPT 21.21 • *Comparing Group Means on Latent Variables*

Latent mean comparison. With [measurement] invariance established, we estimated latent mean differences to determine if Thais had a higher level of mindfulness than Americans. To test for differences, the latent mean for the reference group (American) is fixed to zero and freely estimated in the other group [with] selection of one group as the reference group an arbitrary decision—the estimated differences between the groups are the same either way. . . . The latent mean MAAS difference between Americans (0.000) and Thais (0.003) was not statistically significant ($t = 0.038; p = ns$), indicating that Thais and Americans do not significantly differ in mindfulness as measured by the MAAS.

Source: Christopher, M. S., Charoensuk, S., Gilbert, B. D., Neary, T. J., & Pearce, K. L. (2009). Mindfulness in Thailand and the United States: A case of apples versus oranges? *Journal of Clinical Psychology, 65*(6), 590–612.

on which these two groups were compared was mindfulness. The results provided no support for the claim that either of these groups is superior to the other in terms of mean score on the mindfulness trait.

The Dental Anxiety SEM Study

Earlier in this chapter, we saw a diagram of the measurement model from a study dealing with dental anxiety. Now, after covering a lot of ground in terms of SEM concepts, I want to show you another diagram from the same study, along with a table that also appeared in the research report.

You should be able to look at the information contained in Excerpt 21.22 and understand what the authors were trying to communicate by this diagram. Be sure

EXCERPT 21.22 • *Diagram of Dental Anxiety SEM Model*

The hypothesized structural model was evaluated with the Chinese data as specified in Figure 2. . . . Of particular interest was the strength of the relationships between the anxiety latent factors (Negative Affectivity NA and Autonomic Anxiety AA) with the 2 dental anxiety latent factors (ADA and TDA).

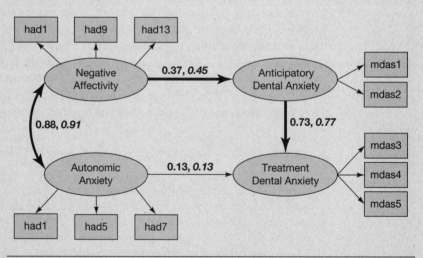

FIGURE 3   Structural model of the relation between negative affectivity, autonomic anxiety, and the two factor version of the MDAS including standardised coefficients: Beijing and Northwest England (italics). Wider arrows denote greater strength of the relationship. Error terms omitted to simplify diagram.

Source: Yuan, S., Freeman, R., Lahti, S., Lloyd-Williams, F., & Humphris, G. (2008). Some psychometric properties of the Chinese version of the Modified Dental Anxiety Scale with cross validation. *Health and Quality of Life Outcomes, 22*(6), 1–11.

to read the figure caption, as it explains why there are two numbers located next to each arrow and why certain of the arrows are thicker than others.

The diagram in Excerpt 21.22 gives no evidence that the researchers were involved in a **model-building** effort. However, the information contained in Excerpt 21.23 shows that three models were created, evaluated, and compared. Desiring their published model to be parsimonious, the researchers presented their initial model rather than either of the other models they developed.

EXCERPT 21.23 • *Consideration of Alternative Models*

Alternative models were also tested. Negative affectivity may influence not only ADA but also TDA. Hence the path NA → TDA was included (Model ii, Table 4) which resulted in a non-significant parameter estimate and little contribution to the overall fit. The further model of AA influencing directly ADA was also tested (i.e. path AA → ADA) (Model iii, Table 4). This path was also redundant.

Table 4: Summary statistics of overall fit for the hypothesized Model (i) with additional paths fitted as indicated by Models ii and iii

Model	χ^2	df	χ^{2diff}	Δdf	RMSEA	GFI	CFI	NFI
i	98.44	40			.056	.964	.979	.966
ii	98.29	39	0.15^{ns}	1	.057	.983	.985	.984
iii	96.93	39	1.51^{ns}	1	.057	.964	.980	.967

Notes: Model i: NA → TDA, AA → ADA, ADA → DTA, NA ↔ AA
Model ii: Model i plus NA → TDA
Model iii: Model i plus AA → ADA
χ^2 difference (χ^{2diff}); root mean square error (RMSEA); goodness of fit index (GFI); comparative fit index (CFI); normative fit index (NFI); ns = non significant ($p > .05$).

Source: Yuan, S., Freeman, R., Lahti, S., Lloyd-Williams, F., & Humphris, G. (2008). Some psychometric properties of the Chinese version of the Modified Dental Anxiety Scale with cross validation. *Health and Quality of Life Outcomes, 22*(6), 1–11.

Two Final Comments

As we come to the end of this chapter, I consider it highly important to offer two final warnings. One concerns the issue of causality. The other deals with unseen models.

Good-fitting structural equation models do not prove cause-and-effect relationships. A good-fitting model is simply one that, based on the data at hand, is *plausible*. Give researchers credit when they acknowledge this. When a structural model is found to have good fit, causation can be inferred only if other important conditions are met. Specifically, to infer a cause-and-effect relationship, a researcher should be able to show that the variables of interest are correlated, that the cause precedes the effect in time, and that other explanations for a cause–effect relationship are ruled out. This third task—of ruling out other possible explanations for a relationship—is often a very difficult process.

Even if an extensive model-building effort has produced a model that has a good fit, there is no guarantee that it is the very best model that could be generated. Other unseen and untested models are out there, and it is conceivable that one of them might be superior to the one generated in a researcher's SEM study. As with the conclusions reached by means of others statistical procedures, the insights generated through an SEM study should be viewed as tentative.

Review Terms

Causal paths	Model fit
Endogenous variable	Model building
Exogenous variable	Model invariance
Indicator variable	Model respecification
Latent variable	Model trimming
Mahalanobis distance measure	Modification indices
Manifest variable	Observed variable
Mardia's test	Nested model
Measurement error	Parceling
Measurement model	Residual error
Mediator variable	Structural model

The Best Items in the Companion Website

1. An interactive online quiz (with immediate feedback provided) covering Chapter 21.
2. Five misconceptions about structural equation modeling.
3. One of the best passages from Chapter 21: "Causal plausibility, not causal proof."

To access chapter outlines, practice tests, weblinks, and flashcards, visit the companion website at http://www.ReadingStats.com.

Review Questions and Answers begin on page 531.

Epilogue

The warnings sprinkled throughout this book were offered with two distinct groups of people in mind. The principal reason for raising these concerns is to help those who are on the *receiving* end of research claims. However, these same warnings should be considered by those who are *doing* research. If both parties are more careful in how they interact with research studies, fewer invalid claims will be made, encountered, and believed.

There are two final warnings. The first has to do with the frequently heard statement that begins with these three words, "Research indicates that. . . ." The second is concerned with the power of replication. All consumers of research, as well as all researchers, should firmly resolve to heed the important messages contained in this book's final two admonitions.

First, you must protect yourself against those who use research to intimidate others in discussions (and in arguments) over what is the best idea, the best practice, the best solution to a problem, or the best anything. Because most people (1) are unaware of the slew of problems that can cause an empirical investigation to yield untenable conclusions, and (2) make the mistake of thinking that statistical analysis creates a direct pipeline to truth, they are easily bowled over when someone else claims to have research evidence on his or her side. Do not let this happen to you! When you encounter people who promote their points of view by alluding to research ("Well, research has shown that. . . ."), ask them politely to tell you more about the research project(s) to which they refer. Ask them if they have seen the actual research reports(s). Then pose a few exceedingly legitimate questions.

If the research data were collected via mailed questionnaires, what was the response rate? No matter how the data were collected, did the researchers present evidence as to the reliability and validity of the data they analyzed? Did they attend to the important assumptions associated with the statistical techniques they used? If they tested null hypotheses, did they acknowledge the possibility of inferential error when they rejected or failed to reject any given H_0? If their data analysis produced one or more results that were significant, did they distinguish between statistical and practical significance? If you ask questions such as these, you may find that the person who first made reference to what research has to say may well become a bit more modest when arguing his or her point of view. And never, ever forget that you not only have a *right* to ask these kinds of questions, you have an *obligation* to do so (presuming that you want to be a discerning recipient of the information that comes your way).

Second, be impressed with researchers who replicate their own investigations—and even more impressed when they encourage others to execute such replications. The logic behind this admonition is simple and shines through if we consider this little question: Who are you more willing to believe, someone who demonstrates something once, or someone who demonstrates something twice? (Recall that a correlation matrix containing all bivariate rs among seven or more variables is more likely than not to be accompanied by the notation $p < .05$ *even if all null hypotheses are true,* unless the level of significance is adjusted to compensate for the multiple tests being conducted. Similarly, the odds are greater than $50-50$ that a five-way ANOVA or ANCOVA will produce a statistically significant result *simply by chance,* presuming that each F's p is evaluated against an alpha criterion of .05.)

It is sad but true that most researchers do not take the time to replicate their findings before they race off to publish their results or present them at a convention. It would be nice if there were a law stipulating that every researcher had to replicate his or her study before figuratively standing up on a soapbox and arguing passionately that something important has been discovered. No such law is likely to appear on the books in the near future. Hence, *you* must protect yourself from overzealous researchers who regard publication or being a convention speaker as more important than replication. Fortunately, there *are* more than a few researchers who delay making any claims until they have first checked to see if their findings are reproducible. Such researchers deserve your utmost respect. If their findings emanate from well-designed studies that deal with important questions, their discoveries may bring forth improvements in your life and the lives of others.

Review Questions

CHAPTER 1

1. Where is the abstract usually found in a journal article? What type of information is normally contained in the abstract?
2. What information usually follows the review of literature?
3. If an author does a good job of writing the method section of the research report, what should a reader be able to do?
4. The author of this chapter's model article used the term *participants* to label the people from whom data were collected (see Excerpt 1.5). What is another word that authors sometimes use to label the data suppliers?
5. If a researcher compares the IQ scores of 100 boys with the IQ scores of 100 girls, what would this researcher's dependent variable be?
6. What are three ways authors present the results of their statistical analyses?
7. Will a nontechnical interpretation of the results usually be located in a research report's results section or in its discussion section?
8. What is the technical name for the bibliography that appears at the end of a research report?
9. If a research report is published, should you assume that it is free of mistakes?
10. Look again at these five parts of the model article: the statement of purpose (see Excerpt 1.3), the researcher's hypothesis (see Excerpt 1.4), the final paragraph of the results section (see Excerpt 1.8), the first sentence of the discussion section (see Excerpt 1.10), and the abstract (see Excerpt 1.1). With respect to this study's purpose and findings, how many of these five sentences are consistent with one another?

CHAPTER 2

1. What does each of the following symbols or abbreviations stand for: N, M, s, Mdn., Q_3, SD, R, σ, Q_2, s^2, Q_1, σ^2, μ?

2. If cumulative frequency distributions were to be created for each column of data of Excerpt 2.2, what would be the cumulative frequency for women in the age group 41–50?

3. Each of several people from your home town is asked to indicate his or her favorite radio station, and the data are summarized using a picture containing vertical columns to indicate how many people vote for each radio station. What is the name for this kind of picture technique for summarizing data?

4. True or False: In any set of data, the median is equal to the score value that lies halfway between the high and low scores.

5. Which one of these two terms means the same thing as negatively skewed?
   a. Skewed left
   b. Skewed right

6. If the variance of a set of scores is equal to 9, how large is the standard deviation for those scores?

7. If the standard deviation for a set of 30 scores is equal to 5, how large do you think the range is?

8. What measure of variability is equal to the numerical distance between the 25th and 75th percentile points?

9. Which of the following three descriptive techniques would let you see each and every score in the researcher's data set?
   a. grouped frequency distribution
   b. stem-and-leaf display
   c. box-and-whisker

10. True or False: The distance between the high and low scores in a data set can be determined by doubling the value of the interquartile range.

CHAPTER 3

1. Following are the quiz scores for five students in English (E) and History (H).

   Sam: E = 18, H = 4
   Sue: E = 16, H = 3
   Joy: E = 15, H = 3
   John: E = 13, H = 1

   Within this same group of quiz-takers, what is the nature of the relationship between demonstrated knowledge of English and history?
   a. *high–high, low–low*
   b. *high–low, low–high*
   c. little systematic tendency one way or the other

2. If 20 individuals are measured in terms of two variables, how many dots will there be if a scatter diagram is built to show the relationship between the two variables?

3. Which of the following five correlation coefficients indicates the weakest relationship?
   a. $r = +.72$
   b. $r = +.41$
   c. $r = +.13$
   d. $r = -.33$
   e. $r = -.84$

4. In Excerpt 3.6, what are the numerical values of the two highest correlations?

5. What is the name of the correlational procedure used when interest lies in the relationship between two variables measured in such a way as to produce each of the following?
   a. two sets of raw scores
   b. two sets of ranks (with ties)
   c. two sets of truly dichotomous scores
   d. one set of raw scores and one set of truly dichotomous scores

6. What does the letter s stand for in the notation r_s?

7. If a researcher wanted to see if there is a relationship between people's favorite color (e.g., blue, red, yellow, orange) and their favorite TV network, what correlational procedure would you expect to see used?

8. True or False: If a bivariate correlation coefficient turns out to be closer to 1.00 than to 0.00, you should presume that a causal relationship exists between the two variables.

9. If a correlation coefficient is equal to.70, how large is the coefficient of determination?

10. True or False: If a researcher has data on two variables, there will be a high correlation if the two means are close together (or a low correlation if the two means are far apart).

CHAPTER 4

1. The basic idea of reliability is captured nicely by what word?

2. What is the name of the reliability procedure that leads to a coefficient of stability? To a coefficient of equivalence?

3. Regardless of which method is used to assess reliability, the reliability coefficient cannot be higher than —— or lower than ——.

4. Why is the Cronbach alpha approach to assessing internal consistency more versatile than the Kuder–Richardson 20 approach?

5. True or False: If the split-half and Kuder–Richardson 20 procedures are applied to the same set of test scores, both procedures will yield the same reliability estimate.

6. True or False: As reliability increases, so does the standard error of measurement.

7. What might cause the correlation coefficient used to assess concurrent or predictive validity to turn out *low* even though scores on the new test are *high* in accuracy?

8. Persuasive evidence for discriminant validity is provided by correlation coefficients that turn out close to
   a. $+1.00$
   b. 0.00
   c. -1.00
9. Should reliability and validity coefficients be interpreted as revealing something about the measuring instrument, or should such coefficients be interpreted as revealing something about the scores produced by using the measuring instrument?
10. True or False: If a researcher presents impressive evidence regarding the reliability of his or her data, it is safe to assume that the data are valid too.

CHAPTER 5

1. In which direction does statistical inference move: from the population to the sample, or from the sample to the population?
2. What symbols are used to denote the sample mean, the sample variance, and the sample value of Pearson's correlation? What symbols are used to denote these statistical concepts in the population?
3. True or False: If the population is abstract (rather than tangible), it is impossible for there to be a sampling frame.
4. In order for a sample to be a probability sample, what must you or someone else be able to do?
5. Which of the following eight kinds of samples are considered to be probability samples?

cluster samples	simple random samples
convenience samples	snowball samples
purposive samples	stratified random samples
quota samples	systematic samples

6. If you want to determine whether a researcher's sample is a random sample, which of these two questions should you ask?
   a. Precisely how well do the characteristics of the sample mirror the characteristics of the population?
   b. Precisely how was the sample selected from the population?
7. True or False: Studies having a response rate lower than 30 percent are not allowed to be published.
8. The best procedure for checking on a possible nonresponse bias involves doing what?
   a. Comparing demographic data of respondents and nonrespondents.
   b. Comparing survey responses of respondents and a sample of nonrespondents.
   c. Comparing survey responses of early versus late respondents.

9. If randomly selected individuals from a population are contacted and asked to participate in a study, and if those who respond negatively are replaced by randomly selected individuals who agree to participate, should the final sample be considered a random subset of the original population?

10. Put the words *tangible* and *abstract* into their appropriate places within this sentence: If a researcher's population is —————, the researcher ought to provide a detailed description of the sample, but if the researcher's population is —————, it is the population that ought to be described with as much detail as possible.

CHAPTER 6

1. True or False: Sampling errors can be eliminated by selecting samples randomly from their appropriate populations.

2. If many, many samples of size n are drawn randomly from an infinitely big population, and if the data from each sample are summarized so as to produce the same statistic (e.g., r), what would the resulting set of sample statistics be called?

3. The standard deviation of a sampling distribution is called the _____.

4. True or False: You can have more faith in a researcher's sample data if the standard error is large (rather than small).

5. The two most popular levels of confidence associated with confidence intervals are ——— and ———.

6. If the confidence interval reported in Excerpt 6.4 had been a 99 percent CI, the upper end of the CI would have been:
   a. lower than 56
   b. higher than 56
   c. equal to 56

7. One type of estimation is called *interval estimation;* the other type is called ————— estimation.

8. True or False: When a researcher includes a reliability or validity coefficient in the research report, such a coefficient should be thought of as a point estimate.

9. Which type of interval is superior to the other, a confidence interval or a standard error interval?

10. Excerpt 6.6 contains a confidence interval built around a Pearson's correlation coefficient. Does the sample value of r lie precisely in the middle of the CI?

CHAPTER 7

1. What another way to express the null hypothesis in Excerpt 7.1?

2. Suppose a researcher takes a sample from a population, collects data, and then computes the correlation between scores on two variables. If the researcher

wants to test whether the population correlation is different from 0, which of the following would represent the null hypothesis?

    a. $H_0: r = 0.00$

    b. $H_0: r \neq 0.00$

    c. $H_0: \rho = 0.00$

    d. $H_0: \rho \neq 0.00$

3. True or False: If the alternative hypothesis is set up in a nondirectional fashion, this decision will make the statistical test one-tailed (not two-tailed) in nature.

4. The null hypothesis is rejected if the sample data turn out to be (consistent/ inconsistent) with what one would expect if H_0 were true.

5. Which level of significance offers greater protection against Type I errors, .05 or. 01?

6. Does the critical value typically appear in the research report?

7. If a researcher sets $\alpha = .05$ and then finds out (after analyzing the sample data) that $p = .03$, will the null hypothesis be rejected or not rejected?

8. If a researcher's data turn out such that H_0 cannot be rejected, is it appropriate for the researcher to conclude that H_0 most likely is true?

9. If a null hypothesis is rejected because the data are extremely improbable when compared against H_0 (with $p = .00000001$), for what reason might you legitimately dismiss the research study as being totally unimportant?

10. True or False: Even if the results of a study turn out to be statistically significant, it is possible that those results are fully insignificant in any practical sense.

CHAPTER 8

1. Is it possible for a researcher to conduct a study wherein the result *is* significant in a statistical sense but *is not* significant in a practical sense?

2. What are the two popular strength-of-association indices that are similar to r^2?

3. Statistical power equals the probability of not making what kind of error?

    a. Type I

    b. Type II

4. What kind of relationship exists between statistical power and sample size?

    a. direct

    b. indirect

    c. power and sample size are unrelated

5. The statistical power of a study must lie somewhere between —— and ——.

6. What are the numerical values for small, medium, and large effect sizes (as suggested by Cohen) when comparing two sample means?

7. If a study is conducted to test $H_0: \mu = 30$ and if the results yield a confidence interval around the sample mean that extends from 26.44 to 29.82, will H_0 be rejected?

8. When the Bonferroni adjustment is used, what gets adjusted first?
   a. H_0
   b. H_a
   c. α
   d. p
9. If a researcher wants to use the nine-step version of hypothesis testing instead of the six-step version, what three additional things must he or she do?
10. If the researcher's sample size is too ———, the results can yield statistical significance even in the absence of any practical significance. However, if the sample size is too ———, the results can yield a nonsignificant result even though the null hypothesis is incorrect by a large margin.
    a. small; large
    b. large; small

CHAPTER 9

1. If a researcher reports that a sample correlation coefficient turned out to be statistically significant, which of the following most likely represents the researcher's unstated null hypothesis?
   a. H_0: $\rho = -1.00$
   b. H_0: $\rho = 0.00$
   c. H_0: $\rho = +1.00$
2. If a researcher reports that "$r(58) = 2.61, p < .05$," how many pairs of scores were involved in the correlation?
3. When a researcher checks to see if a sample correlation coefficient is or is not significant, the inferential test most likely will be conducted in a (one-tailed/two-tailed) fashion.
4. Suppose a researcher has data on five variables, computes Pearson's r between every pair of variables, and then displays the rs in a correlation matrix. Also suppose that an asterisk appears next to three of these rs, with a note beneath the table explaining that the asterisk means $p < .05$. Altogether, how many correlational null hypotheses were set up and tested on the basis of the sample data?
5. In the situation described in question 4, how many of the rs would have turned out to be statistically significant if the Bonferroni technique had been applied?
6. Is it possible for a researcher to have a test–retest reliability coefficient of .25 that turns out to be statistically significant at $p < .001$?
7. A confidence interval built around a sample correlation coefficient leads to a retention of the typical correlational null hypothesis if the CI overlaps which of the following numbers?
   a. -1.0
   b. $-.50$
   c. 0.00

   d. $+.50$
   e. $+1.00$
8. Is it possible for r^2 to be low (i.e., close to zero) and yet have $p < .01$?
9. True or False: To the extent that the p-value associated with r is small (e.g., $p < .01$, $p < .001$, $p < .0001$), the researcher more confidently can argue that a cause-and-effect relationship exists between the two variables that were correlated.
10. Attenuation makes it (more/less) likely that a true relationship will reveal itself through the sample data by means of a statistically significant correlation coefficient.

CHAPTER 10

1. If 20 eighth-grade boys are compared against 25 eighth-grade girls, should these two comparison groups be thought of as correlated samples or as independent samples?
2. In the null hypothesis of an independent-samples t-test comparison of two group means, what kind of means are referred to?
   a. sample means
   b. population means
3. If the *df* associated with a correlated-samples t-test is equal to 18, how many pairs of scores were involved in the analysis?
4. Based on the information contained in the following ANOVA summary table, the researcher's calculated value would be equal to what number?

Source	df	SS	MS	F
Between groups	1	12		
Within groups	18	54		

5. If a researcher uses an independent-samples t-test to compare a sample of men with a sample of women on each of five dependent variables, and if the researcher uses the Bonferroni adjustment technique to protect against Type I errors, what does he or she adjust?
   a. each group's sample size
   b. each t-test's calculated value
   c. the degrees of freedom
   d. the level of significance
6. True or False: Whereas strength-of-association indices *can* be computed in studies concerned with the mean of a single sample, they *cannot* be computed in studies concerned with the means of two samples.
7. Suppose a researcher compares two groups and finds that $M_1 = 60$, $SD_1 = 10$, $M_2 = 55$, and $SD_2 = 10$. Based on this information, how large would the estimated effect size be? According to Cohen's criteria, would this effect size be considered small, medium, or large?

8. If a researcher uses sample data to test the homogeneity of variance assumption in a study involving two independent samples, what will the null hypothesis be? Will the researcher hope to reject or fail to reject this null hypothesis?

9. If the measuring instrument used to collect data has less than perfect reliability, the confidence interval built around a single sample mean or around the difference between two sample means will be (wider/narrower) than would have been the case if the data had been perfectly reliable.

10. Suppose a one-way analysis of variance is used to compare the means of two samples. Also suppose that the results indicate that $SS_{Total} = 44$, $MS_{Error} = 4$, and $F = 3$. With these results, how large was the sample size, assuming both groups had the same n?

CHAPTER 11

1. If a researcher uses a one-way ANOVA to compare four samples, the statistical focus is on (means/variances), there will be ——— (how many) inferences, and the inference(s) will point toward the (samples/populations).

2. In a one-way ANOVA involving five comparison groups, how many independent variables are there? How many factors?

3. If a one-way ANOVA is used to compare the heights of three groups of first-grade children (those with brown hair, those with black hair, and those with blond hair), what is the independent variable? What is the dependent variable?

4. For the situation described in question 3, what would the null hypothesis look like?

5. Based on the information contained in the following ANOVA summary table, what is the numerical value for SS_{Total}?

Source	df	SS	MS	F
Between groups	4			3
Within groups			2	
Total	49			

6. Which of these two researchers would end up with a statistically significant finding after they each perform a one-way ANOVA?
   a. The F-value in Bob's ANOVA summary table is larger than the appropriate critical F-value.
   b. The p-value associated with Jane's calculated F-value is larger than the level of significance.

7. Suppose a one-way ANOVA comparing three sample means (8.0, 11.0, and 19.0) yields a calculated F-value of 3.71. If everything about this study remained the same except that the largest mean changed from 19.0 to 17.0, the calculated value would get (smaller/larger).

8. Suppose a researcher wants to conduct 10 one-way ANOVAs, each on a separate dependent variable. Also suppose that the researcher wants to conduct these ANOVAs such that the probability of making at least one Type I error is equal to. 05. To accomplish this objective, what alpha level should the researcher use in evaluating each of the F-tests?

9. A one-way ANOVA is *not* robust to the equal variance assumption if the comparison groups are dissimilar in what way?

10. Is it possible for a one-way ANOVA to yield a statistically significant but meaningless result?

CHAPTER 12

1. Which term more accurately describes Tukey's test: planned or post hoc?

2. What are the differences among these three terms: post hoc comparison, follow-up comparison, a posteriori comparison?

3. If a one-way ANOVA involves five groups, how many pairwise comparisons will there be if the statistically significant omnibus F is probed by a post hoc investigation that compares every mean with every other mean?

4. Will a conservative test procedure or a liberal test procedure more likely yield statistically significant results?

5. True or False: If three sample means are $M_1 = 60$, $M_2 = 55$, and $M_3 = 50$, it is impossible for the post hoc investigation to say $M_1 > M_2 > M_3$.

6. Which kind of comparison is used more by applied researchers, pairwise or nonpairwise?

7. True or False: When conducting post hoc investigations, some researchers use the Bonferroni technique in conjunction with t-tests as a way of dealing with the inflated Type I error problem.

8. True or False: Whereas regular t-tests and the one-way ANOVA's omnibus F-test have no built-in control that addresses the difference between statistical significance and practical significance, planned and post hoc tests have been designed so that only meaningful differences can end up as statistically significant.

9. If a researcher has more than two comparison groups in his or her study, it (would/would not) be possible for him or her to perform a one-degree-of-freedom F test.

10. True or False: In a study comparing four groups (A, B, C, and D), a comparison of A versus B is orthogonal to a comparison of C versus D.

CHAPTER 13

1. If a researcher performs a univariate 3 × 3 ANOVA, how many independent variables are there? How many dependent variables?

2. How many cells are there in a 2 × 4 ANOVA? In a 3 × 5 ANOVA?

3. Suppose the factors of a 2 × 2 ANOVA are referred to as Factor A and Factor B. How will the research participants be put into the cells of this study if Factor A is assigned while Factor B is active?

4. How many research questions dealt with by a two-way ANOVA are concerned with main effects? How many are concerned with interactions?

5. Suppose a 2 (gender) × 3 (handedness) ANOVA is conducted, with the dependent variable being the number of nuts that can be attached to bolts within a 60-second time limit. Suppose that the mean scores for the six groups, each containing 10 participants, turn out as follows: right-handed males = 10.2, right-handed females = 8.8; left-handed males = 7.8, left-handed females = 9.8; ambidextrous males = 9.0, ambidextrous females = 8.4. Given these results, what are the main effect means for handedness equal to? How many scores is each of these means based on?

6. True or False: There is absolutely no interaction associated with the sample data presented in question 5.

7. How many different mean squares serve as the denominator when the F-ratios are computed for the two main effects and the interaction?

8. True or False: You should not expect to see a post hoc test used to compare the main effect means of a 2 × 2 ANOVA, even if the F-ratios for both main effects turn out to be statistically significant.

9. How many simple main effects are there for Factor A in a 2 × 3 (A × B) ANOVA?

10. True or False: Whenever a two-way ANOVA is used, there is a built-in control mechanism that prevents results from being statistically significant unless they are also significant in a practical sense.

CHAPTER 14

1. If you see the following factors referred to with these names, which one(s) should you guess probably involve repeated measures? (Select all that apply.)
   a. treatment groups
   b. trial blocks
   c. time period
   d. response variables

2. How does the null hypothesis of a between-subjects one-way ANOVA differ from the null hypothesis of a within-subjects one-way ANOVA?

3. If a 2 × 2 ANOVA is conducted on the data supplied by 16 research participants, how many individual scores would be involved in the analysis if both factors are between subjects in nature? What if both factors are within subjects in nature?

4. If the two treatments of a one-way repeated measures ANOVA are presented to 20 research participants in a counterbalanced order, how many different presentation orders will there be?

5. True or False: Because the sample means of a two-way repeated measures ANOVA are each based on the same number of scores, this kind of ANOVA is robust to the sphericity assumption.

6. If eight research participants are each measured across three levels of factor A and four levels of factor B, how many rows (including total) will there be in the ANOVA summary table? How many df will there be for the total row?

7. How many null hypotheses are typically associated with a two-way mixed ANOVA? How many of them deal with main effects?

8. If each of 10 males and 10 females is measured on three occasions with the resulting data analyzed by a two-way mixed ANOVA, how many main effect means will there be for gender, and how many scores will each of these sample means be based on?

9. Suppose the pretest, posttest, and follow-up scores from four small groups (with $n = 3$ in each case) are analyzed by means of a mixed ANOVA. How large would the interaction F be if it turned out that $SS_{Groups} = 12$, $SS_{Total} = 104$, $MS_{Error(w)} = 2$, $F_{Groups} = 2$, and $F_{Time} = 5$?

10. True or False: One of the nice features of any repeated measures ANOVA is the fact that any statistically significant result is guaranteed to be significant in a practical sense as well.

CHAPTER 15

1. ANCOVA was developed to help researchers decrease the probability that they will make a (Type I/Type II) error when they test hypotheses.

2. What are the three kinds of variables involved in any ANCOVA study?

3. In ANCOVA studies, is it possible for something other than a pretest (or baseline measure) to serve as the covariate?

4. Suppose the pretest and posttest means for a study's experimental (E) and control (C) groups are as follows: $M_{E(pre)} = 20$, $M_{E(post)} = 50$, $M_{C(pre)} = 10$, $M_{C(post)} = 40$. If this study's data were to be analyzed by an analysis of covariance, the control group's adjusted posttest mean might turn out equal to which one of these possible values?
   a. 5
   b. 15
   c. 25
   d. 35
   e. 45

5. For ANCOVA to achieve its objectives, there should be a (strong/weak) correlation between each covariate variable and the dependent variable.

6. True or False: Like the analysis of variance, the analysis of covariance is robust to violations of its underlying assumptions so long as the sample sizes are equal.

7. One of ANCOVA's assumptions states that the ——— variable should not affect the _____ variable.

8. ANCOVA works best when the comparison groups (are/are not) formed by random assignment.

9. In testing the assumption of equal regression slopes, does the typical researcher hope the assumption's null hypothesis will be rejected?

10. True or False: Because ANCOVA uses data on at least one covariate variable, results cannot turn out to be statistically significant without also being significant in a practical sense.

CHAPTER 16

1. In a scatter diagram constructed in conjunction with a bivariate regression analysis, which of the two axes will be set up to coincide with the dependent variable?

2. In the equation $Y' = 2 + 4(X)$, what is the numerical value of the constant, and what is the numerical value of the regression coefficient?

3. In bivariate regression, can the slope end up being negative? What about the Y-intercept? What about r^2?

4. True or False: In bivariate regression, a test of H_0: $\rho = 0$ is equivalent to a test that the Y-intercept is equal to 0.

5. In multiple regression, how many X variables are there? How many Y variables?

6. True or False: You will never see ΔR^2 reported among the results of a simultaneous multiple regression.

7. In stepwise and hierarchical multiple regression, do the beta weights for those independent variables entered during the first stage remain fixed as additional independent variables are allowed to enter the regression equation at a later stage?

8. In binary logistic regression, the dependent variable is (dichotomous/continuous) in nature.

9. An odds ratio of what size would indicate that a particular independent variable has no explanatory value?

10. In logistic regression, does the Wald test focus on individual ORs or does it focus on the full regression equation?

CHAPTER 17

1. True or False: When the sign test is used, the null hypothesis says that the sample data will contain an equal number of observations in each of the two response categories, thus yielding as many pluses as minuses.

2. Which test is more flexible, the sign test or the binomial test?

3. What symbol stands for chi square?

4. Suppose a researcher uses a 2×2 chi square to see if males differ from females with regard to whether they received a speeding ticket during the previous year. Of the 60 males in the study, 40 had received a ticket. The sample data would be in full agreement with the chi-square null hypothesis if —— of the 90 females received a ticket.

5. How many degrees of freedom would there be for a chi-square comparison of freshmen, sophomores, juniors, and seniors regarding their answers to the question: "How would you describe the level of allegiance to your school?" (The response options are low, moderate, and high.)

6. Whose name is often associated with the special chi-square formula that carries the label *correction for continuity*?

7. McNemar's chi-square test is appropriate for (two/more than two) groups of data, where the samples are (independent/correlated), and where the response variable contains (two/more than two) categories.

8. If a pair of researchers got ready to use a one-factor repeated measures ANOVA but then stopped after realizing that their data were dichotomous, what statistical test could they turn to in order to complete the data analysis?

9. True or False: Techniques for applying the concept of *statistical power* to tests dealing with frequencies, percentages, and proportions have not yet been developed.

10. Can confidence intervals be placed around sample percentages?

Chapter 18

1. Why do researchers sometimes use nonparametric tests with data that are interval or ratio?

2. The median test is used with (independent/correlated) samples.

3. If the median test is used to compare two samples, how many medians will the researcher need to compute based on the sample data?

4. A Mann-Whitney U test is designed for situations where there are —— (how many) samples that ——— (do/do not) involve repeated measures.

5. Which of the test procedures discussed in this chapter is analogous to the correlated-samples t-test considered earlier in Chapter 10? Which one is analogous to the one-way ANOVA considered in Chapter 11?

6. Which of the nonparametric tests involves a calculated value that is sometimes symbolized as χ_r^2?

7. True or False: The large-sample versions of the Mann–Whitney, Kruskal–Wallis, and Wilcoxon tests all involve a calculated value that is labeled z.

8. Are random samples important to nonparametric tests?

9. True or False: Because they deal with ranks, the tests considered in this chapter have lower power than their parametric counterparts.

10. The term *distribution free* (should/should not) be used to describe the various nonparametric tests discussed in this chapter.

CHAPTER 19

1. What is the statistical focus in a MANOVA?
   a. Correlations
   b. Standard deviations
   c. Means
   d. Frequencies
2. Compared to an ANOVA, a MANOVA has two or more _____.
   a. dependent variables
   b. independent variables
   c. levels in each factor
   d. factors
3. True or False: If a one-way MANOVA is used to compare a sample of men against a sample of women in terms of the participants' speed of running 100 yards on a track and their speed of swimming 100 yards in a pool, the multivariate null hypothesis would be $\mu_{run} = \mu_{swim}$ for men and $\mu_{run} = \mu_{swim}$ for women.
4. Is there a multivariate analogue to the univariate analysis of covariance?
5. Is Wilks' lambda used very often to test the MANOVA null hypothesis?
6. Multivariate test statistics, such as λ, are typically converted into
   a. z-scores
   b. t-values
   c. F-values
7. What do the first two letters stand for in the acronym NPMANOVA?
8. True or False: No statistical techniques have been developed as yet to assess the practical significance of results from a MANOVA investigation.
9. True or False: The only option for performing a post hoc investigation after a MANOVA has yielded a significant result is to perform univariate tests.
10. Should the Bonferroni adjustment procedure be used in a post hoc investigation involving univariate tests if the initial MANOVA or MANCOVA has produced a significant multivariate result?

CHAPTER 20

1. The statistical part of factor analysis begins with an examination of
   a. the mean scores on the observed variables
   b. the standard deviation of scores on each observed variable
   c. the correlations among the observed variables

2. The number of factors identified at the end of a factor analysis usually is _____ than the number of observed variables.
   a. smaller
   b. larger
3. True or False: The outcome of a factor analysis is clear and good if each observed variable has a high correlation with each of the factors.
4. Factor analysis is often used in studies in which the researchers want to assess which kind of validity?
   a. Content validity
   b. Predictive validity
   c. Construct validity
5. The two main types of factor analysis are called _____ factor analysis and _____ factor analysis.
6. The two main types of factor analysis (can/cannot) be used within the context of the same study.
7. True or False: Factor rotation takes place before (not after) factor extraction?
8. A factor is considered to be worth retaining is its eigenvalue is
   a. small
   b. large
9. How are the results of a parallel analysis summarized?
   a. In a graph
   b. In an *F*-value
   c. In a χ^2-value
10. True or False: Tests of *model fit* are used both in exploratory factor analysis and in confirmatory factor analysis.

CHAPTER 21

1. True or False: In structural equation modeling, the terms *manifest variable* and *latent variable* refer to the same thing.
2. True or False: In structural equation modeling, the terms *indicator* and *observed variable* refer to the same thing.
3. How are the findings of SEM typically summarized?
   a. In a table containing means, standard deviations, and indices of skewness
   b. In a diagram containing boxes, ovals, and arrows
   c. In a single scatter diagram containing dark and light data points
4. Which kind of variable is considered to have an effect on some other variable?
   a. Exogenous
   b. Endogenous
5. In SEM, the two types of error are called measurement error and _____.
6. True or False: If model fit is tested in an SEM study, and if the researcher hopes for a good fit, the desired result will be a nonsignificant (i.e., fail-to-reject) result.

7. In an SEM study, an examination of the "measurement model" involves assessing the reliability of the _____ variables.
   a. manifest
   b. latent
8. A statistic that assesses whether one or more causal paths are missing from the model is called the "_____ index."
   a. completeness
   b. modification
   c. orthogonal
   d. Sherlock
9. What is a latent variable called if it functions as both an exogenous variable and an endogenous variable?
10. True or False: Because of the complexity and sophistication of SEM, a tight model fit proves causality.

Answers to Review Questions

CHAPTER 1

1. The abstract is usually found near the beginning of the article. It normally contains a condensed statement of the study's objective(s), participants, method, and results.
2. Statement of purpose
3. Replicate the investigation
4. Subjects
5. IQ (i.e., intelligence)
6. In paragraphs of text, in tables, and in figures
7. In the discussion section
8. References
9. No
10. All five

CHAPTER 2

1. Size of the data set, mean, standard deviation, median, upper quartile point, standard deviation, range, standard deviation, middle quartile point (or median), variance, lower quartile point, variance, mean
2. 420
3. Bar graph
4. False
5. a
6. 3
7. 20
8. Interquartile range
9. b
10. False

CHAPTER 3

1. a
2. 20
3. c
4. .41 and −.39
5. **a.** Pearson's r
   b. Kendall's tau
   c. phi
   d. point biserial
6. Spearman
7. Cramer's V
8. False
9. .49
10. False (Correlation says *nothing* about the two means!)

CHAPTER 4

1. Consistency
2. Test-retest reliability; parallel-forms reliability (or alternate-forms reliability or equivalent-forms reliability)
3. 1.0; 0.0
4. Cronbach's alpha is not restricted to situations where the data are dichotomous
5. False
6. False
7. Poor measurement of the criterion variable
8. b
9. The score obtained by using the measuring instrument
10. False

CHAPTER 5

1. From the sample to the population
2. M, s^2, r, μ, σ^2, ρ
3. True
4. Assign a unique ID number to each member of the population
5. Cluster samples, simple random samples, stratified random samples, and systematic samples
6. b
7. False
8. b
9. No
10. abstract; tangible

CHAPTER 6

1. False
2. A sampling distribution
3. standard error
4. False
5. 95 percent; 99 percent
6. b
7. point
8. True
9. A confidence interval
10. No

CHAPTER 7

1. H_0: $\mu_1 = \mu_2$
2. c
3. False
4. inconsistent
5. .01
6. No
7. Rejected
8. No
9. A silly null hypothesis
10. True

CHAPTER 8

1. Yes
2. Eta squared and omega squared
3. b
4. a
5. 0, 1.0

6. Small = .20, medium = .50, large = .80
7. Yes
8. c
9. Specify the effect size, specify the desired power, and determine (via formula, chart, or computer) the proper sample size.
10. b

CHAPTER 9

1. b
2. 60
3. two-tailed
4. 10
5. Most likely none of them
6. Yes
7. c
8. Yes, if the sample size is large enough
9. False
10. less

CHAPTER 10

1. Independent samples (because two groups with different ns cannot be correlated)
2. b
3. 19
4. 4
5. d
6. False
7. .50; medium
8. H_0: $\sigma_1^2 = \sigma_2^2$; fail to reject
9. wider
10. 5

CHAPTER 11

1. means; one; populations
2. 1; 1
3. hair color; height
4. H_0: $\mu_1 = \mu_2 = \mu_3$
5. 114
6. Bob
7. smaller
8. .005
9. Group size
10. Yes

CHAPTER 12

1. Neither. It depends on whether the researcher who uses the Tukey test first examines the ANOVA F to see if it is okay to compare means using the Tukey test.
2. Nothing; they are synonyms.
3. 10
4. A liberal test procedure
5. True
6. Pairwise
7. True
8. False
9. would
10. True

CHAPTER 13

1. 2; 1
2. 8; 15
3. Participants will be randomly assigned to levels of Factor B from within each level of Factor A.
4. 2; 1
5. The main effect means would be equal to 9.5, 8.8, and 8.7 (for right-handed, left-handed, and ambidextrous individuals, respectively). Each would be based on 20 scores.
6. False
7. 1
8. True
9. 3
10. False

CHAPTER 14

1. b, c, d
2. They do not differ in any way.
3. 16; 64
4. 2
5. False
6. 8; 95
7. 3; 2
8. 2; 30
9. 2
10. False

CHAPTER 15

1. Type II
2. Independent, dependent, and covariate (concomitant) variables

3. Yes
4. e
5. strong
6. False
7. independent; covariate
8. are
9. No
10. False

CHAPTER 16

1. The vertical axis (i.e., the ordinate)
2. The constant is 2; the regression coefficient is 4.
3. Yes; yes; no
4. False
5. At least two; just one
6. True
7. No
8. dichotomous
9. 1.0
10. Individual ORs

CHAPTER 17

1. False. (The null hypothesis is a statement about population parameters, not sample statistics.)
2. The binomial test
3. χ^2
4. 60
5. 6
6. Yates
7. two; correlated; two
8. Cochran's Q test
9. False
10. Yes

CHAPTER 18

1. Because researchers know or suspect that the normality or equal variance assumptions are untenable, especially in situations where the sample sizes are dissimilar.
2. independent
3. 1
4. 2; do not
5. The Wilcoxon matched-pairs signed-ranks test; the Kruskal-Wallis one-way ANOVA of ranks
6. Friedman's two-way analysis of variance of ranks
7. False. (The Kruskal–Wallis test, when conducted with large samples, yields a calculated value symbolized as χ^2.)

8. Yes
9. False
10. should not

CHAPTER 19

1. c
2. a
3. False
4. Yes
5. Yes
6. c
7. Nonparametric
8. False
9. False
10. Yes

CHAPTER 20

1. c
2. a

3. False
4. c
5. exploratory; confirmatory
6. can
7. False
8. b
9. a
10. False

CHAPTER 21

1. False
2. True
3. b
4. a
5. residual error
6. True
7. b
8. b
9. A mediator variable
10. False

Credits

1.1 through 1.11

Perceptual and motor skills by J. J. Annesi. Copyright 2009 by AMMONS SCIENTIFIC, LTD. Reproduced with permission of AMMONS SCIENTIFIC, LTD. in the format Textbook via Copyright Clearance Center.

2.1

McDermott, R. J., Nickelson, J., Baldwin, J. A., Bryant, C. A., Alfonso, M., Phillips, L. M., *et al.* (2009). A community–school district–university partnership for assessing physical activity of tweens. *Preventing Chronic Disease.* 6(1). Contributed to the public domain by the U.S. Centers for Disease Control and Prevention.

2.2

Giannoglou *et al.*, Difference in the topography of atherosclerosis in the left versus right coronary artery in patients referred for coronary angiography. *BMC Cardiovascular Disorders* (published by BioMed Central), 2010, 10:26.

2.3

Chyung, S. Y. (2007). Age and gender differences in online behavior, self-efficacy, and academic performance. *Quarterly Review of Distance Education, 8*(3), 213–222.

2.4

Radat, F., Lantéri-Minet, M., Nachit-Ouinekh, F., Massiou, H., Lucas, C., Pradalier, A., *et al.* (2008). The GRIM2005 study of migraine consultation in France: III: Psychological features of subjects with migraine. *Cephalalgia, 29,* 338–350. Reprinted with permission of Sage Publications.

2.5

Faseru *et al.*, Design, recruitment, and retention of African-American smokers in a pharmacokinetic study. *BMC Medical Research Methodology* (published by BioMed Central), 2010, 10:6.

2.7

Sloma *et al.*, Knowledge of stroke risk factors among primary care patients with previous stroke or TIA: a questionnaire study. *BMC Family Practice* (published by BioMed Central), 2010, **11**:47.

2.17

Tilson, Validation of the modified Fresno Test: assessing physical therapists' evidence based practice knowledge and skills. *BMC Medical Education* (published by BioMed Central), 2010, **10**:38.

2.20

Exner *et al.*, Worry as a window into the lives of people who use injection drugs: a factor analysis approach. *BMC Harm Reduction Journal* (published by BioMed Central), 2009, **6**:20.

3.1

Bago *et al.*, The Trunk Appearance Perception Scale (TAPS): a new tool to evaluate subjective impression of trunk deformity in patients with idiopathic scoliosis. *Scoliosis* 2010, **5**:6.

3.6

Elwood, S., et al. (2009). "The Incubation Effect: Hatching a Solution," *Creativity Research Journal*. 21(1): 11. Reproduced with permission of the publisher, Taylor & Francis, Ltd.

3.7

van Osch *et al.,* Action planning as predictor of health protective and health risk behavior: an investigation of fruit and snack consumption. *International Journal of Behavioral Nutrition and Physical Activity* 2009, **6**:69.

6.3

Reproduced from the Journal of Experimental Social Psychology, 46, Turning the knots in your stomach into bows: Reappraising arousal improves performance on the GRE, by Jamieson, J. P., Mendes, W. B., Blackstock, E., and Schmader, T., 208–212, 2010, with permission from Elsevier.

9.12

Ljoså *et al.* Shiftwork in the Norwegian petroleum industry: overcoming difficulties with family and social life – a cross sectional study. *Journal of Occupational Medicine and Toxicology* 2009, **4**:22.

10.11

COMMUNITY COLLEGE JOURNAL OF RESEARCH & PRACTICE. Copyright 2008 by TAYLOR & FRANCIS INFORMA UK LTD—JOURNALS. Reproduced with permission of TAYLOR & FRANCIS INFORMA UK LTD—JOURNALS in the format Textbook via Copyright Clearance Center.

10.15

Grammatikopoulos *et al.*, The Short Anxiety Screening Test in Greek: translation and validation. *Annals of General Psychiatry* 2010, **9**:1.

11.1 and 11.5

Akbulut, Y. (2007). Effects of multimedia annotations on incidental vocabulary learning and reading comprehension of advanced learners of English as a foreign language. *Instructional Science, 35*(6), 512. Reprinted by permission of the publisher, Springer.

12.14

Wilkerson, T. W. (2009). An exploratory study of the perceived use of workarounds utilized during the prescription preparation process of pharmacies in Alabama. Unpublished doctoral dissertation, Auburn University, Auburn, Alabama. Reproduced with permission of the author.

12.15

Drenowatz *et al.*, Influence of socio-economic status on habitual physical activity and sedentary behavior in 8- to 11-year old children. *BMC Public Health* (published by BioMed Central), 2010, 10:214.

13.8

Chen, L. J., Ho, R. G., & Yen, Y. C. (2010). Marking strategies in metacognition-evaluated computer-based testing. *Journal of Educational Technology & Society*, *13*(1), 246–259. Reproduced with permission of IFEST: the International Forum of Educational Technology and Society.

13.11

Dixon, L. J. (2009). The effects of betrayal characteristics on laypeople's ratings of betrayal severity and conceptualization of forgiveness. Unpublished doctoral dissertation. University of Tennessee, Knoxville. Reproduced with permission of the author.

14.6

Sohlberg, M. M., Fickas, S., Hung, P.-F., & Fortier, A. (2007). A comparison of four prompt modes for route finding for community travellers with severe cognitive impairments. *Brain Injury, 21*(5), 531–538. Reproduced by permission of the publisher (Taylor & Francis Group, http://www.informaworld.com).

14.17

TESOL QUARTERLY : A JOURNAL FOR TEACHERS OF ENGLISH TO SPEAKERS OF OTHER LANGUAGES AND OF STANDARD ENGLISH AS A SECOND DIALECT by K. J. Hartshorn. Copyright

Index

Note: Roman numbers refer to page numbers; bold numbers refer to excerpt numbers.